T0328295

GENE-ENVIRONMENT INTERACTIONS IN PSYCHIATRY

GENE-ENVIRONMENT INTERACTIONS IN PSYCHIATRY

Nature, Nurture, Neuroscience

BART ELLENBROEK
School of Psychology, Victoria University of Wellington
Wellington, New Zealand

JIUN YOUN
School of Psychology, Victoria University of Wellington
Wellington, New Zealand

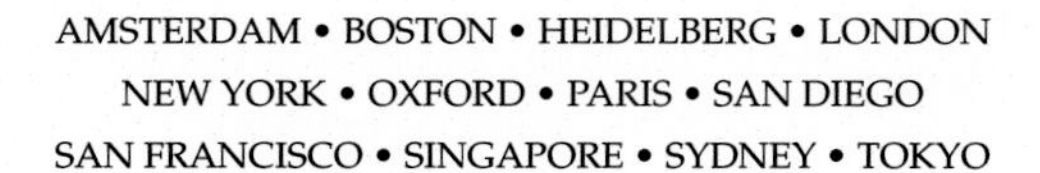

Academic Press is an imprint of Elsevier

Academic Press is an imprint of Elsevier
125 London Wall, London EC2Y 5AS, United Kingdom
525 B Street, Suite 1800, San Diego, CA 92101-4495, United States
50 Hampshire Street, 5th Floor, Cambridge, MA 02139, United States
The Boulevard, Langford Lane, Kidlington, Oxford OX5 1GB, United Kingdom

Library of Congress Cataloging-in-Publication Data
A catalog record for this book is available from the Library of Congress

British Library Cataloguing-in-Publication Data
A catalogue record for this book is available from the British Library

ISBN: 978-0-12-801657-2

For information on all Academic Press publications
visit our website at https://www.elsevier.com/

Publisher: Mara Conner
Acquisition Editor: Natalie Farra
Editorial Project Manager: Kathy Padilla
Production Project Manager: Julia Haynes
Designer: Maria Inês Cruz

Typeset by Thomson Digital

For Inessa, Arabella, and Robin.

It is sobering to realize how profound the effects of environmental factors can be on mental health, by themselves and more importantly by interacting with the genetic makeup of an individual. However, it is at the same time a hopeful realization, especially as parents, that an environment can be protective as well, hence we dedicate this book to our children.

Contents

II

GENE-ENVIRONMENT INTERACTIONS IN PSYCHIATRIC DISORDERS

6. Drug Addiction

7. Affective Disorders

8. Childhood Disorders

9. Schizophrenia

10. Conclusions and the Road Ahead

Preface

The idea for this book came several years ago when we moved from Europe to New Zealand to work at Victoria University of Wellington and one of us started teaching a fourth year honors course entitled *"nature, nurture, neuroscience"* within the school of psychology. The aim of the course was to convey to the students both the fundamental aspects of gene environment interactions as well as teach them our current knowledge on the role genetic and environmental factors play in shaping our brain and in the aetiology of major psychiatric disorders. While preparing for that course and during discussions with the students it became clear that there was no textbook available that covers both the fundamental and the more applied aspects central to the course. Therefore, after discussions with Natalie Farra from Elsevier we decided to write this book.

The book consists of two separate parts, with the first part focussing on the fundamental aspects. Chapter 1 gives a relatively concise overview of the history of genetics, starting with the ground-breaking experiments by Gregor Mendel and leading up to the discovery of the structure of DNA by James Watson and Francis Crick. The chapter finishes with a brief description of the history behavioral genetics. Chapter 2 give an overview of the current knowledge of molecular genetics, including the structure and function of DNA and RNA. The second part of this chapter focusses on genetic alterations. As alterations in the structure of DNA are at the heart of all psychiatric disorders, it is important to describe the different types of genetic mutations. Much of our knowledge on the structure and function of our brain derives from animal research. Therefore, chapter 3 focusses on animal modelling. After a general introduction on animal research and animal modelling, the chapter also discusses ways to alter the genetic make-up of animals. Chapters 4 and 5 then discuss the role of environmental factors. Whereas chapter 4 describes how environmental factors affects behavior, Chapter 5 details the mechanisms underlying these effects. Within this chapter we focus predominantly on two factors, namely how environmental factors alter the development of the brain and secondly how environmental factors influence epigenetic mechanisms. Epigenetics is the branch of research that focusses on how alterations in the three dimensional shape of DNA alters the expression of genes.

After this general introduction, the second part consists primarily of a description of several major psychiatric disorders and the role genetic and environmental factors play: drug addiction (Chapter 6), affective disorders (Chapter 7); childhood disorders (Chapter 8) and schizophrenia (Chapter 9). The basic structure of these chapters is the same, starting with a short description of the diagnostic criteria, symptoms and epidemiology, the chapter then continues with a description of the neurobiology and treatment. After this, we describe our current knowledge on the etiology focusing on the genetic factors and how these factors are moderated by specific environmental factors. The chapters then conclude with a description of the most often used animal models, again starting with a description of how disease specific features can be modeled, followed by the most important genetic models including models based on gene–environmental interactions. It is perhaps important to note that our aim was not to be all inclusive. Given the enormous number of papers being published in the field, this is already an almost impossible task. Moreover, it would make the book incomplete the moment it is published. Rather we aimed to include the most important current ideas and findings within each of the disorders discussed. Likewise, we deliberately left out a description of individual environmental factors that have been implicated in each disorder (and animal models based on only environmental challenges) focusing rather on the interaction between genetic and environmental factors.

The last chapter of the book is meant as a summary chapter, as well as a look forward. As will become clear in the individual chapters, although we have made important progress, the role of genetic and environmental factors in our major psychiatric disorders is still largely a mystery. The chapter therefore summarizes the reasons for this and formulates possible alternative approaches.

Writing this book has been a major undertaking for us and in the course of this we have learned a great deal. Fortunately, we have had many people help us, some very concrete by reading and commenting on specific chapters, others by listening to us, discussing with us or by simply being patient with us when we (again) did not meet a deadline. *"Still writing the book eh?"* was a phrase we often heard (and admittingly used ourselves too). We would therefore like to thank all these people for their help and patience. We would like to thank all the students that have participated in the psyc 444 *"Nature, Nurture, Neuroscience"* course over the years whose questions and discussions have helps us in writing this book. Special thanks goes to Anne Arola, Alana Oakly, Peter Ranger, Charlotte Gutenbrunner, and Kris Nielsen whose discussions and critical comments have contributed greatly to the final version of this book. We also would like to thank our friends and colleagues who were so kind to critically read several of the chapters, especially Edwin Cuppen who commented on Chapter 2, Tim Karl on Chapter 3 and Clare Stanford who took the time to read through Chapter 8. Laura Anderson, Michaela Pettie, Katherine Vlessis and Matthew Westbury helped us with proofreading several chapters. We are grateful to Natalie Farra and Kathy Padilla from Elsevier, San Diego for their help and their patience with us while we were writing the book.

Finally, we hope our book will be helpful for both (under)graduate and postgraduate students that are interested in the field of gene-environment research. However, given the nature of our text, we think it also contains a wealth of information useful for scientists and psychiatrists that study how genetic and environmental factors interact to shape our brain and behavior.

Bart Ellenbroek and Jiun Youn
Wellington Jan. 2016

SECTION I

GENERAL INTRODUCTION

CHAPTER

1

Introduction

INTRODUCTION

It seems reasonable to state that the era of genetics started in February and March in 1865, when Gregor Mendel gave two presentations at the monthly meeting of the "Naturforschenden Vereines" in Brünn (present day Brno in the Czech Republic). In his two-part paper entitled "*Ueber Planzenhybriden*" (Experiments on Plant Hybrids), Mendel described the results of his experiments on peas and laid the foundation for our understanding how traits are passed on from one generation to the next (Mendel, 1866).

Of course scientists had wondered about this for many hundreds if not thousands of years, but this had never culminated in a coherent theoretical framework. Indeed, it was not even clearly understood which characteristics were inherited from parents to children and which were not. Aristotle, around 300 BC (Box 1.1 for a timeline of important events) already pointed out that some characteristics (both physical and behavioral, such as gait) may reappear in the offspring while others (such as the loss of limbs or other body parts) were not. Interestingly, although he acknowledged that he lacked a clear hypothesis of inheritance, Aristotle did suggest that what is inherited is actually the potential of producing specific characteristics, or in modern day terminology, genes represent risk factors that increase the vulnerability for developing specific (behavioral) traits. Charles Darwin on the other hand, in his book "*The variation of animals and plants under domestication*" reported a case in which all the children of a man who lost his little finger were born with deformed fingers. He also suggested that many Jewish children were born without the foreskin of their penis due to the fact that their parents were circumcised. For a more detailed description of the history of genetics, the reader is referred to the excellent book by Sturtevant (2001).

Given the uncertainty about which characteristics are and which are not heritable, it is not surprising that the overarching idea was that the characteristics of a child were randomly determined from the range of homologous traits of the parents. This idea is (now) generally referred to as blending inheritance. However, in the 18th and 19th century, it was already well known that the blending inheritance idea had several significant shortcomings. First of all, as illustrated in Fig. 1.1, if every child represented a blend between the two "extreme" traits of the parents, over the generations, the range on the extremes would

Gene-Environment Interactions in Psychiatry. http://dx.doi.org/10.1016/B978-0-12-801657-2.00001-X

BOX 1.1

A CHRONOLOGY OF MAJOR MILESTONES IN GENETICS

±300 BC	Aristotle on reproduction and inheritance
1859	Darwin and Wallace publish on the principle of evolution
1866	Mendel publishes "Über Pflanzenhybriden"
1866	Haeckel proposes that the cell nucleus contains the hereditary elements
1869	Miescher publishes on "nuclein" an acidic compound in the nucleus
1881	Kossel identifies nuclein as a nucleic acid and isolates the five bases that make up DNA and RNA.
1900	Correns, deVries, and Tschermak rediscover Mendel's work
1902	Boveri and Sutton propose that chromosomes are responsible for inheritance
1905	Stevens identifies the X and Y chromosome as determinants for the sex of an individual
1911	Morgan describes linkage between genes depends on the nearness on the chromosome
1928	Griffith publishes his paper on the transformation of Pneumococcus
1944	Avery shows that the transforming particle of Griffith is DNA
1952	Chargaff shows that the ratio of A-T and C-G in DNA is close to 1
1952	Hershey and Chase prove that DNA alone is responsible for heredity
1953	Watson and Crick publish their report on the double helix structure of DNA
1957	Crick proposes the central dogma (DNA $\rightarrow$ RNA $\rightarrow$ protein) and speculates that three nucleotides code for a single amino acid
1972	Berg creates the first piece of recombinant DNA
1982	Human insulin based on recombinant DNA enters the market
1983	Mullis invents the PCR method
2001	The complete sequence of the human genome is published
2002	The complete sequence of the mouse genome is published
2004	The complete sequence of the rat genome is published

become smaller and smaller and resulting in more and more similar individuals. The second obvious shortcoming of the blending theory is the failure to explain how certain traits can disappear for one or more generations and then reappear again. For instance, blue eyes or blond hair sometimes disappears for several generations in a family before reappearing again.

Although these shortcomings were clearly recognized, scientific research into heritable traits was not very widespread and usually confined to breeding two different species together and looking at their offspring. As the offspring of such breeding experiments is usually infertile, the genetic information obtained is limited to a single generation.

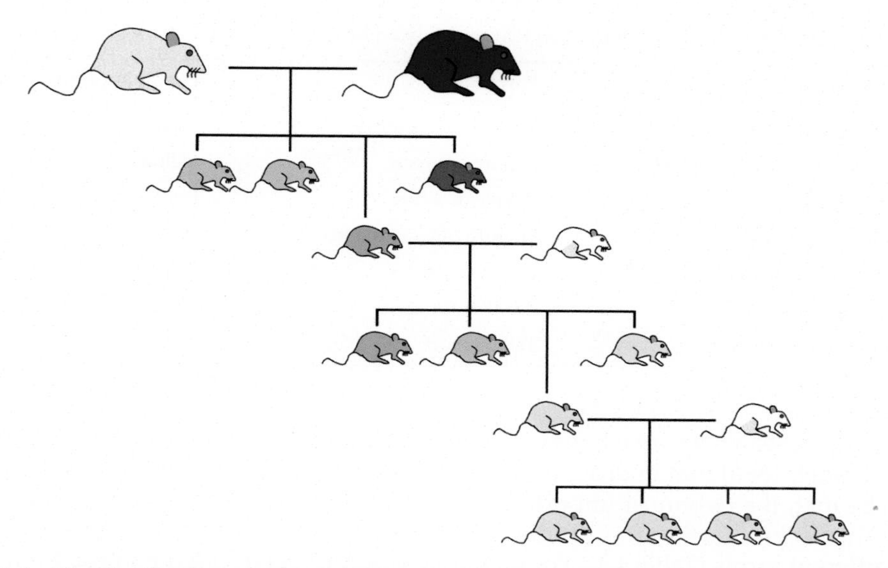

FIGURE 1.1 **The blending theory.** The blending theory of inheritance proposed that the characteristics of the offspring were a blend between the characteristics of each of the parents. The ultimate consequence of this theory is that over time all the offspring will become similar.

GREGOR JOHANN MENDEL

Gregor Mendel was born Johann Mendel, the son of Anton and Rosine Mendel on the 20th of July 1822 in the little town of Heisendorf in Silesia (now known as Hynˇice, in the Czech Republic, close to the borders with Slovakia and Poland). Although, as the only son among two sisters, he was destined to take over the farm of his father (which had been in the family for over 130 years) his parents allowed him to enter the University of Olomouc to study physics and philosophy. Upon the recommendation of his physics teacher, Mendel entered the Augustinian abbey of St Thomas in Brno, and took the name Gregor. This allowed him to study without having to pay for it. In 1851, he was sent to the University of Vienna to receive a more formal training (where his professor of physics was the famous Christian Doppler, well known for his discovery of the change in frequency of a wave moving towards an observer, now known as the Doppler effect). He returned to the abbey in 1851 to teach physics and in 1867 when he became abbot.

Even though he predominantly studied physics and had a keen interest in astronomy and meteorology (he was the founder of the Austrian Meteorological Society in 1865), he is, of course, best known for his studies on heredity. Although he originally started his research in mice, he switched to pea, mainly because his bishop was uncomfortable that one of his monks was studying sexual behavior in mice. In retrospect, this switch was probably the best thing that could have happened to Mendel, as he would unlikely have been able to deduce the principles of heredity now laid down in his two laws from studying mice.

TABLE 1.1 The Characteristics of the Pea Plants Analyzed by Gregor Mendel (1866)

Characteristic traits	Forms	Dominant trait
Shape of the mature seeds	Round or wrinkled	Round
Color of the seed	Yellow or green	Yellow
Color of the flower	Purple or white	Purple
Shape of the mature pod	Smooth or constricted	Smooth
Color of the immature pod	Green or yellow	Green
Position of the flowers	Axial or terminal	Axial
Height of the principal stem	Tall or short	Tall

In fact, his research approach was brilliant, as he pointed out in the very beginning of his paper (Mendel, 1866) that in order to understand the basis of inheritance, several conditions need to be met, the two most important ones being: (1) the offspring in subsequent generation are themselves fertile, and (2) the characteristics are easily recognizable and exist only in two different forms (Table 1.1). For these reasons, Mendel decided on peas from the genus *Pisum* (especially *Pisum sativum*). He also, serendipitously, selected characteristics that were themselves independent of each other. In addition to this, Mendel showed an extraordinary patience and energy, cultivating and analyzing around 29,000 pea plants, and in doing so, laid the foundation for our understanding of the principles of heredity.

MENDEL'S FIRST LAW OF INHERITANCE: THE LAW OF SEGREGATION

The first observation Mendel made when he started his experiments was that all of the first generation (F1 from Filial 1) hybrids showed only one of the characteristics traits, which he called the dominant trait (Table 1.1). As a particularly interesting example he showed that all the F1 hybrids of the tall (around 15 cm) and the short (2.5 cm) plants were themselves between 15 and 17 cm tall. This was in clear conflict with the blending model, which would predict that the offspring would be around 8–9 cm.

However, more interesting were the results of the next (F2) generation. When the F1 hybrids were subsequently crossed, both of the original traits were found again, and always in a ratio of roughly 3:1. For example of the 253 hybrids Mendel analyzed for the shape of the seeds, 5474 were round, while 1850 were wrinkled (even though the parent plants only had round seeds). This, again, was in obvious violation of the blending model.

On the basis of these results, Mendel concluded that "*In Pisum it is beyond doubt that for the formation of the new embryo a complete union of the elements of both reproductive cells must take place. How could we otherwise explain that among the offspring of the hybrids both original forms reappear in equal numbers and with all their peculiarities?*" (p. 41). Formulated slightly differently this leads to Mendel's first law of inheritance: "*Each individual has two elements of heredity which segregate during reproduction, with each offspring inheriting one element from each parent.*" Moreover, this is sometimes referred to as Mendel's third law, one of the two

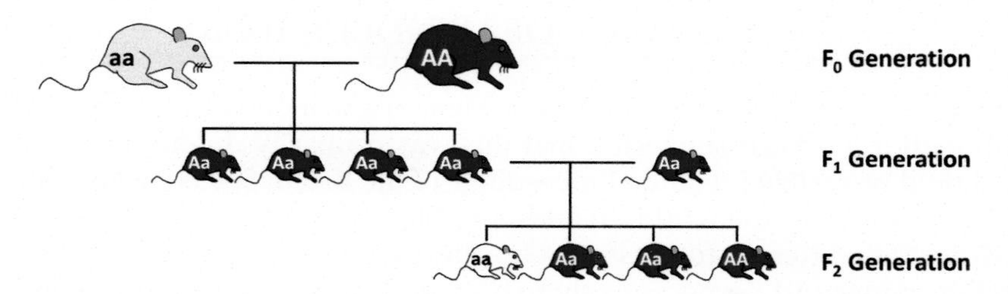

FIGURE 1.2 **Mendel's first law.** Mendel's first law states that each individual inherits two elements one from each parent. In the figure these are indicated by A and a. In many cases one of the element dominates, in our example the dark green coat colour of the rat (indicated by a capital A). As all animals in the first generation (F1) inherit one capital A from the dark green coloured father, all the offspring will be dark green coloured. However, in the second (F2) generation, 25% of the animals will obtain both a lower case "a" from each parent and retain the original white colour again. Thus, in contrast to the blending theory of Fig. 1.1, Mendel's theory can explain why certain characteristics may not occur for a generation but return afterwards.

elements may dominate the other. If this is the case, only the dominant trait will be visible. This is illustrated in Fig. 1.2 showing how this leads to the ratio of 3:1. As the nondominant element reappeared in the F2 generation, Mendel referred to this as the recessive element. As we shall see in Chapter 2, with a few exceptions, Mendel's laws are still valid as long as the studied traits are inherited by a single element. Although Mendel reported on studies with beans (*Phaseolus vulgaris* and *P. nanus*), he was careful enough to report that he only did preliminary studies with small number of these plants. Nonetheless in most cases the same results were obtained as with *Pisum*: One independent element was inherited from each of the parents.

MENDEL'S SECOND LAW OF INHERITANCE: THE LAW OF INDEPENDENT ASSORTMENT

In addition to experiments in which only one characteristic was varied, Mendel also crossed plants with multiple different characteristics, for instance plants that differed in both the shape and the color of the seeds. As he studied seven different characteristics (Table 1.1), this could lead to in theory $2^7 = 128$ different combinations, if all the characteristics were independent of each other. In his experiments, he reported that indeed all 128 different combinations were found. This led to Mendel's second law of inheritance: "*Each element is inherited independently from other elements of inheritance.*" Although his experimental data are completely in line with this law, we now know that this is certainly not true for all traits. In fact, many traits do inherit together, as we will discover in Chapter 2.

During his experiments Mendel carefully mirrored his experimental conditions, that is, ensuring that the dominant and recessive elements were either present in the male or the female plants. By doing so, he could conclude that both parents are equal in their transfer of one element of inheritance. Again, we now how clear reason to believe, that there are also some exceptions in which the male element is more important than the female, or vice versa.

THE LEGACY OF MENDEL'S WORK

It took until the turn of the century before Mendel's works was rediscovered and appreciated. Although it has been suggested that this was primarily due to the 'obscurity' of the journal, it should be realized that the Proceedings of the Brünn Society were sent out to more than 120 libraries all over the world. In addition, Mendel himself sent copies to Nägeli (professor in Münich) and Kerner (professor in Innsbruck). However, Nägeli seemed not to have understood the paper and Kerner is likely to have never even read it (Sturtevant, 2001). Until 1900, Mendel's work was only cited four times and it seems more likely that people did not understand it or failed to see its significance.

The rediscovery was independently done by the Dutch Hugo de Vries the Germans Carl Correns, and Erich von Tschermak. All three reached the same conclusions that F2 hybrids showed the characteristics of the parent plants in a ratio of 3:1. Especially de Vries recorded a series of quite different genera of plants and concluded that the 3:1 ratio probably hold for all discontinuous variations. As Sturtevant emphasizes, all three were quick to point out that it was in fact Mendel who first developed this idea. As de Vries wrote *"This [Mendel's] memoir, very beautiful for its time, has been misunderstood and then forgotten."*

However, it was Williams Bateson who probably became the most vocal supporter of Mendel. According to his wife, he read Mendel's work on the train from Cambridge to London in May 1900, on his way to give a paper at the Royal Horticultural Society. Apparently he was so impressed that he immediately incorporated Mendel's ideas as well as de Vries' confirmation. After that he became a fervent supporter of Mendel's theory and built an active research group in Cambridge. Together with others they set out to answer the most important questions that arose from Mendel's work: how generally applicable are his findings (beyond peas and beans); how are so-called compound characteristics explained; and how widespread is the phenomenon of dominance. With respect to the first, it was Bateson and Cuenot in France who definitely proved that Mendel's principles also applied to animals (fowl in the case of Bateson and mice in the case of Cuenot). Bateson also introduced many of the words we are now so familiar with, including "genetics," "zygote," homozygote, and heterozygote. Bateson also introduced the terms F1 and F2 to indicate the different generations of hybrids.

CRITICISM OF MENDEL'S WORK

Although these principles are now uniformly accepted in the scientific community, Mendel's work and especially his experimental evidence has also been criticized. This criticism concentrates on two different aspects of his work. First of all it has been suggested that the observed ratio of 3:1 of several of the characteristics is too perfect, especially in small sample sizes. Secondly, the fact that the seven characteristics that Mendel analyzed (Table 1.1) are all located on different chromosomes (which enabled Mendel to formulate his law of independent segregation) was considered suspicious.

One of the most famous critics of Mendel's work was Ronald Fisher, who contributed greatly to quantitative genetics by introducing the concept of variance and developing mathematical tools for reconciling Mendelian genetics with population based quantitative traits. In a now (in)famous paper published in 1936 he carefully analyzed the entire Mendel paper and came to the shocking conclusion "*It remains a possibility among others that Mendel was deceived by some assistant who knew too well what was expected. This possibility is supported by independent evidence that the data of most, if not all of the experiments have been falsified so as to agree closely with Mendel's expectation (Fisher, 1936)*. This was, according to Fisher, most apparent for the results of two series of experiments, which were in line with Mendel's alleged erroneous expectation, rather than with Fisher's (alleged) correct interpretation. However, in recent years it has been convincingly argued that, in fact, Fisher's analysis is wrong, or at least, that his criticism of Mendel's results was unfounded (Novitski, 2004a, 2004b; Hartl and Fairbanks, 2007).

THE SEARCH FOR THE MENDELIAN ELEMENTS

Around the time of the rediscovery of Mendel's work, significant progress was also made in identifying the physical location of the genetic information. It was already well known by the end of the 19th century that each species had the same number of chromosomes in all its cells. Moreover, the number of chromosomes was usually even, suggesting that equal numbers came from the sperm and the eggs. And it was generally assumed that, somehow the chromosomes contained the hereditary information. In an interesting series of experiments Boveri showed in 1902 that in sea urchins excess sperm could lead to cells with three sets of chromosomes. Although these cells could divide several times, the resulting embryos were all quite abnormal, suggesting that a specific number of chromosomes is crucial for normal development.

However, it was thought at that time that after the resting phase of the cell, prior to cell division, the chromosomes formed one single thread, a "spireme," which then split into different individual chromosomes. It was therefore assumed that the chromosomes were connected end to end. Most importantly, however, it was at that time thought that all chromosomes were identical, and the notion that chromosomes were present in pairs was completely unknown. It was up to the cytologists to finally resolve these issues. Although there were several people close to the correct solution, it was finally Walter Sutton who in 1902 showed that in the grasshopper, chromosomes occur in clearly distinguishable pairs. He finished his paper with the remark "*I may finally call attention to the probability that the association of paternal and maternal chromosomes in pairs and their subsequent separation during the reducing division may constitute the physical basis of the Mendelian law of heredity.*"

However, there were some problems with this general idea. Most importantly, it was well known that the total number of chromosomes was fairly small. In humans, there were 23 pairs of chromosomes (Fig. 1.3) and although the number of chromosomes differ between different species (Table 2.1), it was usually relatively small. However, Mendel's second law of inheritance (the law of independent assortment) states that all traits and inherited independently. As one would assume there are more than 23 traits in humans, this must imply that there are more genes on a single chromosome, and that somehow the chromosomes must split and recombine. It was de Vries who suggested that during meiosis genes were freely exchanged between homologous regions. It was already known that during meiosis the number of chromosomes halved in each cell, but that there were actually two divisions during meiosis. In the first division the number of chromosomes doubles and in the subsequent second division the resulting four cells each have half of the original number of chromosomes.

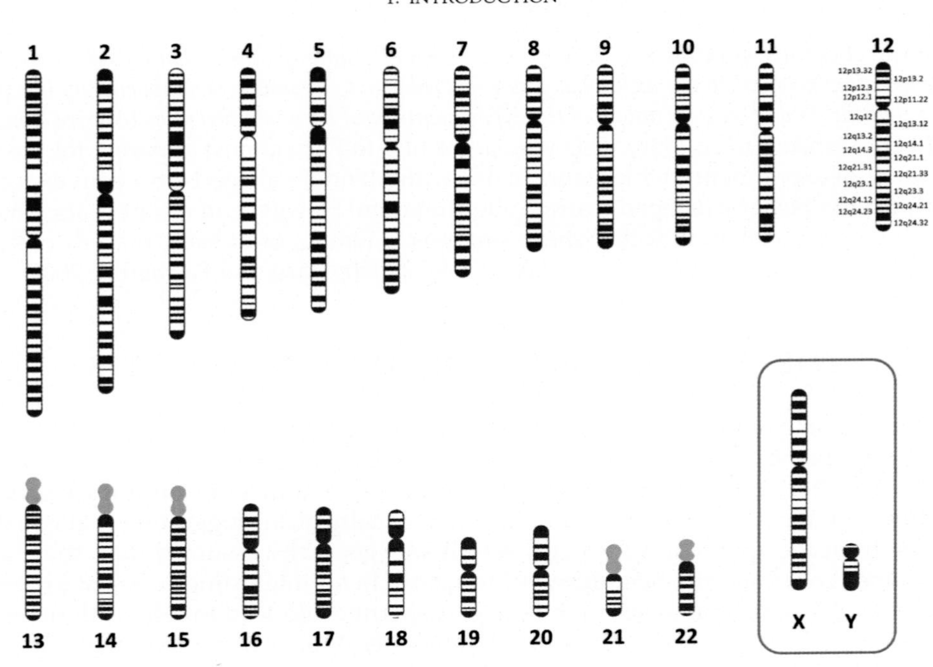

FIGURE 1.3 **The human chromosomes.** Like most other species, humans have two sets of chromosomes, consisting of autosomes (22 pairs in the case of humans) and sex chromosomes. Females have the X chromosomes, males have one X and one Y chromosome. Each chromosome has a short (p) and a long (q arm) and using special staining techniques distinct bands can be visualized, allowing for detailed mapping of specific characteristics. The banding pattern start from the centromere (p1, p2, p3, q1, q2, q3) with subbands receiving additional numbers). An example is given for chromosome 12 (top right corner).

It was Janssens who subsequently explained the necessity of this two-step process. He showed that longitudinally paired meiotic chromosomes each undergo a split, giving a quadripartite structure made up of two daughter strands of each of the original members of the pair. During the first meitotic division, two strands pass to each daughter cell (Sturtevant, 2001). Moreover, Janssens suggested that during this process occasionally an exchange took place between two strands (Fig. 1.4). He also suggested that the two strains need not always be sister strands. Thus this explains why there need to be two meiotic divisions (since only two of the four strands undergo an exchange at any one level). Moreover, it could explain the law of independent assortment, as genes may freely move from one chromosome to another.

At around the same time, however, the first indications that the law of independent assortment was sometimes violated and that, in fact, different characteristics may actually be genetically linked. Correns was among the first to show linkage, in his experiments with two strains on Matthiola (stock), one with coloured flowers and seeds, and hoary leaves and stems, and one with white flowers and seeds, and smooth flowers and stems (Correns, 1900). The F1 had coloured flowers and hoary leaves. In the F2 he expected to find many different combination resulting from independent segregation, but instead he found only the original F0 combinations in the now well-known ratio of 3:1. This was a clear indication that flower colour and leaves surface were linked to each other. Given the fact that combinations of white flower and

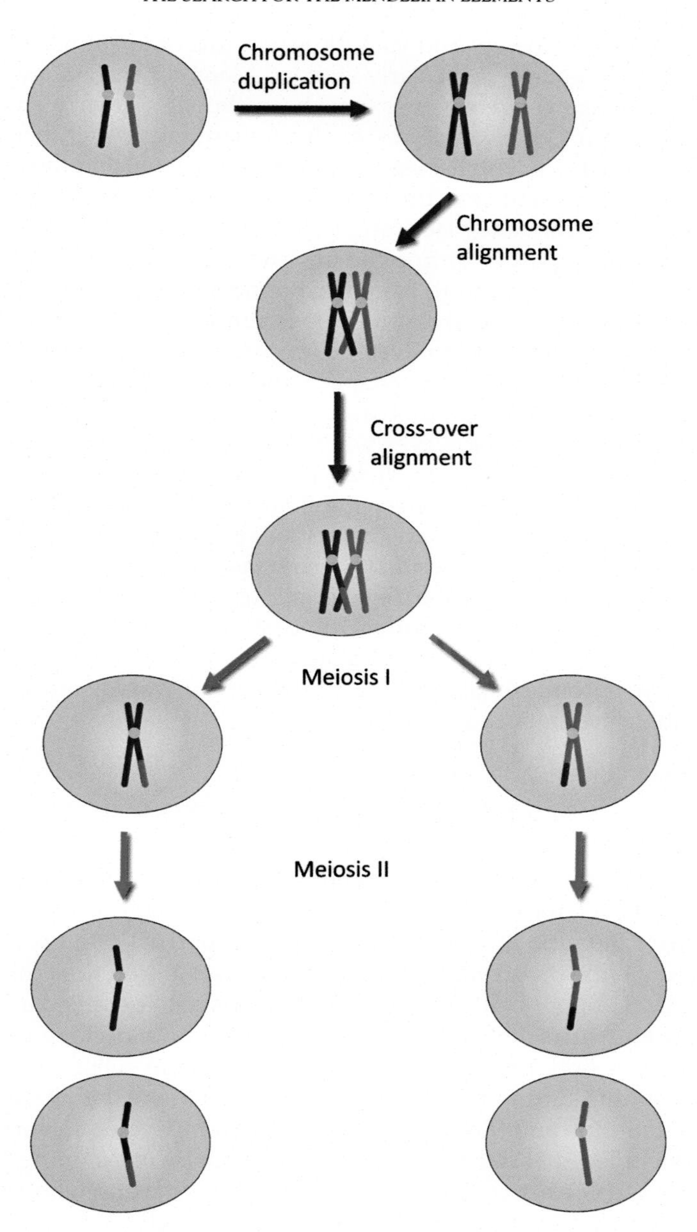

FIGURE 1.4 **Meiotic division.** Meiotic divisions lead to the formation of gametes (male or female germ cells). Several different stages can be identified, the most important of which are illustrated in the figure. First chromosomal duplication takes place, after which the homologous chromosomes line up close together. This stage is followed by the cross-over stage, during which part of one chromosome is exchanged with the homologous part of the second chromosome. Next during the first meiotic division (meiosis I) the cells divide in two each carrying both an original parental copy as well as a mixed (crossed-over) copy. During the subsequent second meiotic division (meiosis II), each of the cells divides again and the haploid cells (carrying only a single set of chromosomes) gametes are formed.

hoary leaves and coloured flowers and smooth leaves did occur in nature (confirming they are the result of different genes), this clearly violated Mendel's second law.

Since then, such linkage between genes have been shown regularly and the principle has proven very useful in our genetic analyses, such as quantitative trait loci and genome wide association studies (which we will discuss in more detail in later chapters). Using Fig. 1.4, the difference between independent assortment and linkage can be easily demonstrated. During meiosis cross-over events can take place (on an average 2–3 cross-over events occur between each pair of human chromosomes during the first meitotic division. During this meitotic division, the two chromosomes begin to form a synaptosomal complex, aligning very closely together, allowing for an exchange of chromosomal information. Crossing over (also referred to as homologous recombination) involves the physical breaking of the two copies of the DNA molecule (see below and the next chapter) and rejoining of the homologous (similar) paternal and maternal ends. Through this process an almost unlimited number of genetically different gametes can be formed. However, whether the genes from the paternal and maternal chromosomes cross-over depends on the distance between the genes on the chromosomes. As crossing over is in principle a random effect, the chances that two genes will end up separated after a cross-over is larger if the genes are farther away from each other on the chromosome. The theory of cross-over was first described by Thomas Morgan (relying heavily on Janssens work discussed above). In his honor the distance between genes on the same chromosomes is now measured in Morgan, or more commonly in centimorgan (cM). It is important to realize that this is not actually an absolute physical distance (as in cm). Rather 1 cM is defined as the distance between two chromosomal positions for which the average number of chromosomal crossovers (in a single generation) is 0.01.

THE SEX CHROMOSOMES

Theories about what determines the sex (gender) of an individual were numerous even in Aristotle's time. However, most of these theories had little scientific basis and none convincingly explained why the ratio of males to females is around 1:1 in all species. The first suggestion that genes may be involved in determining the sex differences came from studies in the grasshoppers (McClung, 1901). Earlier work in insects had identified a so-called accessory chromosome (now known as the "X"-chromosome) which only divided in one of the meiotic divisions and thus was only present in two of the four sperm cells, a finding confirmed by several others, including Sutton and McClung. Unfortunately, the analysis of female cells proved difficult and Sutton and McClung counted 22 chromosomes in females and 23 in male cells, and thus concluded that the X-chromosome was male-determining (Sturtevant, 2001). Several years later the correct conclusion was drawn by Stevens who showed that in the beetle (Tenebrio), female cells have two X chromosomes, while male cells have an X and a Y. These findings were quickly replicated in several other kinds of animals, although controversy existed for some time as it is not a universal principle. In fact, in birds (as well as some reptiles and insects) the situation is reversed. Here the sex chromosomes are generally referred to as Z and W and it is the females that is heterozygous (ie, ZW) while the males are homozygous (ZZ). In addition, other sex chromosome systems exist, as well as several systems for determining sex which are not genetically based. For instance, certain reptiles including alligators, certain turtles, and the very ancient tuatara (now only found in New Zealand) use a temperature-dependent system. Within this system two different variant are known: in variant I extreme temperatures on one side (ie, high) produce males and on the other females, while in variant II both extremes produce animals from one sex and only at middle temperatures animal from the other sex are produced.

BEYOND MENDEL: THE HISTORICAL DEVELOPMENT OF MOLECULAR GENETICS

With his two laws, Mendel laid the foundation of the fundamental principles of inheritance. However, it took almost another 100 years before the molecular mechanisms underlying these principles were finally elucidated. It started with the discovery by Friedrich Miescher of a long molecule from the nucleus of pus cells which he called nuclein (Dahm, 2008). Although he submitted his paper in 1869, the editor (Hoppe Seyler who was also the head of his department) apparently withheld the paper for 2 years as he repeated some of the observations he thought improbable (a very special form of peer review). Unfortunately, some of the DNA preparations Miescher worked on were contaminated with proteins (protamine, which will be discussed in subsequent chapters). So when in 1889, Richard Altmann finally managed to isolate the two components, he called the nonprotein substance nucleic acid, failing, however, to realize that it was not a subcomponent of nuclein, but the exact same substance (Altmann, 1889; Dahm, 2008). This molecule, he noted, was peculiar since it was very large but while containing a lot of phosphorus it completely lacked sulfur. And although the structure of this substance was far from elucidated, already in 1896 Wilson hypothesized that "*And thus we reach the remarkable conclusions that inheritance may, perhaps, be effected by the physical transmission of a particular chemical compound from parents to offspring*" (Wilson, 1896).

It was Albrecht Kossel, also a former research assistant of Hoppe Seyler in Strasbourg, who identified two different forms of nucleus acids from thymus and yeast cells. Consequently, they were referred to as "thymus nucleic acid" and "yeast nucleic acid," now known as deoxyribonucleic acid (DNA) and ribonucleic acid (RNA) respectively. In addition, he identified the building blocks of nucleic acids (the purines and pyrimidines, sugars, and phosphates), as well as the histones, proteins tightly connected to DNA (Dahm, 2008). For his work on nucleic acids Kossel received the Nobel Prize for physiology and Medicine in 1910.

In the meantime, it was found by Levine that both forms of nucleic acid contained adenine, cytosine, and guanine, while DNA also contained thymine, but in RNA this was replaced by uracil. Levine also showed that the sugars in DNA and RNA were deoxyribose and ribose respectively (Levene, 1919). However, the exact composition was largely unclear. In fact it was thought that each DNA molecule contained equimolar amounts of the four bases, which led to the so-called "tetranucleotide hypothesis." According to the hypothesis, the basic unit of DNA consists of a group of one of the four bases.

A major breakthrough, although it was not recognized as such at the time, were the experiments performed by Fred Griffith (Griffith, 1928). He studied the effects of several different strains of Pneumococcus bacteria. It was already known that some of the strains, such as the rough, type II strains had lost their pathogenicity, while others, such as the smooth, type III

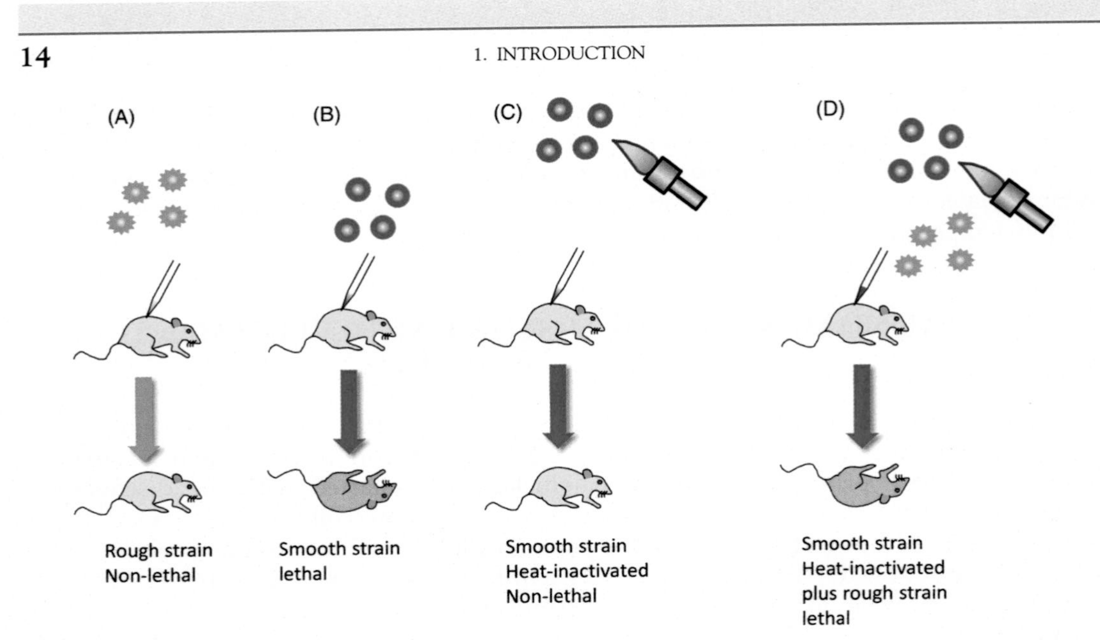

FIGURE 1.5 **Griffith "transformation" experiment.** During these experiments Griffith first of all confirmed that some strains of Pneumococcus such as the rough strain were nonlethal for animals: (A) while others such as the smooth strain ere (B). He also confirmed that heating the smooth strain killed the bacteria making them nonlethal (C). In the crucial experiment (D) he showed that mixing the heat-inactivated smooth strain and the alive (but harmless) rough strain led to the death in his animals. Thus, he proposed that some aspect of the dangerous smooth strain could pass onto the harmless strain (even though the dangerous strain was itself already dead) and 'transform' it into a dangerous strain.

strains were still very pathogenic (Fig. 1.5). In his experiments, Griffith first confirmed that injecting the smooth but not the rough Pneumococcus strain into mice led to the death of all the animals. Moreover, he showed that when he heat-inactivated the smooth strain, these bacteria lost their pathogenicity and all animals survived after inoculation. In the crucial experiment, he then showed that when he combined the heat-inactivated smooth strain with the normal (nonpathogenic) rough strain, the mice died, and from their bodies he could isolate virulent type III bacteria. This was the first example of, what became known as, *"transformation."* However, the mechanism behind this was completely unknown, and Griffith himself suggested that the rough strain must somehow have "ingested" the pathogenic factor of the smooth strain, which he referred to as *"pabulum."* In fact it took another 16 years before Oswald Avery et al. (1944) solved the mystery. Using a systematic approach, they successively removed all the individual components of the smooth strain to identify which of the components were responsible for the *"transformation."* They showed that even after removing proteins, lipids and polysaccharides, the extract still retained the ability to transform the rough strains. They subsequently moved on to show that it was DNA that was responsible for transforming the rough strain. This then explained Griffith's results: heat inactivation left the DNA within the chromosomes (more or less) intact. Part of this DNA was then incorporated into the DNA of the rough strain transferring the lethality of the original smooth strain. Unfortunately, Avery's results and interpretation were fiercely attacked by one of his direct colleagues, Alfred Mirsky, who was convinced that DNA was much too simple a molecule to be able to be responsible for transferring genetic information. Instead, he reasoned, it must be one or more proteins. It has even been suggested that his fierce opposition persuaded the Nobel committee to deny Avery the Nobel Prize (White, 2002). However, Mirsky was proven wrong, and in 1952 Alfred Hershey and Martha Chase proved that it was DNA and only DNA that contained the genetic information (Hershey and Chase, 1952). Their ingenious experiments involved the bacteriophage T2, a virus that infects bacteria thereby forcing the bacteria to produce new bacteriophages. As the T2 bacteriophage consists of only two components, they selectively labelled the DNA with radioactive ^{32}P, and (in a separate batch) the proteins with ^{35}S. Next they allowed both type of T2 to infect bacteria, and subsequently analyzed the progeny of the bacteriophages. The results showed that the progeny contained radioactive ^{32}P but not ^{35}S, thereby convincingly showing that only DNA is transferred from the bacteriophage to the bacteria and on to the next generation. For his work, Hershey would subsequently receive the Nobel prize for medicine and physiology in 1969 (together with Max Delbrück and Salvador Luria). In the conclusion of their original paper, Hershey and Chase were very careful with their interpretation stating that "*This protein probably has no function in the growth of intracellular phage. The DNA has some function. Further chemical inferences should not be drawn from the experiments presented.*"

The final answer to the question how such a (relatively simple) molecule could perform the most important function in any living creature came only one year later with the famous paper by James Watson and Francis Crick (Watson and Crick, 1953).

The story of the race to identify the structure of DNA has been told in many popular books (including "the double helix" by James Watson himself). One of the most interesting and entertaining descriptions can be found in Michael White's book "*Rivals: conflict as the fuel of science*" (White, 2002). In the next chapter we will discuss the structure of DNA in much more detail. Suffice to say that the Watson & Crick paper marks the end of the search for the molecule responsible for transmitting the genetic information, while at the same time marking the beginning of the molecular genetic revolution.

A BRIEF HISTORY OF BEHAVIORAL GENETICS

We want to finish this introductory chapter with a brief overview of the history of behavioral genetics. Behavioral genetics has been described as having a long past but a short history (Loehlin, 2009). Indeed, that animals and humans inherit specific traits from their parents was already known in ancient Rome, Greece, and Egypt. This was perhaps most explicitly voiced by Plato, who suggested that like dogs and birds, the breeding of elite class of humans should be based on the principle of matching the best with the best (Loehlin, 2009). However, Plato also recognized that in addition to a good ancestry, proper education was also very important, thus already emphasizing the interaction between nature and nurture. The nature nurture discussion can also be found in Shakespeare's the Tempest "*a devil, a born devil, in whose nature nurture never sticks.*" The British philosopher John Locke was probably one of the most explicit proponents of the nurture side of the nature – nurture debate. He suggested that humans were basically born with a blank mind which was then "filled in" by experience.

With his famous work "on the origin of species" and even more so with "the descent of man," Charles Darwin concluded that humans (and their behavior) were governed by the same evolutionary principles as all other animals. This was made even more explicit by Darwin's younger cousin Francis Galton (1822–1911). In contrast to Locke and his followers, he was a clear proponent of the "nature over nurture" hypothesis stating that, as long as nurture was within the normal limits, nature's influence on our behavior was overwhelming. Galton is probably most (in)famous for the introduction of the term "eugenics" (from the Greek words "*eugenes*," meaning well-born and "*genos*," meaning race). Eugenics aims to improve the human race by the promotion of a higher reproduction of people with "desirable" properties, while at the same time limiting the reproduction of people with "undesirable" properties. Although Galton predominantly supported voluntary eugenics, his ideas have been a constant source of heated debate (Graves, 2001). Galton's eugenics rapidly spread to the United Kingdom and United States championed by people like Charles Davenport and Henry Goddard. The latter became famous with his book "*The Kallikak family: A study in the heredity of feeble-mindedness*" (Goddard, 1912), in which he described the genealogy of Martin Kallikak. The name Kallikak is a pseudonym, a combination of the Greek words "kallos" (beautiful) and "kakos" (bad). Martin was a respectable man (a revolutionary war hero) married with a Quaker woman with which he had a number of "wholesome" children without any signs of mental retardation. However, on his way back from the battle, Martin had sex with a "feeble minded girl" and a single (feeble minded) son (who later became known as "old horror") was born who went on to father 10 children, many of who also showed signs of mental retardation. From these data Goddard concluded that intelligence, sanity and even morality were genetically determined and hence that every effort should be undertaken to prevent feeble minded individuals from procreating.

The ideas proposed by eugenics culminated in the racial theories that developed in Nazi Germany during the 1930–40. Although these theories were predominantly developed by figures such as Alfred Rosenberg, he was guided by psychiatrists such as Ernst Rüdin and his colleagues in Munich. As Propping formulated it "*what had begun as a romantic ideal of eugenics ended as an uncompassionate delusion of the need to attain a cleansed race*" (Propping, 2005).

It can't be a surprise that after this horrible period in our history the pendulum swung back in the direction of nurture over nature. This was probably best illustrated by the behaviorists such as John Watson and his student B. F. Skinner illustrated by Watson famous quote "Give me a dozen healthy infants, well-informed, and my own specified world to bring them up in and I'll guarantee to take any one at random and train him to become any type of specialist I might select–doctor, lawyer, artist, merchant-chief, and yes, even beggar-man and thief, regardless of his talents, penchants, tendencies, abilities, vocations and race of his ancestors" (Watson, 1930).

Although this rather extreme view has by now been abandoned again, behavioral genetics still regularly leads to controversy, both in the general media as well as in the scientific community. Especially studies that found genetic differences in intelligence often lead to heated public debate. Examples are Jensen's studies suggesting that there are genetically determined differences in IQ between Caucasian and African Americans (Jensen, 1969), and Herrnstein and Murray's book "*The bell curve.*" Although the authors only devoted a small part of their book on racial differences in cognition, it was this aspect that received most attention in the popular media.

Perhaps more interesting for our purpose is the continued discussion in the scientific literature about behavior genetics, especially between developmental biologists and geneticists. The main controversy is illustrated by Richard Lerner, in a paper in which he asks: "*Why do we have to keep reinterring behavior genetics or other counterfactual conceptualizations of the role of genes in behavior and development? Why is it still necessary to continue to drive additional nails into the coffin of this failed approach*" (Lerner, 2006). The main point of controversy is related to the influence of environmental factors. According to Lerner, "*genes always function through interaction with the context*" (p. 338), and therefore studying genes in isolation is useless. As we will see in later chapters, we feel that this is a very important point, and may indeed be one of the reasons why genetic findings in mental disorders are notoriously difficult to replicate. However, as exemplified in a rebuttal of Lerner's argument, Matt McGue convincingly argues that Lerner seems to give an incorrect description of behavioral genetics (McGue, 2010). It would, indeed, be difficult to find a behavioral geneticist these days who does not believe that genes function through an interaction with the context. In fact, it is the aim of this book to illustrate how important the interaction between genetic and environmental factors is and how it is essential to study the two in unison, rather than as individual contributing factors.

References

Altmann, R, 1889. Ueber Nucleinsaeuren. Arch. f. Anatomie u. Physiol. 1, 524–536.

Avery, O.T., Macleod, C.M., McCarty, M, 1944. Studies on the chemical nature of the substance inducing transformation of pneumococcal types: induction of transformation by a deoxyribonucleic acid fraction isolated from *Pneumococcus* type III. J. Exp. Med. 79, 137–158.

Correns, C, 1900. Mendels Regel ueber das Verhalten der Nachkommenschaft der Rassenbastarde. Berichte Der Deutschen botanischen Gesellschaft 18, 158–168.

Dahm, R., 2008. Discovering DNA: Friedrich Miescher and the early years of nucleic acid research. Hum. Genet. 122, 565–581.

Fisher, R.A., 1936. Has Mendel's work been rediscovered? Annals Sci. 1, 115–137.

Goddard, H.H., 1912. The Kallikak Family: a Study in the Heredity of Feeble-Mindedness. MacMillan, New York.

Graves, J.L.J., 2001. The Emperor's New Clothes: Biological Theories of Race at the Millennium. Rutgers University Press, New Brunswick.

Griffith, F, 1928. The significance of pneumococcal types. J. Hygiene 27, 113–159.

Hartl, D.L., Fairbanks, D.J., 2007. Mud sticks: on the alleged falsification of Mendel's data. Genetics 175, 975–979.

Hershey, A.D., Chase, M., 1952. Independent functions of viral protein and nucleic acid in growth of bacteriophage. J. General Physiol. 36, 39–56.

Jensen, A.R., 1969. How much can we boost IQ and scholastic achievement. Harvard Education. Rev. 39, 1–123.

Lerner, R.M., 2006. Another nine-inch nail for behavioral genetics! Hum. Dev. 49, 336–342.

Levene, P.A., 1919. The structure of yeast nucleic acid: IV. Ammonia hydrolysis. J. Biol. Chem. 40, 415–424.

Loehlin, J.C., 2009. History of behavior genetics. In: Kim, Y.-K. (Ed.), Handbook of Behaviour Genetics. Springer, New York, pp. 3–11.

McClung, C.E., 1901. Notes on the accessory chromosome. Anatomischer Anzeiger 20, 220–226.

McGue, M., 2010. The end of behavioral genetics? Behav. Genetics 40, 284–296.

Mendel, G., 1866. Versuche ueber Pflanzenhybriden. Verhandlungen Naturforschender Vereines Bruenn 10, S3–S47.

Novitski, C.E., 2004a. Revision of Fisher's analysis of Mendel's garden pea experiments. Genetics 166, 1139–1140.

Novitski, E., 2004b. On Fisher's criticism of Mendel's results with the garden pea. Genetics 166, 1133–1136.

Propping, P., 2005. The biography of psychiatric genetics: from early achievements to historical burden, from an anxious society to critical geneticists. Am. J. Medical Genetics Part B 136B, 2–7.

Sturtevant, A.H., 2001. A history of genetics. Cold Spring Harbor Laboratory Press, New York.

Watson, J.B., 1930. Behaviorism. University of Chicago Press, Chicago.

Watson, J.D., Crick, F.H., 1953. Molecular structure of nucleic acids; a structure for deoxyribose nucleic acid. Nature 171, 737–738.

White M, 2002. Rivals: Conflict as the Fuel of Science. Vintage.

Wilson, E.B., 1896. The Cell in Development and Heredity, first ed. The Macmillan company, New York.

CHAPTER

2

The Genetic Basis of Behavior

INTRODUCTION

"We wish to suggest a structure for the salt of deoxyribose nucleic acid (D.N.A.)" is the first sentence of the famous paper by James Watson and Francis Crick (Watson and Crick, 1953). It forms the culmination of a race between several groups, to identify the molecular structure of the molecule responsible for transferring genetic information from one generation to the next. The most important players were, in addition to Watson and Crick, Linus Pauling at Caltech and Maurice Wilkins and Rosalind Franklin in London.

The basic ingredients of DNA were known since the beginning of the century: its acidic nature, phosphate, the sugars (deoxyribose in DNA and ribose in RNA, see Fig. 2.1) and the four bases: the purines adenine [A] and guanine [G] and the pyrimidines cytosine [C] and thymine [T], the latter being replaced by the pyrimidine uracil [U] in RNA (Fig. 2.1). Perhaps the most important characteristic of DNA was reported in 1952 by Erwin Chargaff, an Austrian biochemist who had fled to the USA during the Nazi era. Although it had long been known that different DNA molecules contain different amounts of the four bases, Chargaff found that the ratio of A to T and C to G was almost always close to 1. This is now known as the Chargaff rule, and suggested that somehow A and T (as well as C and G) were closely related (Chargaff et al., 1952). The final piece of the puzzle came when Maurice Wilkins showed James Watson the (now famous) X-ray crystallographic photo 51. The photo was taken by Raymond Gosling while working as a PhD student for Rosalind Franklin. Apparently Franklin wasn't aware of this, but to Watson and Crick the diffraction pattern provided valuable evidence that DNA has a helical structure.

THE MOLECULAR STRUCTURE OF DNA

DNA molecules are large polymers (Fig. 2.2) consisting of a sugar moiety linked together by covalent phosphodiester bonds. The sugar moiety is a pentose (5-membered ring) called deoxyribose (as the hydroxyl group of ribose on the 2′ position is removed), which is linked to the phosphate groups either on the 3′ or the 5′ position, thus forming long linear chains.

Gene-Environment Interactions in Psychiatry. http://dx.doi.org/10.1016/B978-0-12-801657-2.00002-1

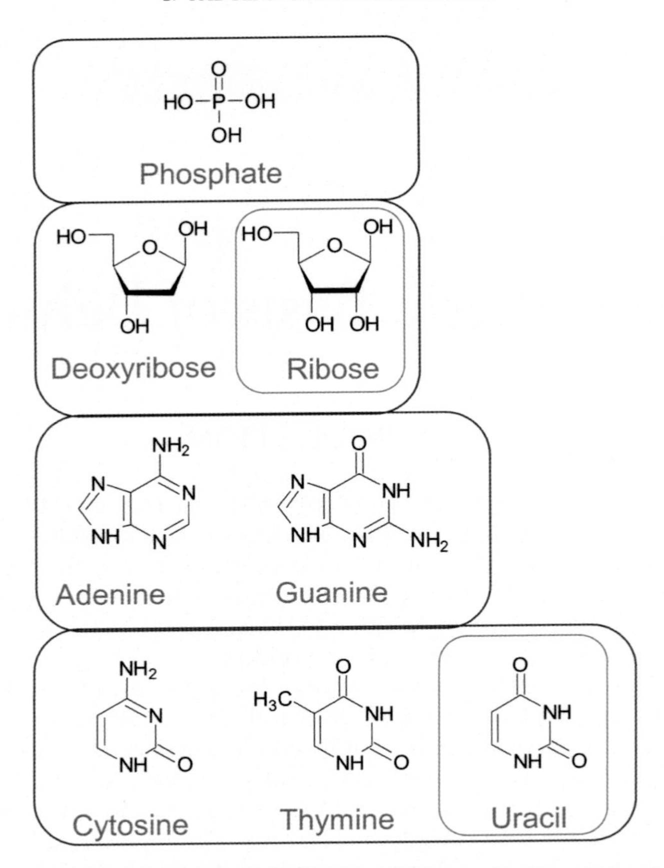

FIGURE 2.1 **Basic building blocks of DNA and RNA.** The backbone of DNA consists of a phosphate coupled to a deoxyribose sugar, while in RNA the backbone consists of a phosphate coupled to a ribose sugar. In addition, both DNA and RNA contain the purines adenine and guanine, while DNA also contains the pyrimidines cytosine and thymine. In RNA, thymine is replaced by uracil.

Due to this configuration, each DNA strand is said to have a 5′ to 3′ polarity, the relevance of which will be discussed later. In addition to the phosphate binding to the 3′ and 5′ position, deoxyribose is substituted at the 1′ with one of the nucleotide bases.

The beauty of DNA lies in the fact that the chemical structure of each nucleotide base allows for binding to one (and only one) of the other nucleotide bases through hydrogen bonds. Hydrogen bonds are (fairly weak) electrostatic bonds between polarized parts of a molecule such as between nitrogen (N) and oxygen (O). As is illustrated in Fig. 2.2, adenine and thymine can form two hydrogen bonds with each other, while adenine and guanine can form three hydrogen bonds. This then results in a double stranded form of DNA (which is much more stable than the single stranded form) and at the same time explains Chargaff's rule.

The stability of DNA is determined by the strength of the covalent binding between the different components, as well as the hydrogen bonds between the nucleotide bases on opposite

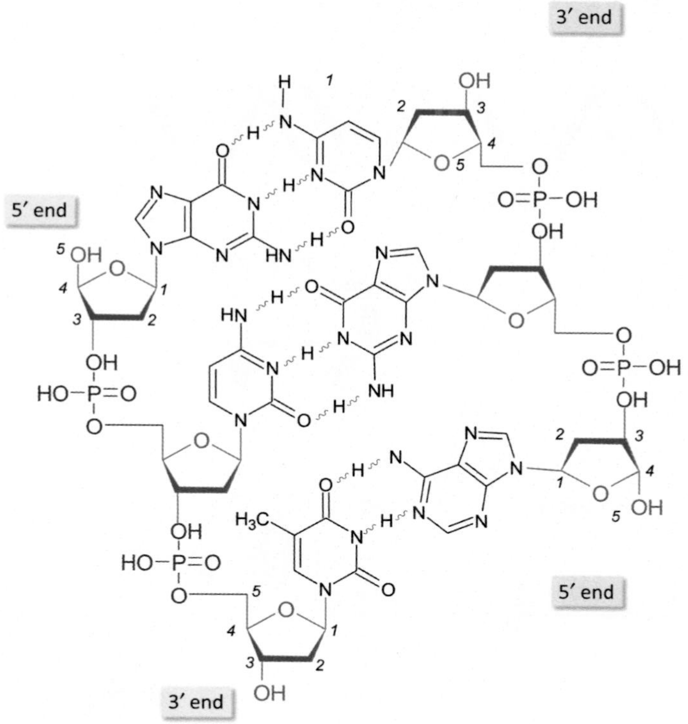

FIGURE 2.2 **The basic structure of DNA.** DNA consists of a backbone of phosphate groups connecting deoxyribose sugars at the 3′ and 5′ positions. In addition, one of the four nucleobases is attached to the 1′ position of the deoxyribose. Due to the chemical structure of the nucleobases, adenine can only pair with thymine as they can form two hydrogen bonds. Likewise, cytosine and adenosine pair with each other through three hydrogen bonds. Since the backbones run in opposite directions they are said to be antiparallel.

strands. Although hydrogen bonds are relatively weak as mentioned previously, DNA molecules generally have millions of nucleotide bases which together form a formidable force. Due to the large number of phosphate groups, DNA is a highly negatively charged molecule and therefore dissolves well in water. DNA is further stabilized by globular proteins called histones. These proteins are now known to play a very important role in determining the overall functioning of DNA (which will be discussed further in Chapter 5).

The double-stranded DNA generally takes the form of a double helix (Fig. 2.3A) and several different forms have been identified: A-DNA, B-DNA, and Z-DNA. A- and B-DNA are right-handed helixes (in which the helix spirals clockwise) while Z-DNA is a left-handed (counter clockwise) spiral. The difference between A- and B-DNA is the number of base-pairs (bp) per turn (10 vs 11). Under physiological conditions, the vast majority of DNA adopt the B-form with a pitch (ie, the distance occupied by a single turn of the helix) of 3.4 nm (Ghosh and Bansal, 2003). The helical structure of DNA is further characterized by a major and a minor groove.

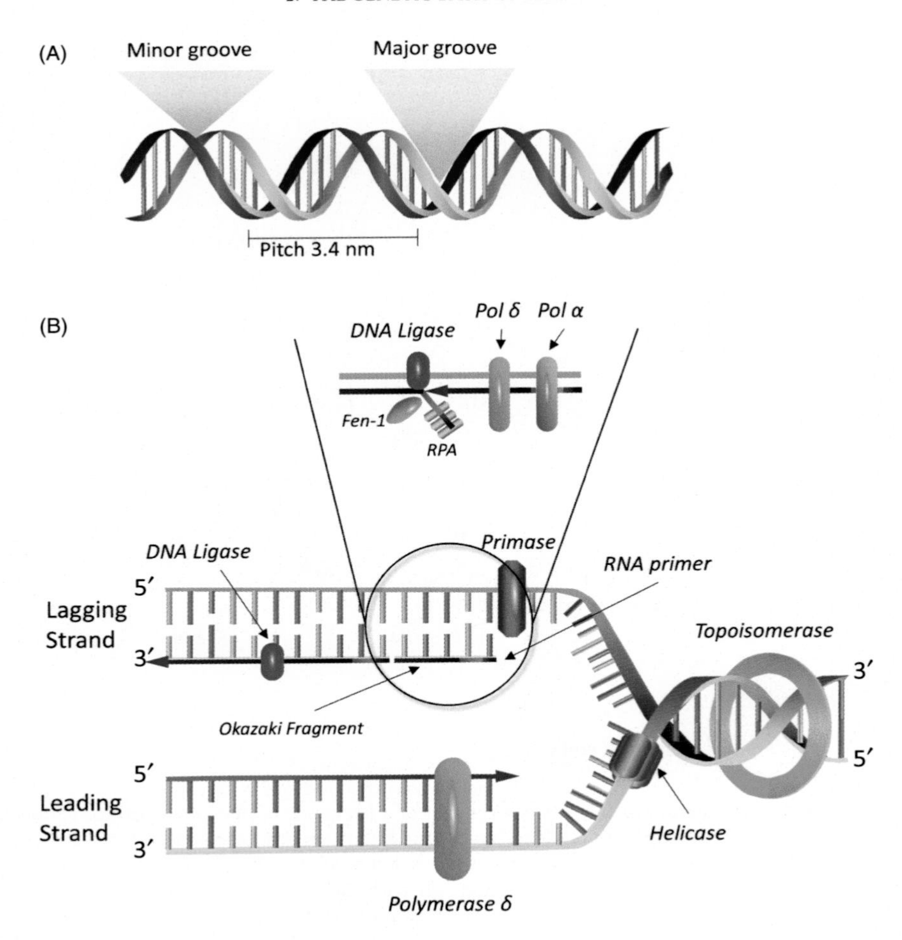

FIGURE 2.3 **DNA replication.** A: DNA forms a double helix, characterized by a minor and a major groove. The most common form of DNA has a pitch of 3.4 nm (length of a single turn of the double helix). B: DNA replication starts with the unwinding of the double helix after which each of the two strands is replicated. The process involves a number of different steps and different proteins. See the text and Box 2.1.

As discussed before, each DNA molecule has one terminal sugar in which the 3′ carbon is not linked to a phosphate (the 3′ end), and similarly one terminal sugar on the other side where the 5′ carbon is not attached to a phosphate (the 5′ end). The two strands of the double helix are said to be antiparallel because they always associate (anneal) in such a way that the 5′ → 3′ direction of one DNA strand is opposite to that of its partner. As both strands are complementary and thus essentially contain the same information, it is customary to write the DNA sequence of only a single strand. For this, we normally use the 5′ → 3′ direction as this is the direction in which DNA is synthesized and in which RNA molecules are synthesized. When a dinucleotide is described, it is customary to include a p (to indicate the phosphate bond). Thus CpG refers to a cytosine covalently linked to a guanine of the same strand.

DNA SYNTHESIS AND REPLICATION

The unique structure of DNA with its complementary strands also explains one of the great mysteries of genetics; namely how information can be faithfully transferred from one cell and one generation to the next. Indeed, Watson and Crick already stated at the end of their paper *"It has not escaped our notice that the specific pairing we have postulated immediately suggests a possible copying mechanisms for the genetic material"* (Watson and Crick, 1953). Box 2.1 summarizes the main proteins involved in DNA replication.

The process of DNA replication starts with the unwinding of the double helix (Fig. 2.3) by so-called DNA helicase proteins, (a class of motor proteins) as they travel along the phosphodiester backbone. There are a large number of helicases, and 95 different nonredundant proteins have been identified in the human genome (Umate et al., 2011), of which 64 are RNA helicases (aimed at unwinding RNA) and 31 are DNA helicases (aimed at unwinding DNA). The existence of so many different helicases can be explained by the many different processes that helicases are involved in, such as DNA replication, transcription, translation, and mismatch repair (see later). In addition, helicases are involved in processes such as recombination and ribosome biogenesis.

BOX 2.1

DNA REPLICATION

DNA replication is a tightly controlled process following several distinct steps:

1. DNA replication is initiated at many different sites within the molecule (origins of replication) by the binding of the origin replication complex.
2. This multi-protein complex leads to the recruitment of several additional proteins and ultimately to the binding of DNA helicase to the leading strand of DNA.
3. DNA helicase functions to unwind the DNA and separate the two strands.
4. To prevent overwinding of the DNA in front of the separation, the enzyme topoisomerase functions to relax the DNA.
5. To initiate the synthesis of the daughter DNA strands, DNA primase attaches a short RNA primer sequence to the original DNA strands.
6. In the leading strand, DNA polymerase δ attaches new nucleotides to the growing daughter strand.
7. In the lagging strand, short Okazaki fragments are formed. They are initiated by the DNA primase induced attachment of an RNA primer.
8. To this primer, DNA polymerase α binds a short sequence of nucleotides, after which DNA polymerase δ attaches the remaining nucleotides.
9. RPA and Fen-1 then attach to the RNA/DNA stretch to allow the exact removal of the RNA primer plus the DNA stretch synthesized by polymerase α.
10. DNA ligase then connects the 5′ phosphate group to the free 3′ –OH group of the deoxyribose sugar.

DNA unwinding takes place at the replication fork within the so-called replisome complex, with hexameric helicases forming a ring around the phosphodiester backbone of DNA (McGlynn, 2013). This is thought to be important to ensure tight binding of the helicase to the DNA molecule. Although the principle of helicase directed unwinding is similar in all organisms, subtle differences exist. Thus in bacteria, the hexamer consists of 6 identical helicases and binds to the 5′ → 3′ strand, while in other organisms the hexamer consists of different helicases and travels in the 3′ → 5′ direction. While moving along the phosphodiesterase backbone of DNA, the helicase complex uses ATP to break-up the hydrogen bonds that keep the two strands of the DNA helix together. The number of base pairs unwound by a single ATP molecule is not precisely known, but in bacteria there is evidence that two base pairs are split by every molecule of ATP.

In addition to helicases another group of enzymes is essential for the unwinding of DNA, the so-called topoisomerases. As the DNA unwinds and opens up due to the actions of DNA helicases, the DNA tends to become overwound ahead of the replication fork, which would severely inhibit (and ultimately stop) further unwinding and separation of the strands. Topoisomerases function to release the tension by temporarily cutting the phosphodiester backbone, allowing the DNA strand to uncoil and release the tension. Two different families of topoisomerases can be delineated: type I cut only one single strand on DNA, either at the 5′ phosphate (type IA) or at the 3′ phosphate (type IB) while type II topoisomerases temporarily cut both strands (Champoux, 2001). Unfortunately, the terminology of topoisomerases is slightly confusing, as several different forms of topoisomerase III exist as well, which actually belong to the type IA subfamily.

After the combined action of topoisomerases and helicases, the two strands of DNA are separated allowing each strand to replicate, leading to two daughter molecules of DNA. As each of these consists of one of the original strands and one newly synthesized strand, the process is referred to as semiconservative. The process of synthesis of the new strands of DNA is critically dependent on the enzyme DNA polymerase. This enzyme uses the four different nucleotide triphosphates to add a single nucleotide monophosphate (the two additional phosphates are discarded) to the growing DNA strand. As the base-pair rule strongly favors A to bind to T and C to G, the new daughter molecule is an identical copy of the original (although some replication errors can occur, see later). There are over 120 different DNA polymerases, generally subdivided into three groups, of which the classical DNA-directed DNA polymerases are the most important ones. However, all polymerases share two fundamental properties: [1] they can only add nucleotides to an existing chain of DNA; [2] they can only attach nucleotides to the 3′ –OH group of deoxyribose. Both these properties pose a serious problem in DNA replication. First of all, if polymerases can only add nucleotides to an existing chain, how does replication start? This problem is solved by attaching a short sequence to the single stranded DNA. This short stretch is attached by a primase enzyme and is in fact a short stretch of RNA rather than DNA, and is generally referred to as an RNA primer. The RNA primer then provides the free 3′ –OH group for DNA polymerase to attach nucleotides. The RNA primer is eventually removed by a so-called ribonuclease enzyme, after which DNA polymerase fills in the gap. The gap is then closed by DNA ligase enzymes, which catalyze the phosphodiester bond between adjacent 3′ –OH and 5′ -phosphate groups.

The second problem in DNA replication is illustrated in Fig. 2.3. Since the DNA strands are antiparallel, only one of the new strands will have a free 3′ –OH group, while the other must have a free 5′ –OH group. If DNA polymerase can only attach new nucleotides to a free 3′ –OH group, how can the second strand be copied? The answer is provided by the pioneering work of Reiji Okazaki and his colleagues. They showed that DNA replication is in fact a discontinuous process, with short stretches of DNA being formed, now referred to as Okazaki fragments. Each of these stretches starts again with an RNA primer attached by a primase, to which DNA polymerase can then attach nucleotides. In bacteria only one DNA polymerase (polymerase III) is active in the synthesis, while a second polymerase (polymerase I) removes the RNA primers and allows ligases to close the gap between the individual fragments. In contrast, in eukaryotic cells two different polymerases are involved in the synthesis of Okazaki fragments. After the RNA primer (around 12 bases) is attached by the primase PriSL, the first stretch of DNA (around 25 bases) is synthesized by polymerase α, while the remainder is then synthesized by polymerase δ. In addition, as neither of these DNA polymerases can remove the RNA primer by themselves, several nucleases and the single stranded binding protein RPA are required to remove the primer. There are further differences between bacteria and eukaryotic cells, including the speed of Okazaki fragment formation (about 1000 bases per second in bacteria versus only about 50 bases per second in eukaryotic cells) and the length of the fragments (about 1000–2000 bases in bacteria versus 150–250 bases in eukaryotic cells). As a result, DNA replication is much faster in bacteria than in eukarya, although the reasons for this are still unclear (Forterre, 2013). Although it has been suggested that the involvement of a second polymerase might represent a further stage in the evolution, this seems unlikely. In fact apart from slowing down DNA replication, the introduction of polymerase α has the major drawback that it lacks endonuclease activity which is essential for proofreading. Proofreading is important to ensure that the DNA molecule is replicated faithfully. As polymerase δ (and its bacterial counterpart polymerase III) can excise incorrect nucleotides (through its endonuclease activity) and insert correct nucleotides at the same time, the error rate of polymerase α is up to 100 times higher. It seems that the mechanism for Okazaki fragment maturation has been designed to compensate for this. Thus the polymerase δ of the previous Okazaki fragment synthesizes exactly the same number of bases to allow the Fen-I endonuclease to remove not only the RNA primer, but also the part of the DNA synthesized by polymerase α, allowing the ligase to close the gap between the two fragments.

Given the complexity of the formation and maturation of the Okazaki fragments, and the slow actions of the primases (compared to DNA polymerases) the two daughter strands are synthesized at different rates and are therefore referred to as the leading (ie, the continuously synthesized 5′ → 3′ strand) and the lagging (ie, Okazaki fragments 3′ → 5′ strand). However, as the speed of synthesis is so different between the two strands, the primases also act as a stopping signal, to prevent the lagging strand from falling too much behind the leading strand.

A final important aspect of DNA replication that we have not yet discussed is the origin of replication. Given the length of the DNA molecules (in humans up to 250 million base pairs), replication of a single DNA molecule would be extremely slow. Thus in most organisms, DNA replication starts at multiple sites within a single DNA molecule. These sites are referred to as origins of replication and there can be hundreds if not thousands of these sites per DNA molecule. The exact nature (ie, base pair composition) of these sites varies enormously between species (and even within one species), but they all can bind to the so-called origin recognition complex. This is a six-subunit complex which is remarkably conserved in eukaryotic cells and the originator of replication. The binding of the origin recognition complex triggers the subsequent binding of several other proteins including Cdc6 and Mcm2-7 leading to the so-called prereplicative complex. There is evidence to support the idea that Mcm2-7 (which is a hexameric protein complex) functions as a helicase, unwinding the DNA (Bell and Dutta, 2002).

THE STRUCTURE OF THE GENOME

The genome is defined as the total amount of genetic material of a species (see Box 2.2 for a summary of the most important terms used in this chapter). As can be expected this differs dramatically between different species. Although more complex organisms generally have

BOX 2.2

DEFINITION OF THE IMPORTANT TERMS

Allele	An alternative version of a gene. Diploid organisms carry two alleles of each gene (although they may be identical)
Alternative splicing	The process whereby one gene (and one mRNA molecule) can produce multiple different proteins (isoforms)
Autosome	All chromosomes with the exception of the sex chromosomes
Chromatin	A complex of macromolecules consisting of DNA, RNA and proteins
Codon	A sequence of three nucleotides coding for a single amino acid
Diploid	Cells (or organisms) with two copies of all autosomes
DNA	Deoxyribonucleic acid, the chemical substance that contains the genetic material
Dominance	An allele that produces a certain phenotype even in a heterozygous individual
Exon	A part of the DNA that codes for a specific (part of) a protein
F_1, F_2	The offspring in first and second generation after mating of the F_0
Gamete	Mature reproductive cell (sperm in males, egg or ovum in females)
Gene	The basic unit of inheritance that codes for a specific product. Through the process of alternative splicing, more than one isoform of a protein can be formed from a single gene
Genome	The total DNA sequence in an organism
Genotype	The genetic composition of an organism or at a specific gene location
Heterozygosity	The presence of different alleles for the same gene on the two chromosomes
Histones	Globular proteins involved in packaging DNA (see Chapter 5)
Homozygosity	The presence of two identical alleles for the same gene on the two chromosomes

Intron	A part of the DNA that does not code for a protein
mRNA	Messenger RNA, a single strand molecule involved in DNA transcription
Nucleobase	The basic building blocks of DNA and RNA: adenine, guanine, thymine, cytosine and uracil
Nucleoside	Organic molecules consisting of a nucleobase connected to a sugar: ribose to form ribonucleosides or deoxyribose to form deoxyribonucleosides
Nucleosome	A segment of DNA (about 147 bp) wrapped around a central core of 8 histone molecules
Nucleotide	Organic molecules consisting of a nucleoside connected to a phosphate group. They form the basic unit of DNA and RNA.
Phenotype	A functional characteristic of an organism that results from genetic factors (often in combination with specific environmental factors)
RNA	Ribonucleic acid, a group of chemical substances involved in transcription, translation and regulation of genetic information
rRNA	Ribosomal RNA, critical for the formation of ribosomes which assists in DNA translation
tRNA	Transfer RNA, a macromolecule involved in protein formation in ribosomes

larger genomes, there is not a simple linear relationship. For instance, the genome of mammals is about 10^9–10^{10} bases, while some amphibians and birds have genomes close to and above 10^{11} bases. In all species the genomic information is stored in the chromosomes, with the number of chromosomes again being very different between different species. Table 2.1 list a number of different species and their chromosomes, again illustrating that there is no linear relationship between the complexity of the organism and the number of genes. It is important to realize that whereas in eukaryotic cells the vast majority of chromosomes are located in the nucleus, some chromosomes are also present in the organelles, most notably the mitochondria in animals and chloroplasts in plants. These organellar chromosomes are mostly circular and are much more tightly packed with genes (see later). As such, the organellar chromosomes resemble bacterial DNA and it has been suggested that mitochondria (and chloroplasts) were originally prokaryotic cells that were encapsulated by eukaryotic cells forming a so-called endosymbiotic relationship (Thiergart et al., 2012).

Whereas the genome is the sum of all genetic material, a gene can be defined as the molecular unit of heredity. It is generally identical to a stretch of DNA that codes for a single polypeptide (protein) chain. In most organisms, genes include both coding stretches of DNA (so-called exons) but also intervening noncoding areas (so-called introns) as well as several regulatory stretches of DNA. In the next section we will describe the structure of genes in more detail. As Table 2.1 shows, the human genes are encoded in 23 pairs of chromosomes, 22 pairs of autosomes plus the X and Y sex chromosomes. The two most often used animal species in biomedical research, mice and rats, have 20 and 21 pairs respectively. This implies that the same

TABLE 2.1 Chromosomal Composition of Several Interesting Species

Species	Official name	No. of chromosome
Adders tongue	*Ophiglossum reticulatum*	1260
Shrimp	*Penaues semiculcatus*	90
Pigeon	*Columbidae*	80
Dog	*Canis lupus*	78
Cow	*Bos primigenius*	60
Elephant	*Loxodonta africana*	56
Zebrafish	*Danio rerio*	50
Chimpanzee	*Pan troglodytes*	48
Human	*Homo sapiens*	46
Rat	*Rattus norvegicus*	42
Mice	*Mus muculus*	40
Cat	*Felis catis*	38
Fruit fly	*Drosophila melanogaster*	8
Jack jumper ant	*Myrmecia pilosula*	2

genes are located on different chromosomes in different species. For instance, the gene for the serotonin transporter [a protein that is crucially involved in regulating the extracellular concentration of the neurotransmitter serotonin and which has been implicated in several psychiatric disorders (see Section 2 of the book)] is located on chromosome 17 in humans, on chromosome 11 in mice and chromosome 10 in rats. Conserved synteny refers to the colocalization of genes on chromosomes in different species. It is generally accepted that during evolution the genome has become rearranged due to, for instance, chromosomal translocation. It follows that the larger the conserved synteny, the more closely related the two species are (in evolutionary terms).

DNA TRANSCRIPTION

The expression of the genetic information stored in DNA involves two more steps namely transcription (during which information from DNA is passed on to RNA) and translation (during which information from RNA is passed on to proteins). Since virtually all organisms use exactly the same flow of information, the principle is commonly known as the "central dogma" of molecular biology (Fig. 2.5). It was first proposed by Francis Crick in a lecture in 1956 and later reiterated in a paper in Nature in 1970 (Crick, 1970). In this paper Crick already acknowledged that the simplest model DNA → RNA → protein (Fig. 2.4A) is likely not correct and does not account for RNA replication (which occurs in many viruses) and for the flow of information from RNA to DNA (Fig. 2.4B). This latter process occurs in so-called retro-viruses but also in so-called retrotransposons.

DNA transcription starts with the binding of the enzyme RNA polymerase to a specific sequence of DNA called the promotor region. However, in order to be able bind to DNA and

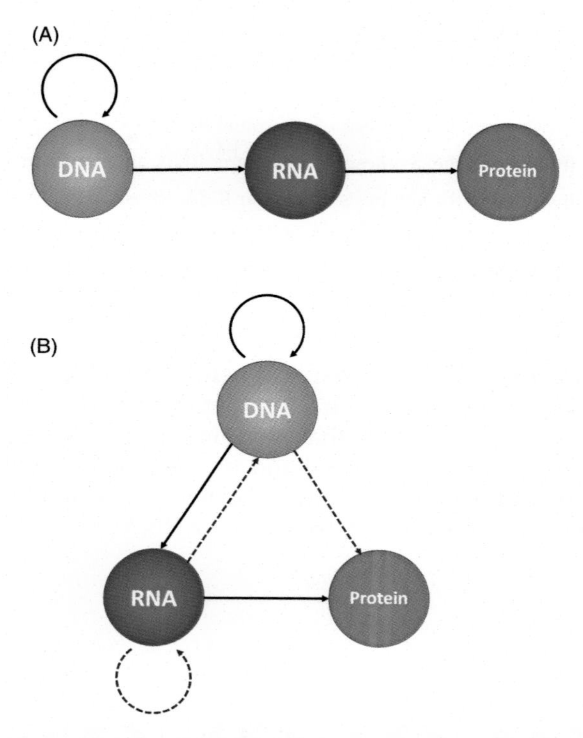

FIGURE 2.4 **The central dogma of molecular biology.** (A) In its traditional form the central dogma states that the flow of information is from DNA to RNA to proteins, while acknowledging that DNA replicates itself. (B) In 1970, Crick adjusted the central dogma slightly, acknowledging that in certain circumstances the flow of information goes from RNA to DNA and RNA can replicate itself too. There is even some evidence, albeit in very rare circumstances, that information can flow directly from DNA to proteins.

initiate transcription, the RNA polymerase requires several other specific proteins to bind [so-called transcription factors (TF)] which allows for a differential expression of genes per cell. We will discuss this in the next section. There are several different RNA polymerases, the most important for gene transcription being RNA polymerase II (Table 2.2). As RNA polymerases contain intrinsic helicase activity, the binding to DNA automatically unwinds the double helix. Another crucial difference is that whereas DNA polymerase requires a primer, RNA polymerase can start from scratch. Normally, the primary RNA nucleotide binds to the $3' \rightarrow 5'$ template strand of DNA in antiparallel fashion (Fig. 2.5). Thus, chain elongation occurs by adding new nucleotides to the free 3′ prime –OH group of the ribose sugar. As with DNA, this implies that RNA is also polarized, although as only a single strand of RNA is generally synthesized, most commonly only the $5' \rightarrow 3'$ RNA molecule exists, with the 5′ end containing a free triphosphate group and the 3′ end containing a free ribose hydroxyl group. Since chain elongation follows the same basic hydrogen-bond rule that governs DNA replication, the RNA strand is a copy of the $5' \rightarrow 3'$ strand of DNA. For that reason, this strand of DNA is usually referred to as the sense strand, and the template $3' \rightarrow 5'$ strand is referred to as the antisense strand. However, it is important that the RNA is not an exact copy of the sense

TABLE 2.2 The Different RNA Polymerases and their Function

Species	Function
RNA polymerase I	Involved in the synthesis of ribosomal RNA
RNA polymerase II	Involved in the synthesis of messenger RNA, small nuclear and micro RNA
RNA polymerase III	Involved in the synthesis of transfer RNA, ribosomal RNA and small nuclear RNA
RNA polymerase IV	Involved in the synthesis of small interfering RNA in plants
RNA polymerase V	Involved in the synthesis of small interfering RNA in plants

DNA strand. In addition to having a ribose-phosphate (rather than a deoxyribose-phosphate) backbone, RNA contains the pyrimidine uracil rather than thymine (opposing adenine). The reason why RNA contains uracil rather than thymine is not known. However, as thymine and deoxyribose are more stable than uracil and ribose, one theory has proposed that over the course of the evolution the primary source of genetic information changed from RNA to DNA. In line with this, many viruses still only contain RNA and as mentioned earlier, many rely on RNA replication.

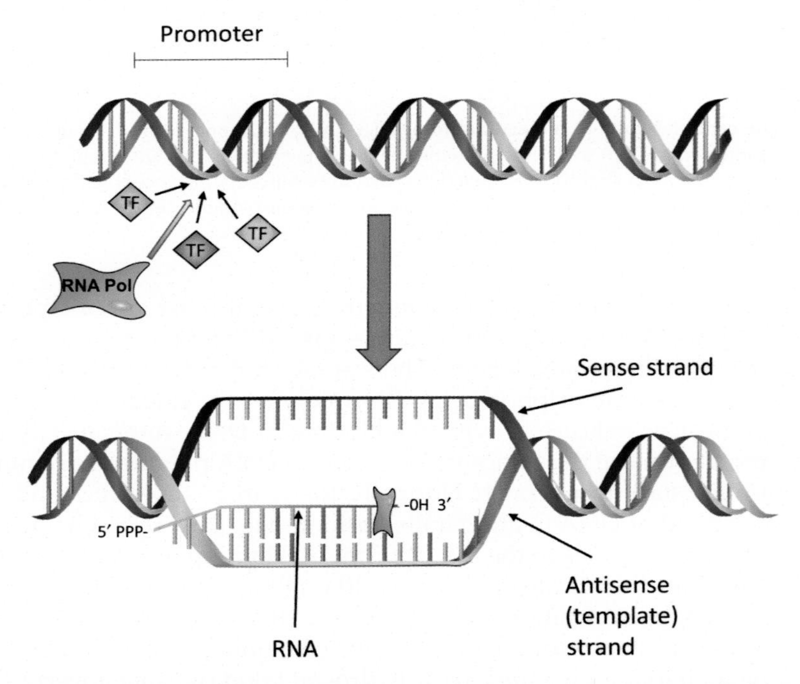

FIGURE 2.5 **DNA transcription I.** Like DNA replication, DNA transcription starts with the unwinding of the double helix. This occurs when RNA polymerase II binds to the promoter region of the gene in addition to specific TFs. In contrast to DNA replication however, the RNA resulting from DNA transcription is a single strand, only formed from the anti-sense (3′ to 5′) strand of DNA. For more details, see the text.

THE STRUCTURE OF GENES

While the genetic information is more or less the same in each cell in an organism (although there are exceptions but this will be discussed later), there are large differences in which genes are actually expressed and which are not. For instance, whereas all cells contain the genetic code for voltage gated Na^+ ion channels, only neurons express them, and therefore only these cells can actively initiate and propagate action potentials. But even within the neuronal cells different genes (and their protein products) are differentially expressed. For example the enzyme dopamine-β-hydroxylase converts dopamine into noradrenaline. So in cells that use noradrenaline as a neurotransmitter, this enzyme is highly expressed. However, dopamine is, in itself also a neurotransmitter (involved in among others Parkinson's disease, schizophrenia and drug addiction). Therefore, in dopaminergic cells, dopamine-β-hydroxylase cannot expressed. Thus, each cell must have tightly controlled mechanisms in place to determine which genes are and which are not expressed.

This differential gene expression is governed by the interaction between RNA polymerase, general TFs, and additional regulatory proteins. As mentioned earlier, although RNA polymerase II is responsible for the actual initiation and elongation of the RNA chain, a variable number of enzymes called TFs need to bind to the promoter region of DNA in order for the transcription process to start. Moreover, additional proteins can enhance or inhibit the transcriptional process. The combination of RNA polymerase II, the TFs and regulatory proteins together are referred to as the RNA polymerase II holoenzyme. These transcription and regulatory proteins are generally referred to as *trans*-acting elements, as they are coded by genes on chromosomes different from the gene they regulate. They not only bind to the promoter region of the gene they regulate but also to other parts of the chromosome such as enhancers or silencers, both short DNA sequences that are usually located upstream of the gene they regulated. When regulatory proteins bind to enhancers, they alter the three-dimensional structure of the DNA, enhancing interaction of the TFs or RNA polymerase binding to the promoter region, resulting in enhanced transcription. Silencers are similarly short stretches of DNA upstream of the promoter region. However, binding of specific regulatory proteins to silencers inhibit the transcription of the gene. Since the promoter region, enhancers, and silencers are all located on the same DNA as the gene they regulate, they are referred to as *cis*-acting elements. It is important to realize that in contrast to the promoter region which is mostly located directly upstream of the gene, enhancers and silencers may be located a considerable distance from the gene they regulate.

Transcription starts with TFs binding to the promoter region of the gene. A large number of different promoter regions have been identified which in vertebrates were originally subdivided into two classes: TATA and CpG types, and in mammals in TATA box enriched and CpG-rich promoters. TATA box enriched promoters are characterized by a DNA sequences consisting of TATAAA while the CpG type promoters contain a variable number of CG repeats. Analysis of mammalian DNA has found that the classical TATA box promoters actually represent a minority of the mammalian promoters (Carninci et al., 2006). A recent, more detailed investigation of promoters in several different species (*Homo sapiens*, *Drosophila melanogaster*, *Arabidopsis thaliana*, and *Oryza sativa*) has identified 10 different clusters (Gagniuc and Ionescu-Tirgoviste, 2012), with CG based promoter again being the most common ones in humans, while the other AT based promoters being more common in the other three species.

During the first step of gene transcription (Fig. 2.6), the TATA box (about 25–30 base pairs upstream of the start of the gene) is recognized by TF TFIID, which consists of the TATA binding protein (TBP) and several transcription associated factors (TAF). This complex then recruits TFIIB, allowing RNA polymerase and TFIIF to bind, after which several additional TFs bind and transcription starts. It is important to realize that these TFs are also involved in transcription of genes that do not contain a TATA box, hence they are usually referred to as general TFs.

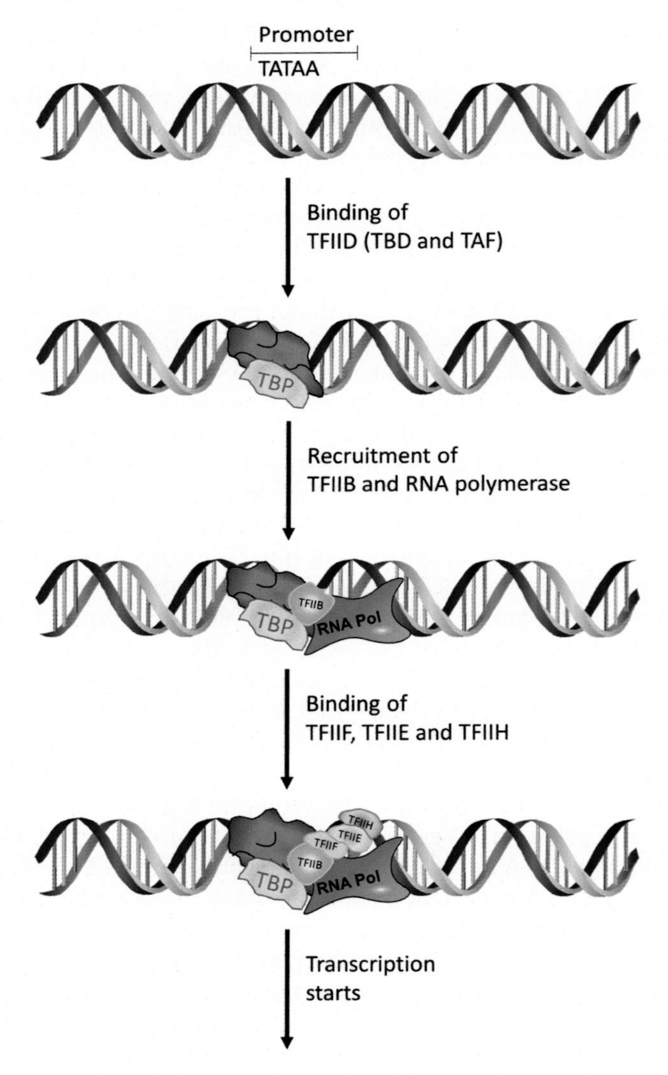

FIGURE 2.6 **DNA transcription II.** DNA transcription critically depends on the binding of a series of TF to the promotor region of a gene.

mRNA PROCESSING

The result of RNA transcription with RNA polymerase II is a single stranded RNA sequence complementary to the entire length of the gene, usually referred to as the primary (RNA) transcript or precursor messenger RNA (pre-mRNA). This transcript, in most cases, contains both protein-coding and noncoding segments (Fig. 2.7), so-called exons and introns respectively. The primary RNA transcript therefore has to undergo a series of splicing reactions in which the RNA strand is cut on the boundaries between exons and introns (splice junctions), after which the separate exons are fused together again. In the vast majority of cases, an intron starts with a GT and ends with an AG. We therefore often speak of the GT – AG (or in the case of RNA GU – AG) rule. However, although GT and AG are essential for splicing they are not sufficient by themselves. In addition, introns generally have a branch site, consisting of an adenosine located approximately 18–40 nucleotides upstream of the 3′ splice site (ss), and usually followed by a polypyrimidine tract. The introns are removed in a two-step transesterification process: in the first step the 2′ –OH of the adenosine in the branch site carries out a nuclear attack on the 5′ ss, resulting in the cleavage at this site. Subsequently the free 5′ end ligates onto the adenosine leading to form the so-called lariat structure. In the second step, the 3′ end is attacked by the free 3′ –OH of the exon leading to the ligation of the 5′ of exon 1 and the 3′ of exon 2, thereby releasing the exon. Splicing is catalyzed by a large complex of proteins and small RNAs called the spliceosome. In fact, in eukaryotic cells there are two different spliceosomes: the more ubiquitous U2-dependent and the much less abundant U12-dependent spliceosome. The conformation and composition of the spliceosome are flexible and can change rapidly, allowing it to be both accurate and flexible (Will and Luhrmann, 2011; Matera and Wang, 2014). The U2-dependent spliceosome consists of 5 different so-called ribonucleoproteins (snRNP): U1, U2, U5, and U4/U6 as well as numerous other non snRNP proteins. Each snRNP consists of a small nuclear RNA (or two in the case of U4/U6) a common set of seven Sm proteins and a variable number of additional proteins. This complexity is essential to ensure a perfect orientation of the splicing. Similar to the DNA double helix, the spliceosome complex forms many relatively weak interactions with the splice junction, resulting in a strong, highly precise splicing event. For instance, the RNA part of U1 binds to the conserved 5′ ss via the normal base-pairing rule, while the RNA part of U2 recognized the branch point (again binding via the base-pairing rule).

After the splicing out of the introns, two additional changes usually occur in the mRNA (Fig. 2.7). First, a 7-methylguanoside (m^7G) is linked to the first 5′nucleotide. It is thought that this capping process serves several important functions, including protecting mRNA from exonuclease attack, facilitating the transfer of mRNA from the nucleus to the cytoplasm, and enhancing the attachment of the mRNA to the 40S subunit of the ribosomes, which is essential for the final translation process (see later). Secondly, at the 3′ end of mRNA about 200 adenylate monophosphate (AMP) residues are added by the enzyme poly(A) polymerase. This poly(A) tail is thought to serve similar functions as the 5′ capping process: facilitation of the transport from the nucleus to the cytoplasm, stabilizing the mRNA molecule in the cytoplasm and enhancing the binding to the ribosomes.

An important aspect of mRNA processing is alternative splicing, which refers to the inclusion (and exclusion) of different exons in the final mRNA. It has been suggested that about 95% of all mammalian genes undergo alternative splicing. As a result, about 20,000 human

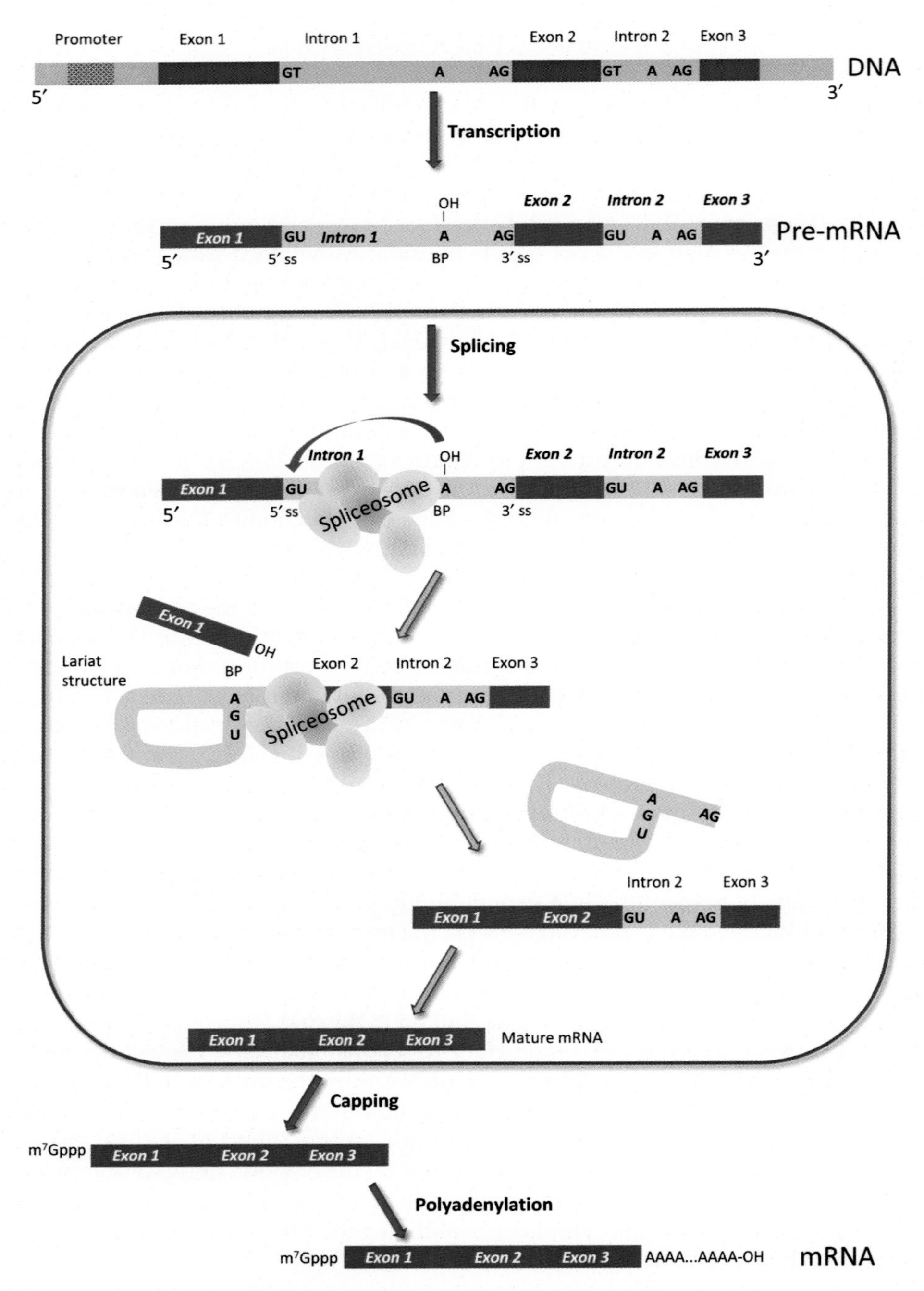

FIGURE 2.7 **mRNA processing.** DNA transcription leads to pre-mRNA which contains all the exons and the introns. Subsequently, through recruitment of the spliceosome the introns are excised leading to the mature mRNA. However, in most cases mRNA is further processed by adding a 7 methylguanoside to the 5 side and a long AMP tail on the 3′ end.

protein coding genes can lead to between 250,000 and 1 million different proteins (de Klerk and 't Hoen, 2015). An analysis of 15 different human cell lines showed that a single gene can lead to 25 different mRNAs, with up to 12 expressed in a single cell (Djebali et al., 2012). It is important to realize that isoforms are not always equally expressed. Some isoforms may indeed be very rare (although this does not necessarily mean they are less important: a small change in a crucial pathway may lead to big effects). Several different mechanisms can underlie alternative splicing. For example, the BDNF gene undergoes significant alternative splicing due to alternative transcription initiation. Thus at least 11 different BDNF transcripts have been identified as a result of different promoters (Aid et al., 2007). In addition, alternative splicing can involve an alternative order of exons, or the exclusion of specific exons. Alternative splicing involves the recruitment of specific RNA-binding proteins that are not part of the normal spliceosome, but can enhance or suppress splicing sites (Witten and Ule, 2011).

mRNA TRANSLATION

The final step in gene expression is the DNA translation, leading to the formation of proteins, molecules that consists of sequences of amino acids. Although there are 23 proteinogenic (protein-forming) amino acids, only 20 are found in all living species and are therefore referred to as standard amino acids (Fig. 2.8). Of the remaining three, pyrrolysine only occurs in some archaea and one bacterium, and N-formylmethionine occurs in bacteria, mitochondria, and chloroplasts. The last of the proteinogenic amino acids is selenocysteine, which has been found in many noneukaryotes as well as most eukaryotes. However, in contrast to the 20 standard amino acids, selenocysteine only occurs in about 25 human proteins.

In order for the DNA translation process to take place, the mature mRNA is transported out of the cell nucleus into the cytoplasm, where it binds to ribosomes. These are large complex organelles, consisting of both proteins and a special form of RNA called ribosomal RNA (transcribed by RNA polymerase I and III, rather than II). Ribosomes consists of two different subunits, a large subunit (50S in prokaryotes and 60S in eukaryotes) and a small subunit (30S and 40S respectively). The term S refers to the Svedberg unit, an indication of the rate of sedimentation, rather than actual size. Ribosomes can be found freely around the cytoplasm or bound to the endoplasmatic reticulum (forming the rough endoplasmatic reticulum). These ribosomes are involved in different types of protein synthesis.

The process of RNA translation is illustrated in Fig. 2.9 and starts with the formation of the ribosome around the mRNA (Fig. 2.9A). In most cases this initiation starts around the 5′ m^7G cap and the first part of exon 1. Since this part is normally not translated into amino acids, it is usually referred to as the 5′ untranslated region (5′–UTR, a similar region occurs at the 3′ site). Once the ribosomes are bound to the mRNA, a special type of RNA binds to the start of the mRNA. This start sequence is referred to as the initiation codon recognition sequence and often has the sequence GCCPuCCAUGG (known as the Kozak sequence with translation starting at the AUG, where Pu can code for either of the two purines adenine or guanine). This special type of RNA is referred to as tRNA (or transfer RNA, transcribed by RNA polymerase III) and has a very characteristic cross-like structure. An amino acid is attached to the 3′ site of tRNA through the enzyme aminoacyl tRNA synthetase, leading to a covalent ester binding of the amino acid to the 2′ or 3′ –OH group of the terminal adenosine

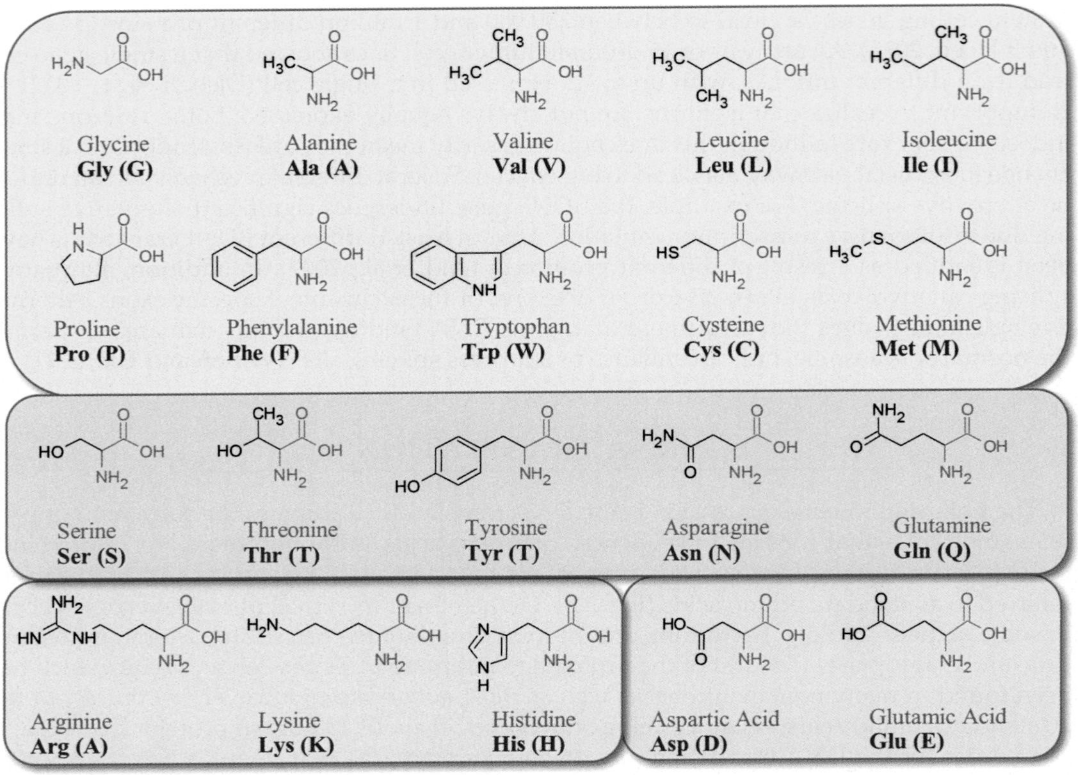

Glycine **Gly (G)** | Alanine **Ala (A)** | Valine **Val (V)** | Leucine **Leu (L)** | Isoleucine **Ile (I)**

Proline **Pro (P)** | Phenylalanine **Phe (F)** | Tryptophan **Trp (W)** | Cysteine **Cys (C)** | Methionine **Met (M)**

Serine **Ser (S)** | Threonine **Thr (T)** | Tyrosine **Tyr (T)** | Asparagine **Asn (N)** | Glutamine **Gln (Q)**

Arginine **Arg (A)** | Lysine **Lys (K)** | Histidine **His (H)** | Aspartic Acid **Asp (D)** | Glutamic Acid **Glu (E)**

FIGURE 2.8 **The major protein forming amino acids.** There are 20 major protein forming amino acids that can be subdivided into 4 groups: neutral nonpolar amino acids (yellow box), neutral polar amino acids (green box), positively charged amino acids (blue box), and negatively charged amino acids (orange box).

nucleotide. If the amino acid is attached to the 2′ –OH it will ultimately be transferred to the 3′ –OH through a process called transesterification. The result of this process is a so-called aminoacyl-tRNA. However, given that there are multiple amino acids (and hence multiple aminoacyl-tRNAs), the question is how the ribosome determines which amino acids should be attached to each other. This is determined by the nucleotide sequence of RNA (and ultimately the DNA). Given that we have 20 standard amino acids, only a sequences of 3 (or more) nucleotides would have enough information to code for all amino acids. As there are 4 different nucleotides, a "sequence" of one would only code for 4 different amino acids, and a sequence of 2 for $4^2 = 16$. A sequence of 3 could code for $4^3 = 64$ different combinations, more than required for the 20 standard amino acids. It was again Francis Crick and his colleagues who first proposed that amino acids were coded by a triplet (codon) of nucleotides (Crick et al., 1961). As is evident from Fig. 2.10 there is considerable redundancy in the system and many amino acids are coded for by a number of different triplets. It was long thought that the nucleotide codes for the different amino acids were identical in all life forms and it therefore became known as the "universal code". However, we now know that although highly

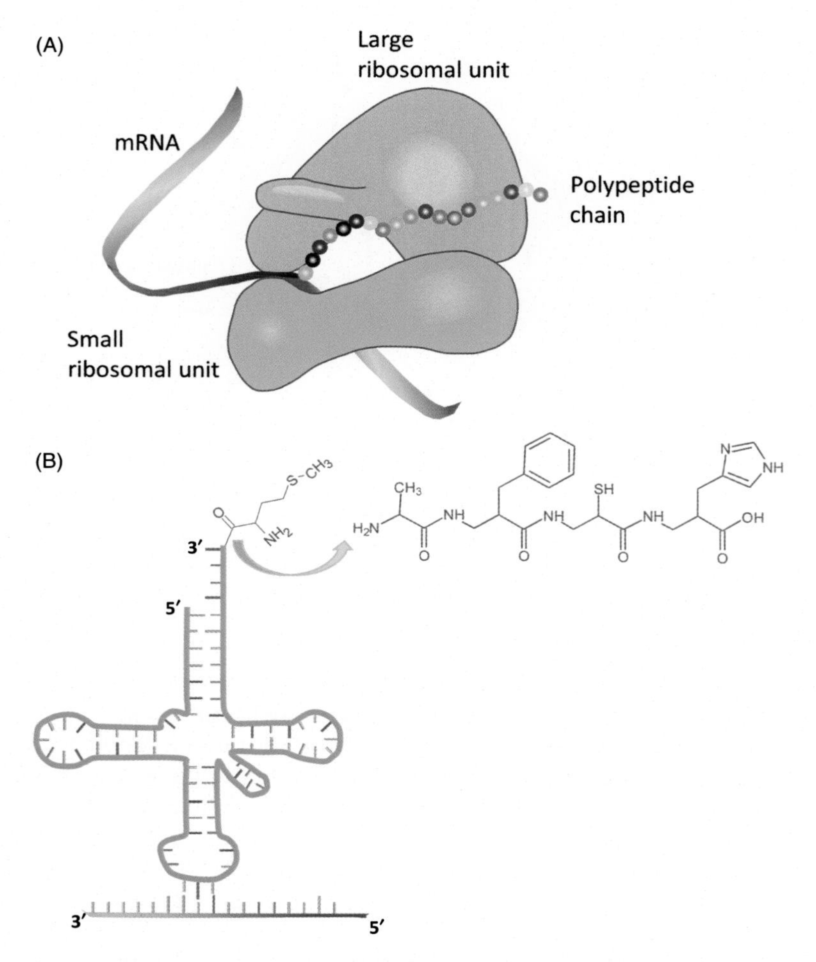

FIGURE 2.9 **RNA translation.** Through RNA translation, the nucleotide sequence of DNA is translated into amino acids. Translation starts with the binding of ribosomes to mRNA (A). Subsequently, tRNA binds to the triplet codons of mRNA. Each tRNA is covalently bound to one specific amino acid (based on the triplet codon) and thus each subsequent tRNA adds one amino acid to the growing amino acid chain until a stopcodon is reached.

conserved, there are some differences between species and also between cytoplasmic and mitochondrial codes. For instance, the codon ATA codes for Isoleucine in cytoplasmic mRNA, but for Methionine in mitochondrial mRNA.

The process of DNA translation therefore requires the aminoacyl-tRNA to bind to the RNA strand based on the triplet (codon) nucleotide sequence. Hence the tRNA sequence holds what is known as an anti-codon sequence. Although based on the triplet code one would expect 64 different aminoacyl-tRNAs, in reality there are only about 30 types of cytoplasmic and 22 mitochondrial aminoacyl-tRNA. The explanation for this is that the general base pair rule for binding of tRNA to mRNA is strict for the first two nucleotides of the triplet codon, but relaxed for the last one. For instance, a G in the third position of the anticodon of tRNA

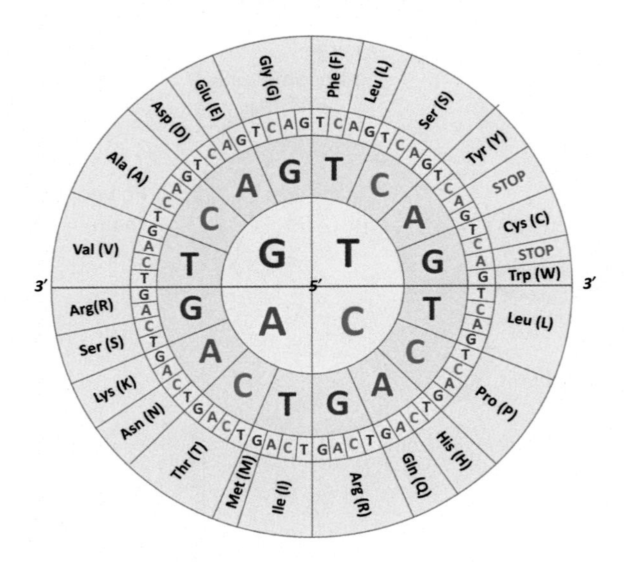

FIGURE 2.10 **The triplet code.** The triplet code of DNA (and thus of RNA) determines the amino acid sequence. Illustrated here is the DNA triplet code, with the first letter of the codon (5′) in the center, the second codon in the middle ring and the third and final codon in the outer ring (3′).

can bind to both C and U on the mRNA strand. Likewise, a U can bind to both A and G on the mRNA strand. This is generally referred to as the **wobble hypothesis**.

As the aminoacyl-tRNA binds to the mRNA, the 3′ terminal amino acid is linked to the growing chain of amino acids by the formation of a peptide bond. The ribosome then repositions itself to allow the next aminoacyl-tRNA to bind to the next codon and the process repeats itself. The end of protein synthesis is signaled by one of the so-called stop-codons (UAA, UAG or UGA for RNA, or TAA, TAG or TGA for DNA see Fig. 2.10). These codons do not have a corresponding tRNA but they are recognized by the release factor eRF1. This protein, together with the ribosome dependent GTP-ase eRF3 leads to the release of the peptide chain from the ribosome. The released polypeptide chain can then undergo a series of post-translational modifications (such as phosphorylation, methylation, glycosylation etc.) and adopt its final three dimensional structure.

GENETIC ALTERATIONS

The correct replication, transcription and translation of DNA is so fundamental to the survival of the species that it is accompanied by numerous quality control and repair mechanisms. It would be beyond the scope of this book to describe this in detail, but some of the mechanisms are being exploited in the creation of genetically altered animals (see Chapter 3). Nonetheless, in spite of these extensive quality control and repair mechanisms, incorrect DNA replication can occur and can lead to more or less serious problems, depending on the nature of the replication error. In addition, environmental toxins and infections can cause genetic alterations. In the remainder of the chapter we will discuss the most important genetic

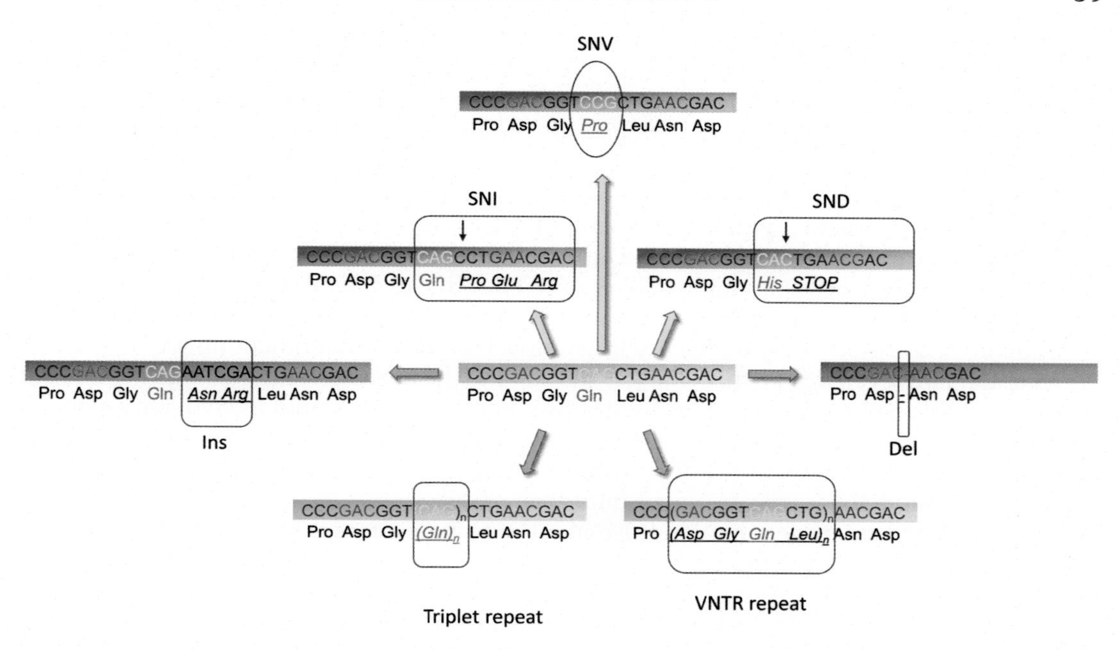

FIGURE 2.11 **The most important genetic alterations.** The most important genetic alterations include single nucleotide variants (SNV) in which a single nucleotide is changed. Single nucleotide insertions (SNI) and deletions (SND) generally lead to frameshift alterations (if they occur in exons) and involve the insertions or deletion of a single nucleotide (upper part of the figure). Tandem repeats can either take the form of a triplet repeat, in which one specific triplet is repeated, or variable number tandem repeat (VNTR) which involves repetition of larger nucleotide sequences (lower part of the figure. Finally, in the middle part of the figure, insert (ins) and deletions (del) involve the insertion or deletion of large nucleotide sequences.

alterations. They are illustrated in Fig. 2.11. There is no clear consensus about how genetic variations should be classified and subdivided (Ku et al., 2010), however, we have chosen to subdivide them based on the length of the nucleotide change, cognizant of the fact that this also, is quite arbitrary.

SINGLE NUCLEOTIDE VARIATIONS

The simplest variations occur at the single nucleotide level (illustrated at the top part of Fig. 2.11) and include single nucleotide variants (SNV), insertions (SNI) and deletions (SND), A single nucleotide variant (SNV) is defined as a change at the nucleotide level where only one single nucleotide is exchanged for another (Fig. 2.11).

SNVs are very common and are usually subdivided into three different types:

- *Silent or synonymous mutations*: These are mutations in which the single nucleotide change either occurs in a noncoding segment of DNA, or where the change does not lead to a different amino acid. Remember that the genetic code contains a large degree of redundancy. As an example (Fig. 2.11) if the SNV would change the original triplet AAA to AAG, the resulting amino acid would still be Lysine.

- *Missense or nonsynonymous mutations*: These mutations, on the other hand, lead to a change in the amino acid of the peptide. What the functional consequence of such an amino acid change is, is difficult to predict. It can lead to a loss of function, a gain of function or no change in function, depending on the amino acid change itself, as well as the position of the amino acid change. For instance, most neurotransmitter receptors belong to the family of the so-called G-protein coupled receptors. These receptors all have the same basic structure with 7 transmembrane domains. As the membrane of cells is a hydrophobic environment, the amino acids that make up these transmembrane domains all belong to the nonpolar amino acids such as Leucine or Valine (Fig. 2.9). An SNV which changes AUC to CUC leads to an exchange of Leucine for Isoleucine in a transmembrane domain, and would likely have very little impact on the overall functioning of the receptor. However, a SNV that changes the same AUC to AGC would change the neutral hydrophobic Isoleucine to a polar hydrophilic Serine, which would likely have much greater effect on the functioning. Indeed, we induced exactly such a SNV in the third transmembrane domain of the dopamine D1 receptor in Wistar rats, which led to a virtually complete loss of function (Smits et al., 2006).
- *Nonsense mutation*: these mutations also involve a single change in the triplet code. However, rather than changing one amino acid for another, the mutation leads to premature stop-codon. As a result, the protein either appears in a truncated form or gets completely degraded through a process known as nonsense mediated mRNA decay. This process is part of the quality control system of the cell and metabolizes nonsense RNA in order to prevent potentially dangerous truncated proteins from being formed (Baker and Parker, 2004). Nonsense mutations almost invariably lead to a (complete) loss of function, basically creating what is known as a knock-out (see Chapter 3). As an example, we induced a nonsense mutation in the serotonin transporter by changing the codon from UGC (cysteine) to UGA (STOP). Detailed analysis showed that through nonsense mediated mRNA decay, no mRNA and no protein was formed, creating a serotonin transporter knock-out rat (Homberg et al., 2007).

In addition to these SNVs, single nucleotoide insertions (SNI) and deletions (SND) can occur anywhere within the DNA. These changes can lead to frameshift mutations, depending on the number of nucleotides deleted or inserted, as the reading frame (the codons) shifts either to the 3′ or 5′ end of the mRNA. As a result, from the mutation onwards virtually all the amino acids change (Fig. 2.11) and premature stopcodons may also occur. As can be imagined, such a frameshift mutation almost invariably leads to a loss of function (unless the mutation is found close to the terminal amino acid in which case the consequences of the frameshift might not be too dramatic).

TANDEM REPEAT SEQUENCES

A second class of genetic variations are the so-called tandem repeat sequences. This refers to the insertion (or in rarer cases deletion) of a repeat sequence within a gene. We generally distinguish between two different type of tandem repeats, namely trinucleotide (triplet) repeats or variable number of tandem repeats. Both are illustrated in the lower part of Fig. 2.11.

TRIPLET REPEATS

These genetic alterations involve the insertion of triplet codes (as opposed to single nucleotides). Although this does not lead to a frameshift this can have detrimental effects, especially when large numbers are inserted. The triplet repeat can occur in both protein coding as well as noncoding (including the 5′-UTR) regions and have been implicated in at least 22 hereditary disorders (La Spada and Taylor, 2010), including Fragile X (where a CGG repeat occurs in the 5′ –UTR) and Huntington's disease (where CAG repeat in exon 1 leads to the insertion of additional glutamine residues). In the majority of these disorders (11 out of 22) the dysfunction involve a CAG repeat in the coding region of the gene, leading to the insertion of glutamine (Fig. 2.10) and are therefore known as poly-glutamine or poly(Q) disorders. However, other sequences such as CTG, CGG and GCC can also be replicated.

The most accepted mechanism thought to underlie the development and elongation of triplet repeats involves the development of semistable hairpins and subsequent slippage of the DNA (Mirkin, 2007). As illustrate in Fig. 2.12, semistable hairpins can spontaneously form during DNA duplication, especially with triplet CXG or GXC (where X can be any nucleotides), as the C from the first triplet can form a stable bond with the G of the second triplet and vice versa. Although in the case of the CAG, the A of the first triplet only weakly binds to the A of the second triplet, the hairpin is still stable enough, especially when there are multiple triplets. A characteristic of such semistable hairpins structures is that they often lead to a similar, out of register realignment of the complementary repetitive strands, leading to so-called 'slip-outs' which can also fold into hairpins structures (Mirkin, 2007; Galka-Marciniak et al., 2012). Such hairpin formations can lead to a stalling of DNA polymerases and can ultimately lead to a reduction or an expansion of the repeats (depending on whether the stable hairpin is formed on the nascent or the template strand).

An important characteristic of triplet repeat disorders is genetic anticipation, meaning that the disorder starts progressively earlier and is progressively more severe in subsequent generations. This is due to the increase in the number of repeats with each generation. Studies with several triplet repeat disorders have shown that a threshold exists, below which triplets are fairly stably transmitted. However, once the triplet has crossed the threshold, the triplet can increase massively from generation to generation. For example, in Huntington's disease, CAG repeats below 35 are fairly normal, generally stable and do not lead to any symptoms. However, CAG repeats above 35 are inherently unstable and lead to the characteristic symptoms of Huntington's disease. A major challenge in triplet repeat disorders is understanding the switch between below-threshold stable and above threshold unstable triplet repeats (Lee and McMurray, 2014).

VARIABLE NUMBER TANDEM REPEATS (VNTRs)

VNTR are structural regions of the DNA where a short sequence of nucleotides (longer than 3) is repeated a variable number of times in tandem. VNTRs are commonly subdivided into microsatellites (repeat sequences shorter than 5 nucleotides) and minisatellites (repeat sequences larger than 5 nucleotides) and, like triplet repeats, are thought to be due to DNA slippage errors during DNA replication. As the number of repeats is very individually

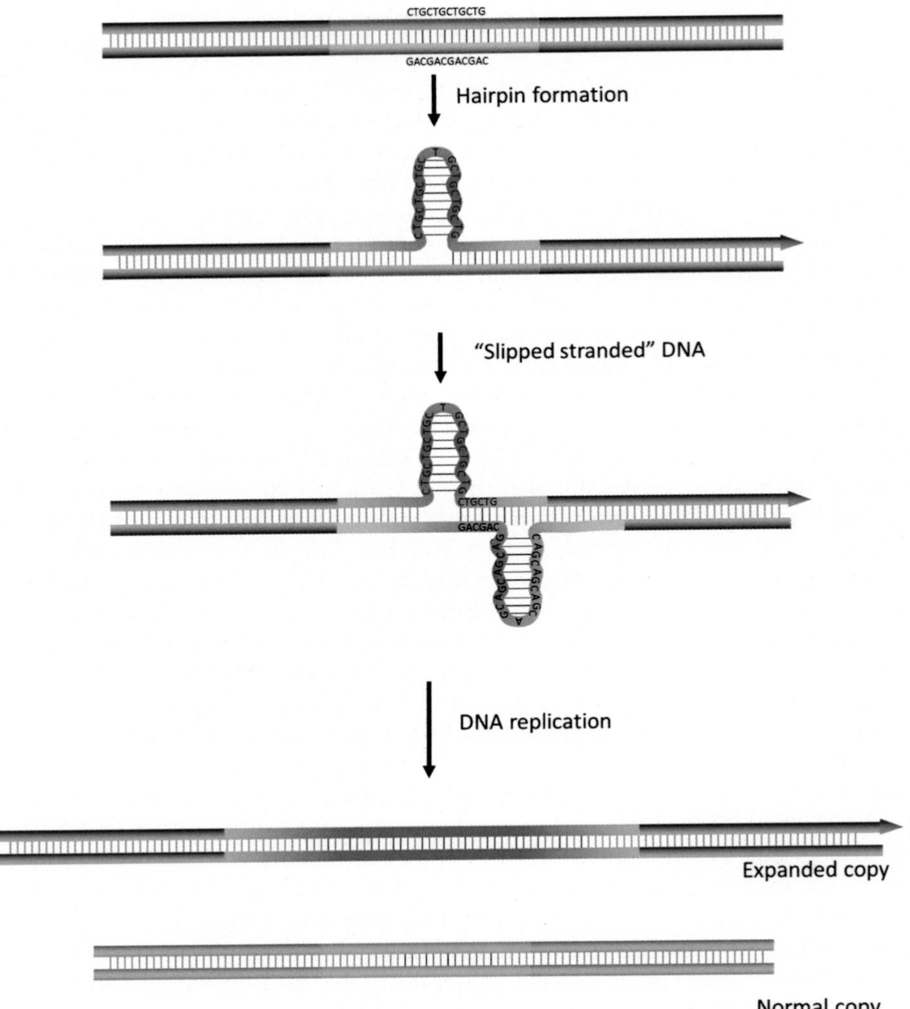

FIGURE 2.12 **The most likely mechanism underlying triplet repeats.** Although the development of triplet repeats is not yet fully understood, the most likely mechanism involves the formation of stable hairpin formations within the DNA. This then leads to an out of register realignment leading to so-called "slip outs" leading to hairpin structure in the complimentary DNA strand and triplet replication.

determined (and generally differs between the maternal and paternal copy), analysis of VNTRs is often used in forensic and paternal identity research. In neuroscience, VNTRs occur in several important genes. Often studied examples are the 40 nucleotide sequence in the 3′ UTR of the dopamine transporter (Vandenbergh et al., 1992) that can be repeated between 3 and 11 times, with the 9- and 10- repeat being the most common and the 17 nucleotide sequence in intron 2 of the serotonin transporter that is repeated 9, 10 or 12 times (Lesch et al., 1994). In several studies, the 12-repeat has been linked to increased aggression and mood disorders (Haddley et al., 2012). Finally, an often studied VNTR is within exon 3 of the dopamine D4 receptor, with 2 to 10 repeats of a 48 nucleotide stretch (Van Tol et al., 1992). This VNTR has been associated with drug addiction and ADHD (see Chapters 6 and 8).

SHORT INSERTIONS AND DELETIONS (INDELS)

As the words imply, these genetic variations refer to the insertions or deletions of short nucleotide sequences as indicated in the middle of Fig. 2.11. A general, albeit arbitrary, rule of thumb is that insertions or deletion have sizes between 100 and 1000 nucleotides, while everything above 1000 is referred to as a copy number variation. However, this does not encompass the large number of indels with sizes below 100 nucleotides. It has been estimated that the human genome contains about 1.6 to 2.5 million indels, although a full scale analysis has yet to be undertaken. Indeed, a recent study identified almost 2,000,000 nonredundant indels with about equal numbers of insertions and deletions (Mills et al., 2011), however, the sizes of the indels were between 1 and 10,000 nucleotides, thus including SNI, SND, and CNVs, although the majority were <100 nucleotides. These indels were found on all chromosomes with a frequency of 1 indel per almost 1600 nucleotides. More than 800,000 indels were found in known genes, although only 2123 indels actually affect exons. However, more than 39,000 indels were found in promoter regions of known genes and could thus affect gene expression.

In this respect, probably the best studied genetic variations in neuroscience is an indel within the promoter region of the serotonin transporter (SERT), the so-called 5-hydroxytryptamine transporter linked polymorphic region (5-HTTLPR). This region contains a 44 nucleotide insertion/deletion region leading to a large (l) and short (s) variant (Lesch et al., 1996). Although it was originally thought that the s-allele conferred lower SERT activity, further detailed analysis has identified an additional SNV within the long variant (L_A and L_G) with the L_A also having low SERT activity (Murphy et al., 2008). Hence, it is now thought that the presence of the L_G genotype confers high SERT activity.

COPY NUMBER VARIATIONS (CNVs)

CNVs also represent insertion/deletion alterations. However, they are generally defined as that are larger than 1000 nucleotides. It has been estimated that up to 12% of the human genome is subject to CNVs and it can occur both during meiosis as well as during mitosis, as evidenced by the fact that monozygotic twins can have different CNVs (Bruder et al., 2008). It has been suggested that CNVs are as important as SNVs in determining individual differences. Although it was originally thought that CNVs can be advantageous as they induce redundancy and can therefore protect against mutations, much of the variation in CNVs is now considered to be detrimental, as they are involved in a multitude of diseases, including various cancers (Valsesia et al., 2013) and CNS disorders (Morrow, 2010).

A change in CNV requires a change in chromosomal structure joining previously separate strands of DNA. Detailed analysis of these junctions shows that most CNVs are located in areas with high homology within the genome (Hastings et al., 2009). This suggests that CNVs are the result of abnormal homologous recombination. In addition, CNVs can occur in areas with only a limited degree of homology (often referred to as microhomology). In such cases, CNVs are unlikely due to aberrant homologous recombination.

Homologous recombination was already discussed in chapter 1 as the mechanisms underlying the crossing-over event that occurs during meiosis. It requires extensive DNA sequence identity (up to 300 nucleotides in mammalian cells (Liskay et al., 1987)), as well as a specific strand breaking protein (Rad51 in eukaryotic cells). HR underlies many DNA repair processes by using identical sequences to repair a double stranded DNA. When the process involves exchange of the homologous chromosomal position in the sister chromatid no structural change will take place. However, if the exchange involves a homologous segment from another chromosome (a process referred to as nonallelic homologous recombination NAHR) CNVs can occur. NAHR induces an unequal crossing over leading to a duplication in one, and a deletion of a DNA segment in the other chromosome. Homologous replication is also important for repairing broken replications and, if not performed accurately, can not only lead to duplications or deletions, but also to translocations (ie, exchange of DNA sequences between different chromosomes) or inversions (in which a stretch of DNA is reversely inserted in the same chromosome). These latter two structural changes (translocations and inversions) are generally known as copy neutral variations, in contrast to deletions and inversions, as the total number of nucleotides per chromosome does not change.

In the field of psychiatry, CNVs have received increased attention, especially in autism and schizophrenia (see Chapters 8 and 9) and it has been suggested that both disorders are prone to de novo CNVs. In this respect, one of the most studied CNVs is the 22q11.2 CNV, involving a deletion of 1.5 to 3 million basepairs and more than 25 different genes. Individuals with this deletion show symptoms reminiscent of schizophrenia, autism, anxiety and ADHD (Jonas et al., 2014). Another often studied CNV is a deletion of 15q11-13. The interesting aspect of this CNV is that the functional consequences of this CNV depend on whether the deletion is on the maternal or paternal copy. If the deletion comes from the mother, the child will suffer from Angelman syndrome, whereas if the deletion is inherited from the father, the child will suffer from Prader-Willi syndrome. The reason for this difference lies in the fact that the genes of the paternal and maternal copies are differentially expressed (Mabb et al., 2011).

CHROMOSOMAL VARIATIONS

The last group of genetic variations we will discuss are alterations in the number of chromosomes. Under normal circumstances, an individual obtains a single set of chromosomes from each parent, thus leading to a so-called diploid genome. As discussed more extensively in Chapter 1, this results from the separation of the chromosomes during meiosis, when sperm and egg cells become haploid (ie, carry only one set of chromosomes) and the subsequent fusion of the gamete cells during fertilizations.

However, in certain circumstances during meiosis the chromosomes fail to separate, leading to gametes with two sets of chromosomes or no chromosomes. When such a gamete is subsequently fertilized, the resulting cells will contain an uneven number of chromosomes (generally referred to as aneuploidy leading to either monosomy or trisomy). In the vast majority of cases, such embryos are not viable and in fact they are likely to constitute the majority of miscarriages. However, a few exceptions exist, especially when the smaller autosomal chromosomes or the sex chromosomes are involved. The most common autosomal trisomies are trisomy 21 (Down syndrome, occurring about 1 in 1000 live births) and trisomy 18 (Edwards syndrome, about 1 in 6000 live births), although cases of trisomy 13, 9, 8, and 22 have also been described. Trisomies can also occur in the sex chromosomes, leading to disorders such as triple X (XXX), XXY (Klinefelter Syndrome) or XYY. These sex chromosome trisomies all occur roughly in about 1 in 1000 live births, and compared to the autosomal trisomies present a much more subtle phenotype. In fact, many females with XXX or males with XYY often are not diagnosed at all.

Monosomies, that is, missing one copy of a gene appears much more detrimental than having an additional copy. As a result, with the exception of Turner's syndrome (X0) no other monosomies appear viable. Partial monosomy (perhaps more accurately referred to as CNVs) such as "cri du chat" (deletion of the end of the short arm of chromosome 5) and 1p36 syndrome (deletion of the end of the short arm of chromosome 1) have been described however. Turner's syndrome occurs when only an X chromosome is present (in the absence of another X or Y) and is found in about 1 in every 2000 live births (although about 99% of all cases result in spontaneous termination in the first trimester). The symptoms are quite varied from patient to patient and often include short stature, gonadal dysfunction and sterility. However, additional health problems including congenital heart disease, diabetes and cognitive deficits are also frequently observed.

References

Aid, T., Kazantseva, A., Piirsoo, M., et al., 2007. Mouse and rat BDNF gene structure and expression revisited. J. Neurosci. Res. 85, 525–535.

Baker, K.E., Parker, R., 2004. Nonsense-mediated mRNA decay: terminating erroneous gene expression. Curr. Opinion Cell Biol. 16, 293–299.

Bell, S.P., Dutta, A., 2002. DNA replication in eukaryotic cells. Annual Rev. Biochem. 71, 333–374.

Bruder, C.E., Piotrowski, A., Gijsbers, A.A., et al., 2008. Phenotypically concordant and discordant monozygotic twins display different DNA copy-number-variation profiles. Am. J. Hum. Genet. 82, 763–771.

Carninci, P., Sandelin, A., Lenhard, B., et al., 2006. Genome-wide analysis of mammalian promoter architecture and evolution. Nature Genetics 38, 626–635.

Champoux, J.J., 2001. DNA topoisomerases: structure, function, and mechanism. Ann. Rev. Biochem. 70, 369–413.

Chargaff, E., Lipshitz, R., Green, C., 1952. Composition of the desoxypentose nucleic acids of four genera of sea-urchin. J. Biol. Chem. 195, 155–160.

Crick, F., 1970. Central dogma of molecular biology. Nature 227, 561–563.

Crick, F.H., Brenner, S., Watstobi S Rj, et al., 1961. General nature of genetic code for proteins. Nature 192, 1227-&.

de Klerk, E., 't Hoen, P.A., 2015. Alternative mRNA transcription, processing, and translation: insights from RNA sequencing. TIG 31, 128–139.

Djebali, S., Davis, C.A., Merkel, A., et al., 2012. Landscape of transcription in human cells. Nature 489, 101–108.

Forterre, P., 2013. Why are there so many diverse replication machineries? J. Mol. Biol. 425, 4714–4726.

Gagniuc, P., Ionescu-Tirgoviste, C., 2012. Eukaryotic genomes may exhibit up to 10 generic classes of gene promoters. BMC Genomics, 13.

Galka-Marciniak, P., Urbanek, M.O., Krzyzosiak, W.J., 2012. Triplet repeats in transcripts: structural insights into RNA toxicity. Biol. Chem. 393, 1299–1315.

Ghosh, A., Bansal, M., 2003. A glossary of DNA structures from A to Z. *Acta crystallographica*. Section D. Biol. Crystallograph. 59, 620–626.

Haddley, K., Bubb, V.J., Breen, G., et al., 2012. Behavioural genetics of the serotonin transporter. Curr. Topics Behav. Neurosci. 12, 503–535.

Hastings, P.J., Lupski, J.R., Rosenberg, S.M., et al., 2009. Mechanisms of change in gene copy number. Nature Rev. Genetics 10, 551–564.

Homberg, J.R., Olivier, J.D.A., Smits, B.M.G., et al., 2007. Characterization of the serotonin transporter knockout rat: a selective change in the functioning of the serotonergic system. Neuroscience 146, 1662–1676.

Jonas, R.K., Montojo, C.A., Bearden, C.E., 2014. The 22q11.2 deletion syndrome as a window into complex neuropsychiatric disorders over the lifespan. Biol. Psychiatry 75, 351–360.

Ku, C.S., Loy, E.Y., Salim, A., et al., 2010. The discovery of human genetic variations and their use as disease markers: past, present and future. J. Hum. Genet. 55, 403–415.

La Spada, A.R., Taylor, J.P., 2010. Repeat expansion disease: progress and puzzles in disease pathogenesis. Nat. Rev. Genetics 11, 247–258.

Lee, D.Y., McMurray, C.T., 2014. Trinucleotide expansion in disease: why is there a length threshold? Curr. Opinion Genetics Dev. 26, 131–140.

Lesch, K.P., Balling, U., Gross, J., et al., 1994. Organization of the human serotonin transporter gene. J. Neural Trans. General Sec. 95, 157–162.

Lesch, K.P., Bengel, D., Heils, A., et al., 1996. Association of anxiety-related traits with a polymorphism in the serotonin transporter gene regulatory region. Science 274, 1527–1531.

Liskay, R.M., Letsou, A., Stachelek, J.L., 1987. Homology requirement for efficient gene conversion between duplicated chromosomal sequences in mammalian cells. Genetics 115, 161–167.

Mabb, A.M., Judson, M.C., Zylka, M.J., et al., 2011. Angelman syndrome: insights into genomic imprinting and neurodevelopmental phenotypes. Trends Neurosci. 34, 293–303.

Matera, A.G., Wang, Z., 2014. A day in the life of the spliceosome. Nat.ure Rev. Mol. Cell Biol. 15, 108–121.

McGlynn, P., 2013. Helicases at the replication fork. Adv. Exp. Med. Biol. 767, 97–121.

Mills, R.E., Pittard, W.S., Mullaney, J.M., et al., 2011. Natural genetic variation caused by small insertions and deletions in the human genome. Genome Res. 21, 830–839.

Mirkin, S.M., 2007. Expandable DNA repeats and human disease. Nature 447, 932–940.

Morrow, E.M., 2010. Genomic copy number variation in disorders of cognitive development. J. Am. Acad. Child Adolesc. Psychiatry 49, 1091–1104.

Murphy, D.L., Fox, M.A., Timpano, K.R., et al., 2008. How the serotonin story is being rewritten by new gene-based discoveries principally related to SLC6A4, the serotonin transporter gene, which functions to influence all cellular serotonin systems. Neuropharmacology 55, 932–960.

Smits, B.M.G., Mudde, J.B., van de Belt, J., et al., 2006. Generation of gene knockouts and mutant models in the laboratory rat by ENU-driven target selected mutagenesis. Pharmacogenetic. Genomic. 16, 159–169.

Thiergart, T., Landan, G., Schenk, M., et al., 2012. An evolutionary network of genes present in the eukaryote common ancestor polls genomes on eukaryotic and mitochondrial origin. Genome Biol. Evol. 4, 466–485.

Umate, P., Tuteja, N., Tuteja, R., 2011. Genome-wide comprehensive analysis of human helicases. Commun. Integrat. Biol. 4, 118–137.

Valsesia, A., Mace, A., Jacquemont, S., et al., 2013. The growing importance of CNVs: new insights for detection and clinical interpretation. Frontiers Genetics 4, 92.

Van Tol, H.H., Wu, C.M., Guan, H.C., et al., 1992. Multiple dopamine D4 receptor variants in the human population. Nature 358, 149–152.

Vandenbergh, D.J., Persico, A.M., Hawkins, A.L., et al., 1992. Human dopamine transporter gene (DAT1) maps to chromosome 5p15.3 and displays a VNTR. Genomics 14, 1104–1106.

Watson, J.D., Crick, F.H., 1953. Molecular structure of nucleic acids; a structure for deoxyribose nucleic acid. Nature 171, 737–738.

Will, C.L., Luhrmann, R., 2011. Spliceosome structure and function. Cold Spring Harbor Perspect. Biol. 3, a003707.

Witten, J.T., Ule, J., 2011. Understanding splicing regulation through RNA splicing maps. TIG 27, 89–97.

CHAPTER

3

Animal Modelling in Psychiatry

INTRODUCTION

Research in animals, especially rodents (rats and mice) has been crucial for advancing our understanding of the functioning of the human body. Likewise the development of new drugs would be impossible without animal models, both for detecting potential therapeutic effects as well as for identifying side effects and toxic liability. Nonetheless, especially in the area on CNS disorders, the development of animal models has proven particularly difficult, predominantly in relation to the translational value for clinical practice. In line with this, developing new drugs for psychiatric and neurological disorders has proven particularly difficult. An analysis of the success rates of the major pharmaceutical companies a decade ago showed that from all drugs that started in clinical phase I trials (see Box 3.1 for a description of the various stages of drug development) only 8% make it to registration in the CNS area (Kola and Landis, 2004). More recent analyses have shown that the situation has unfortunately not improved (Arrowsmith and Miller, 2013). A careful analysis shows that the failure is mostly due to lack of therapeutic efficacy (Arrowsmith and Miller, 2013), suggesting indeed that our current animal models do not translate well into clinical practice. As an example, the selective positive allosteric modulator of the metabotropic glutamate 2 and 3 receptors AZD8529 was found to act like an antipsychotic in seven different animal models for antipsychotic activity. However, clinical phase II trials against placebo and risperidone failed to show a significant improvement in the patients treated with AZD8529 and further development was dropped (Cook et al., 2014).

There are several obvious reasons why animal models for brain disorders are notoriously difficult. First and foremost, the brain is undoubtedly the most complex organ in our body. Not only does it contain about 80 billion neurons (and about an equal number of glial cells (Lent et al., 2012)), but each of these neurons connect with about 1,000–10,000 other neurons forming extremely complex networks. Thus, whereas we can (more or less) infer the function of the heart, liver or kidney by studying a single cell from each organ, studying a single neuron does not tell us much about how external stimuli are processed, how thought processes are formed, or how behavior is initiated. Another major obstacle in animal modelling is that, again in contrast to most other organs, the brain shows large species differences. Although at a fundamental (molecular) level the differences are likely to be less, on a more

Gene-Environment Interactions in Psychiatry. http://dx.doi.org/10.1016/B978-0-12-801657-2.00003-3

BOX 3.1

THE MAIN PHASES OF DRUG DEVELOPMENT

Drug development is usually subdivided in two phases, a discovery (nonclinical) phase and a development (clinical) phase. Whereas the developmental phase is traditionally subdivided in well-defined clinical phases, the subdivision of the discovery phase is less clear. This is, in part, due to the fact that different approaches have been used in drug discovery, such as the phenotypic vs the target based approach (Sams-Dodd, 2005). Using the latter, currently more popular, approach we can distinguish 4 different discovery phases (Paul et al., 2010):

Target to hit	During this phase a large number of chemical substances (often from a compound library) are screened for a specific target (ie, a receptor or enzyme). Those that show an appreciable effects are termed "hits."
Hit to lead	In this phase the "hits" are further chemically altered to improve the (in vitro) efficacy and selectivity. The most promising candidates are termed "leads"
Lead optimization	In the phase the "leads" are further optimized in terms of in vivo profiling, pharmacokinetics etc.
Preclinical phase	In this final discovery phase the best compound (or compounds) is (are) tested for their safety and potential toxicological side effect profile.

In the developmental phase of drug discovery, the most successful candidate from the discovery phase will be administered in humans:

Phase I	During this phase the candidate drug is given to healthy volunteers to investigate the pharmacokinetic properties and to identify potential side effects
Phase II	In this phase the candidate drug is given to a selected group of patients to investigate whether the drug has therapeutic properties and to identify potential side effects.
Phase III	In this phase the candidate drug is tested in large groups of patients, usually in multi-center trials, often investigating different patient groups (acute vs chronic).
Launch	In this final phase the results of all the discovery and development phase studies are compiled in a portfolio which is then submitted to the regulatory office in order to obtain registration as a novel drug.

global level there are clear and important differences. This is seen both in gross anatomy (the cortex is highly folded in humans but smooth in rats and mice and the prefrontal cortex is much larger in humans for instance) but also in functioning. An obvious but far from trivial difference between rodents and humans is that rodents walk on four legs. Thus the brain areas involved in motor coordination are obviously different. Likewise, in contrast to humans, mice and rats rely heavily on olfactory cues and thus the olfactory bulb as well as the other areas of the brain that process olfactory information are much more prominent in rodents than in humans. On the other hand, rodents rely much less on visual cues than humans do.

In spite of these (and other which we will discuss later on) limitations, animal research offers a number of very important benefits. First of all, rodents such as rats and mice have a very short life span. Their pregnancy lasts about 3 weeks and within about 2 months they are (young) adults (see for a comparison between rodent and human development Chapter 4) and in about another two months more they can give birth to the next generation. Much more important, however, is the fact that it is much easier to keep experimental conditions as constant as possible, varying only a single variable. This applies to environmental, genetic and other factors. Thus, using for instance inbred strains of rats or mice, we can assure there is a very strong genetic similarity between all animals (although recent studies have found that inbred rats or mice are not 100% identical as is often thought). In addition, as we will discuss later on in this chapter it is relatively easy to selectively modify the genetic make-up of mice and rats and induce SNVs, CNVs (see Chapter 2), or other mutations and study the functional consequences of these. Similarly, we can selectively manipulate one aspect of the environment, by injecting drugs or exposing animals to specific environmental challenges (this will be discussed further in Chapter 4).

The importance of selectively manipulating one (or more) of the experimental factors cannot be overstated. It allows us to study the causal relationship between these factors and the behavior or other functional outcomes. In human studies this is very difficult for practical as well as ethical and financial reasons. Although it is possible to study the (short term) outcome of a pharmacological or environmental manipulation, most human studies rely on establishing a correlative relationship. For example, if we want to investigate whether a genetic variation in the dopamine D2 receptor is involved in alcoholism, the traditional approach in humans is to collect blood or saliva from large number of individuals that suffer from alcoholism and from controls and study whether in the group of patients individuals with the short allele are overrepresented. However, this does not automatically establish a causal relationship as many other potentially relevant variables cannot be controlled for. The approach in animals is more straightforward, we can use rats with and without the same genetic mutation in the dopamine D2 receptor and give them access to alcohol. If the mutation is indeed linked to alcoholism, animals with the mutation would drink significantly more than animals that do not have the mutation.

Studies focused on the long term consequences of early environmental challenges are even more difficult in humans. While it is relatively easy to expose rats or mice to environmental challenges, in humans we mostly rely on retrospective information from specific questionnaires. Although some studies have now used a prospective approach, these studies are very expensive, and take a very long time, especially as major psychiatric disorders such as depression, anxiety and schizophrenia generally develop after puberty. Moreover, even in prospective studies, we cannot, for obvious ethical reasons, selectively expose individuals to environmental challenges.

A final important advantage of animal studies is the accessibility of the brain. Although modern in vivo imaging techniques, such as functional magnetic resonance imaging (fMRI) and magnetoencephalography (MEG) and more classical techniques as electroencephalography (EEG) allow us to investigate changes in regional brain activity in humans they are of limited use in analyzing changes in neurochemical processes. Although Single Photon Emission Computer Tomography (SPECT) does allow us to study changes in, for instance, neurotransmitter receptor occupancy, its spatial and temporal resolution are relatively poor. In contrast, in vivo techniques such as microdialysis and in vivo voltammetry allow us to study neurotransmitter changes in very small regions of the animal brain and with a temporal resolution in the order of minutes or even less. In addition, ex vivo techniques such as cFos allow us to look at changes in neuronal cell activation at a cellular level.

ANIMAL MODELS

A very broad definition of an animal model is *"the preparation in one species in order to study phenomena in another species"*. According to this definition testing a new veterinary product for horses in a rat or a mouse would be considered an animal model. In the present context (and indeed the vast majority of animal models described in the scientific literature), however, animal models are developed to study the human species. Another often used definition was originally coined by McKinney states: *"In the case of animal models for human psychopathology one seeks to develop syndromes in animals which resemble those in humans in certain ways in order to study selected aspects of human psychopathology"* (McKinney, 1984). Although intuitively, this seems a valid and comprehensive definition, it is in fact limited to only one specific group of animal models usually referred to as simulation models. However, there are other classes of animal models as well. One useful way to subdivide animal models is illustrated in Fig. 3.1 and is based on the principle aim and focus of the model. Before discussing these models in more detail, it is important to understand that the subdivision is somewhat artificial, and often one animal model can have multiple aims, and hence belong to different classes depending on how the model is used. This is indicated by the arrows linking the various models together.

Screening Models

The principle aim of screening models, sometimes referred to as screening tests or models with pharmacological isomorphism (Matthysse, 1986), is to predict the pharmacological properties in humans based on their properties in animals. Thus, the main focus of such models is on pharmacotherapy. In its simplest form, they are based on the premise that if two drugs produce similar effects in animals, they will also produce similar effects in humans. In other words, in these models, a new drug is compared to a well-known drug, generally referred to as the gold standard. The general steps in the development of screening models are illustrated in Fig. 3.2. Essential for this class of animal models is a strong interaction between clinical and preclinical research, the starting point being a drug with

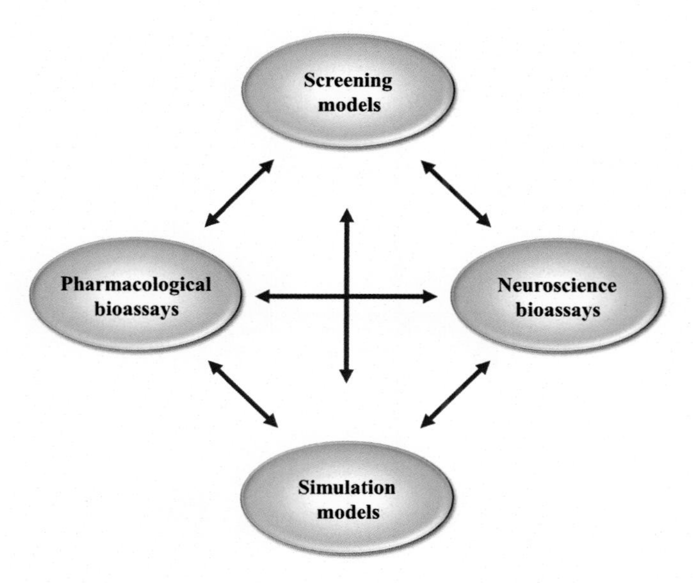

FIGURE 3.1 **Animal models.** Although there are several different ways to subdivide animal models, based on techniques, on species or disorders, a more theoretical subdivision is on the basis of the research question addressed. Using this approach, we can distinguish between four different classes of animal models, which are discussed in the text in more detail. It is important to realize, however, that these models often intersect and should not be viewed as completely complimentary.

well-known clinical properties. Although there are no explicit criteria for selecting the gold standard, the more information we have about the drug's clinical effects, the better the drug is suited. Likewise, as all drugs induce side effects, and we are primarily interested in developing a model for the therapeutic effects, the less side effects a drug has, the better is it suited to function as a gold standard. Moving from the clinical to the preclinical domain, we next test the drug in an animal. Important in the selection of the experimental set-up is that the drug induces a clear, discernible effect, ideally on an interval or ratio scale, so that we can use parametric tests to evaluate statistical differences (which are generally considered to be more powerful than nonparametric tests). Although sometimes an experimental paradigm is chosen that is (or is thought to be) related to the clinical effects, this is by no means necessary. In fact, it could actually obscure the validity of the screening model. For instance, one of the classical screening models for antipsychotic drugs is reversal of the apomorphine induced stereotypy, based on the idea that stereotyped behavioural patterns are often seen in patients with schizophrenia. However, a closer inspection of the model shows that it is actually not a very good model for the therapeutic effects at all (Ellenbroek, 1993). The experimental paradigms may involve changes in normal "spontaneous", drug or lesion induced behavior, and may be focused on acute behavior or learned paradigms. Although an acute paradigm allows for a faster throughput of new substances, with the right controls, learned paradigms can also be very helpful. Another well-known screening model for antipsychotics is disruption of active avoidance. In this paradigm, animals are first trained in a two compartment shuttle box, where they learn to avoid a weak electric shock by

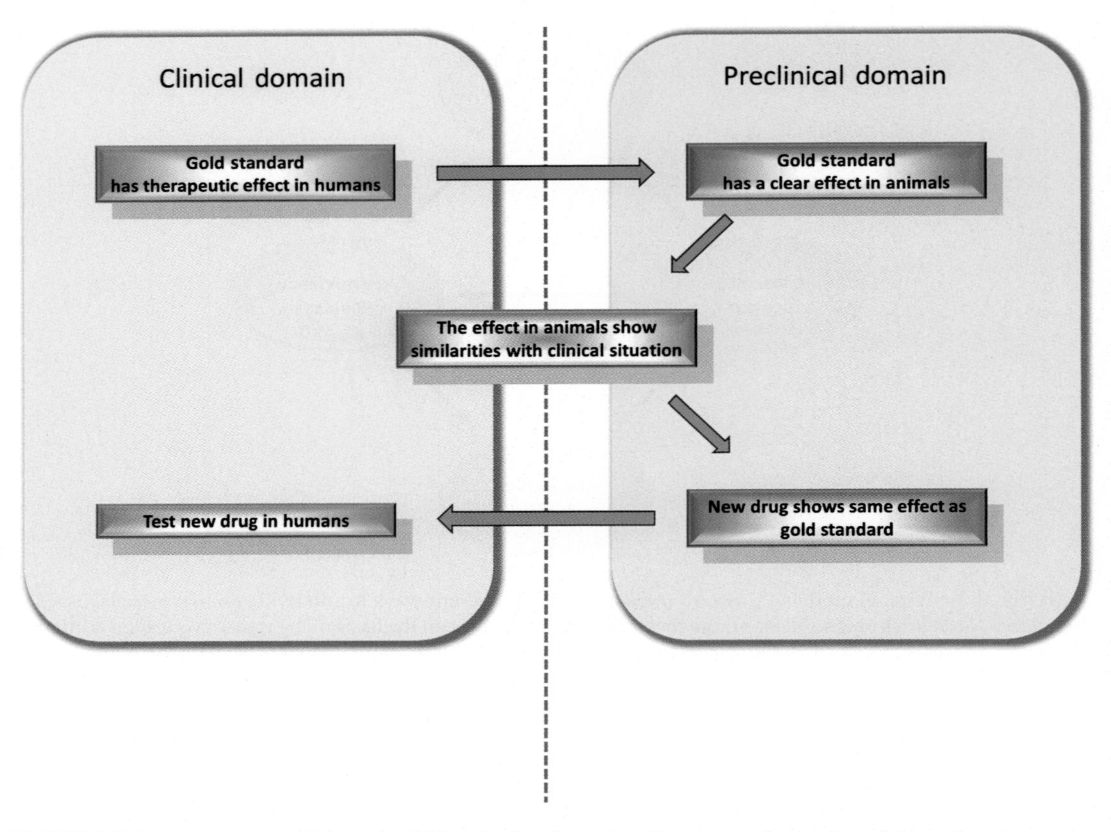

FIGURE 3.2 **Screening models.** Essential for the development of any type of animal model is a close interaction between the clinical (human) and the preclinical (animal) research domain. This is illustrated here for screening models. These models are based on comparing new drugs with existing well-described drugs (usually referred to as the Gold Standard). After establishing a specific change in animals, new drugs are investigated to see whether they induce the same changes in animals. If so, it is assumed that the new drug will also induce the same changes in humans, which can subsequently be tested in clinical trials.

jumping from one compartment to another once a tone or light is turned on. Although this takes some time, once performance has reach constant levels (usually > 85% accuracy), the animals can repeatedly be used for testing, as long as some wash-out period is taken into consideration.

The next step in screening model development is probably the most important one, often referred to as the validation phase. As mentioned before every drug has a number of different effects in humans, and this phase is meant to ensure that the parameter modelled in animals is as closely related to the desired clinical effect as possible. As illustrated in Fig. 3.2, this step requires a combination of clinical and preclinical knowledge. More specifically a series of validation criteria need to be developed based on clinical evidence. It would be beyond the scope of this chapter to discuss these criteria in any great detail and obviously they are different for the different classes of drugs, for example, screening models for antidepressants

(Willner, 1984) have (partly) different criteria from screening models for antipsychotics (Ellenbroek, 1993). Nonetheless, several general validation criteria can be identified:

- Drugs from different chemical classes should be effective in the model. In all therapeutic classes (ie, antipsychotics, antidepressants, anxiolytics etc.) substances from different chemical classes are effective. If a screening model is to be useful, its effect should not be limited to only one, or a few of the chemical classes.
- Related to this, the usefulness of an animal model increases when fewer false negatives are found. False negatives are defined as drugs that do not act like the gold standard in the model, but are therapeutically effective in the clinic.
- In addition, the usefulness of the model improves when the number of false positives is low. A false positive is a drug which acts like the gold standard in the model, but does not show the same therapeutic effect in humans. Obviously, it is impossible to evaluate all potential false positives as there are thousands of substances, many of which have not been properly evaluated in clinical trials. Hence, this criterion usually focuses on a well-defined set of drugs that either show structural (chemical) similarity with the gold standard or that is known to interfere with general motor control/coordination (such as drugs that cause sedation).
- A last general criterion is that there is a positive correlation between the doses used in the clinic and those in the model. It is important to realize that this correlation does not necessarily have to be absolute. The dose required depends strongly on the way the drug is metabolized, and it is well known that species differences can exist in metabolism. Moreover, drug metabolism also depends in part on the route of administration, and while in rodents intraperitoneal and subcutaneous injections are most common, in humans the oral route is by far the most preferred route of administration. These species differences are not only related to the speed of metabolism but may even involve the production of active metabolites (ie, metabolites that by themselves have a therapeutic effect) in one species but not in another. Nonetheless, the usefulness of a model will be reduced if for instance drug A would be 1000-fold more active than drug B in humans, but 1000 times less active in rats.

Once the usefulness of the animal model has been established, novel drugs can be evaluated. If such novel drugs produce the same (or very similar) effects as the gold standard in the model, they can be further evaluated in clinical tests (once it has been established that the new drug is safe of course, see Box 3.1). The ultimate test of the quality of the screening model of course is determined by how successful such novel drugs are in clinical trials.

Screening models, in spite of their (relative) simplicity, have been very successful in drug development research in the past 50 years. Indeed the vast majority of drugs in the field of CNS therapeutics have been identified on the basis of such screening models. However, there are a number of inherent limitations to screening models. The most important limitation is that all drugs are compared to a gold standard, and hence all drugs will be like the gold standard. In other words, such models create "more of the same" kind of drugs. They may be more potent, have a longer duration of action or perhaps less side effects, but a major breakthrough such as a drug with a completely new mode of action is unlikely to be identified in such models. Another obvious limitation of screening models is that it is very difficult to develop drugs for a condition where so far treatment has been limited. For example, we have had very limited success in developing effective drugs for autism or Huntington's disease and thus there are no proper gold standards to which we can compare novel therapeutic options. As a result of these limitations [in combination with stricter rules from the regulatory authorities such as the Food and Drug Administration (FDA) in the USA and the European Medicines Agency (EMEA) in Europe] screening models have lost their prominent position in the last decades.

Screening models do, however, have another purpose, namely as a model to understand the mechanism of action of therapeutically effective drugs. Although most drugs such as antipsychotics and antidepressants have been in clinical use for a long period of time, their exact mechanism of action is still not completely understood. If we have a screening model which faithfully mimics the therapeutic effects of a specific class of drugs (or even a single drug), we can use it to probe the neurobiological substrate underlying this effect. Strictly speaking we would then be using the test not as a screening model but more as a neuroscience bioassay (as discussed later).

Pharmacological Bioassays

As discussed in many papers, the development of new drugs has seen an important shift in strategy in the last 20 years from a more phenotypical to a more target-based approach (Sams-Dodd, 2005). A phenotypical approach, in this respect, refers to strategy where drugs are primarily selected on the basis of their physiological effect, that is, drugs that increase heart rate, or decrease active avoidance, irrespective of which neurobiological mechanism underlies this physiological effect. A target-based approach, on the other hand, focuses primarily on the neurobiological mechanism of action, that is, drugs that selectively activate the dopamine D_1 receptor or inhibit the histamine H_3 receptor. The advantages of the target-based approach are obvious. Not only do we already know the primary mechanism of action of the newly developed drugs, but most importantly from a drug discovery point of view, identifying and optimizing drugs using a target-based approach is much faster. The basic strategy is to artificially insert the DNA for the protein of choice (ie, the dopamine D_1, histamine H_3 receptor etc.) in a cell line and grow them in vitro. Using specific read-outs (such as increase in cAMP production or decrease in Ca^{2+} concentration) we can then test whether our drugs indeed have the desired action. Often, as most drugs need to be tested in rats or mice as well, cell lines are prepared that express the human, the rat and/or the mouse protein. For this strategy to work optimally it is important that the cells do not express these (and many other receptors) under normal circumstance, and therefore laboratories often use special cell lines such as the Chinese Hamster Ovary (CHO) or the Human Embryonic Kidney (HEK) cells. Using this strategy coupled with special robotics allows for a very high throughput (up to 100,000 compounds per day).

However, target-based approaches have several inherent limitations as well. First of all, as we have very limited knowledge of the underlying pathology for most disorders (and especially CNS related diseases), the selection of the "correct" target is extremely difficult. Secondly, although the target based approach allows for (relatively) easy selection of drug with high selectivity for a single target, it is much more complicated to develop drugs that bind to multiple targets. In this respect it is important to realize that many very effective drugs do bind to multiple sites, such as most of the second generation antipsychotics (like olanzapine, clozapine, and asenapine). These two limitations are inherent to the target based approach and cannot be addressed properly unless one changes the approach itself (Sams-Dodd, 2005). There is another limitation to the target based approach, due to the fact that the cell lines used are rather artificial. Moreover, since drugs are selected in an in vitro assay, there is a need to investigate whether the drugs are also active on the same target in vivo.

Pharmacological bioassays are the class of models focused on testing drugs with a known biochemical mode of action. As with screening models, the type of behavior that is used is not essential and it can even be a physiological change such as an increase in heart rate, or a characteristic change in EEG. However, if a drug is developed for a CNS disorder, it is important that the read-out parameter is related to changes in the CNS, to ensure that the drug is able to cross the blood brain barrier. Table 3.1 list some of the more commonly used pharmacological bioassays.

A special form of a pharmacological bioassay is the drug discrimination paradigm. This approach can be used for instance when a specific class of drugs does not induce any overt changes in behavior or physiology. In this paradigm, rats are trained in a two lever operant chamber where they have to learn to press the correct lever in order to obtain a food reward (Koek, 2011). Which of the two levers are coupled to the reward depends on whether the animal receives a specific drug or only the vehicle solution. Thus, when we for instance are developing selective dopamine D_1 receptor agonists, rats need to learn that when they receive a standard D_1 agonist, pressing the left lever will be rewarded, whereas when they receive vehicle, pressing the right lever will give access to reward. In other words, the rats are trained to discriminate the interoceptive cue (the "feeling") of a D_1 agonist from vehicle. Once animals have learned this, we can challenge them with newly selected drugs, which show high in vitro activity for the D_1 receptor. If the drugs are also active in vivo, the animals will experience a similar interoceptive cue and will press the left lever. In principle this procedure works with any drug that affects the brain, although some drugs have much stronger discriminative properties than others. Although this procedure requires the rats to be trained, once the animals are trained they can be used for a prolonged period of time and can be tested with many different drugs, as long as an adequate period of retraining with the original drugs is maintained in the periods in between novel drug testing.

TABLE 3.1 Some of the Often used Pharmacological Bioassays

Receptor	Assays
$5\text{-}HT_{1A}$	8-OHDPAT induced lower lip retraction
$5\text{-}HT_{2A}$	5-MeODMT induced head twitches
$5\text{-}HT_{2C}$	DOI induced deficits in prepulse inhibition
CB_1	Win-55,212 induced hypothermia, analgesia, catalepsy and hypolocomotion
D_2	Haloperidol induced catalepsy
D_3	7-OHDPAT induced yawning
H_3	R-alpha-methylhistamine induced drinking behavior
NK_1	Foot tapping (only seen in gerbils)
NMDA	Phencyclidine induced hyperactivity

Neuroscience Bioassays

This is a very large class of animal models, which we will only briefly discuss as they are not the most relevant ones for the current discussion. In these models, the neurobiology is the central focus and the aim of these models is to identify the role of specific brain structures or neurotransmitters in behaviors or other physiological processes. Thus, these classes of models would aim to answer questions like: "Does the medial prefrontal cortex play a role in fear extinction?" or "What is the involvement of the histamine H_3 receptor in exploratory behavior?" However, the output parameter does not necessarily have to be a behavior, as long as it is an *in vivo* parameter. Thus, a model which assesses the role of the 5-HT_{1A} receptor in the regulation of the firing of dopaminergic cells would equally constitute a neuroscience bioassay.

As mentioned above, screening models can be used to probe the neurobiological mechanisms underlying the therapeutic actions of drugs. Likewise, simulation models (see next paragraph) can be used as neuroscience bioassays. Thus, the same animal experimental set-up can be used for different purposes and hence fall into different classes of models.

Simulation Models

The last class of animal models and for our current purpose the most important one is the class of the simulation models. These models aim to mimic the human disorder in an animal as closely as possible. They have also been referred to as animal models with construct validity or models with cross-species psychological processes (Willner, 1984; Matthysse, 1986; Ellenbroek and Cools, 1990). In the most optimal situation, a simulation model mimics the human aetiology, which should then lead to the same pathology and symptoms in animals. However, developing simulation models is a very difficult task particularly for CNS disorders for several reasons. First of all for the vast majority of CNS related illnesses neither the aetiology nor the pathology is known. Thus, at best, we can only try to mimic a hypothesized aetiology. Additionally, the majority of symptoms seen in patients with psychiatric disorders are not behavioural in nature, and can only be determined by an interview with the patient. Hence such symptoms (like for instance auditory hallucinations or suicidal ideation) cannot be determined in animals. Secondly, the symptoms can be very different within the same patient population. As we will see in later chapters, most patients are diagnosed on the basis of a checklist of symptoms with individuals needing a certain minimum number of symptoms to receive the diagnosis. As an example see Box 7.1 (Chapter 7) where the symptoms for the diagnosis of major depression are described (according to the Diagnostic and Statistical Manual, DSM V). Since a simulation model aims to mimic "the disorder" one would in a first instance expect that it should encompass all nine symptoms. However, only five are sufficient for the diagnosis of major depression. In other words, one patient can have symptoms 1, 3, 5, 7 and 9, while another can have symptoms 2, 4, 6, 8 and 9. It seems highly unlikely that both patients have the same aetiology and pathology as they differ so widely in their clinical symptoms. From a modelling perspective this raises the question of whether we need two different models (or, many more if we are to mimic all possible combinations of 5 out of 9). However, even if we were to try and incorporate all the symptoms in Box 7.1, we are faced with another important problem: patients with major depression can have significant weight gain or loss (symptom 1c), insomnia or hypersomnia (symptom 1d) and psychomotor retardation or agitation (symptom 1e). It is obvious that within the same simulation model an animal cannot show both weight gain and weight loss. In other words, it is impossible to mimic all the symptoms of the disorder (even if they lend themselves to be assessed in animals). This clearly suggests that the currently used diagnostic systems (such as DSM V) lead to a very heterogeneous group of patients with widely varying symptoms, pathology and aetiology, and has led many researchers to suggest that rather than trying to model the entire disease, it is more worthwhile to model specific aspects of the disease, such as specific symptoms, or, on a more basic level, specific biomarkers or endophenotypes (Almasy and Blangero, 2001). We will discuss these approaches in more detail in Chapter 10. However, as these approaches are still relatively new, we will focus in the remainder of this section on the traditional simulation models that aim to model specific (aspects of) disorders.

As discussed in previous publications, there are several different strategies how to develop a simulation model, depending on the amount of information available (Ellenbroek and Cools, 2000; Ellenbroek et al., 2000; Ellenbroek, 2010). As with screening models, developing, optimizing and validating simulation models requires a continuous interaction between clinical and preclinical research. Fig. 3.3 shows a general approach to developing

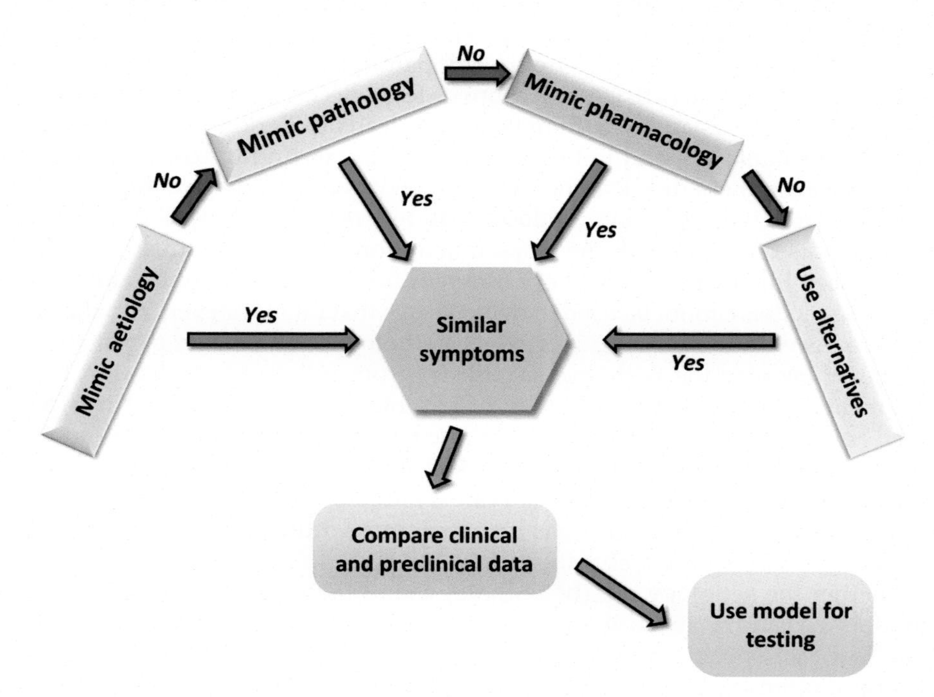

FIGURE 3.3 **Simulation models.** Given our limited knowledge on the cause and pathology of psychiatric disorders, developing simulation models has been challenging. Several different approaches have been attempted. These approaches have first attempted to mimic the (presumed) aetiology or pathology. When such approaches fail, one can try to induce symptoms through pharmacological means, or finally, use alternative approaches such as selective breeding. Once an approach has been identified, it is important to compare the outcome with our clinical knowledge to establish the validity of the model before using it for testing.

simulation models. The highest category of simulation models is that in which the aetiology of the disease is mimicked. As mentioned before, very few models exist that encapsulate the aetiology completely, as for most disorders this is not known. Exceptions are monogenetic disorders, such as Huntington's disease, which results from a triplet repeat (see Chapter 2) in exon 1 of the huntingtin gene. This realization has led to a large number of different mice and rat models, differing in whether the species-specific huntingtin gene is altered, or whether the human mutated huntingtin gene (or only exon 1) is inserted into the genome of the rat or mouse. All these models have advantages as well as disadvantages, but none have been shown to completely replicate the human condition. In the case of psychiatric disorders, aetiologically-based simulation models necessarily have to be limited to mimicking only a part of the aetiology (for instance one specific genetic mutation that has been linked to the disorder or one specific environmental effect). We will discuss such models in more detail in Chapters 6–9.

If it is not possible to mimic the aetiology (either because it is completely unknown or so complicated that only very limited information is available), the next category of simulation models aims to mimic the pathology. Such models have been quite successful in neurological disorders, such as Parkinson's disease. In this disorder there is a fairly selective lesion of the dopaminergic cells within the substantia nigra pars compacta (although pathological inclusions such as Lewy bodies are also prominent in several brain areas). There are several neurotoxic substances such as 6-hydroxydopamine (6-OHDA) and 1-Methyl-4-Phenyl-1,2,3,6,-TetrahydroPyridine (MPTP) that more or less selectively destroy dopaminergic cells. Administering these substances to animals indeed leads to Parkinson's disease-like pathology in the brain of animals, and in line with the symptoms in humans, to rigidity and bradykinesia. However, these drugs do not lead to Lewy bodies and more importantly, in most cases the drugs are given only once in doses that destroy 80 – 90% of all dopaminergic cells. In Parkinson's disease cell death develops much more slowly and thus the compensatory mechanisms seen in patients will be very different from those seen in the animal models. In addition, these models cannot be used to detect drugs that retard or even stop the progression of the disease. As with the aetiological models, pathological models for psychiatric disorders are rare, partly because the pathology is still far from fully understood, but also because our current thinking about the pathology of psychiatric disorders is that it is much more distributed across the central nervous system. As we will discuss later in this book, it is now generally thought that most psychiatric disorders are due to disturbances in the communication between brain regions, rather than to one dysfunctional structure or neurotransmitter. Obviously such network disconnections are much more difficult to mimic that a single neurotransmitter or brain region dysfunction.

When neither the aetiology nor the pathology lend themselves for modelling in a simulation model, we can try and use a pharmacological approach. In this approach we make use of the fact that specific drugs exist that can mimic (aspects) of a disorder in healthy volunteers and often exacerbate existing symptoms in patients. Examples of such drugs are listed in Table 3.2. Based on the same principle as the screening models, the idea is that if these drugs induce certain (aspects of) psychiatric disorders in humans, they will also do so in animals. Although in most cases this is true to a certain extent, this category of simulation models has an inherent weakness. In humans, these drugs induce only acute effects of short duration (although the effects of reserpine can last for a few days, as it depletes cells

TABLE 3.2 Examples of Drugs that can Induce Psychiatric Symptoms in Humans

Drug	Effect
Amphetamine	Psychotic symptoms
CCK	Panic attacks
Cocaine	Addictive behavior (also holds true for other drugs of abuse)
Phencyclidine	Psychotic symptoms
Reserpine	Depressive symptoms

of monoamines, which take time to replenish), whereas psychiatric disorders are likely to develop over prolonged periods of time, and by definition symptoms have to persist for a significant period of time (weeks to months). Thus such pharmacological models can, at best, mimic the acute exacerbation of the disorder. In addition, most of the drugs mentioned in Table 3.2 only mimic a subset of symptoms. Amphetamine, for instance, has often been used as a pharmacological simulation model for schizophrenia, but in humans predominantly induces hallucinations and delusions, but none of the negative symptoms (such as blunted affect and apathy).

When all the previously mentioned approaches are unsuccessful, some alternatives are still possible. Such models usually start by investigating one (or more) of the symptoms of the disorder we are interested in. One approach is to compare different strains of rats or mice to investigate whether any of them "spontaneously" show this symptom. An example is the Brown Norway rat (Feifel and Shilling, 2013) which was found to show at least one schizophrenia-like deficit (ie, a reduction in prepulse inhibition of the acoustic startle response). Alternatively, one can look for spontaneously occurring symptoms within one strain, and assuming this occurs, selectively breed these symptom-exhibiting animals to produce selection lines. One example of this approach is the development of high and low anxiety lines (Liebsch et al., 1998). These animals were originally selected from an outbred Wistar strain of rats based on their response on the elevated plus maze. This is an often used animal test to detect anxiety-like behavior consisting of two open arms and two arms with high walls. The animals have to choose between the relative safety of two enclosed arms and the tendency to explore all four arms (including the more aversive open arms). The selection led to a group of animals that spent a relatively short time on the open arms (high anxiety) and a group of animals that spent relatively long time on the open arms (low anxiety). Subsequent breeding of animals led to the High Anxiety Behavior (HAB) and Low Anxiety Behavior (LAB) rat lines. Interestingly, subsequent testing of these animals in other models for anxiety confirmed that HAB rats are indeed in general more anxious than LAB rats.

A special case of this alternative strategy for simulation models is the repurposing of existing strains or selection lines. With this we mean that rats that were originally selected on the basis of one specific behavior or pharmacological response, are subsequently found to represent an animal model for another disorders. One of the most well-known examples is the Flinders Resistant and Flinders Responsive line. Originally selected on the basis of their response to the cholinesterase inhibitor DFP, they are now regarded as an animal model for depression (Overstreet and Wegener, 2013).

VALIDATING ANIMAL MODELS

One of the most important stages in the developing of an animal model is the validation stage. We already touched upon this in the description of the screening models. However, validation is a complex process and can involve a series of different criteria. We generally distinguish three different levels of validation:

1. *Predictive validity*: This level of validity is focused on how well the model predicts the clinical effects of drugs. Obviously this validity is closely related to screening models (where we have already discussed this is some detail) but it is equally important in simulation models. In addition to the general criteria that we discussed above (false positives/negatives etc.) we often identify additional criteria depending on the animal model. For instance it is well known that antidepressants only work after chronic treatment and that anticholinergic drugs inhibit the neurological side effect but not the therapeutic effects of antipsychotics.
2. *Face validity*: This level of validity focuses on the similarity between the model and the disorder at the level of the symptoms. Again several different face validity criteria can be identified, including false positives and false negatives. In this case, false positives would refer to symptoms that occur in the model, but not in the patients, while the reverse holds true for false negatives, these are symptoms that occur in the patients, but not in the model. As we discussed in the previous section, it is highly unlikely (if not impossible) to completely model all symptoms of a disorder, thus it is impossible to completely fulfil this set of criteria. It is important to realize that face validity is not only related to the traditional symptoms (as they are assessed using DSM V) but also other signs and biomarkers (when they are available). Likewise, certain (sets of) symptoms may not be required for the official diagnosis, but may still be prominent in certain diseases. As an example, most patients with schizophrenia suffer from cognitive deficits, such as disturbances in working memory and/or executive functioning. Although the presence of these symptoms is irrelevant for the diagnosis, a good animal model should encompass these features.
3. *Construct validity*: This level of validity is generally regarded as the highest and most important one. It focuses on the similarity in underlying psychopathological construct. In order words, whether or not the model faithfully represents the aetiology and pathology of the disorder. We have already touched upon the relative lack of knowledge regarding the aetiology and pathology of CNS disorders, and hence this validity is the most difficult to investigate. In general, we assume some degree of validation if the model at least in part shows similarity with the clinical situation. For instance, we know that a subgroup of patients with schizophrenia have a mutation in a gene called *DISC1* (disrupted in schizophrenia 1). Thus, a mouse model in which this gene is disturbed can be regarded as a model with a certain degree of construct validity, in spite of the fact that the majority of patients with schizophrenia do not have a mutation in the *DISC1* gene.

The description of the different sets of validating criteria already shows that not all animal models aim to fulfil all the criteria. Thus screening models are unlikely to fulfil any aspect of face or construct validity. Likewise the pharmacological and neuroscience bioassays are often not validated for any of the above criteria. Nonetheless, it is important, especially for the simulation models, that these criteria are assessed to ascertain the translational value of the model is.

GENETIC MODULATION IN ANIMALS

In Chapter 2, we discussed a number of different genetic alterations that can occur in humans. Although similar alteration can (and do) "spontaneously" occur in animals too, they are in general too random and unpredictable to be useful in the development of simulation models for CNS disorders. Fortunately, if we want to model genetic factors in simulation models a plethora of techniques are now available. Traditionally genetic approaches in animal modelling have been subdivided into two categories, forward and reverse genetics (Fig. 3.4).

Forward Genetic Approaches

Forward genetics refers to the process used to identify genes responsible for a particular phenotype. In other words, the starting point is the phenotype and from thereon we move forward to the genes responsible. Before the advent of genetic manipulation, this was the method of choice. It generally started with the identification of differences among animals (either different strains or different individuals within a single strain of rats or mice) in response to a challenge. This challenge can be a specific behavioural setup (such as exploration of an open field), a specific learning paradigm (such as active avoidance) or the response to a particular drug (such as the response to the dopaminergic agonist apomorphine). Subsequently, the extremes of the populations are selectively bred and the offspring again tested. As an example, many years ago, we had found that within a Wistar rat population roughly about 40% was extremely sensitive to the stereotyped gnawing response induced by apomorphine (ie, these rats gnawed more than 500 times within 45 minutes). On the other hand another 40% were completely insensitive (ie, gnawed less than 10 times in the same 45 minutes). We selectively bred these so-called apomorphine susceptible (APO-SUS) and unsusceptible

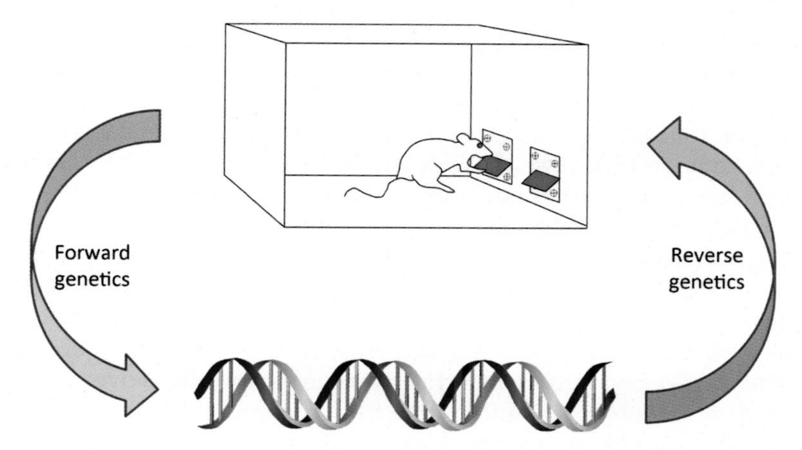

FIGURE 3.4 **Genetic modelling in animals.** Broadly speaking there are two different approaches to studying the relationship between genetic factors and the phenotype. In the forward genetic approach the starting point is a difference in phenotype between animals, and from there moving forwards to identifying the gene(s) involved. In the reverse genetic approach, the starting point is a genetic alteration (generally induced by specific random or targeted techniques), and from there move backwards towards the phenotype.

TABLE 3.3 Some Examples of Selection Lines and their Principle Characteristics

Abbreviation	Name	Principle selection criterion
APO-SUS/UNSUS	Apomorphine susceptible/unsusceptible	Apomorphine induced stereotyped gnawing behavior
FSL/FRL	Flinders sensitive/resistant	Response to cholinesterase inhibitor DFP
HAB/LAB	High anxiety/low anxiety behavior	Open arm behavior in elevated plus maze
MR/MNR	Maudsley reactive/nonreactive	Defecation in a novel open field
RHA/RLA	Roman high/low avoidance	Active avoidance behavior
SHR	Spontaneous Hypertensive	High blood pressure
TMB/TMD	Tryon maze bright/dull	Spatial maze learning

(UNSUS) rats, carefully testing each generation for their apomorphine response (Ellenbroek and Cools, 2002). This approach showed that indeed the apomorphine induced gnawing response had a strong genetic component and after 20 generations we found that 80% of the APO-SUS rats scored more than 500, and 80% of the APO- UNSUS rats scored less than 10 gnawing counts over a 45-min period. Similar selection procedures have been used by others and have led to a large number of selection lines (see Table 3.3 for some examples).

Once the selection lines are stable, the process to find the underlying gene(s) can start, often using a technique known as Quantitative Trait Loci (QTL) analysis. In short this technique uses specific markers to identify regions on the DNA (the chromosomes) that differ in a quantitative trait (ie, a phenotype). The principle behind this linkage technique was already discussed in Chapter 1. During meiosis parts of each chromosome cross over from the paternal chromosome to the maternal chromosome. As a result the genes segregate independently (Mendel's second Law). However, the closer two genes are on the same chromosome, the smaller the chance that the cross over happens between these two genes, hence they are considered to be linked. By using genetic markers (that show high polymorphism), we can see which markers are linked to the quantitative trait, and it is thus inferred that the gene (or more likely genes) responsible for the trait is (are) located near these markers. More detailed analyses of these regions can then be undertaken to narrow down the region and find the relevant genes.

A special technique for identifying genes responsible for behavior is the use of consomic strains (also referred to as chromosome substitution lines). These are strains derived from two original strains that show clear differences in one or more phenotypes. The procedure to develop consomic strains is illustrated in Fig. 3.5 and starts with breeding an F1 generation from a cross between the original lines. Next, this F1 generation is crossed back to one of the parent strains (in this case the red strain). Although cross-over between the parent chromosomes often occurs, there are occasions where a chromosome escapes cross over. In that instance we will have animals with one pure chromosome from the original blue strain (in our example chromosome 4) and mixed blue/red chromosomes (where cross over did occur). By repeatedly backcrossing to the original red strain, continuously selecting only those animals where chromosome 4 did not undergo a cross over event, we can selectively breed in the red chromosomes and ultimately end up with animals that have the complete genetic information from the red strain, with the exception of one chromosome. Cross breeding these animals

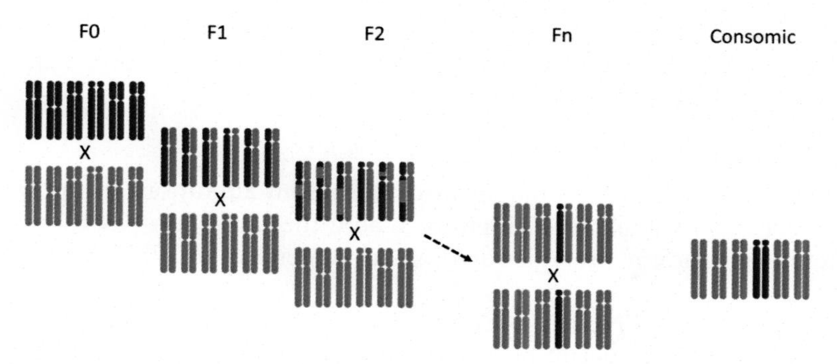

FIGURE 3.5 **The development of consomic strains.** The aim of consomic strain is to develop an animal that has one intact chromosome of one strain of a background of another strain. It is based on the fact that although crossing-over events can occur during meiosis, in some cases this does not happen. Thus when rats of two strains (blue and red) are crossed the F1 generation will have one chromosome of each of the original strains. When this strain is mated (backcrossed) with one of the original strains (the red strain in this case), most of the F2 chromosomes will be either red (from the red parent) or a mixture of red and blue (due to the cross-over events). However, if one chromosome does not show a cross-over event, a fully retained blue chromosome will be present. These animals will be selectively backcrossed again, selecting in each generation only those animals that have a fully retained blue chromosomes. By backcrossing for many generations, the entire chromosome will be from the red strain with the exception of the one blue chromosome. When two such animals are then mated a homozygous consomic strain for this specific chromosome has been developed.

with each other will give us animals with two copies of the blue chromosome 4 on a completely red background. Using this procedure, it has been possible to generate panel of consomic or chromosome substitution strains (CSS) for both rats and mice (Cowley et al., 2004; Wahlsten, 2012). Martien Kas and his coworkers (de Mooij-van Malsen et al., 2009) used the chromosome substitution panel between C57BL/6J and A/J mice to study exploratory and anxiety-related behavior and found that the original C57Bl/6J mice showed significantly less anxiety than CSS15 and CSS19 (in which chromosome 15 resp. 19 was substituted with the corresponding A/J chromosome). Subsequent more detailed analysis led to the identification of one specific gene on chromosome 15 that was related to avoidance behavior.

With the advent of molecular genetic techniques, such as the microarrays chips, it has now become possible to directly link individual genes to phenotypic differences. Although different types of microarrays exist, they are generally based on the PCR method (see Chapter 2), but allow for the simultaneous identification of thousands of genes. For instance, we used the cDNA microarray technique to identify gene expression differences between the APO-SUS and APO-UNSUS rats and found a very large difference in one specific gene, the *Aph-1B* gene. Subsequent analysis showed that this difference was due to a copy number variation: Whereas APO-UNSUS rats have 3 copies of the *Aph-1B* gene, APO-SUS rats have only 1 or sometimes 2 (Coolen et al., 2005).

Reverse Genetic Approaches

In the reverse genetics approach, we study the functional consequences of genetic alterations. In the last several decades our toolbox for manipulating genes has increased enormously and it would be beyond the scope of this book to describe all the techniques is detail.

However, as many different techniques have been used in the development of genetic simulation models, we will discuss the most relevant ones. In general, the techniques can be subdivided into random and targeted techniques.

Random Techniques

As the word implies, these techniques lead to random alterations in the DNA and hence require significant DNA analysis to identify where the alteration has taken place. Two of the most often used random strategies are ENU mutagenesis and Transposable Elements (or transposon).

N-ethyl-N-nitrosourea (ENU) is a chemical compound with a high mutagenicity (it is sometimes referred to as a supermutagen). It acts as an alkylating agent, transferring its ethyl group to either the nitrogen or oxygen of one of the nucleotide bases. Fig. 3.6 shows how ENU transfers its ethyl group onto a thymine, leading to a mismatch with the adenine in the complementary strand. There are two possible consequences of ENU: either the DNA mismatch repair mechanisms complete remove the faulty ethyl-nucleotide base and restores the original thymine adenine pair, or both the ethyl-nucleotide base and its complementary base are excised and replaced by another pair leading to a point mutations. Studies in both rats and mice (Justice et al., 1999; Smits et al., 2006) have shown that ENU primarily attacks A/T pairs, although C/G pairs can also be mutated (Fig. 3.6). Moreover, ENU predominantly targets spermatogonial stem cells that lead to sperm cells. The strategy is therefore to treat male rats or mice and mate them with wild type (genetically unaltered) females. All the F1

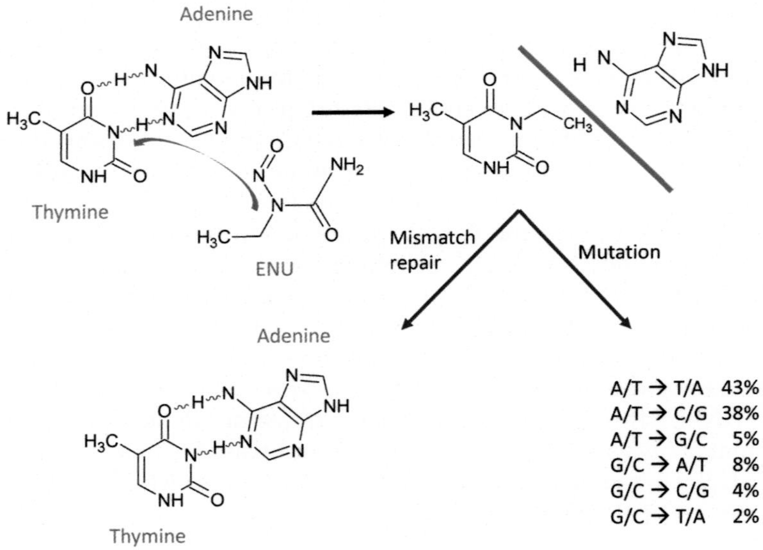

A/T → T/A	43%
A/T → C/G	38%
A/T → G/C	5%
G/C → A/T	8%
G/C → C/G	4%
G/C → T/A	2%

FIGURE 3.6 **ENU mutagenesis.** N-ethyl-N-nitrosourea (ENU) is a chemical mutagen often used in animal research. It acts by alkylating the nucleobases in the DNA, transferring an ethyl group. This interferes with the normal base pairing and generally leads to mismatch repair. This can either be complete and the original (T-A) basepair is restored, or it can lead to a mutation. Although ENU is considered a random mutagen, it has a clear preference for A/T base pairs as is evident from the table in the lower right hand corner.

offspring can then be genotyped to identify the mutated gene. As ENU mutagenesis is a random technique it is impossible to predict which genes are affected and how they are affected. Therefore, the usual strategy is to select a number of target genes that are of particular interest and sequence those, to investigate whether a mutation can be found. As we discussed in the previous chapter, the mutations can either be a silent mutation (leading to no change in the amino acid sequence), a missense mutation (leading to a single amino acid change) or a nonsense mutation (leading to a premature stopcodon). Studies in rats have shown that ENU produces about 1 mutation in 1 million basepairs (Smits et al., 2006), although it differs somewhat from strain to strain. Extrapolating this, it was calculated that about 50,000 offspring would be necessary to obtain a 96% chance of having a missense mutation in each gene (assuming an average gene of about 1250 basepairs). Obviously, this is not a very feasible strategy. Fortunately, other more targeted strategies are available.

Transposable elements (or transposons) are mobile genetic units, first identified by Barbara McClintock in maize (McClintock, 1950). Transposons are DNA elements that can change their location within the genome, which can lead to an insertion mutation. Initially transposons were used to create insertional mutagenesis in plants and invertebrates, as active transposons were thought to be absent in higher vertebrates. However, in 1997 a fish transposon was identified, and although dormant for 15 million years the researchers were able to activate it and called it "Sleeping Beauty" (Luft, 2010). Subsequently other transposon systems such as PiggyBac have also gained widespread popularity. Transposon mutagenesis is illustrated in Fig. 3.7 and requires a two-step process (Jacob et al., 2010). It is based on the assumption that the random insertion of a transposon into a gene is likely to cause a null mutation by disrupting the transcribed mRNA at the site of insertion. In order to optimize this system, the transposon is usually linked to a so-called gene-trap insertion. In its simplest form, a gene-trap consists of a splice acceptor, a reporter gene and a poly(A) tail. When this gene-trap is inserted into a gene, the splice acceptor will force a fusion between the first part of the original gene and the reporter gene, thus disrupting the normal mRNA and subsequent protein formation. By combining this gene-trap with the transposon vectors, we can ensure the gene-trap can randomly insert into other genes. However, for transposon to move along the genome, they need to be activated using an enzyme called transposase. The standard procedure is therefore to create two transfected animals, one carrying the transposon linked to the gene-trap, and one carrying the transposase gene. Once these animals are mated up, the offspring (seed males) will have mobilized (jumping) transposons in the spermatogonial stem cell. By outcrossing them with wild type females, and subsequent DNA analysis, we can identify the gene that is disrupted. Compared to the ENU mutagenesis technique, the transposon techniques have several advantages but also disadvantages. Technically, the transposon technique is more complicated and involves the generation of two different transgenic animals (which also implies that special laboratories are required). As ENU is a mutagenic but not a transgenic technique (ie, no "foreign" DNA is inserted into an animal) the procedures and the subsequently generated animals can be housed in conventional laboratories. On the other hand, using special reporter genes in the gene-trap cassette, identifying which gene(s) are affected by the transposon insertion can be much easier and therefore widespread DNA sequences is not necessary. Moreover, only a few transposon insertions are created in each animal (as compared to many point mutations in ENU mutagenesis). Therefore, it is easier to isolate the mutation from potential

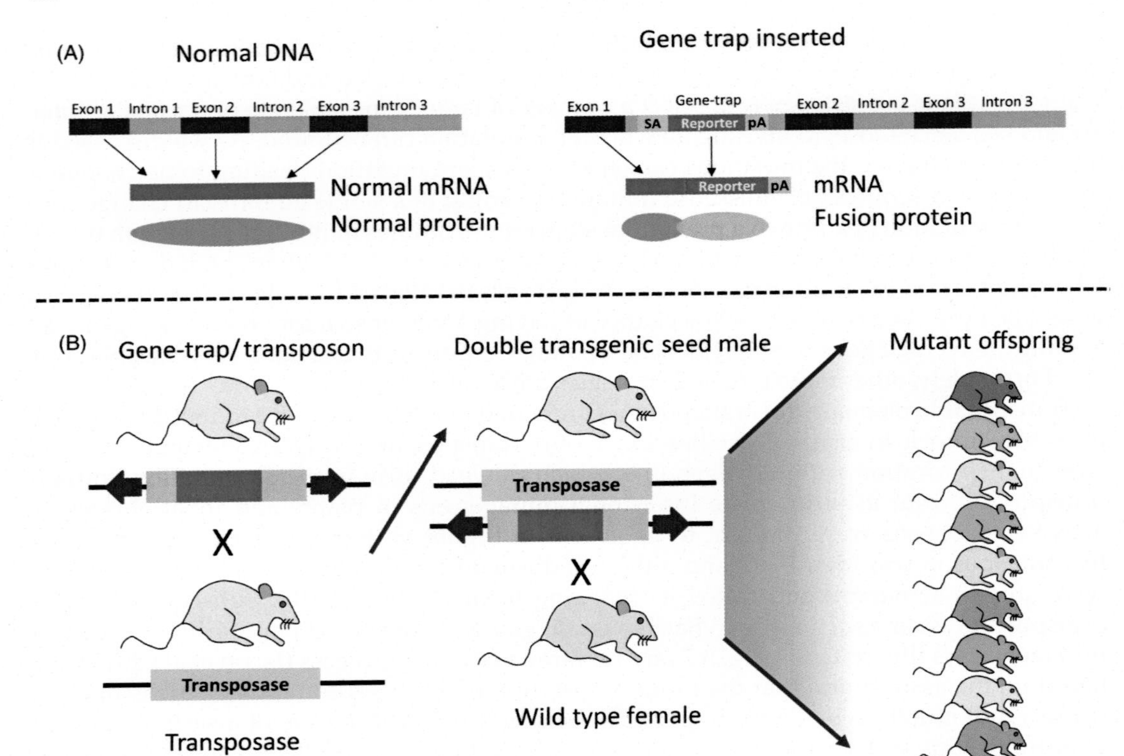

FIGURE 3.7 **The principle of transposon induced mutagenesis.** Transposons are movable element that can, under the correct circumstances move from one gene location to the next. The principle behind the mutation (A) is that when the transposon is relocated to another position, it will likely cause a null (or nonsense)-mutation as it interferes with the normal mRNA formation. To increase the chance, the transposon usually contains a splice-acceptor (SA), which "forces" the spliceosome to create a fusion protein consisting of the upstream elements of the original gene (in this case only exon 1) and the transposon reporter protein. For the transposon system to work, the transposon vector needs to be activated by a so-called transposases enzyme. Hence, the typical strategy is to develop two types of animals: one in which the gene-trap/transposon is inserted in the genome and one which contains the DNA for the transposase enzyme. By breeding these rats together a so-called "seed male" is created with an activated transposon system. By breeding this seed male with wild-type females, a large number of offspring can be created that all differ in the site of transposon insertion.

background mutations that might segregate with it. Finally, whereas ENU mutagenesis can induce many missense and silent mutations, transposon mutagenesis almost exclusively lead to nonsense (ie, knock-out) mutations.

Before moving to more targeted genetic approaches, it is important to realize that both the ENU and transposon mutagenesis techniques can also be used in combination with forward genetics. As both are random techniques, and hence we don't know which genes are affected, the F1 generation can also be studied for phenotypical differences before investigating the genetic factor(s) involved. One complicating factor here is that all F1 animals only have heterozygous SNPs, and thus only strong effects can be identified.

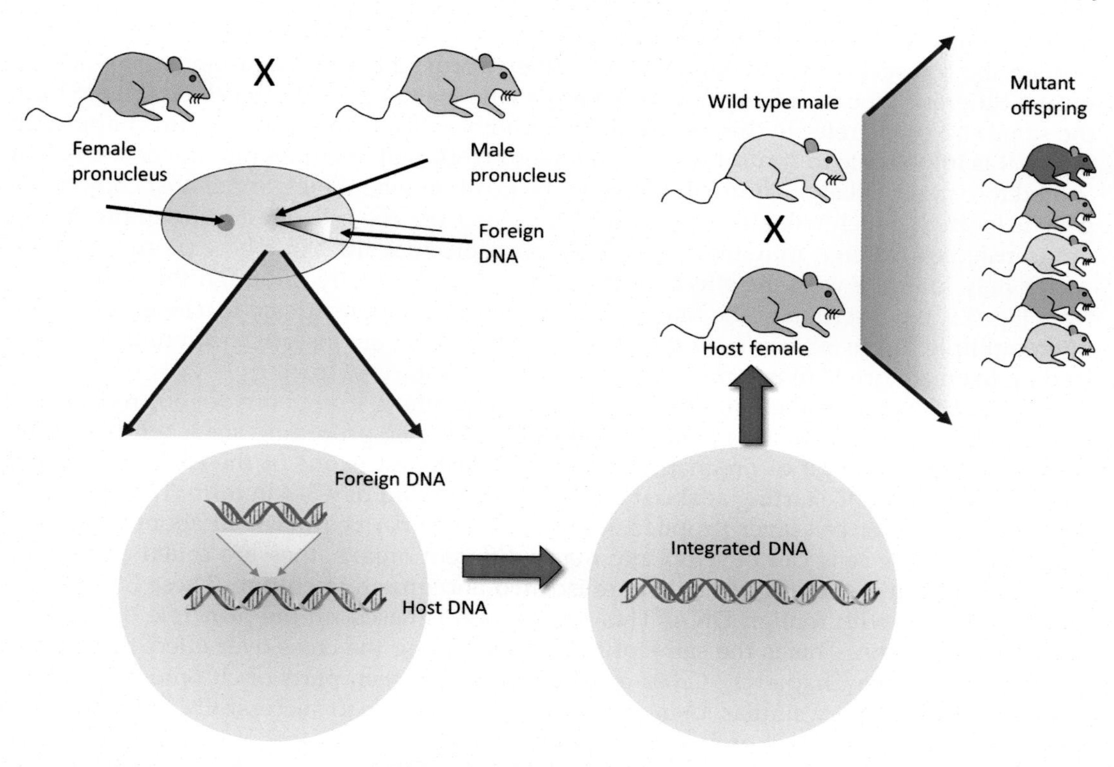

FIGURE 3.8 **The principle of direct DNA microinjections.** In the direct microinjection technique, foreign DNA is directly injected in a pronucleus of a fertilized egg cells. As the male pronucleus is larger than the female pronucleus, this one is usually selected for the injection. Once the DNA is integrated into the host DNA, the transfected female is mated with the wild type males to produce offspring with the specific genetic transfection.

Targeted Techniques

Although the random techniques offer a number of advantages (such as speed of mutagenesis and a relative easy technology, especially ENU mutagenesis), the most important limitation is that it is impossible to predict which gene will be affected, and thus if we are interested in one specific gene, these techniques are unlikely to be very helpful. Fortunately, several additional techniques have been developed which can be used to alter one specific gene. Such targeted approaches can be subdivided into direct and indirect techniques.

In the direct DNA microinjection technique (Fig. 3.8), foreign DNA is directly injected into a fertilized oocyte. When an oocyte (an immature egg cell) is fertilized, the male and female pronuclei exist for a short period of time, before they merge into a single nucleus. During this stage, DNA can be injected into the male pronucleus (because it is larger than the female pronucleus). This DNA can then integrate into the host nucleus at specialized sites (called nicks) and the fertilized transgenic cells are re-implanted into pseudopregnant females. This technique is quite efficient in mice (up to 50% of microinjected mouse oocytes can lead to transfected germline cells although the efficiency is much less in other species). The DNA integration can happen immediately within the pronucleus, but it is more common that it does not happen until after one or two cell divisions. In that case only some of the cells will contain the transgene and others do not. In other words, the individual mouse is made up of two different type of cells with a different genetic make-up. As both cells originate from the same strain, we refer to this as mosaicism. Another disadvantage of the direct approach is that it is impossible to control where the foreign DNA will integrate into the host DNA. It is therefore highly likely that it will not be at the position where the normal host copy of the gene is located, which will affect its expression. Moreover, the cell, in addition to producing the transfected (foreign mutated) gene(s) will also produce the same host (normal) genes, which may interfere with the effects of the mutated gene. Finally, although this is not clear from Fig. 3.8, the insertion of the foreign DNA is generally not restricted to a single copy, but often multiple (up to 50) copies of the same DNA construct are inserted head to tail (CNVs) leading to an abnormal overexpression of the gene(s) in the construct.

In order to overcome these limitations, an indirect method, using homologous recombination, has been developed. This is illustrated in Fig. 3.9. The procedure starts with isolating cells from the blastocyst of pregnant female. The blastocyst is one of the earliest stages of embryo development, starting at about day 5 in humans and day 3–4 in mice. The blastocyst consists of an outer cell layer (trophectoderm) and an inner cell mass (of about 20 cells in mice). Since these cells can develop into many different organs, they are called embryonic stem (ES) cells. It is these ES cells that are isolated and grown in culture. These ES cells are then transfected with foreign DNA. This transfection is based on the principle of homologous recombination. This is the same principle underlying the cross-over effect that occurs during meiosis (see Chapter 1). Cross-over can occur between parts of chromosomes if the DNA stretches are very similar. Using this principle, in order to successfully transfer a cell, the foreign DNA is flanked on both sides with a stretch of DNA that is identical to the sequences of the host DNA. This ensures that the foreign DNA is inserted at the correct location. Moreover, in contrast to the direct approach, this also significantly reduces the risk of the insertion of multiple copies (unless of course this is the objective of the transfection). By using this technique, we can insert a SNP, a premature stopcodon or a variety of other genetic alterations. Next the cells are placed back into a blastocyst, which are then implanted in a pseudopregnant female. This blastocyst will then develop into an adult. As the blastocyst contains both normal cells (from the host mother) and transgenic cells from the donor mother (from a different strain) the resulting animals is called a chimera (having DNA from two different strains of animals). When the foreign DNA is also found in the gametes, it will transfer to the next generation, and from there on homozygous transgenic animals can be bred. This technique is very powerful but also has several limitations. First of all homologous recombination is actually a very rare event, and thus many cells need to be treated. One strategy to identify which cells are successfully transfected is to combine the transgene with the gene for Neomycin phosphotransferase. This protein protects cells for the lethal effects of antibiotics such as neomycin. Thus, when the cell culture is treated with neomycin, all cells will die, except those that are successfully transfected. A second and more serious limitation is that the technique only works if the ES cells that are transfected, can develop into all other cells, and especially into gametes (ie, egg or sperm cells). In other words, the ES should be pluripotent or omnipotent. For a long time, such omnipotent ES cells were only found in a few strains of mice, especially the S129 mice. For that reason, the vast majority of transgenic animals are mice. As the S129 mice are not the best strains for behavioural studies, the transgenic animals

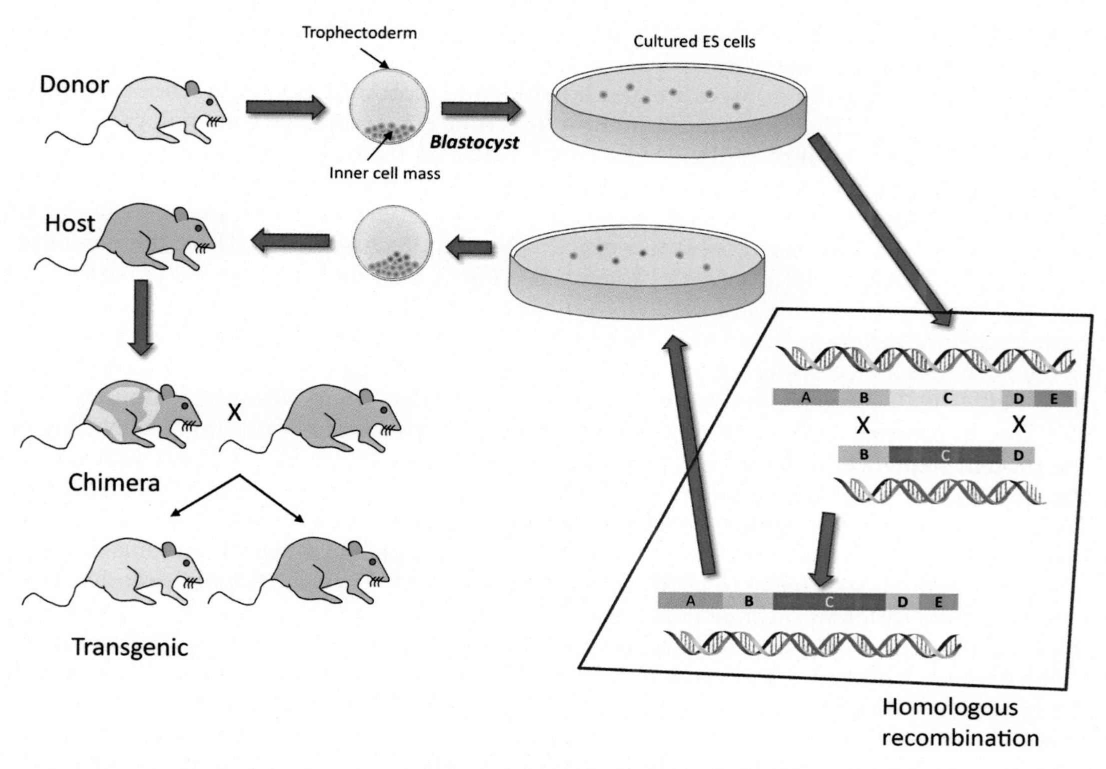

FIGURE 3.9 **The principle of genetic engineering using homologous recombination.** In this technique, pluripotent (or omnipotent) embryonic stem cells are isolated from the inner cell mass of blastocyst. These cells are subsequently transfected with the gene of interest. This is done by using a double stranded DNA with very specific overlapping sequences (see lower right hand), and is based on the principle of homologous recombination, which also underlies the cross-over events occurring during meiosis (see Chapter 1). If such a cross-over event occurs, the endogenous section of DNA is exchanged for the foreign DNA. These transfected cells are then transplanted back into the blastocyst of a host female. As these offspring will consist of cells from both the donor and the host, they are referred to as chimeras. If the transfection is also found in the gametes, subsequent breeding of these chimeras will lead to the development of a transgenic animal.

are usually back-crossed onto behaviorally more conducive animals such as the C57B/6 mice. Fortunately, in more recent years, procedures have been developed to induce omnipotent ES cells in other mice strains and even other species, such as rats as well (Blair et al., 2011). Essential for this success was the recognition that the key players in determining whether an ES cell will retain its pluripotency are subtly different between mice and rats. Thus, in mice a cocktail of three inhibitors (3i) in the culture media was found to keep the cells in a basal pluripotent state: 1. The FGF receptor inhibitor SU5402; 2. The MEK activation inhibitor PD184352; and 3. The GSK3 inhibitor CHIR99021. Although this cocktail also was somewhat effective in rat ES cells, it was subsequently found that the combination of a more effective inhibition of MEK (using PD0325901) along with CHIR99021 but without SU5402 (2i) led to much more stable pluripotent ES cells, especially when LIF (Leukemia Inhibitory Factor) was added (Jacob et al., 2010). Overall, 30%–60% of the blastocyst explants will grow and propagate under these circumstances.

Although the targeted approach using homologous recombination is much more precise than the direct pronuclear injection technique, it is time consuming and technologically challenging. Moreover, until recently, it was only feasible in mice. Several alternative targeted approaches have therefore been developed, predominantly for producing knock-out animals, but these techniques can also be used to produce knock-ins (where specific genes are inserted).

The Zinc finger nuclease (ZFN) technique is a direct technique based on the combined effect of two different DNA binding proteins (see Fig. 3.10). Nucleases are enzymes that can selectively sever the double stranded DNA molecule. The most often used endonuclease *Fok*I requires dimerization in order to induces a DNA double stranded break. As *Fok*I has relatively little specificity, applying just *Fok*I would lead to a (large) number of random breaks. Zinc Finger (ZF) proteins, are proteins with a characteristic loop structure, stabilized by a Zn^{2+} ion. In contrast to *Fok*I, ZF proteins are highly selective in their binding. Depending on the protein sequence in the α-helix section of the molecule, they will bind only to a specific triplet. Thus, by engineering different ZF proteins and combining multiple ZF proteins with *Fok*I, we can selectively guide the endonuclease to cut the DNA double bond only in one specific location. Since we need two different ZFNs an average of 4 to 6 ZF proteins on either side gives enough selectivity to produce a very precise cut. Once the double stranded DNA is broken, two different mechanisms are available to the cell to repair the break: Homology dependent repair (HDR) and Nonhomologous end joining (NHEJ). Whereas HDR leads to a faithful repair (and hence a nonmutated DNA), NHEJ is much less accurate and can lead to the insertion or deleting of a single nucleotide at the repair site, thus leading to a frame-shift and in most cases a nonsense protein. Although this technique has predominantly been used to induce knock-outs, it has been adapted to produce other types of mutations as well (Geurts and Moreno, 2010).

In addition to the ZNF techniques, several alternative techniques have also been developed, most of which are based on the same principles: that is, a two component process combining DNA binding components with endonucleases. TALEN, transcription activator-like effector nucleases are artificially generated fusions between DNA binding domains of transcription activator-like (TAL) effectors and DNA cleavage domains of endonucleases (often *Fok*I). TAL effectors are proteins secreted by the gram negative bacteria *Xanthomonas* spp. but it was not until the TAL effector-DNA code was uncovered that their application in gene editing could be fully realized (Wright et al., 2014). One of the most recently published methods is the CRISPR/CAS technique (Doudna and Charpentier, 2014). This technique is based on bacterial defense mechanisms against viruses and plasmids. CRISPS (clustered regularly interspaced short palindromic repeats) constitute an RNA that normally binds to the viral DNA and its associated endonuclease proteins Cas subsequently cleaving the double helix, thus countering the viral attack. In contrast to ZFN and TALEN, the CRISPS/Cas system uses guide RNAs to selectively bind to DNA (rather than proteins). Thus, by selectively engineering the guide RNAs, it is possible to precisely arrange the Cas protein (usually Cas9) over the gene of interest. The guide RNAs are about 80 nucleotides, with the final 20 being used to direct the RNA to the target of interest. Although the guide RNA and Cas9 protein can be expressed separately, it is more efficient to place them together in a single plasmid (a circular stretch of DNA). Its ease and versatility combined with its effectiveness in many different

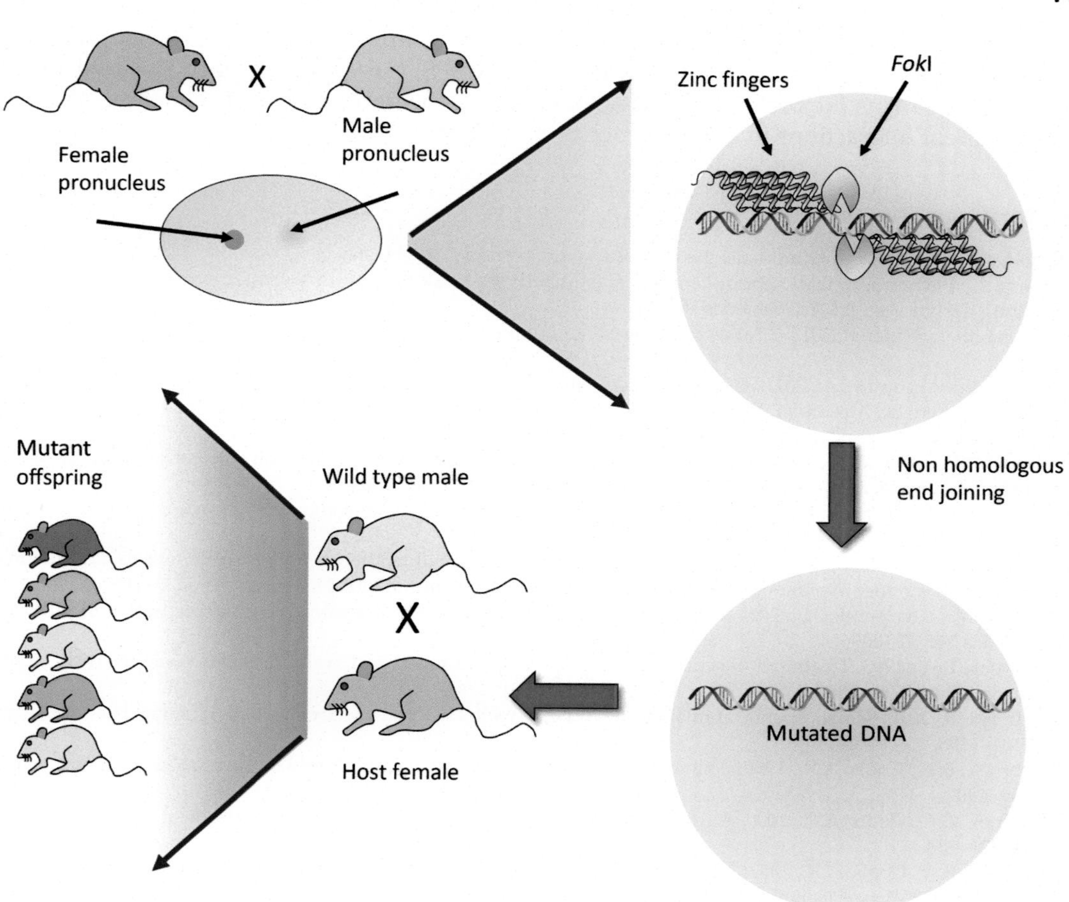

FIGURE 3.10 **The principle of genetic engineering using zinc finger nucleases.** Zinc finger nucleases are chemically designed molecules that consist of two specific types of proteins: zinc finger proteins and nucleases. Zinc finger proteins are unique in that they can selectively bind to specific triplets. By combining several different zinc fingers (6 are illustrated in the figure) the sequence specificity can be very high (ie, they bind selectively to a unique stretch of 18 bases. Nucleases, on the other hand, cut double stranded DNA. However, this cutting requires two nucleases that attack both strands at the same time. Thus by chemical engineering, zinc finger nucleases can be developed to cut only one specific site in the entire genome. Although DNA repeat mechanisms will aim to repair this double stranded break, the most common mechanism of repair is the so-called nonhomologous end joining which often induces a deletion, leading to a frameshift mutation. Zinc finger nucleases are generally injected into the male pronucleus of fertilized egg cells and the mutated DNA can then be transferred to the offspring.

species (including nonhuman primates and human cells) has made the CRISPS/Cas9 system one of the most exciting new developments in gene editing (Doudna and Charpentier, 2014).

In summary, in this section we have discussed the extensive genetic toolbox that is now at our disposal to study the role of genes in regulating behavior and in the development of psychiatric disorders. Originally, the mouse was the preferred species for genetic manipulation, although the rat has been the model organism of choice for CNS disorders. However, the elucidation of the complete rat genome (Gibbs et al., 2004) and the developments of newer techniques (including ZNF, TALEN and CRISPS/Cas9) have opened the door for the return of the rat as the preferred species for studying the (long-term) consequences of gene – environment interactions in neuroscience research (Abbott, 2004).

References

Abbott, A., 2004. Laboratory animals: the renaissance rat. Nature 428, 464–466.
Almasy, L., Blangero, J., 2001. Endophenotypes as quantitative risk factors for psychiatric disease: rationale and study design. Am. J. Med. Genetics 105, 44.
Arrowsmith, J., Miller, P., 2013. Trial watch: phase II and phase III attrition rates 2011-2012. Nat. Rev. Drug Discov. 12, 569.
Blair, K., Wray, J., Smith, A., 2011. The liberation of embryonic stem cells. PLoS Genetics 7, e1002019.
Cook, D., Brown, D., Alexander, R., et al., 2014. Lessons learned from the fate of AstraZeneca's drug pipeline: a five-dimensional framework. Nat. Rev. Drug Discov. 13, 419–431.
Coolen, M.W., van Loo, K.M.J., van Bakel, N.N.H.M., et al., 2005. Gene dosage effect on g-secretase component aph-1b in a rat model for neurodevelopmental disorders. Neuron 45, 497–503.
Cowley, Jr., A.W., Roman, R.J., Jacob, H.J., 2004. Application of chromosomal substitution techniques in gene-function discovery. J. Physiol. 554, 46–55.
de Mooij-van Malsen, A.J., van Lith, H.A., Oppelaar, H., et al., 2009. Interspecies trait genetics reveals association of Adcy8 with mouse avoidance behavior and a human mood disorder. Biol. Psychiatry 66, 1123–1130.
Doudna, J.A., Charpentier, E., 2014. Genome editing. The new frontier of genome engineering with CRISPR-Cas9. Science 346, 1258096.
Ellenbroek, B.A., 1993. Treatment of schizophrenia: a clinical and preclinical evaluation of neuroleptic drugs. Pharmacol. Ther. 57, 1–78.
Ellenbroek, B., 2010. Schizophrenia: animal models. Encyclopedia of Psychopharmacology. Springer, Heidelberg, pp. 1181–1186.
Ellenbroek, B.A., Cools, A.R., 1990. Animal models with construct validity for schizophrenia. Behav. Pharmacol. 1, 469–490.
Ellenbroek, B.A., Cools, A.R., 2000. Animal models for the negative symptoms of schizophrenia. Behav. Pharmacol. 11, 223–233.
Ellenbroek, B.A., Cools, A.R., 2002. Apomorphine susceptibility and animal models for psychopathology: genes and environment. Behav. Genetics 32, 349–361.
Ellenbroek, B.A., Sams Dodd, F., Cools, A.R., 2000. Simulation models for schizophrenia. In: Ellenbroek, B.A., Cools, A.R. (Eds.), Atypical Antipsychotics. Birkhauser Verlag, Basel, pp. 121–142.
Feifel, D., Shilling, P.D., 2013. Modelling schizophrenia in animals. In: Conn, P.M. (Ed.), Animal Models for the Study of Human Disease. Elsevier, Amsterdam, pp. 727–755.
Geurts, A.M., Moreno, C., 2010. Zinc-finger nucleases: new strategies to target the rat genome. Clin. Sci. (London) 119, 303–311.
Gibbs, R.A., Weinstock, G.M., Metzker, M.L., et al., 2004. Genome sequence of the Brown Norway rat yields insights into mammalian evolution. Nature 428, 493–521.
Jacob, H.J., Lazar, J., Dwinell, M.R., et al., 2010. Gene targeting in the rat: advances and opportunities. TIG 26, 510–518.
Justice, M.J., Noveroske, J.K., Weber, J.S., et al., 1999. Mouse ENU mutagenesis. Hum. Mol. Genetics 8, 1955–1963.
Koek, W., 2011. Drug-induced state-dependent learning: review of an operant procedure in rats. Behav. Pharmacol. 22, 430–440.
Kola, I., Landis, J., 2004. Can the pharmaceutical industry reduce attrition rates? Nat. Rev. Drug Discov. 3, 711–715.
Lent, R., Azevedo, F.A., Andrade-Moraes, C.H., et al., 2012. How many neurons do you have? Some dogmas of quantitative neuroscience under revision. Eur. J. Neurosci. 35, 1–9.
Liebsch, G., Montkowski, A., Holsboer, F., et al., 1998. Behavioural profiles of two Wistar rat lines selectively bred for high or low anxiety-related behavior. Behav. Brain Res. 94, 301–310.
Luft, F.C., 2010. Sleeping beauty jumps to new heights. J. Mol. Med. (Berlin) 88, 641–643.
Matthysse, S., 1986. Animal models in psychiatric research. Prog. Brain Res. 65, 259–270.

McClintock, B., 1950. The origin and behavior of mutable loci in maize. Proc. Natl. Acad. Sci. USA 36, 344–355.
McKinney, W.T., 1984. Animal models of depression: an overview. Psychiatric Dev. 2, 77–96.
Overstreet, D.H., Wegener, G., 2013. The flinders sensitive line rat model of depression—25 years and still producing. Pharmacol. Rev. 65, 143–155.
Paul, S.M., Mytelka, D.S., Dunwiddie, C.T., et al., 2010. How to improve R&D productivity: the pharmaceutical industry's grand challenge. Nat. Rev. Drug Discov. 9, 203–214.
Sams-Dodd, F., 2005. Target-based drug discovery: is something wrong? Drug Discov. Today 10, 139–147.
Smits, B.M.G., Mudde, J.B., van de Belt, J., et al., 2006. Generation of gene knockouts and mutant models in the laboratory rat by ENU-driven target selected mutagenesis. Pharmacogenetics Genomics 16, 159–169.
Wahlsten, D., 2012. The hunt for gene effects pertinent to behavioral traits and psychiatric disorders: from mouse to human. Dev. Psychobiol. 54, 475–492.
Willner, P., 1984. The validity of animals models of depression. Psychopharmacology 83, 1–16.
Wright, D.A., Li, T., Yang, B., et al., 2014. TALEN-mediated genome editing: prospects and perspectives. Biochem. J. 462, 15–24.

CHAPTER

4

The Environmental Basis of Behavior

INTRODUCTION

In the previous chapters, we discussed how alterations in the genetic code can affect behavior in animals including humans, and there is no doubt that genetic factors play an essential role in the aetiology of major psychiatric disorders. Indeed, in the chapters in part II we will discuss this in more detail. However, it is equally clear that none of the major psychiatric disorders are actually caused by genetic factors, but that these factors only increase the vulnerability. The most compelling evidence in favor of this comes from family, twin and adoption studies.

The aim of family studies is to investigate whether a specific phenotype (ie, a specific behavior) occurs more often in certain families, and based on the occurrence over generation, how this condition is genetically transmitted. Fig. 4.1 shows two classical examples of family trees, where only the phenotype or trait (ie, whether the individual is affected or not) is indicated (in red). A comparison of the two traits shows that there are clear differences. The trait in Fig. 4.1A occurs much more often and is seen in each generation, while the trait in Fig. 4.1B occurs more rarely, and even disappears in some generations, only to reappear in the next. By careful analysis, we can deduce the pattern of inheritance from the occurrence of the trait. The trait displayed in Fig. 4.1B is similar to that originally found by Mendel in his experiments with peas (see Chapter 1). Thus in the second generation all individuals are identical to one of the parents, while in the second generation the famous 1:3 ratio is found: 1 individual has the original characteristic of one grandparent while the other 3 have the characteristic of the other grandparent (and of the parents). As Mendel explained this type of inheritance occurs when the trait is recessive, that is it only occurs if an individual inherits the recessive gene from both parents. As the trait in Fig. 4.1A is very different, the trait must be a dominant trait. However, the exact genotype of the parents is slightly more difficult to deduce. As in the first generation not all individuals are affected. The original affected male (1) and female (2) parent cannot be homozygous for the trait, as this would mean that all the offspring should also be affected. Thus detailed analysis of the family trees can help us determine the genotype. With the advent of modern genetic analyses, we can nowadays also identify the genotypes directly on the basis of DNA analysis, and thus confirm that trait A is indeed a dominant trait (see Fig. 4.2A)

Gene-Environment Interactions in Psychiatry. http://dx.doi.org/10.1016/B978-0-12-801657-2.00004-5

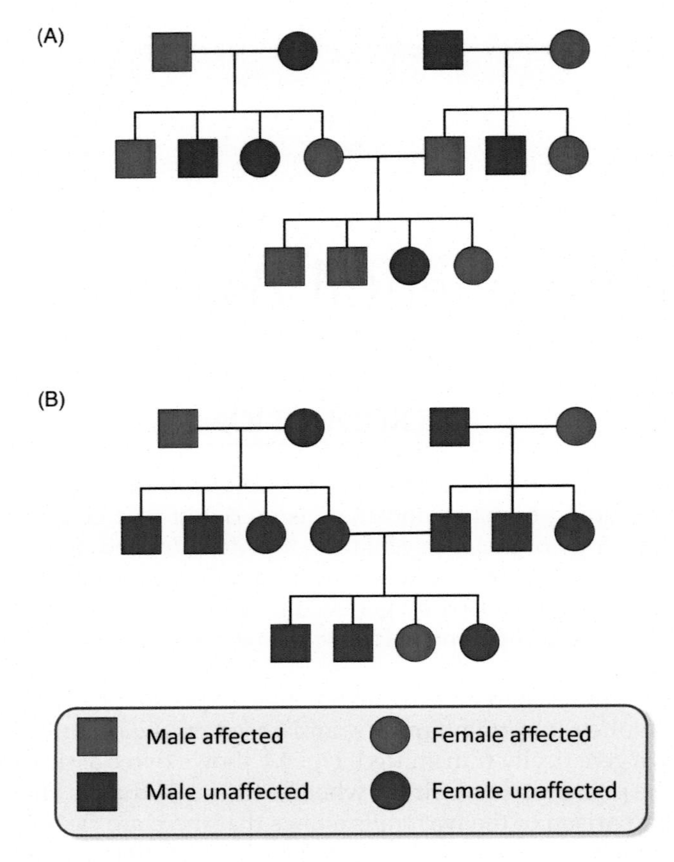

FIGURE 4.1 **Family trees.** Family trees can help identify the mode of inheritance. In fig A, we see that there many more affected individuals (in red), while in B there are only a few, and in the second generation, there is actually no affected individual. This suggests that the trait in A is a dominant and in B a recessive trait.

as one copy of the A is sufficient to display the trait and with both originally affected parents being heterozygous for the trait, while trait B is a recessive trait and only occurs when two copies of the recessive gene b are present (Fig. 4.2B).

Analyses of family trees are relatively simple for monogenetic traits as they normally follow Mendel's first law. However, there are a few exceptions, some of which we have already encountered in Chapter 2. If a de novo (a new) mutation occurs, Mendel's first law is violated (as the genetic information in the offspring is not 50% identical with each parent. However, once the mutation has been inserted, it will be transmitted in a Mendelian fashion in the next generations. Trisomies and monosomies (ie, the offspring inheriting either three or only one chromosome from the parents), on the other hand are in direct violation of Mendel's law as are triplet repeats. As we saw in Chapter 2, triplet repeats (once they breached the threshold) have a tendency to lengthen with each generation (leading to genetic anticipation, see Chapter 2) and thus the offspring will have different genetic information. A special case

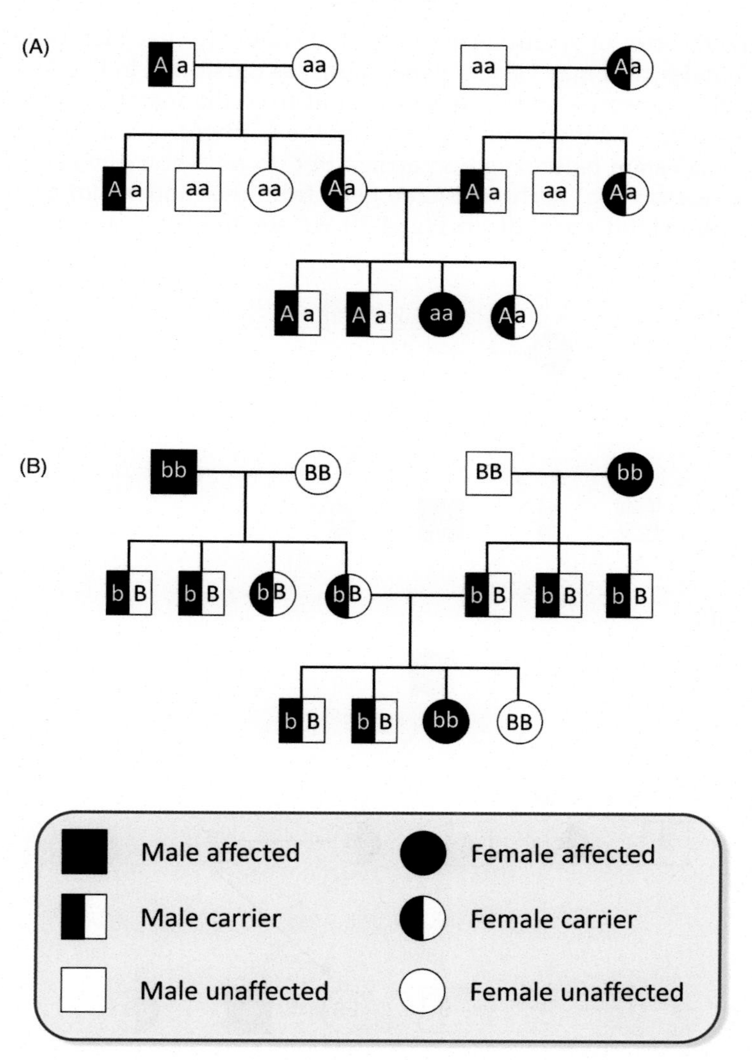

FIGURE 4.2 **Family trees with genotyping.** Assuming the traits in Fig. 4.1A and B are due to a single gene, we can use the family tree to deduce the genotypes of each of the family members, confirming that trait A is indeed a dominant trait (as a single copy is enough to lead to a phenotype, while trait B is a recessive trait only occurring if both b traits are present.

occurs when the gene responsible for a specific trait is located on the sex chromosomes. Although in theory, such genes could be both on the X or the Y chromosome, in most cases sex chromosome related traits originate from the X chromosome. This is not surprising if we realize that whereas the human X-chromosome has about 2000 genes, the human Y chromosome only has about 450. This clear difference between the X and Y chromosomes also explains why traits linked to these chromosomes have a more complicated pattern of inheritance. The two main characteristics of X-chromosome related traits are that they occur more often in males than in females, and may "skip a generation," as illustrated in Fig. 4.3. Fig. 4.3A shows the example of the transmission of a recessive trait from the maternal side. As each female receives a recessive (a) X-chromosome gene from the mother but a dominant (A) gene from the father, all females are heterozygous carriers but do not show the trait. However, as all males receive a recessive (a) X-chromosome gene from the mother but a Y chromosome of the father, all will exhibit the trait. In Fig. 4.3B we see the "skip a generation" effect that

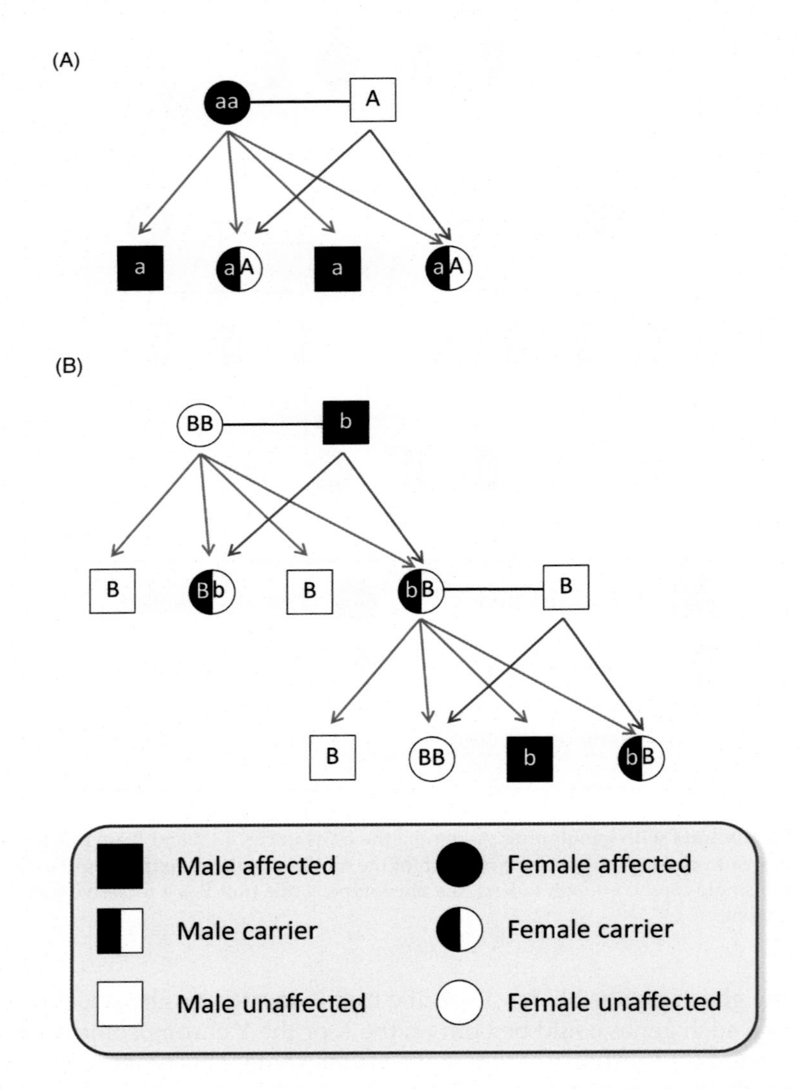

FIGURE 4.3 **X-linked mode of inheritance.** Because in contrast to the autosomes the sex chromosomes (X and Y) are differently expressed in males and females, the inheritance of X- and Y-linked traits is also different, as illustrated in A and B. X-linked inheritance has two main characteristics, namely it occurs more often in males than females (A) and it can "skip a generation" (B).

occurs when the X-linked trait is transmitted from the father. If the mother is homozygous for the dominant trait, all the sons and daughters will receive one dominant (B) gene from the mother and either a Y chromosome of the father (sons) or a recessive (b) gene (daughters) and therefore none will display the trait (although all daughters will be carriers). However, in the next generation, sons (from female carriers) will have a 50% chance of inheriting the trait, while the daughters will be 50% healthy and 50% heterozygous carriers (and therefore none will display the trait).

Even though X-linked genetic traits are slightly more complicated than autosomal linked traits, inheritance analyses for monogenetic disorders from family trees is fairly simple and can generally be done even without molecular genetic analyses. Unfortunately, most traits and disorders are not determined by a single gene, but by multiple genes, which makes the relation between genotype and phenotype much more complicated. As illustrated in Fig. 4.4, monogenetic traits have only three different genotypes and phenotypes (though in practice, as one of the alleles is usually dominant only two phenotypes may be apparent). With 2 genes involved, the number of genotypes becomes 9 and the number of phenotypes 5 (based on the principle that each gene has only additive effects). With three different genes involves we obtain 64 genotypes and 7 phenotypes. It is clear from Fig. 4.4 that with an increasing number of genes involved, the trait becomes more and more normally distributed. This explains why many traits have a strong genetic component (such as cognitive capabilities or height) and yet are normally distributed in the population.

Given that the mode of inheritance becomes virtually impossible to deduce from family studies if more than one gene is involved, researchers have looked for other methods, such as twin and adoption studies. The rationale behind twin studies (Fig. 4.5) is based on the fact that monozygotic (identical) twins share more or less 100% of their genes, while dizygotic (fraternal) twins share, on average 50% of their genes (just as normal siblings). Thus, if a trait (or disorder) has a genetic basis, the risk that both twins have the same trait should be higher in monozygotic twins than in dizygotic twins. This shared risk is referred to as concordance rate. Fig. 4.6 shows mono- and dizygotic concordance rates for a number of different psychiatric disorders. The data clearly show that for all these disorders the concordance rate is significantly higher in monozygotic twins compared to dizygotic twins. However, perhaps equally important, none of the concordance rates in monozygotic twins equals 100%, indicating that nongenetic factors also play an important role. The concordance rates allow us to calculate the heritability. Heritability refers to the proportion of the total phenotypic variance that is attributable to genetic variation. However, the exact heritability is difficult to calculate (see Box 4.1), especially since the contribution of nongenetic (environmental) factors is difficult to quantify and isolate from the genetic factors. For instance, in calculating twin heritability the assumption is usually that both monozygotic and dizygotic twins share the same environment. However, this is not necessarily the case. In fact, the uterine environment of monozygotic twins is usually more similar than for dizygotic twins, as illustrated in Table 4.1. In addition, the postnatal environment of monozygotic twins is often also more similar than of dizygotic twins. Indeed in a recent study, equal-environment assumption in twin studies in schizophrenia was assessed and the results clearly indicated that social adversity was much more strongly correlated in monozygotic than dizygotic twins, in line with the idea that shared environment is actually stronger in monozygotic twins (Fosse et al., 2015). Hence, the exact contribution of genetic and environmental factors in twin studies is difficult to ascertain.

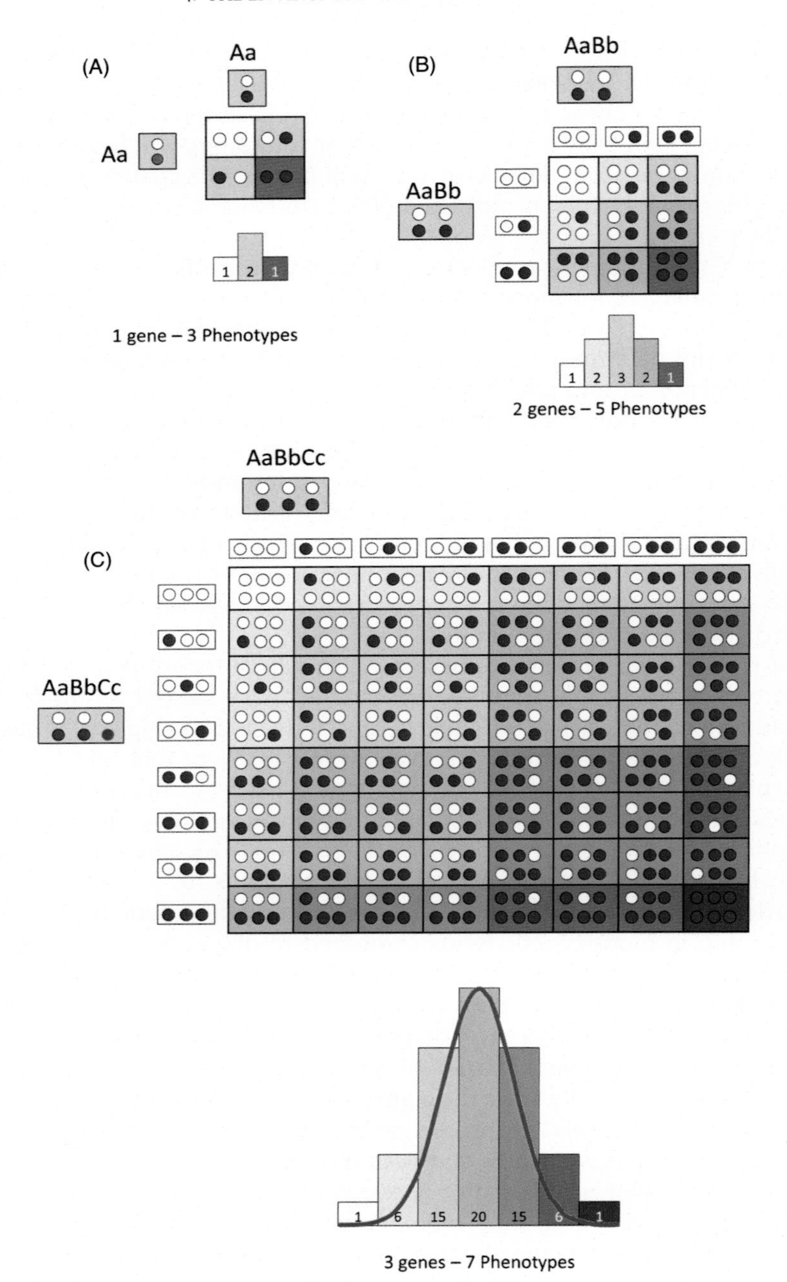

FIGURE 4.4 **The relationship between genotypes and phenotypes.** (A) In the simplest situation where a trait is determined by only one gene, three different phenotypes are possible. (B) In the case of two genes the number of possible phenotypes is 5 and (C) In the case of three genes this number increases to 7. More importantly, as indicated by the histograms, the phenotypes become more and more normally distributed.

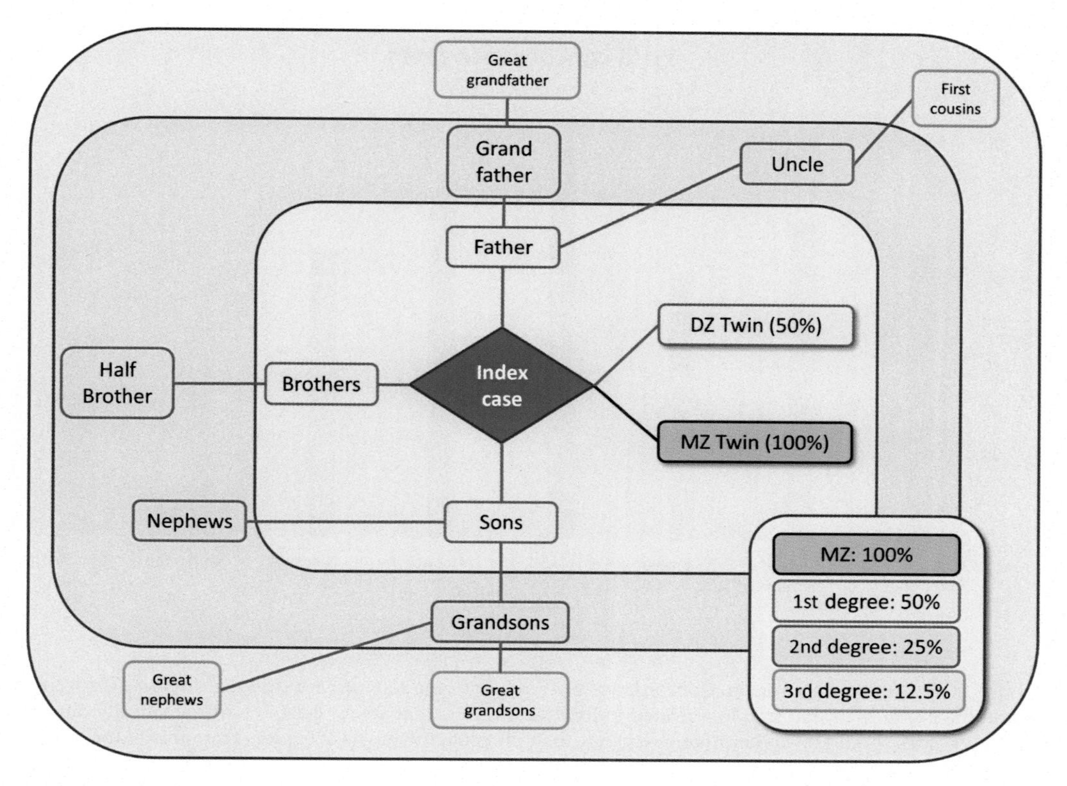

FIGURE 4.5 **Family relationships.** Given that most of the traits in humans, and certainly those related to brain functioning are determined by multiple genes, family studies are essential to determine the heritability of a trait, especially if we can obtain information of a wide number of family members of the first, second and third degree.

Adoption studies are a kind of "natural" experiment to separate genetic and environmental factors, although it is certainly not a randomized controlled experiment. Indeed, one of the main criticisms of adoption studies, is that they are likely to involve only extreme cases of psychopathology in the parents, that is so severe that the offspring is at risk. Moreover, given that most adoption agencies have very strict screening procedures for the host families, they generally do not constitute an average environmental influence either. They are, in general, older, better educated and wealthier than the average (and certainly than the biological) parents. Moreover, the decision to adopt is a very intentional one, in general made after extensive deliberations. In addition, adoption studies have often relied on official records of the parents' medical history, rather than a personal diagnosis of the parents by the research team itself. Finally, the prenatal and the postnatal environment until adoption will still be shared by the child and the biological parents (Lynskey et al., 2010).

In spite of these limitations, many long-term adoption studies have been reported and many prospective studies are still running. Many of these come from Scandinavian countries, as these excel in their health record keeping. For instance the Danish Adoption Register covers 14,425 adoptions between 1924 and 1947 (Petersen and Sorensen, 2011a). This register has the advantage

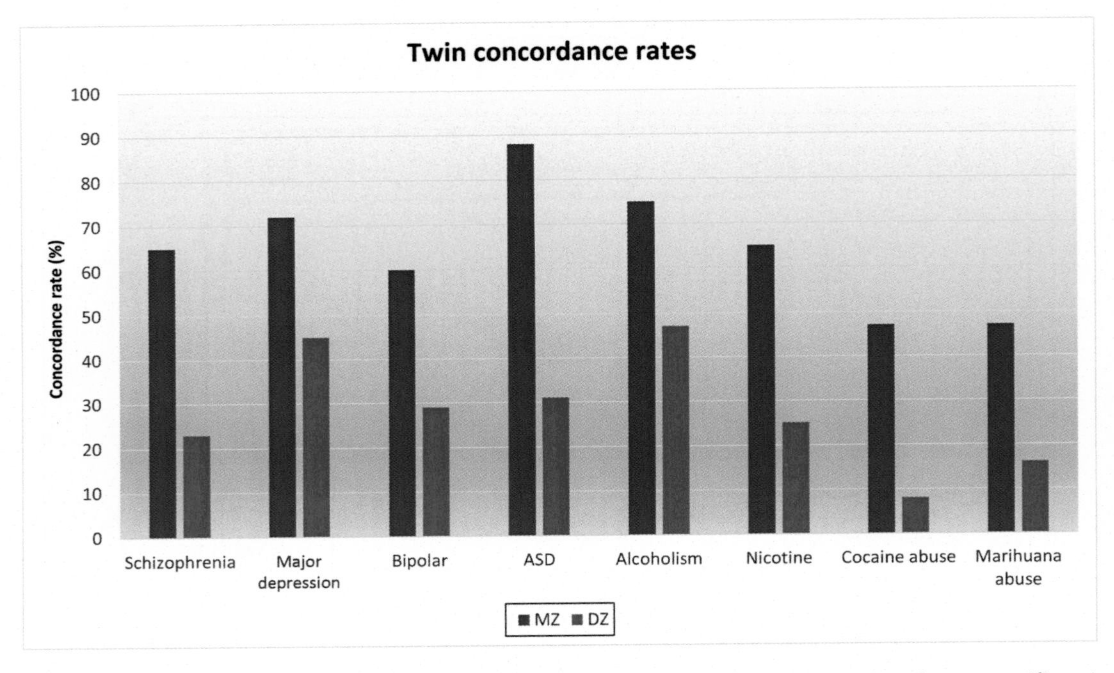

FIGURE 4.6 **Concordance rates.** Concordance rates refers to the risk of an individual to show a specific trait based on the presence of that trait in a related individual. In traits that are (at least in part) genetically determined concordance rates should be higher in monozygotic than dizygotic twins. As is evident from this figure, this is true for many psychiatric disorders.

that around 90% of the children were transferred to their foster parents within 2 years after birth, thus reducing the shared environmental factor. Although initially this register was used in the area of schizophrenia, it has now become a major source for many different phenotypes, including criminality, IQ, psychopathy and mortality (Petersen and Sorensen, 2011b). A similarly large register exists in Sweden with over 18,000 adoptees (as well as over 79,000 biological and 47,000 adoptive relatives). In a recent study this register was used to investigate criminal behavior. The authors found that criminal behavior in either one of the biological or one of the adoptive parents increased the risk of violent behavior in the adoptee, thus suggesting both genetic and environmental factors (Kendler et al., 2014). Interestingly, a closer look at the risk factors showed that whereas some were significant for both adoptive and biological parents (criminal behavior, alcohol abuse and maternal divorce), others were specific only for adoptive parents (drug abuse) or for the biological parents (low education). Likewise, whereas alcohol abuse in biological siblings increased the odds risk of criminal behavior in the adopted child, alcohol abuse in the adoptive siblings did not. Thus, although both genetic and environmental factors seem to play a role in criminal behavior, there are differences in the extent specific environmental factors affect criminality. The authors also distinguished between violent and nonviolent criminal behavior but very few differences were found between these two categories. Another recent adoption study also suggested that genetic and environmental factors are also involved in alcohol abuse (McGue et al., 2014). In this study the authors investigated the similarity in drinking patterns

BOX 4.1

GENETIC HERITABILITY

Twin and adoption studies allow for a separation of genetic and environmental factors. However, as discussed elsewhere in the text, this separation is not complete. Twins do share a womb, which is more identical in monozygotic than dizygotic twins, and even postnatally, monozygotic twins are treated more similarly than dizygotic twins (especially if the latter is a mixed gender twin). Nonetheless, twin studies have been used to estimate the heritability of specific traits.

Heritability is defined as the percentage of the trait variance that can be attributed to genetic factors. A distinction is often made between additive genetic factors (referred to as narrow-sense heritability estimates h^2) or total genetic factors (broad sense heritability estimates H^2).

If we assume that a trait (T) results from a genotype (G) and an environment (E) and the variance is given by σ^2 then in its simplest form:

$$\sigma_T^2 = \sigma_G^2 + \sigma_E^2 \qquad [1]$$

We can then calculate H^2 as $H^2 = \frac{\sigma_G^2}{\sigma_T^2}$

However, both the genetic and the environmental variance can be subdivided into different components. Thus the genetic variance is the sum of the additive (A), dominance (D) and interactive (I) effects, while the environmental variance can be subdivided in individual (I) and shared (S) environment (and generally a residual term is added as well for any environmental influences not accounted for).

As narrow sense heritability only relates to additive effects we calculate this as

$$h^2 = \frac{\sigma_A^2}{\sigma_T^2}$$

One major shortcoming of equation [1] is that it assumes a lack of gene * environment interaction, which, of course is generally not in line with our findings. Therefore a more correct description would be:

$$T = G + E + G^*E$$

And the variance will then be:

$$\sigma_T^2 = \sigma_G^2 + \sigma_E^2 + 2\sigma_{G,E}^2 + \sigma_{G*E}^2 \qquad [2]$$

In which G, E is the gene environment covariance and G^*E the gene * environment interaction. In most heritability calculations these terms are ignored because they are difficult to calculate, leading to an inflation of either of the first two terms.

TABLE 4.1 Characteristics of the Intrauterine Environment in Monozygotic (MZ) and Dizygotic (DZ) Twins

Placenta	Chorion	Amnion	%DZ	%MZ	Timing of the MZ embryonic split
Separate	Separate	Separate	58	18	Before day 4 after conception
Fused	Separate	Separate	42	19	Between day 4 and 6 after conception
Fused	Fused	Separate	0	64	Between day 6 and day 10 after Conception
Fused	Fused	Fused	0	4	between day 10 and day 14 after conception

between parent and siblings. They found that there was a clear correlation between the drinking patterns of children and their adoptive parents indicating an environmental factor. However, the correlation in drinking pattern was much stronger between nonadopted children and their (biological) parents, indicative of an additional genetic component.

ENVIRONMENTAL FACTORS

The data from family studies clearly indicate that psychiatric disorders have a complicated aetiology, with multiple genes interacting with each other. The relatively low concordance rates in monozygotic twins adds further to this complexity and shows that heritability is less than 1.0 and thus that nongenetic (environmental) factors are also involved. Although the heritability calculated from adoption studies often differs from that calculated from twin studies, the results nonetheless confirm that genetic and nongenetic factors both play a crucial role in the aetiology of psychiatric disorders.

Research into the role of nongenetic factors is considerably more complicated and less "specific" than genetic research. In contrast to genetic factors, environmental factors usually impinge upon the phenotype for only a short period, and often long before the disorder becomes evident or the phenotype is studied. As discussed further, the pre- and early postnatal period is, in this respect, probably the most important period, while many psychiatric disorders do not develop until during or after puberty [(Paus et al., 2008), see also Fig. 8.1]. This implies that in most cases we have to rely on retrospective data, often using questionnaires, which always have the risk of being biased by the knowledge of the present day. In some cases, such questionnaires can be supplemented by hospital, police or other official records, thereby improving the reliability of the information. In only a few cases do we actually have prospective information. An important limitation of such prospective studies is the aforementioned timing effect. If prenatal factors are involved in for instance schizophrenia or depression, such prospective studies will need to span at least 25 to 30 years in order to ensure that the majority of individuals are correctly diagnosed. Although there is evidence, as we shall see in the next chapter that genetic factors can also impinge on the phenotype in a time-critical manner, the genotype itself does not change after birth, and thus identifying the genetic factors in a "retrospective" study is not a significant concern.

When studying the role of environmental factors, we often distinguish between population studies and case studies, sometimes referred to as ecological and birth cohort studies (Brown and Derkits, 2010). In population studies, different populations are compared, exposed to a specific environmental condition, and one that is not. In case studies, individuals are compared that are either exposed or not. Although this latter approach would seem to be the more specific strategy, population studies have the advantage that the exposure to the environmental factor is better documented. For example, the first studies that identified prenatal exposure to influenza as a significant risk factor for schizophrenia were based on a population design. In 1957, a severe influenza pandemic affected millions of people all around the world in a short period of time. Known as the Asian flu, it was first identified in China, reaching Hong Kong in Apr., the United States in Jun. and Europe in Sep./Oct. A study in Finland was the first to provide evidence for a link between this pandemic and schizophrenia, by showing excess births of children that later developed schizophrenia among people in the Uusimaa county in Finland who were exposed during the second trimester of pregnancy (Mednick et al., 1988). This was followed by a similar study from the UK (O'Callaghan et al., 1991). However, these studies used the population design, that is, they compared the entire population of children born in the pandemic area in 1957 with those born in the years before and/or after the pandemic. Although in itself a useful approach, it tends to underestimate the relative risk of environmental factors as it ignores the fact that not all pregnant mothers will actually have been exposed to the influenza virus themselves. Thus only very strong environmental factors can be identified with this method and it may explain why subsequent studies failed to replicate the original finding (Selten et al., 2010). As discussed by Brown and his colleagues, using a case controlled design by analyzing hospital records of identified cases of influenza and other infections during pregnancy has provided more robust support for the hypothesis that prenatal infections enhances the risk for schizophrenia (Brown and Derkits, 2010).

Another important difference between the genetic and environmental factors relates to the "specificity" of the effects. Although the effect of a mutation on the function of the gene is not always exactly known, the normal function of the gene usually is, as well as the molecular pathway or network in which it is involved. It is therefore often possible to predict the functional consequences of a genetic alteration (at least at the molecular level). However, the functional consequences of an environmental effect are much harder to predict and it is often even difficult to exactly pinpoint what aspect of the environmental challenge drives the phenotypical changes. In the above mentioned example of the prenatal influenza infection, it might be the infectious agent itself, it may be the maternal immune response or the treatment of the infection. Likewise, many studies have investigated the long-term consequences of the Dutch hunger winter of 1944–45 (as discussed later). Although the most parsimonious explanation of these findings is that these long-term effects are due to malnutrition, they may also (in part) be due to increased prenatal stress, eating less healthy food or a combination of some of these factors. Thus, isolating the exact factor (or factors) related to the development of psychiatric disorders is difficult, and of course the results are always, at best, correlative and do not allow a firm conclusion on the causal relationship. Fortunately, studies investigating the effects of environmental challenges in animals have provided a wealth of information on the issue of causality. We will discuss some of these studies below.

TIMING

Although the exact nature of the environmental risk factors is still mostly elusive, as discussed above, it has become clear that the time when an individual is exposed to the risk factor is of critical importance for the long-term consequences. We usually distinguish between several different critical periods (see Fig. 4.7). However, even within each of these periods

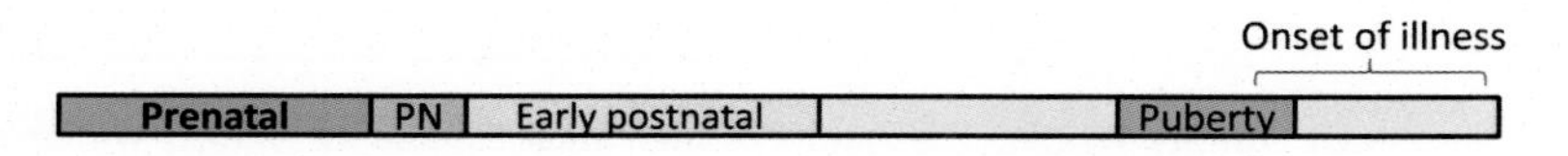

FIGURE 4.7 **The different environmental time windows.** Although environmental factor can affect an organism throughout development, several critical time windows have been identified during which environmental challenges are particularly significant. These are the prenatal period, the perinatal period (PN), the early postnatal period and the period around puberty.

there is increasing evidence that the precise timing of the environmental challenge can determine the nature of the long-term alteration. This evidence comes mainly from research in animals, although there is also epidemiological evidence from human studies available. One of the most compelling sets of data in this respect comes from studies on the Dutch Hunger Winter (also referred to as the Dutch Famine). This took place at the end of the second world-war from about Oct. 1944 until the liberation in May 1945, and resulted from a ban on (heavy) traffic through the country. This was instigated by the German occupying forces as a retaliation to the national railway strike in Sep. 1944 aimed to further the allied forces attempts at liberation the Netherlands. It is considered a unique event as it took place in a modern, well-developed country. Moreover, it was very restricted in time as well as in area. As the main agricultural eastern part of the Netherlands was very self-sufficient, only the western part (around Amsterdam, the Hague and Rotterdam) was affected (in total about 4.5 million people). In Nov. 1944, the average rations dropped to about 4000 kJ and were further reduced to about 2000 kJ in Apr. 1945. Butter had already disappeared by Apr. 1945 and the weekly ration consisted of 400 grams of bread and 1 kg of potatoes. Given these numbers it is not surprising that the death toll was high, with estimates ranging from 18,000 to 22,000.

More important for our present discussion are, however, the long-term consequences of the offspring of mothers pregnant during the hunger winter. As the famine was well localized in time and space, it is possible to determine whether exposure during different periods of development have different long-term effect (Kyle and Pichard, 2006). Inspection of Fig. 4.8 leads to a number of important conclusions. First of all, the Dutch hunger winter affected

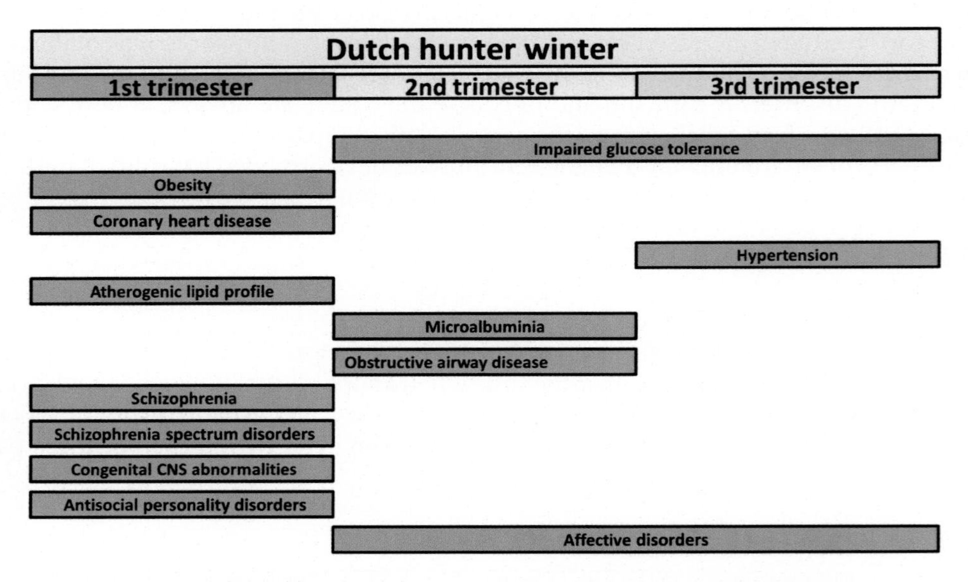

FIGURE 4.8 **The Dutch Hunger Winter.** The results of the Dutch Famine or Hunger winter are an excellent example of how timing can affect the long-term consequences of environmental challenges. The results show that although this environmental challenge had more effects early during pregnancy, it also affected the outcome later during pregnancy and even during early childhood (not shown). Moreover, not only psychiatric disturbances, but also more somatic symptoms can occur.

many different organs, including lungs, heart, blood vessels as well as the brain. Secondly, the timing of the hunger winter is a crucial factor in determining the long term consequences. For instance, whereas the offspring of mothers exposed in the first trimester of pregnancy had an increased risk of developing schizophrenia, those exposed in the second and/or third trimester had an increased risk of affective disorders such as depression or anxiety disorder. Thirdly, the first trimester of pregnancy seems to be the most sensitive period, as most long-term consequences are seen after exposure in this period. Finally, although not shown in Fig. 4.8, exposure during (early) childhood also led to long-term problems, such as breast cancer or endocrinological changes, including as early onset menopause or changes in reproductive functioning (Elias et al., 2003; Elias et al., 2004).

THE PRENATAL PERIOD

The results from the Dutch hunger winter emphasize the role the environment can play during the prenatal period. This is also supported by a large number of other studies. Perhaps one of the most intriguing phenomena in this respect is the season of birth effect. Again, the exact underlying mechanisms is not clear, but for certain diseases, there is an increased risk depending on the date of birth (or, probably more correct) the period of pregnancy. This has probably most often been studied in schizophrenia where studies as early as the 1930s indicated that children born in the first months of the year (in the northern hemisphere) had an increased risk. An influential study in this respect was published in Nature where spring birth increased the risk for schizophrenia by 7% and for bipolar disorder by 9% (Hare et al., 1973). Since then, this has repeatedly been found (although failures to replicate have also been published). Overall, several meta-analyses have confirmed this season of birth effect. For instance pooled data from 27 sites in the northern hemisphere (total of 126,196 patients and 86,605,807 healthy controls) also indicated a 7% increased risk for winter/spring births (Davies et al., 2003). Interestingly, this study found a positive correlation between the odds ratio for the season of birth effect and latitude, indicating that the effect was larger, the closer the site was to the North Pole. It is beyond the scope of this chapter to discuss in detail the various hypothesis developed to explain the season of birth effect in schizophrenia but they include seasonal variations in vitamin D (which is mainly produced by direct sunlight), ambient temperatures, infectious agents, geomagnetic field, and others (Kay, 2004; Tochigi et al., 2004).

A season of birth effect is by no means specific for schizophrenia and bipolar disorders, but has also been found for a number of other psychiatric disorders, including major depression, autism, and obsessive compulsive disorders. As Table 4.2 clearly indicates, a variety of non-CNS related disorder, such as various form of cancer and cardiovascular disorders also show a season of birth effect. It is important to note that not all findings in Table 4.2 are equally robust, and some of the findings have proven difficult to replicate, especially as the effects are sometimes rather small necessitating large numbers of participants. Another important issue to note here is that a season of birth effect does not necessarily implies a prenatal risk factor, and postnatal factors may also contribute to the increased risk.

However more specifically prenatal environmental risk factors have also been identified. We can subdivide these in a number of different categories, such as pharmacological/toxicological factors, infectious factors and more social/emotional risk factors. Much research has

TABLE 4.2 A Summary of Major Findings of a Season of Birth Effect on Disorders. References with a Star (*) Indicate Reviews and/or Meta-Analyses

Disorders	Seasons of birth effect	References
Adult onset glioma	Excess in winter	Efird (2010)*
Schizophrenia	Excess in winter/spring	Davies et al. (2003)*
Bipolar Disorder	Excess in winter/spring	Torrey et al. (1997)*
Schizoaffective disorders	Excess in winter/spring	Torrey et al. (1997)*
Nonmelanoma skin cancer	Excess in winter/spring	La Rosa et al. (2014)
Major Depression	Excess in Spring	Torrey et al. (1997)*
Autism	Excess in Mar.	Torrey et al. (1997)*
Cardiovascular mortality	Excess in Mar./Apr.	Ueda et al. (2014)
Multiple Sclerosis	Excess in Apr./May	Dobson et al. (2013)*
Non-Hodgkin Lymphoma	Excess in spring/summer	Crump et al. (2014)
Type I diabetes (+coeliac disease)	Excess in summer	Adlercreutz et al. (2015)
Obsessive Compulsive Disorder	Excess in summer/autumn	Cheng et al. (2014)
Melanoma skin cancer	Excess in Jan., Mar., and Jun.	La Rosa et al. (2014)
Infectious mortality	Excess in Sep.	Ueda et al. (2014)

focused on the long-term consequences of drug administration during pregnancy, especially after the thalidomide disaster (Kim and Scialli, 2011). Thalidomide, originally marketed as a sedative, was widely used to treat nausea in pregnant women in the late 1950s until it was discovered that it leads to severe limb reduction, but also to congenital heart disease, malformation of the inner ear, and ocular abnormalities. A close examination of the data shows that the effects of thalidomide critically depend on the time of administration (Kim and Scialli, 2011). As exceptional as the thalidomide disaster was, it is now well recognized that many other drugs may induce long-term abnormalities in the offspring when given during pregnancy and requirements for systematic testing of new pharmaceutical products for developmental toxicology were adopted as a direct consequence. Nonetheless, certain developmental abnormalities still occur, for instance as a direct consequence of addictive substance abuse during pregnancy, and in addition subtler effect on the development of the brain and body may occur that may go unnoticed during systematic testing in animals. Because these effects are more subtle, it is often difficult to unequivocally attribute them to the use of specific agents, especially as there are many environmental pollutants and potentially harmful substances that, by themselves, or in combination can affect development.

Prenatal alcohol exposure is probably one of the most prevalent risk factors for somatic behavioural and brain abnormalities. Depending on the severity (and probably the amount of alcohol consumed) different terms have been used to describe the condition with Fetal alcohol spectrum disorder (FASD) representing the entire continuum of the disorder and Fetal alcohol syndrome (FAS) being the most severe form (Dorrie et al., 2014). FASD and FAS are characterized by abnormalities in the facial region (thin lips, flattened philtrum), growth deficits, and structural and functional CNS changes, such as microcephaly, intellectual disability, language, psychiatric, and cognitive disorders. Other drugs of abuse have been less extensively studied, although there is strong evidence that cocaine and (meth)amphetamine also disrupt fetal development. The same holds true for smoking cigarettes, which is complicated by the fact that cigarette smoking leads to the inhalation of a large number of chemical such as carbon monoxide, cyanide, tar etc. in addition to nicotine itself, and has been associated with short term changes such as low birth weight, preterm birth and sudden infant death. Moreover, long term effects such as behavioural problems, hypertension, type II diabetes and obesity have been linked to maternal smoking (Bruin et al., 2010). In addition, exposure during pregnancy of drugs of abuse has been associated with increased risk of addiction and, in the case of maternal smoking, also of ADHD and Autism Spectrum Disorder (see Chapter 8).

While one can argue that prenatal exposure to drugs like cocaine, alcohol or nicotine are an avoidable risk (although addictive behavior is a very powerful drive as we will see in Chapter 6), many pregnant females are also treated with therapeutic drugs for existing conditions, some of which are increasingly recognized to be able to affect development of the fetus. Examples of such drugs are selective serotonin reuptake inhibitors (SSRIs) and valproic acid (VPA, also known as valproate). SSRIs are a class of antidepressant drugs, which compared to more traditional tricyclic antidepressants and monoamine inhibitors have significantly less side effects in patients (see Chapter 7). However, in recent years it is becoming increasingly clear that SSRI treatment during pregnancy can lead to preterm delivery, smaller birth weight, decrease in cortisol levels and disturbed psychomotor development (El Marroun et al., 2012; Pawluski, 2012; Gur et al., 2013), as well as increased autistic traits (El Marroun et al., 2014). Moreover, several markers for (neuro)development such as Reelin and S100B were significantly lower at birth (Pawluski, 2012; Brummelte et al., 2013). interestingly, in a large Scandinavian study, there was no evidence of an increase in stillbirth of infant mortality (Stephansson et al., 2013), suggesting that the effects of SSRI are a much more subtle than prenatal alcohol or thalidomide treatment. However, on the positive side, SSRI treatment during pregnancy reduced the depressive symptoms in mothers as well as in their offspring (El Marroun et al., 2014; Man et al., 2015), and the offspring actually had higher weight gain than offspring from depressed mothers that were not treated with SSRIs (El Marroun et al., 2012), indicating that whether to use SSRIs during pregnancy or not is a complex question to answer.

Like SSRIs, VPA is another commonly used drug that has been reported to increase the risk of autism in the offspring when used during pregnancy (Bromley et al., 2008; Christensen et al., 2013). VPA was originally used for the treatment of epilepsy, but has also been shown to be effective as a mood stabilizer for the treatment of bipolar disorder and for the treatment of migraine. In recent years it has even been proposed as a treatment for Alzheimer's disease. This surprisingly wide therapeutic effect is paralleled by an equally broad biochemical effect, many of which we do not yet know in all its details.

In addition to pharmacological and toxicological agents, many studies have provided strong evidence that prenatal infections influence the development of the unborn fetus and can have long lasting effects on its behavior and emotional development. Above, we already discussed the influenza pandemic of 1957 and its relation to schizophrenia. A similar pandemic occurred in 1917 and studies again suggested that it led to an increase in schizophrenia. In the wake of the 1957 pandemic–schizophrenia studies, an increased risk for major depression (Cannon et al., 1996; Machon et al., 1997; Selten and Morgan, 2010) and bipolar disorder was also reported (Parboosing et al., 2013). Another disorder which has often been associated with maternal infection is autism or autism spectrum disorder (ASD). This was confirmed in a recent very extensive study from Denmark (Atladottir et al., 2010). Using the Danish National Hospital Registry the authors investigated over 10,000 children with ASD (out of a total of over 1.6 million). Combining the information with hospital records of the parents showed a significantly increased risk of the offspring developing ASD when the mother was exposed to either a viral infection in trimester one or a bacterial infection in trimester two. These data confirm (in line with the data on the Dutch hunger winter) the impotance of the timing of the adverse event. Although it also suggests that there may be differences between different infectious agents, studies in humans are not entirely conclusive, and especially in schizophrenia, several different maternal infections have been linked to an increased risk (Brown and Derkits, 2010). Studies in animals also suggest that although differences may exist between different infectious agents they tend to be small and subtle (as discussed further).

The third and most diffuse category of prenatal adverse events can broadly be termed social or emotional life events. Although technically positive social and emotional life events during pregnancy can have a very beneficial influence on the unborn child as well, most studies have concentrated on adverse (usually stressful) life events, such as the death of a spouse, unwantedness of a pregnancy, maternal anxiety or other, more "external" stressful events, such as natural or man-made disasters. Examples of this latter category include hurricanes, earthquakes, war and such unique events as the Chernobyl nuclear plant accident in 1986 and the terrorist attack on the twin towers in New York in Sep. 2001. Overall there is overwhelming evidence that stressful life events cause long lasting alterations in brain and behavior and enhances the risk of later life illnesses. As an example, a recent study using the Swedish national registries, investigated offspring psychopathology following the death of a first degree relative or spouse, either in the 6 months leading up to pregnancy, during pregnancy or in the first two years after birth (Class et al., 2014). The results are summarized in Fig. 4.9 and emphasize, on the one hand, that bereavement indeed can increase the risk of psychiatric disorders and suicide, but also, on the other hand, that timing is again of crucial importance. Thus, while bereavement stress only during term 3 of pregnancy increases the risk of ADHD, the risk for ASD is increased both after stress during this period and in the second year after birth.

Several studies have also investigated the long term consequences of the stress of a natural disaster. Such investigations have the advantage that they are usually very short and well delineated in space and that there is ample objective evidence of its severity (although objective severity and subjective distress experienced by an individual do not necessary correspond). In a very interesting study, Kinney and colleagues investigated the consequences of 10 severe weather events in Louisiana (USA). The authors found an increased risk of autism as a function of prenatal exposures to storms rated severe by the National Weather Service (Kinney et al., 2008). Interestingly, this increased risk was not only dependent on the timing, with gestational weeks 17–24 and 33–40 being the most sensitive, but also on the severity of the storm and there was even an interaction between these two variable (Fig. 4.10). Thus, in the low risk prenatal period the intensity of the storm did not significantly alter the prevalence of autism. However, in the high risk period the intensity of the hurricane was positively correlated to the prevalence. Similar severe natural conditions that have been studied in detail are the Iowa Flooding (Yong Ping et al., 2015) and Project Ice Storm (Laplante et al., 2008). This latter one refers to the worst natural disaster in Canada, when in Jan. 1998 several severe freezing rain

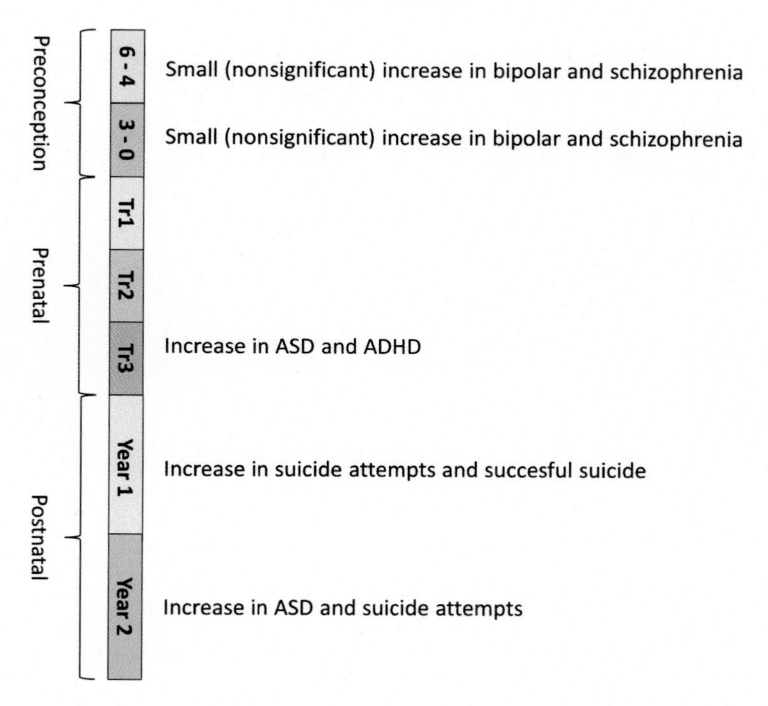

FIGURE 4.9 **The long-term consequences of bereavement.** The results of an extensive study of the Swedish national registries on the long-term effects of bereavement show that, depending on the timing of the bereavement, the incidence of different psychiatric disorders can increase.

storms hit Quebec, depriving 3 million people of electricity for up to 45 days (and killing 30). Recent studies have shown an increased risk of autistic like behavior and childhood asthma (in girl) after prenatal exposure to the ice storm (Turcotte-Tremblay et al., 2014; Walder et al., 2014).

THE PERINATAL PERIOD

The perinatal period is usually restricted to the period of delivery, although the boundaries between prenatal, perinatal and early postnatal are rather fuzzy. The most studied risk factor in this period is obstetric complications, which can take a number of different forms, and include breech position (ie, the baby's head is upwards, rather than downwards in the birth canal), (emergency) caesarean section, preeclampsia (a condition characterized by high blood pressure and large amounts of protein in the urine), bleeding, uterine atony, asphyxia etc. A common factor of most of these conditions is that the fetus during the delivery suffers from a lack of oxygen. The brain consumes about 20% of the body's oxygen (in spite that it only weighs about 2% of the total body weight), therefore a reduction is likely to have a major impact of the function of the brain, especially during development.

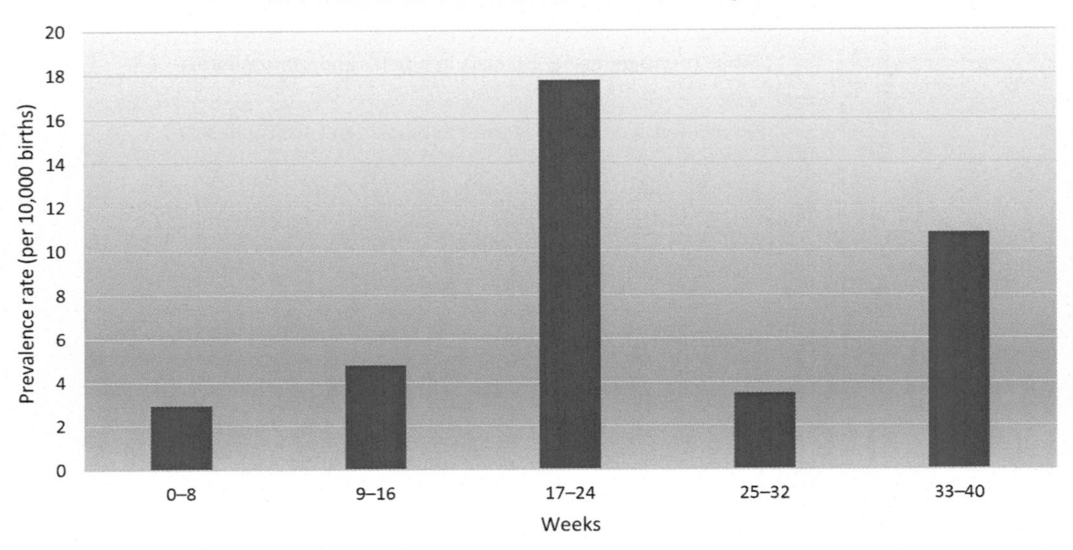

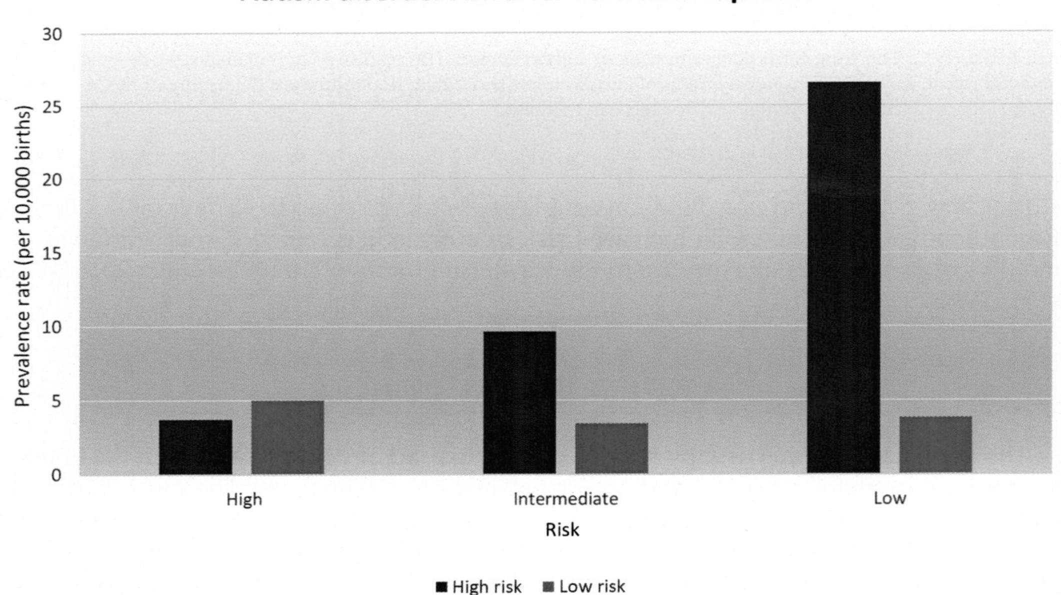

FIGURE 4.10 **The influence of hurricanes on the risk of autism spectrum disorder.** Kinney and colleagues investigated the relationship between severe weather events during pregnancy of the risk for developing autism spectrum disorder in the offspring. The results clearly show that the effects of storm intensity critically depends on the vulnerability window. During the period where the unborn fetus is most vulnerable, severe hurricanes had the strongest effect, while during less vulnerable period, the storm intensity did not influence the risk for autism very much.

TABLE 4.3 The Apgar Scale

Characteristic	Score 0	Score 1	Score 2
Appearance	Blue or pale all over	Blue in extremities	Healthy pink colour
Pulse	Absent	<100 beats per minute	>100 beats per minute
Grimace	No response	Grimace after intense stimulation	Cry after stimulation
Activity	None	Some flexion	Flexed arms/legs that resist extension
Respiration	Absent	Weak/ irregular	Strong, regular

Obstetric complications have been reported to increase the risk of several major psychiatric disorders, including schizophrenia (Clarke et al., 2006), and autism spectrum disorder (Guinchat et al., 2012), while data with respect to bipolar disorder, major depression or ADHD are less convincing (Scott et al., 2006; Adamou et al., 2012).

One of the first assessments of a newborn baby is usually the Apgar score. Originally invented by Virginia Apgar in 1952, the word Apgar can be regarded as a backronym (an acronym created to fit a specific name) as it is also the abbreviation of Appearance, Pulse, Grimace, Activity, and Respiration. Each of these 5 aspects of a newborn baby are scored on a three-point scale ranging from 0 to 2 (Table 4.3) and thus the Apgar score can vary between 0 and 10. The Apgar test is usually done at 1 and 5 minutes after birth, and scores of 7 or higher are considered normal. Scores between 5 and 6 are regarded as relatively low, while scores below 5 are considered critical. Although obstetric complications influence the Apgar scores, it is important to realize that the Apgar score can also be influenced by what happened prior to delivery, during the pregnancy itself. Nonetheless, low APGAR scores have been linked to autism spectrum disorders (Larsson et al., 2005) and schizophrenia. For instance in a recent study the obstetric records of 50 patients with first episode schizophrenia and 50 healthy volunteers were compared. Both the Apgar scores at 1 minutes and at 5 minutes were significantly lower in the group of the first episode patients (Kotlicka-Antczak et al., 2014). In line with this, the records also identified significantly more obstetric complications during the birth of these patients.

THE POSTNATAL PERIOD

The postnatal period roughly equates to the first decade of life and, as with the prenatal period, several different factors have been identified as risk factors for later-life psychiatric disorders. The two most studied factors are early postnatal infections and stressful life events. Although some studies have also investigated the effects of early postnatal drug treatment, this type of research is much less prevalent compared to drug treatment during pregnancy, presumably in part due to the fact that drug treatment (especially those that can affect the brain) is much less common in children than in adults (including pregnant females).

Postnatal, like prenatal, infections have been implicated in several major psychiatric disorders. For instance a recent study investigated pre- and postnatal antecedents of attentional problems in children born very prematurely (between gestational week 23 and 27) and found that tracheal bacterial infection in the first 2–4 weeks postnatally increased the risk for attentional problems at 2 years of age (Downey et al., 2015). Obviously, such very prematurely born children do not represent a normally developing populations, therefore these data should be repeated in normal term births. However, a population-based study in Western Australia also found that discharges because of infections in the first 4 years of life occurred much more often in children subsequently diagnosed with ADHD (Silva et al., 2014). Postnatal infections have also been linked to schizophrenia (Khandaker et al., 2012) and autism (Abdallah et al., 2012).

Stressful life events can take many forms, and similar to what we discussed with prenatal stress, can be due to external factors (such as natural disasters) or more social (predominantly familial) factors. Especially this latter category has been studied in great detail, and we generally distinguish between stressful life events and trauma (mostly restricted to abuse and neglect). Several different rating scales have been developed that focus on different aspect of social stress. Most of these are self-rating scales for adolescents and adults, and thus are retrospective and rely heavily on the objectivity and memory of the participant. This is recognized as an important problem, especially as rating scales are designed by adults and their ideas of what is stressful for children. In order to investigate how children perceive stressful life events and how they correlate with parental information Bailey & Garraldo investigated children between 7 and 12 years, using both an itemized questionnaire as well as a free recall session (Bailey and Garralda, 1990). The study showed that children reported significantly more events and, more importantly, only two of six items scored acceptable levels of agreement between children and parents (birth of a sibling and moving school). Although the numbers of children and parents was relatively small, the lack of agreement on items such as death of a relative, sibling or parent in a hospital or death of a pet (all less than or equal to 20% agreement) is striking and illustrates the need for well-validated questionnaires.

The Childhood Life Events Questionnaire (CLEQ) consists of 13 items including separation/remarriage of parents, death of a parent, friend or sibling, admission to hospital or jail of a parent, suspended from school etc. and is measured as a retrospective self-report questionnaire on a yes/no scale (Upthegrove et al., 2015). Another often used scale for measuring stressful life events is the Childhood Trauma Questionnaire (CTQ). This scale, in contrast to the CLQE, focusses on neglect (emotional or physical) and abuse (emotional, physical or sexual) using a five point Likert scale ranging from 1 *never* to 5 *frequently* (Bernstein et al., 1994). In addition to these and several other validated rating scales (Bailey and Garralda, 1990) other stressful events are occasionally studied, such as problematic teacher-pupil relation (Lang et al., 2013) which was found to be related to higher levels of any psychiatric disorders at 3 year follow-up (especially conduct disorder).

Stressful and/or traumatic life events in childhood have been related to virtually all major psychiatric disorders, including schizophrenia, affective disorders, ADHD and substance abuse disorders, and has also been linked to personality disorders (especially borderline personalities), (attempted) suicide, externalizing disorders and aggressive behavior (Gregorowski and Seedat, 2013; Schmid et al., 2013). However, childhood trauma has also been shown to increase the risk for nonpsychiatric disorders. For instance in a study from Japan (Fuh et al., 2010) it was found that physical maltreatment significantly increased the risk of both depression as well as migraine both with and without aura (see Fig. 4.11).

Similar results have also been reported for other sexual abuse or violence (Cripe et al., 2011).

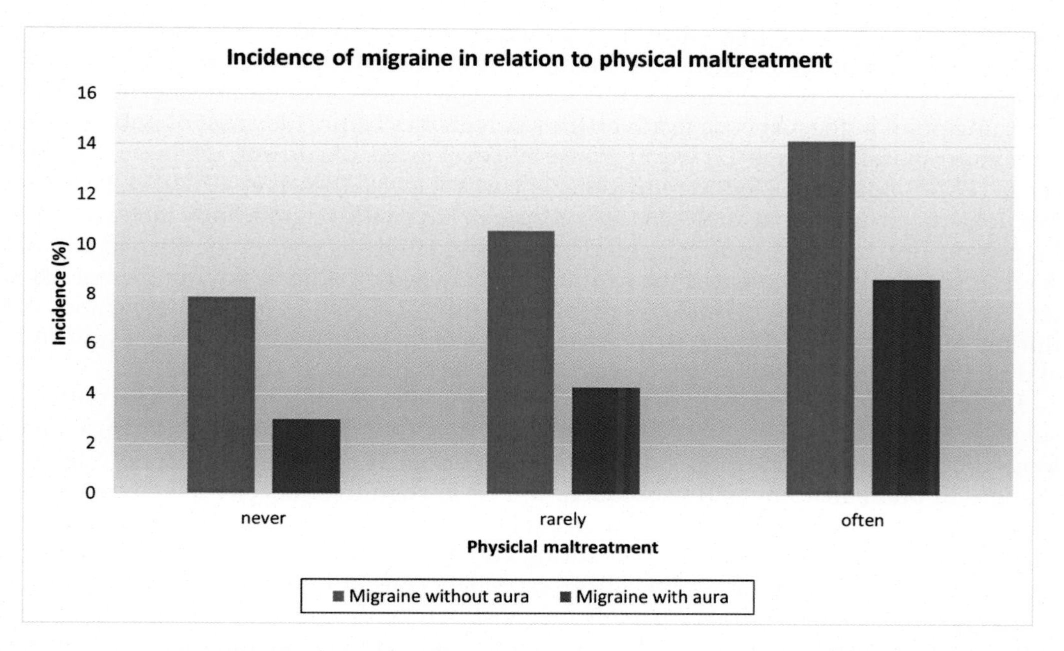

FIGURE 4.11 **Childhood maltreatment and migraine.** Physical maltreatment does not only lead to increased psychiatric symptoms, but also enhances the risk for migraine as is shown in the study of Fuh et al. (2010).

Realizing the role childhood trauma plays in psychopathology it has been proposed to identify "developmental trauma disorder (DTD)" as a separate diagnostic category (in analogy to post traumatic stress disorder). DTD is thought to comprise of three main components: (1) the child (or adolescent) experiences multiple forms of persistent dysregulation in response to traumatic reminders; (2) generalization to other nontraumatic stimuli; (3) behavioural changes centered around anticipation of and attempts to prevent recurrence of the traumatic experience (Gregorowski and Seedat, 2013).

THE ADOLESCENT PERIOD

The adolescent period is another critical time in neuro- and endocrine development. During this period the reproductive organs mature and, consequently, sexuality and sexual behavior develop. In addition, this period is characterized by significant changes in social behavior and attachment, with increasing parent-child conflicts and a shift from familial to peer relationships (Sturman and Moghaddam, 2011). As a result this is a very stressful period in the life of a child. In addition, several functional and structural changes occur in the brain (as discussed in Chapter 5), all emphasizing the sensitivity of this period in development. Adolescence is usually defined as the period between the onset to puberty and adulthood, and roughly equates to 10–17 years in girls and 12–18 in boys.

The adolescence period is also characterized by a much stronger and longer lasting response to stressors compared to adulthood (McCormick and Mathews, 2010), and it is therefore not surprising that stressful experiences during adolescence can have long-term consequences for an individual. A recent report from the TRAILS (TRacking Adolescents Individual Lives Survey) performed in the Netherlands showed that the adolescent period was a turning point in the response to adversities: Adversities before the onset of puberty (prenatal and postnatal until age 11) led to increased basal levels of cortisol, while during adolescence (12–15) adversities led to reductions in cortisol levels (Bosch et al., 2012). In line with this, it has been found that adolescence exposure to adversities significantly increases the risk for personality disorders, especially borderline personalities as well as obesity and psychosis. A recent study also investigated the relationship between modifiable environmental risk factors during adolescence and major depression (Cairns et al., 2014). In this dieting and dating during adolescence were found to increase the risk of major depression, while a healthy diet and relationship with positive peers was found to be significantly protective.

The adolescence period is also a time where many children start to experiment with illegal substances, as well as tobacco and alcohol. Consequently much research has focused on the long-term effects of such drugs use during this period. With respect to alcohol, significant changes in brain anatomy and subsequent to that in cognitive capacity have been reported in adolescents that meet the criteria for alcohol use disorder (Jacobus and Tapert, 2013). Both frequency and quantity of alcohol use during adolescence has also been linked to major depression in adulthood (Cairns et al., 2014), especially consumption during later adolescence (15–18 years). Cannabis or marihuana is the most frequently used illicit drug in adolescents and its use has also been associated with cognitive impairments in adulthood, as well as an increased risk of developing psychotic depressive and anxiety disorders (Cairns et al., 2014; Renard et al., 2014). However, it is also important to emphasize that studies have found that high levels of depressive symptoms enhance the risk of poly-substance abuse. Thus, whether the increased incidence of cognitive impairments or psychiatric symptoms seen in cannabis users are actually causally related to cannabis use, or are an expression of a pre-existing conditions (or indeed a combination of both) remains to be elucidated.

EVALUATING THE ROLE OF ENVIRONMENTAL FACTORS IN ANIMALS

In the previous sections we discussed how environmental challenges can adversely affect the development of behavior and can enhance the susceptibility for psychiatric (and other) disorders in humans. We also discussed that there were several different "vulnerability windows" during which humans are particularly susceptible. Finally, we showed that there is evidence that within each of these vulnerability windows (for instance the prenatal period) the same environmental challenge can have different consequences, depending on the timing of the event. However, studies in humans, as we discussed before, are almost invariably correlative in nature. This most certainly goes for stressful life events, as it would be ethically unacceptable to deliberately expose children or mothers to such stressors. Moreover, even with such well-defined (and objectively quantifiable) life events such as project Ice Storm (as discussed earlier), individuals may be differentially exposed, will have lived in different houses and thus could be more or less protected against the consequences. This all leads to increased variability and makes drawing firm conclusions on the causality between extreme weather events and subsequent psychiatric disorders complicated. Likewise although the relationship between prenatal SSRI treatment and increased risk for autism spectrum disorder may be more causally related (as discussed earlier), it is important to realize that mothers take SSRIs for a reason (ie, they suffer from major depression) and thus it may be difficult to differentiate between the effects of drugs and the maternal condition as such.

Therefore, in order to be able to study the causal relationship between early environmental challenges and changes in brain and behavior, researchers have focused on animal models. To mimic the situation in humans, a large number of different models have been developed differing in the type of challenge as well as the timing of the challenge. With respect to the latter, animal research has the added benefit that we can very precisely time the challenge. Table 4.4 list a number of the most often used environmental manipulations and their (presumed) human equivalent.

With respect to prenatal manipulations, much animal research has focused on studying the long term consequences of maternal infections. Although some studies have directly inoculated pregnant females with influenza virus (or other viruses and bacteria), most studies have used pharmacological tools to mimic a viral or bacterial infection using Polyriboinosinic polyribocytidilic acid (polyI:C) or Lipopolysaccharide (LPS). PolyI:C is a synthetic double stranded RNA which mimics a viral infection, while LPS is a component of the outer cell wall of gram-negative bacteria, and hence mimics a bacterial infection. Although a direct inoculation with bacteria or viruses may represent a more truthful mimicking of the human

TABLE 4.4 Animal Models for Studying the Long-Term Effects of Environmental Challenges

Period	Environmental challenge	Animal manipulation
Prenatal	Influenza infection	Influenza inoculation
	Viral Infection	Poly I:C injections
	Bacterial infection	LPS injections
	Psychological stress	Restraint stress
		Crowded stress
		Variable stress
	Malnutrition	Protein deprivation
	Drug treatment	Drug treatment
Perinatal	Obstetric complications	Perinatal hypoxia
	Caesarean section	Caesarean section
Early Postnatal	Psychological stress	Early handling
		Maternal separation
		Maternal deprivation
Adolescence	Psychological stress	Isolation rearing
		Social stress
		Chronic variable stress
	Drug treatment	Drug treatment

conditions, there are both practical and theoretical reasons why polyI:C or LPS are the preferred methods used to induce a maternal immune response. Practically, as neither polyI:C nor LPS are infectious, the experiments can be performed in a normal laboratory without fear of spreading the infection to other animals or even humans. From a more theoretical point of view, direct PolyI:C or LPS injections have the added benefit that they allow for very precise timing of the immune response. LPS induces an immune response for about 24 h (depending on the dose used), and therefore these components allow to study the effects of timing in a very precise manner. For example, it has been shown that PolyI:C injections during mid-gestation (gestational day 9) in mice decreases exploratory behavior, while administration during late-gestation (day 17) leads to perseverative behavior (Meyer et al., 2006). Likewise, while LPS administration on gestational days 10 and 11 in the rat significantly enhanced the hyperactivity induced by 3,4-methylenedioxy-methamphetamine (MDMA, the active ingredient of ecstasy) in adulthood compared to saline injections on the same gestational days, it produces the opposite effect when injected on gestational days 18/19. As evident from Fig. 4.12 this effect was only apparent in females.

Several different procedures in rats and mice have been developed to mimic the more psychological stress, differing both in the type of stress as well as in the timing. Some studies use classical footshock test (sometimes daily throughout the entire period of pregnancy), while others use subtler forms of stress, such as overcrowding (ie, putting several pregnant females

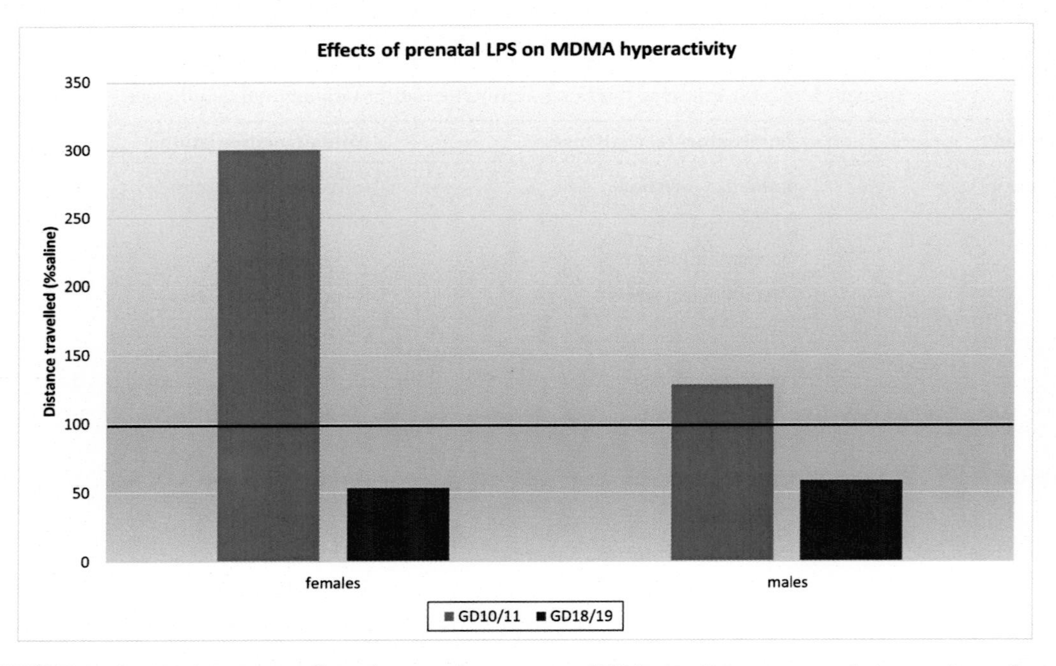

FIGURE 4.12 **The long-term effect of prenatal exposure to LPS in rats.** In a recent study in rats, it was found that the long-term effects of a prenatal exposure to LPS on MDMA induced hyperactivity also critically depends on the timing. Thus, whereas LPS injections on gestational day 10 and 11 leads to hyperactivity after MDMA administration in adult females, exposure to LPS on gestational days 18 and 19 leads to hypoactivity after MDMA administration in adult females.

together in a relatively small environment) or using more unpredictable random stressors. Such stressors are usually quite mild and can involve 24 h light exposure, 8 h exposure to a cold environment, overnight (12 h) food deprivation etc. This procedure is very similar to the chronic mild stress model used to induce a depression-like state in adult rats (Willner, 1997) and has the advantage that the animals cannot adjust to the stressors. This is very important as rats and mice are very resilient and rapidly habituate to stressful situations. However, the procedure has also been criticized as being too "random" and hence difficult to standardize between laboratories and therefore difficult to replicate. Prenatal malnutrition, thought to be at least in part responsible for the long-term effects found after the Dutch Hunger Winter (see Fig. 4.9), has been mimicked by treating mothers with a low protein diet. Normal rodent diets usually contain about 20% casein (phosphoproteins typically found in mammalian milk and the major source of amino acids). Malnutrition is usually induced by lowering the amount of casein to 8 or 6% (Alamy and Bengelloun, 2012). Finally, the long-term effects of drugs and other chemicals (such as pollutants) on (neuro) development have been investigated. Most notably in this respect are the studies done on fluoxetine (Olivier et al., 2011) and valproic acid (Schneider and Przewlocki, 2005), given their purported role in the development of human psychiatric disorders such as autism spectrum disorder (as discussed earlier). In addition, substantial work has been done on studying the effects of drugs of abuse such as cocaine and alcohol. Recent studies have shown that not only treatment of pregnant females affects the development of the brain, but even treatment before pregnancy. Thus in a recent study female rats were treated with cocaine up until 5 days before being mated with experienced males. In adulthood, the male offspring were significantly more sensitive to the psychostimulant effects of cocaine (Sasaki et al., 2014). Interestingly, several recent papers have now also shown that treatment of males with cocaine (Vassoler et al., 2013) or ethanol (Ceccanti et al., 2015) prior to mating with females can affect the offspring, suggesting that such drugs can affect the germline in males. In the next chapter we will discuss mechanisms through which this may be accomplished.

Studies mimicking perinatal environmental risk factors are not as numerous as studies investigating the prenatal period. Nonetheless, caesarean section (with or without an additional period of anoxia) has been shown to affect brain development and especially alter catecholamines, leading to, among others, an increase response to psychostimulants such as amphetamine (Boksa and El Khodor, 2003) and cocaine (Galeano et al., 2013).

Early postnatal manipulations in rats have focused particularly on stressors and can be subdivided into three main categories: early handling, maternal separation and maternal deprivation. In spite of small (but perhaps relevant) methodological variations, early handling generally refers to repeated (typically daily from postnatal day 4 to 14) short-term (typically 15 min) separations of the mother from her pups (Raineki et al., 2014). The procedure is referred to as early handling, as originally, all rats were stroked and handled during the separation. However, subsequent research has shown that this is not essential for inducing long-term changes. In fact many of the behavioural changes in the offspring are due to an increased maternal behavior upon return of the mother. In general, early handling leads to a reduction in anxiety and stress responsivity in adulthood, although the traditional view that early handling only leads to beneficial changes has recently been challenged (Raineki et al., 2014). For instance, early handling has been reported to impair aversive learning such as fear conditioning.

In the literature maternal separation and deprivation are often used interchangeably. However, maternal separation should be reserved for repeated (typically daily from postnatal day

4 to 14) long-term (between 3 and 6 h) separation of the mother from her pups, while maternal deprivation refers to a single (typically 24 h) separation of mother and pups. Maternal separation has traditionally been linked to depression and generalized anxiety disorders as the maternally separated animals showed in adulthood increased stress and anxiety responses, as well as certain depression-like characteristics (Vetulani, 2013). However, maternally separated rats also show an increased sensitivity to the rewarding effect of alcohol (Nylander and Roman, 2013) and altered performance in cognitive tasks (Kosten et al., 2012).

Maternal deprivation, on the other hand, has been linked more to schizophrenia than to depression (Ellenbroek and Riva, 2003) as maternally deprived rats, in adulthood, show behavioural, neurochemical and cognitive deficits reminiscent of patients with schizophrenia. Interestingly, as was observed in several human studies, the effects of maternal deprivation critically depend on the timing, with postnatal day 9 usually being the most sensitive. However, as a recent review showed, maternal deprivation also induces changes in metabolic and immunological parameters and thus may have broader implications (Marco et al., 2015).

Several different paradigms have been developed to study the long-term consequences of environmental challenges during the adolescence period in rodents. In order to mimic the psychological stressors, chronic variable stress paradigms, similar to those described above for the prenatal period, have been developed. In addition, in lieu with the changes in social behavior occurring during adolescence in humans, social defeat and social instability procedures have been described (Holder and Blaustein, 2014). The social defeat model is based on the resident-intruder test often used in aggression research and involves the experimental rat (the intruder) being defeated (usually several times) by a more aggressive rat (the resident). The social instability test is inspired by the idea that peer pressure is stressful to humans (and by analogy to rats). In this model rats are housed in pairs, isolated from 1 hour and upon return housed with another (new) rat. This procedure is repeated daily for 15 days during adolescence after which they are left undisturbed until they are adults. Behavioural, neurochemical and neuroanatomical studies have found substantial alterations in these rats in adulthood (McCormick et al., 2015).

Isolation rearing is an animal model assessing the effects of psychological stress during late childhood/early adolescence. Rats (and mice) are typically weaned at around postnatal day 21 when they are placed together in all male or all female groups. This is important for the establishment of social behavior (play behavior typically starts to develop around this period) and hence, rearing animals without peers leads to strong alterations in social behavior. Typically, isolation rearing starts at the time of weaning and lasts for 8 weeks, thus covering the entire period of adolescence and well into adulthood. Compared to the same length social isolation starting in adulthood, these young isolation reared animals show symptoms of schizophrenia, abnormal social behaviors and an enhanced sensitivity to drugs of abuse (Fone and Porkess, 2008).

Finally, in order to mimic the onset of drug use, and to investigate its functional consequences on brain (and behavioural) development, several different drug treatments have focused on the adolescent period. These include administration of cannabinoid agonists (Viveros et al., 2011), ethanol (Ornelas et al., 2015), methylphenidate and other drugs of abuse. Although differences exist between these different drugs, in general most studies agree that the effects of drugs during adolescence are stronger and longer-lasting than similar treatments in adulthood, although qualitative differences between treatments during adolescence and adulthood have also been described (Jain and Balhara, 2010).

DEVELOPMENTAL DIFFERENCES BETWEEN RODENTS AND HUMANS

A fundamental problem with using rodents to study the long term consequences of mimicking (early) environmental challenges is the difference in developmental speed between the species. Not only is the pregnancy of rats and mice only about three weeks as compared to about 40 weeks in humans, it is now also widely accepted that rats and mice are born in a much more premature state than humans (see Fig. 4.13). Thus, in contrast to humans, the rats and miceare born with their eyelid closed and the outer ears not yet developed. Moreover, they do not yet have fur. On the other hand, after birth rats and mice catch up very quickly and are considered adults around postnatal day 70 (as compared to 17–18 years in humans).

This differences in developmental speed therefore creates a challenge when we want to mimic a certain developmental stage in humans. This challenge is even more apparent when we realize that different processes develop with different speeds between species. For instance, whereas neural tube formation occurs around mid-gestation in rats (gestational day 9–10), it occurs much earlier in humans (week 3–4 of gestation). On the other hand, neurogenesis in rats is virtually complete at birth, while it can extend up to 2.5 years after birth in humans (Semple et al., 2013). Thus the general question "what stage of human development does postnatal day X in rats equates to?" cannot be answered properly, without specifying the developmental process one is interested in. Nonetheless, recent attempts have been made to develop a comprehensive model to describe the timing of neurodevelopmental processes for 18 different species, including rats, mice and humans (Workman et al., 2013), and has led to an online application www.translatingtime.org. Although the model is complex and would be beyond the scope of this book to describe in detail, it basically determines a so-called event score for 271 developmental events in 18 species. An event score can range from 0 to 1, with an earlier age of onset having a lower event score and a later age of onset having a higher score. Using this procedure, the results confirm the idea that rats and mice are born relatively prematurely. This is reflected in the so-called precocial score (an index of how active and mobile an animal is at birth, which is 0.654 for humans but 0.445 for rats and 0.408

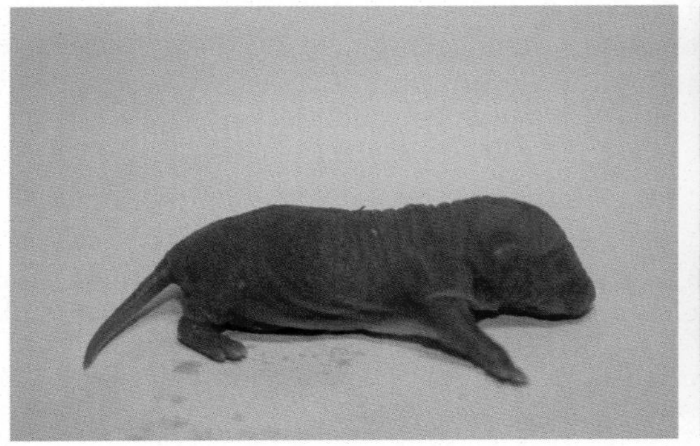

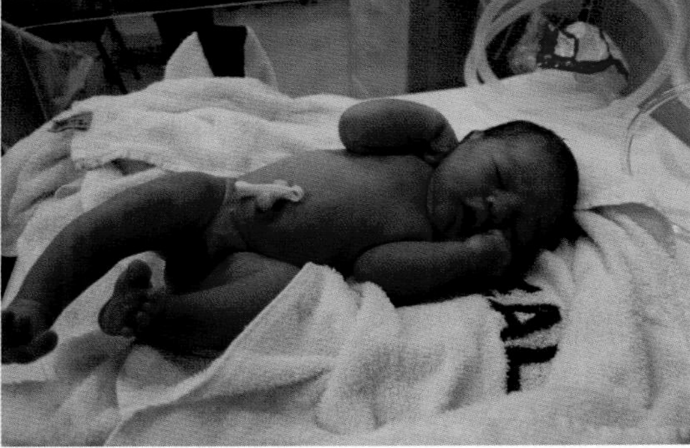

FIGURE 4.13 **The species-differences in developmental speed between rodents and humans.** Compared to humans, rat pups are born in a much more premature state, making it difficult to compare the developmental stages between the two organisms.

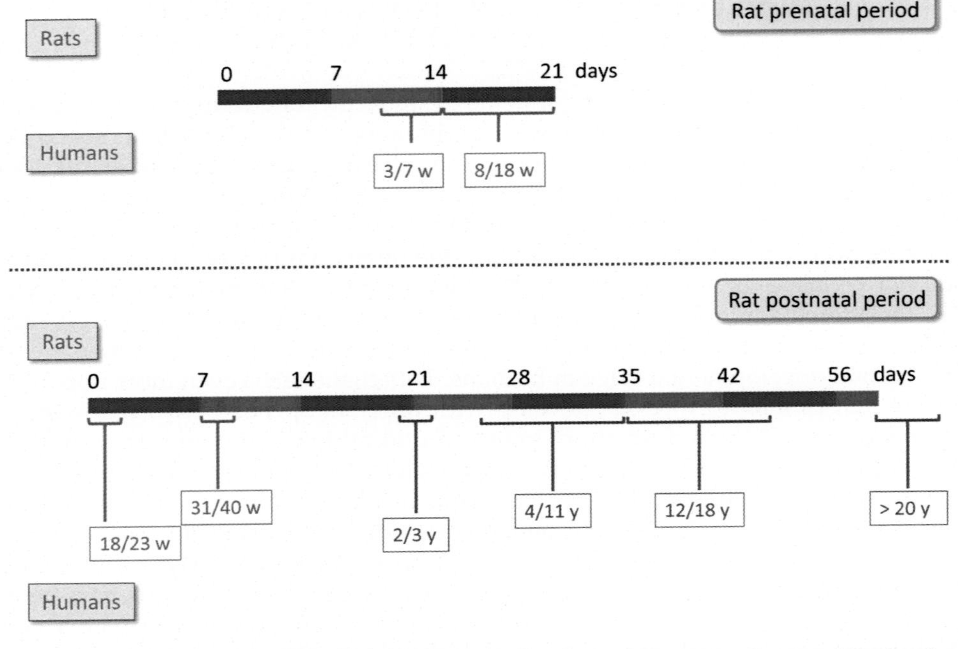

FIGURE 4.14 **Comparative developmental timelines between rats and humans.** In spite of the developmental differences, close inspection of many variables have led to a comparable time line for both the prenatal and postnatal period in rats and humans.

in mice (Workman et al., 2013)). Taken all these data together, a time line for the comparison between rodents and humans can be established. As can be seen in Fig. 4.14, prenatal week 2 in rodents roughly equates to weeks 7 of human pregnancy. Birth in rodents roughly equates to midterm (week 20) in humans and postnatal day 9 - 10 is roughly similar to the period of birth in humans (Semple et al., 2013). However, as mentioned before, these comparisons are only rough guides and differences exist with respect to individual processes. Fig. 4.14 also compares the later postnatal development of rats and humans, emphasizing that adolescence roughly occurs between day 35 and 45 in rats, and (young) adulthood is usually reached around postnatal day 60.

References

Abdallah, M.W., Hougaard, D.M., Norgaard-Pedersen, B., et al., 2012. Infections during pregnancy and after birth, and the risk of autism spectrum disorders: a register-based study utilizing a Danish historic birth cohort. Turk Psikiyatri Dergisi 23, 229–236.

Adamou, M., Russell, A., Sanghera, P., 2012. Obstetric complications in adults with ADHD: a retrospective cohort study. J. Developmental Physical Disabilities 24, 53–64.

Adlercreutz, E.H., Wingren, C.J., Vincente, R.P., et al., 2015. Perinatal risk factors increase the risk of being affected by both type 1 diabetes and coeliac disease. Acta Paediatr. 104, 178–184.

Alamy, M., Bengelloun, W.A., 2012. Malnutrition and brain development: an analysis of the effects of inadequate diet during different stages of life in rat. Neurosci. Biobehavioral Rev. 36, 1463–1480.

Atladottir, H.O., Thorsen, P., Ostergaard, L., et al., 2010. Maternal infection requiring hospitalization during pregnancy and autism spectrum disorders. J. Autism Developmental Disorders 40, 1423–1430.
Bailey, D., Garralda, M.E., 1990. Life events–childrens reports. Social Psychiatry Psychiatric Epidemiol. 25, 283–288.
Bernstein, D.P., Fink, L., Handelsman, L., et al., 1994. Initial reliability and validity of a new retrospective measure of child-abuse and neglect. Am. J. Psychiatry 151, 1132–1136.
Boksa, P., El Khodor, B.F., 2003. Birth insult interacts with stress at adulthood to alter dopaminergic function in animal models: possible implications for schizophrenia and other disorders. Neurosci. Biobehavioral Rev. 27, 91–101.
Bosch, N.M., Riese, H., Reijneveld, S.A., et al., 2012. Timing matters: long term effects of adversities from prenatal period up to adolescence on adolescents' cortisol stress response. The TRAILS study. Psychoneuroendocrinology 37, 1439–1447.
Bromley, R.L., Mawer, G., Clayton-Smith, J., et al., 2008. Autism spectrum disorders following in utero exposure to antiepileptic drugs. Neurology 71, 1923–1924.
Brown, A.S., Derkits, E.J., 2010. Prenatal infection and schizophrenia: a review of epidemiologic and translational studies. Am. J. Psychiatry 167, 261–280.
Bruin, J.E., Gerstein, H.C., Holloway, A.C., 2010. Long-term consequences of fetal and neonatal nicotine exposure: a critical review. Toxicological Sci. Official J. Soc. Toxicol. 116, 364–374.
Brummelte, S., Galea, L.A.M., Devlin, A.M., et al., 2013. Antidepressant use during pregnancy and serotonin transporter genotype (SLC6A4) affect newborn serum reelin levels. Developmental Psychobiol. 55, 518–529.
Cairns, K.E., Yap, M.B., Pilkington, P.D., et al., 2014. Risk and protective factors for depression that adolescents can modify: a systematic review and meta-analysis of longitudinal studies. J. Affect. Disord. 169, 61–75.
Cannon, M., Cotter, D., Coffey, V.P., et al., 1996. Prenatal exposure to the 1957 influenza epidemic and adult schizophrenia: a follow-up study. Br. J. Psychiatry 168, 368–371.
Ceccanti, M., Coccurello, R., Carito, V., et al., 2015. Paternal alcohol exposure in mice alters brain NGF and BDNF and increases ethanol-elicited preference in male offspring. Addict. Biol..
Cheng, C., Lin, C.H., Chou, P.H., et al., 2014. Season of birth in obsessive-compulsive disorder. Depression Anxiety 31, 972–978.
Christensen, J., Gronborg, T.K., Sorensen, M.J., et al., 2013. Prenatal valproate exposure and risk of autism spectrum disorders and childhood autism. JAMA 309, 1696–1703.
Clarke, M.C., Harley, M., Cannon, M, 2006. The role of obstetric events in schizophrenia. Schizophrenia Bull. 32, 3–8.
Class, Q.A., Abel, K.M., Khashan, A.S., et al., 2014. Offspring psychopathology following preconception, prenatal and postnatal maternal bereavement stress. Psychological Med. 44, 71–84.
Cripe, S.M., Sanchez, S.E., Gelaye, B., et al., 2011. Association between intimate partner violence, migraine and probable migraine. Headache 51, 208–219.
Crump, C., Sundquist, J., Sieh, W., et al., 2014. Season of birth and risk of Hodgkin and non-Hodgkin lymphoma. Int. J. Cancer 135, 2735–2739.
Davies, G., Welham, J., Chant, D., et al., 2003. A systematic review and meta-analysis of Northern Hemisphere season of birth studies in schizophrenia. Schizophrenia Bull. 29, 587–593.
Dobson, R., Giovannoni, G., Ramagopalan, S., 2013. The month of birth effect in multiple sclerosis: systematic review, meta-analysis and effect of latitude. J. Neurol. Neurosurgery Psychiatry 84, 427–432.
Dorrie, N., Focker, M., Freunscht, I., et al., 2014. Fetal alcohol spectrum disorders. Eur. Child Adolescent Psychiatry 23, 863–875.
Downey, L.C., O'Shea, T.M., Allred, E.N., et al., 2015. Antenatal and early postnatal antecedents of parent-reported attention problems at 2 years of age. J. Pediatrics 166, 20–U270.
Efird, J.T., 2010. Season of birth and risk for adult onset glioma. Int. J. Environmental Res. Public Health 7, 1913–1936.
El Marroun, H., Jaddoe, V.W., Hudziak, J.J., et al., 2012. Maternal use of selective serotonin reuptake inhibitors, fetal growth, and risk of adverse birth outcomes. Arch. Gen. Psychiatry 69, 706–714.
El Marroun, H., White, T.J.H., van der Knaap, N.J.F., et al., 2014. Prenatal exposure to selective serotonin reuptake inhibitors and social responsiveness symptoms of autism: population-based study of young children. Br. J. Psychiatry 205, 95–102.
Elias, S.G., van Noord, P.A.H., Peeters, P.H.M., et al., 2003. Caloric restriction reduces age at menopause: the effect of the 1944-1945 Dutch famine. Menopause J. North American Menopause Soc. 10, 399–405.
Elias, S.G., Peeters, P.H.M., Grobbee, D.E., et al., 2004. Breast cancer risk after caloric restriction during the 1944-1945 Dutch famine. J. National Cancer Institute 96, 539–546.

Ellenbroek, B.A., Riva, M.A., 2003. Early maternal deprivation as an animal model for schizophrenia. Clinical Neurosci. Res. 3, 297–302.
Fone, K.C.F., Porkess, M.V., 2008. Behavioural and neurochemical effects of post-weaning social isolation in rodents–relevance to developmental neuropsychiatric disorders. Neurosci. Biobehavioral Rev. 32, 1087–1102.
Fosse, R., Joseph, J., Richardson, K., 2015. A critical assessment of the equal-environment assumption of the twin method for schizophrenia. Frontiers Psychiatry 6, 62.
Fuh, J.L., Wang, S.J., Juang, K.D., et al., 2010. Relationship between childhood physical maltreatment and migraine in adolescents. Headache 50, 761–768.
Galeano, P., Romero, J.I., Luque-Rojas, M.J., et al., 2013. Moderate and severe perinatal asphyxia induces differential effects on cocaine sensitization in adult rats. Synapse 67, 553–567.
Gregorowski, C., Seedat, S., 2013. Addressing childhood trauma in a developmental context. J. Child Adolescent Mental Health 25, 105–118.
Guinchat, V., Thorsen, P., Laurent, C., et al., 2012. Pre-, peri- and neonatal risk factors for autism. Acta obstetricia et gynecologica Scandinavica 91, 287–300.
Gur, T.L., Kim, D.R., Epperson, C.N., 2013. Central nervous system effects of prenatal selective serotonin reuptake inhibitors: sensing the signal through the noise. Psychopharmacology 227, 567–582.
Hare, E.H., Price, J.S., Slater, E., 1973. Mental disorder and season of birth. Nature 241, 480–1480.
Holder, M.K., Blaustein, J.D., 2014. Puberty and adolescence as a time of vulnerability to stressors that alter neurobehavioral processes. Front Neuroendocrinol. 35, 89–110.
Jacobus, J., Tapert, S.F., 2013. Neurotoxic effects of alcohol in adolescence. Annu. Rev. Clin. Psychol. 9, 703–721.
Jain, R., Balhara, Y.P., 2010. Impact of alcohol and substance abuse on adolescent brain: a preclinical perspective. Indian J. Physiol. Pharmacol. 54, 213–234.
Kay, R.W., 2004. Schizophrenia and season of birth: relationship to geomagnetic storms. Schizophrenia Res. 66, 7–20.
Kendler, K.S., Larsson Lonn, S., Morris, N.A., et al., 2014. A Swedish national adoption study of criminality. Psychol. Med. 44, 1913–1925.
Khandaker, G.M., Zimbron, J., Dalman, C., et al., 2012. Childhood infection and adult schizophrenia: a meta-analysis of population-based studies. Schizophrenia Res. 139, 161–168.
Kim, J.H., Scialli, A.R., 2011. Thalidomide: the tragedy of birth defects and the effective treatment of disease. Toxicol. Sci. 122, 1–6.
Kinney, D.K., Miller, A.M., Crowley, D.J., et al., 2008. Autism prevalence following prenatal exposure to hurricanes and tropical storms in Louisiana. J. Autism Dev. Disord. 38, 481–488.
Kosten, T.A., Kim, J.J., Lee, H.J., 2012. Early life manipulations alter learning and memory in rats. Neurosci. Biobehavioral Rev. 36, 1985–2006.
Kotlicka-Antczak, M., Pawelczyk, A., Rabe-Jablonska, J., et al., 2014. Obstetrical complications and Apgar score in subjects at risk of psychosis. J. Psychiatric Res. 48, 79–85.
Kyle, U.G., Pichard, C., 2006. The Dutch Famine of 1944-1945: a pathophysiological model of long-term consequences of wasting disease. Curr. Opinion Clinical Nutrition Metabolic Care 9, 388–394.
La Rosa, F., Liso, A., Bianconi, F., et al., 2014. Seasonal variation in the month of birth in patients with skin cancer. Br. J. Cancer 111, 1810–1813.
Lang, I.A., Marlow, R., Goodman, R., et al., 2013. Influence of problematic child-teacher relationships on future psychiatric disorder: population survey with 3-year follow-up. Br. J. Psychiatry 202, 336–341.
Laplante, D.P., Brunet, A., Schmitz, N., et al., 2008. Project Ice Storm: prenatal maternal stress affects cognitive and linguistic functioning in 5 1/2-year-old children. J. Am. Acad. Child Adolescent Psychiatry 47, 1063–1072.
Larsson, H.J., Eaton, W.W., Madsen, K.M., et al., 2005. Risk factors for autism: perinatal factors, parental psychiatric history, and socioeconomic status. Am. J. Epidemiol. 161, 916–925.
Lynskey, M.T., Agrawal, A., Heath, A.C., 2010. Genetically informative research on adolescent substance use: methods, findings, and challenges. J. Am. Acad. Child Adolescent Psychiatry 49, 1202–1214.
Machon, R.A., Mednick, S.A., Huttunen, M.O., 1997. Adult major affective disorder after prenatal exposure to an influenza epidemic. Arch. Gen. Psychiatry 54, 322–328.
Man, K.K.C., Tong, H.H.Y., Wong, L.Y.L., et al., 2015. Exposure to selective serotonin reuptake inhibitors during pregnancy and risk of autism spectrum disorder in children: a systematic review and meta-analysis of observational studies. Neurosci. Biobehavioral Rev. 49, 82–89.
Marco, E.M., Llorente, R., Lopez-Gallardo, M., et al., 2015. The maternal deprivation animal model revisited. Neurosci. Biobehavioral Rev. 51, 151–163.

McCormick, C.M., Mathews, I.Z., 2010. Adolescent development, hypothalamic-pituitary-adrenal function, and programming of adult learning and memory. Prog. Neuropsychopharmacol. Biol. Psychiatry 34, 756–765.

McCormick, C.M., Hodges, T.E., Simone, J.J., 2015. Peer pressures: social instability stress in adolescence and social deficits in adulthood in a rodent model. Dev. Cognitive Neurosci. 11, 2–11.

McGue, M., Malone, S., Keyes, M., et al., 2014. Parent-offspring similarity for drinking: a longitudinal adoption study. Behavior Genetics 44, 620–628.

Mednick, S.A., Machon, R.A., Huttunen, M.O., et al., 1988. Adult schizophrenia following prenatal exposure to an influenza epidemic. Arch. Gen. Psychiatry 45, 189–192.

Meyer, U., Nyffeler, M., Engler, A., et al., 2006. The time of prenatal immune challenge determines the specificity of inflammation-mediated brain and behavioral pathology. J. Neurosci. 26, 4752–4762.

Nylander, I., Roman, E., 2013. Is the rodent maternal separation model a valid and effective model for studies on the early-life impact on ethanol consumption? Psychopharmacol. (Berlin) 229, 555–569.

O'Callaghan, E., Sham, P., Takei, N., et al., 1991. Schizophrenia after prenatal exposure to 1957 A2 influenza epidemic. Lancet 337, 1248–1250.

Olivier, J.D.A., Blom, T., Arentsen, T., et al., 2011. The age-dependent effects of selective serotonin reuptake inhibitors in humans and rodents: a review. Prog. Neuro-psychopharmacol. Biol. Psychiatry 35, 1400–1408.

Ornelas, L.C., Novier, A., Van Skike, C.E., et al., 2015. The effects of acute alcohol on motor impairments in adolescent, adult, and aged rats. Alcohol 49, 121–126.

Parboosing, R., Bao, Y.Y., Shen, L., et al., 2013. Gestational influenza and bipolar disorder in adult offspring. Jama Psychiatry 70, 677–685.

Paus, T., Keshavan, M., Giedd, J.N., 2008. Why do many psychiatric disorders emerge during adolescence? Nature Rev. Neurosci. 9, 947–957.

Pawluski, J.L., 2012. Perinatal selective serotonin reuptake inhibitor exposure: impact on brain development and neural plasticity. Neuroendocrinology 95, 39–46.

Petersen, L., Sorensen, T.I., 2011a. The Danish adoption register. Scandinavian J. Public Health 39, 83–86.

Petersen, L., Sorensen, T.I., 2011b. Studies based on the Danish Adoption Register: schizophrenia, BMI, smoking, and mortality in perspective. Scandinavian J. Public Health 39, 191–195.

Raineki, C., Lucion, A.B., Weinberg, J., 2014. Neonatal handling: an overview of the positive and negative effects. Dev. Psychobiol. 56, 1613–1625.

Renard, J., Krebs, M.O., Le Pen, G., et al., 2014. Long-term consequences of adolescent cannabinoid exposure in adult psychopathology. Frontiers Neurosci. 8, 361.

Sasaki, A., Constantinof, A., Pan, P., et al., 2014. Cocaine exposure prior to pregnancy alters the psychomotor response to cocaine and transcriptional regulation of the dopamine D1 receptor in adult male offspring. Behav. Brain Res. 265, 163–170.

Schmid, M., Petermann, F., Fegert, J.M., 2013. Developmental trauma disorder: pros and cons of including formal criteria in the psychiatric diagnostic systems. BMC Psychiatry 13, 3.

Schneider, T., Przewlocki, R., 2005. Behavioral alterations in rats prenatally exposed to valproic acid: animal model of autism. Neuropsychopharmacology 30, 80–89.

Scott, J., McNeill, Y., Cavanagh, J., et al., 2006. Exposure to obstetric complications and subsequent development of bipolar disorder: systematic review. Br. J. Psychiatry 189, 3–11.

Selten, J.P., Morgan, V.A., 2010. Prenatal exposure to influenza and major affective disorder. Bipolar Disord. 12, 753–754.

Selten, J.P., Frissen, A., Lensvelt-Mulders, G., et al., 2010. Schizophrenia and 1957 pandemic of influenza: meta-analysis. Schizophrenia Bull. 36, 219–228.

Semple, B.D., Blomgren, K., Gimlin, K., et al., 2013. Brain development in rodents and humans: identifying benchmarks of maturation and vulnerability to injury across species. Prog. Neurobiol. 106-107, 1–16.

Silva, D., Colvin, L., Hagemann, E., et al., 2014. Children diagnosed with attention deficit disorder and their hospitalisations: population data linkage study. Eur. Child Adolescent Psychiatry 23, 1043–1050.

Stephansson, O., Kieler, H., Haglund, B., et al., 2013. Selective serotonin reuptake inhibitors during pregnancy and risk of stillbirth and infant mortality. Jama-J. Am. Med. Assoc. 309, 48–54.

Sturman, D.A., Moghaddam, B., 2011. The neurobiology of adolescence: changes in brain architecture, functional dynamics, and behavioral tendencies. Neurosci. Biobehav. Rev. 35, 1704–1712.

Tochigi, M., Okazaki, Y., Kato, N., et al., 2004. What causes seasonality of birth in schizophrenia? Neurosci. Res. 48, 1–11.

Torrey, E.F., Miller, J., Rawlings, R., et al., 1997. Seasonality of births in schizophrenia and bipolar disorder: a review of the literature. Schizophrenia Res. 28, 1–38.

Turcotte-Tremblay, A.M., Lim, R., Laplante, D.P., et al., 2014. Prenatal maternal stress predicts childhood asthma in girls: project ice storm. Biomed. Res. Int. 2014, 201717.

Ueda, P., Bonamy, A.K.E., Granath, F., et al., 2014. Month of birth and cause-specific mortality between 50 and 80 years: a population-based longitudinal cohort study in Sweden. Eur. J. Epidemiol. 29, 89–94.

Upthegrove, R., Chard, C., Jones, L., et al., 2015. Adverse childhood events and psychosis in bipolar affective disorder. Br. J. Psychiatry 206, 191–197.

Vassoler, F.M., White, S.L., Schmidt, H.D., et al., 2013. Epigenetic inheritance of a cocaine-resistance phenotype. Nature Neurosci. 16, 42–U67.

Vetulani, J., 2013. Early maternal separation: a rodent model of depression and a prevailing human condition. Pharmacol. Rep. 65, 1451–1461.

Viveros, M.P., Marco, E.M., Lopez-Gallardo, M., et al., 2011. Framework for sex differences in adolescent neurobiology: a focus on cannabinoids. Neurosci. Biobehav. Rev. 35, 1740–1751.

Walder, D.J., Laplante, D.P., Sousa-Pires, A., et al., 2014. Prenatal maternal stress predicts autism traits in 6 (1/2) year-old children: project ice storm. Psychiatry Res. 219, 353–360.

Willner, P., 1997. Validity, reliability and utility of the chronic mild stress model of depression: a 10-year review and evaluation. Psychopharmacology 134, 319–329.

Workman, A.D., Charvet, C.J., Clancy, B., et al., 2013. Modeling transformations of neurodevelopmental sequences across mammalian species. J. Neurosci. 33, 7368–7383.

Yong Ping, E., Laplante, D.P., Elgbeili, G., et al., 2015. Prenatal maternal stress predicts stress reactivity at 2 (1/2) years of age: the Iowa Flood Study. Psychoneuroendocrinology 56, 62–78.

CHAPTER

5

Environment Challenges and the Brain

INTRODUCTION

In the previous chapter we saw how environmental challenges can have long-term consequences for the development of an individual's behavior and how this can increase the risk for psychiatric disorders. The question that arises is how such (sometimes very) early environmental stimuli can have such profound long lasting effects on an individual. As an example, maternal deprivation (ie, a 24 h separation of the mother from her pups, see previous chapter) of a rat at postnatal day 4 was found to change the stress response and spatial learning for up to 30 months of age (Oitzl et al., 2000; Workel et al., 2001), which is close to the actual life span of the animal. Several studies have shown that the average half-life of proteins in humans is about 1–2 days, although there are exceptions with proteins such as the nucleoporins (proteins involved in the nuclear pore complex that transport substances between the nucleus and the cytoplasm) and the histones [proteins involved in the three-dimensional structure of chromosomes and in gene expression (as discussed later)]. However, these proteins are quite exceptional and for most proteins the lifespan is much shorter, making it unlikely that the long-term changes induced by early environmental challenges are due to a simple increase in protein production induced by the challenge per se. Rather, such long lasting effects require other mechanisms. The two most often discussed mechanisms are: (1) structural alterations in the (developing) brain; (2) epigenetic changes in gene regulation. In the remainder of this chapter, we will discuss these mechanisms in more detail. It is, however, important to realize that these processes are not mutually exclusive and may very well interact with each other.

EARLY ENVIRONMENTAL CHALLENGES AND STRUCTURAL BRAIN CHANGES

The development of the central nervous system in humans follows a very strict pattern and encompasses not only the prenatal period, but extends up to early adulthood. In fact myelination of the prefrontal cortex in humans is not finished until about 20–22 years of age.

Gene-Environment Interactions in Psychiatry. http://dx.doi.org/10.1016/B978-0-12-801657-2.00005-7

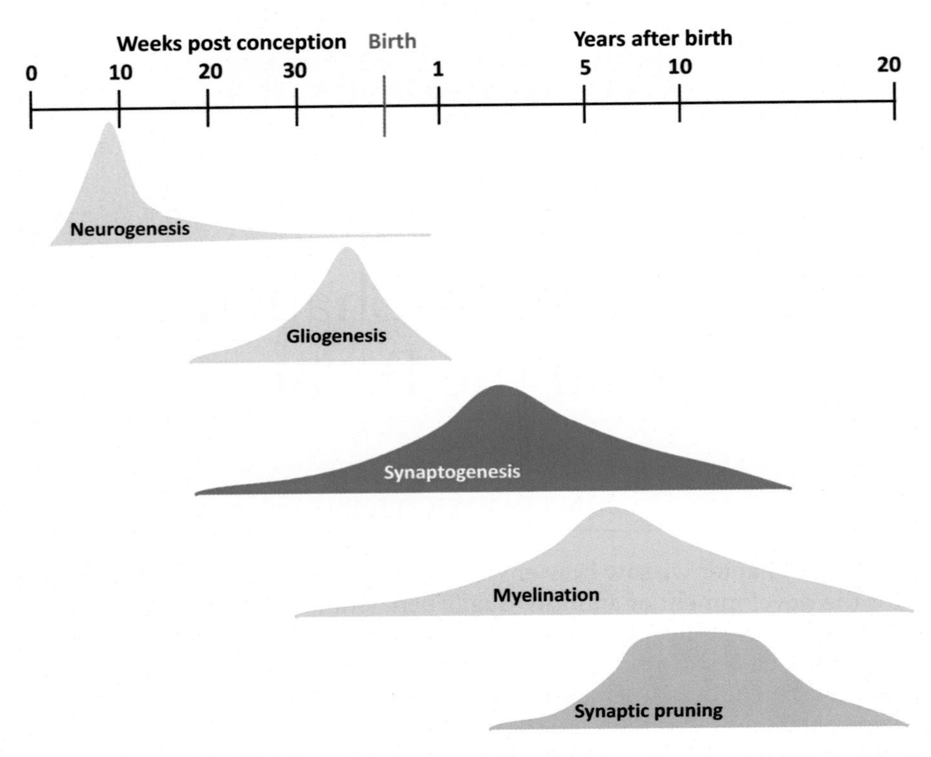

FIGURE 5.1 **Brain development.** Brain development in animals, including humans occurs in distinct phases, starting with neurogenesis in the first few weeks after conception and ending with myelination which continues until about 20 years of age (in humans).

Fig. 5.1 shows a rough timeline of the different developmental processes in humans. Given this protracted period of development it is not surprising that environmental challenges can significantly impact on the structure and function of the brain. Remember that neuronal cells cannot replicate anymore and are therefore called post-mitotic. Thus, when cells die, they are normally not replaced by other cells. Although we do know that neurogenesis (ie, the development of new brain cells from progenitor embryonic stem cells) does occur in specific brain regions (such as the hippocampus), the new cells being generated here play a role in plasticity but not necessarily in replacing dead cells. In other words, if early challenges lead to pathological changes in the brain, the effect is likely to be permanent (although some functional compensation may still occur).

Studies in animals have confirmed that various environmental challenges lead to cell death in the brain, through both necrosis and apoptosis. Apoptosis refers to the process of programmed cell death which in itself is a natural process to regulate the normal number of cells within the body. However, when apoptosis goes awry too many or too few cells may be removed which can lead to abnormal brain connections. Injections of lipopolysaccharide on gestational days 15 and 16, mimicking a bacterial infection, leads to significant reductions in dendritic lengths, branching and spine densities in the hippocampus and prefrontal cortex, some of these effects already observable on postnatal day 10, others more prominent on postnatal 35 or 60 (Baharnoori et al., 2009). Structural alterations have also been seen in rats exposed to prenatal stress (Charil et al., 2010), including reductions in hippocampal volume, dendritic spine density in the anterior cingulate and orbitofrontal cortex (Lupien et al., 2009) and alterations in dendritic arborization in the nucleus accumbens and hippocampus (Martinez-Tellez et al., 2009).

Lack of oxygen, due to perinatal birth complications, is also an obvious risk factor for inducing cell death which was confirmed in a study in sheep where experimentally induced asphyxia near the end of pregnancy led to behavioural changes in the new-born lambs, as well as increased necrosis and apoptosis. In addition, these lambs showed diffuse but widespread grey and white matter lesions (Castillo-Melendez et al., 2013).

Postnatal environmental challenges can also lead to morphological changes in the brain. For instance, maternal deprivation on postnatal day 13 was shown to increase neurodegeneration as evidenced by an increase in Fluoro-Jade C positive neuronal cells (Fluoro-Jade C is a well-established marker for neurodegeneration) in the hippocampus. This increase was seen during the neonatal period shortly after the maternal deprivation and was accompanied by an increase in astroglial cells, which extended into adulthood (Marco et al., 2015). However, structural changes were not restricted to the hippocampus. Reductions in cell proliferation and astrocytes were also observed in the hypothalamus shortly after the deprivation. Shorter term repeated separations of the mother from her litter also affects the morphology of the offspring. Thus 15-min separation (early handling, see Chapter 4) between postnatal days 1 and 12 led to increases in spine density and dendritic lengths in the prefrontal cortex and spine density in the CA1 region of the hippocampus, while dendritic lengths were reduced in the nucleus accumbens. In contrast, 120 min separation during the same postnatal period led to decreases in dendritic lengths in the prefrontal cortex, CA1 area of the hippocampus and nucleus accumbens. Moreover, in this last area spine densities were also significantly reduced (Monroy et al., 2010).

In line with the important role of the adolescent period in shaping the brain (Fig. 5.1) environmental challenges during this period also significantly alter brain morphology and especially maturation. For instance adolescent cannabinoid treatment (with either the synthetic agonist WIN55,212-2 or Δ^9-tetrahydrocannabinol) in rats leads to changes in surviving progenitor cells in the neostriatum and prefrontal cortex, and alterations in the morphology of the hippocampus and prefrontal cortex (Renard et al., 2014). Reductions in dendritic lengths have also been reported after adolescent chronic mild stress, with decreases particularly prominent in (again) the hippocampus and prefrontal cortex, while increases were seen in the total dendritic length in the basolateral amygdala (Eiland and Romeo, 2013).

Studies in humans, while being significantly less in number and in detail of analysis, further support the preclinical findings of altered brain morphology after environmental challenges. So far very few studies have looked at the effect of prenatal stress on brain development in humans, although a large study around project Ice Storm (see Chapter 4) is currently looking at brain structures in children prenatally exposed to severe ice storms in the Quebec region of Canada (Charil et al., 2010). A recent prospective study in healthy volunteers found a relationship between prenatal stress and uncinate fasciculus, the major connection between the limbic system and the prefrontal cortex (Sarkar et al., 2014). Significantly more evidence has linked perinatal birth complications to altered brain morphology. Relatively brief periods of perinatal hypoxia can lead to neuronal death, white and gray matter damage and reduced neuronal growth (Volpe, 2012). Studies in patients with schizophrenia further emphasized the effects of birth complications on brain morphology, showing a direct correlation between perinatal hypoxia and brain volume, cortical surface area and gyrification, but not cortical thickness (Smith et al., 2014).

Childhood maltreatment has also been linked to morphological changes in specific brain areas, including volume reductions in specific hippocampal areas, the left medial prefrontal cortex and other corticolimbic gray matter. Looking more specifically at different forms of maltreatment, physical abuse was associated with reductions in rostral prefrontal, dorsolateral prefrontal and orbitrofrontal cortex and striatum. Physical neglect, on the other hand was associated with reductions in rostral prefrontal cortex and cerebellum. Emotional neglect had the most significant impact on gray-matter affecting many prefrontal cortex areas (rostral, dorsolateral, orbitofrontal and subgenual), hippocampus, amygdala, striatum and cerebellum (Edmiston et al., 2011). These morphological changes seem to lead to deficits in functional connectivity between brain regions, as childhood maltreatment is associated with connectivity deficits between the amygdala and prefrontal cortical areas (Birn et al., 2014) and between different cortical areas (Elton et al., 2014). In an interesting recent paper, adults subjected to childhood maltreatment were subdivided in a vulnerable and resilient group, based on the presence or absence of major depression and/or anxiety disorders. An investigation of the resting-state functional connectivity showed the resilient group were characterized by increased connectivity between the left anterior cingulate and an area between the bilateral lingual gyrus and the occipital fusiform gyrus (van der Werff et al., 2013), suggesting that this increased connectivity might be a protective factor.

Studies focusing on adolescence demonstrated that the effect of stress on brain structure and function can be timing dependent. Thus whereas stress in adolescence or in adulthood leads to an impairment in performance in a go-no-go task, stress in adulthood was accompanied by activation of the dorsolateral prefrontal cortex while stress during adolescence was accompanied by an inhibition of the dorsolateral prefrontal cortex (Rahdar and Galvan, 2014). Hippocampal and cortical structural deficits (reduced volumes, cortical thickness, and activation) have also been reported in relation to adolescent cannabis exposure (Renard et al., 2014).

In summary, there is substantial evidence that different environmental challenges affect brain structure in animal as well as in humans. The results seem to indicate a particular vulnerability of the limbic system, particularly the hippocampus, amygdala and the prefrontal cortex. Several different mechanisms have been proposed to explain how environmental challenge can affect the structure of the brain. The three most important processes are changes in the immune system, in the stress system or in neurotransmitter concentrations.

EARLY ENVIRONMENTAL CHALLENGES AND THE IMMUNE SYSTEM

Several early environmental challenges, such as prenatal stress, prenatal malnutrition (Marques et al., 2013) and prenatal exposure to ethanol (Taylor et al., 2006) and to valproate (Lucchina and Depino, 2014) induce an increase in inflammation. Moreover, as we discussed at length in the previous chapter, maternal viral or bacterial infections also cause long-lasting changes in the offspring. Finally, perinatal hypoxia has been associated with increased inflammatory cell infiltration in the brain, especially in the subventricular zone, the corpus callosum and periventricular and cortical white matter (Castillo-Melendez et al., 2013). This suggests that inflammation may be one common factor in mediating the effects of various environmental challenges on brain structure. However, viruses and bacteria infecting the mother (and antibodies produced against them) are unlikely to pass the blood placental barrier, and, likewise, maternally injected lipopolysaccharide is not transmitted across the blood placental barrier in rats, in spite of the fact that it induces significant changes in the unborn rat pups (Ashdown et al., 2006). This suggests that it is the maternal immune response to an environmental challenge that alters brain development in the unborn offspring.

The maternal immune response leads to an increase in cytokines, chemical messengers that can have both pro- and anti-inflammatory properties (Box 5.1 lists the most important cytokines). Some (IL-2 and IL-6) but not other (IL-1α and TNF) cytokines can pass from the mother to the unborn fetus. In addition, there is some evidence indicating that the placenta itself may be a source of cytokine production which can then be transported to both mother and fetus.

Before discussing the potential effects of increased cytokine levels in fetal brain, it is important to realize that several cytokines are also present in the normal central nervous system, both during development and in adulthood. Indeed, microglia and astrocytes produce cytokines and chemokines as early as week 5 of gestation in humans, suggesting an important role in normal development. In line with this, in fetal sheep brains IL-1β and TNF-α were found as early as gestational day 30 (full term is 150 days) but both had virtually disappeared

BOX 5.1

THE MOST IMPORTANT CYTOKINES

Abbreviations	Names	Functions
IL-1β	Interleukin 1beta	Proinflammatory cytokine produced by the M1 wave of microglia activation
IL-4	Interleukin 4	Stimulates M2 wave of cytokine activation
IL-6	Interleukin 6	Proinflammatory cytokine produced by the M1 wave of microglia activation
IL-10	Interleukin 10	Antiinflammatory cytokine produced by the M2 wave of microglia activation
TNF-α	Tumor necrosis factor alpha	Proinflammatory cytokine produced by the M1 wave of microglia activation
TGF-β	Transforming growth factor beta	Antiinflammatory cytokine produced by the M2 wave of microglia activation

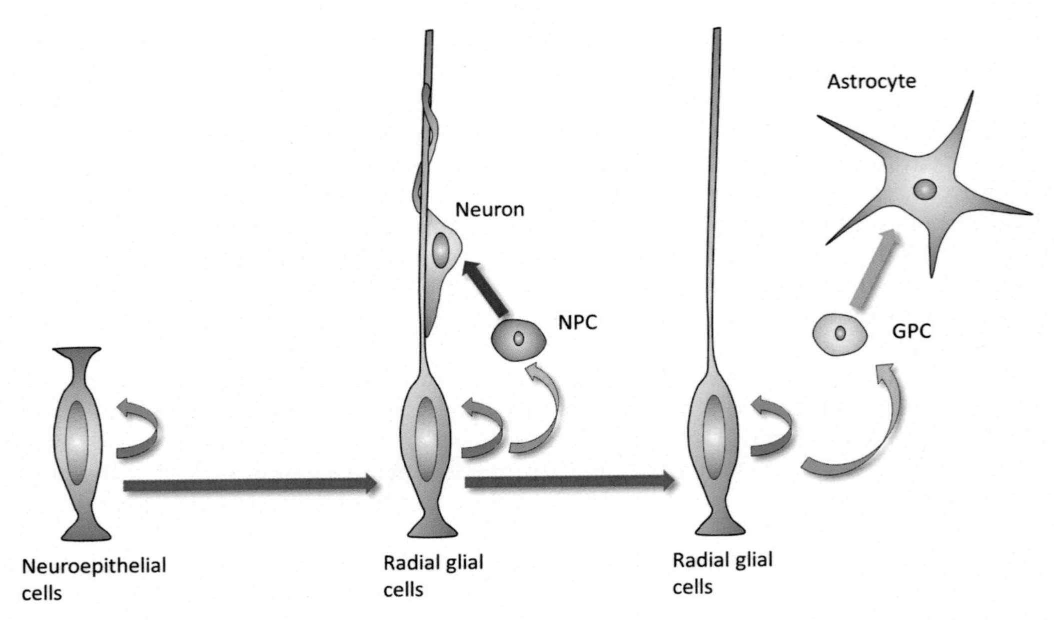

FIGURE 5.2 **Neurogenesis and gliogenesis.** The development of the brain starts with the formation of neuroepithelial cells. These cells develop into radial glial cells, which function both as precursors for neuronal progenitor cells (NPC) as well as providing a scaffolding for the neuronal cells, allowing them to migrate through the brain. Subsequently, radial glial cells can develop into glial progenitor cells (GPC) which subsequently develop into astrocytes and oligodendrocytes.

again around embryonic day 80. Since this coincides with a period of intense synaptogenesis, it was suggested that both cytokines play an important role in this developmental process. However, cytokines are now thought to play a role in virtually all aspect of brain development (Deverman and Patterson, 2009). Most importantly, certain cytokines play a role in the self-renewal of radial glia cells. These cells do not only provide an essential scaffolding along which neuronal cells can migrate in the brain, but also give rise to neurons, astrocytes and oligodendrocytes (see Fig. 5.2). It is thus conceivable that changes in cytokines alter the balance between these three processes.

Perhaps even more importantly, the brain also contains large number of microglia. In contrast to the other cells in the brain, microglia arise from hematopoietic stem cells (which also give rise to all blood cells) and in the adult brain function as immune surveillance cells. Microglia are first detected at the onset of neurogenesis and have multiple functions throughout development (Fig. 5.3). First of all, the central nervous system produces many more neurons than necessary and microglia appear to be involved in removing the ones that are superfluous. In addition, they appear to modulate axon path finding, neurite outgrowth and synaptogenesis and are involved in angiogenesis and astrogliogenesis. Thus microglia play a critical role in shaping the brain during the early development, primarily by releasing specific cytokines and other chemical messengers. Moreover, viral or bacterial infections increase cytokine production in microglia. Microglial activation occurs in two waves: a classical M1 (leading to phagocytosis and production of pro-inflammatory cytokines such as IL1β, IL-6,

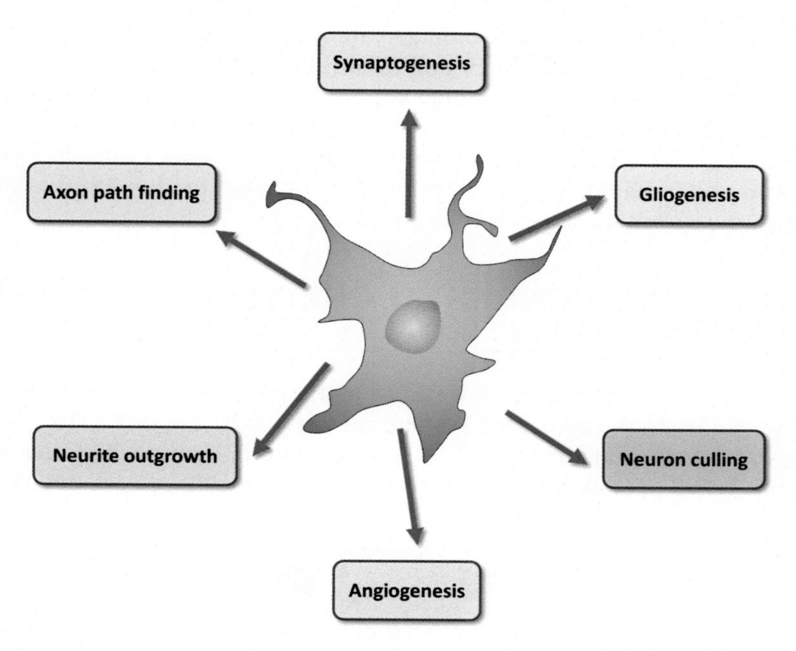

FIGURE 5.3 **The functions of microglia.** In addition to its function as the "immune cells" of the central nervous system, microglial cells also have a number of structural functions, such as gliogenesis, synaptogenesis, and axon path finding.

and TNF-α) and an alternative M2 wave (associated with the release of anti-inflammatory cytokines such as IL-10 and TGF-β).

Taken together, it is not surprising that changes in the levels of cytokines (either directly by increasing levels in the mother or from the placenta, or indirectly by changing the activity of astrocytes or microglia) can have far reaching consequences for the development of the brain. Alterations can be subtle, such as cell displacement as seen for instance in the hippocampus of patients with schizophrenia (Kovelman and Scheibel, 1984) and which have been linked to alterations in the radial glial cells that guide neurons to their final destination (Fig. 5.2). Likewise, a disturbance in the role of microglia in axon pathfinding can be expected to lead to changes in connectivity between different brain structures. Such disconnection is now generally thought to underlie a large number of psychiatric disorders, including autism spectrum disorder, major depression and schizophrenia (Menon, 2013). However, given the microglial role in cell survival, increased cytokine levels may even lead to increased cell death and as a result in reduced brain matter, as has also been described in several brain disorders.

One fundamental question that has yet to be answered satisfactorily is which of the many different cytokines is (or are) responsible for disturbing brain development. Interestingly, it has been shown that overexpression of either the pro-inflammatory cytokine IL-6 or the anti-inflammatory cytokine IL-10 produce very similar behavioural and cognitive effects, leading to the suggestion that it is not a specific cytokine that is involved, but rather an imbalance between the levels of pro- and antiinflammatory cytokines. On the other hand, the finding that many of the long-term consequences of the prenatal viral–mimetic polyI:C can be reproduced by a single injection of IL-6 on gestational day 12 and administration of an IL-6 antibody prevents many of the effects of polyI:C administration seems to indicate that IL-6 is largely responsible for these effects, an idea supported by the finding that most behavioural effects of polyI:C are absent in mice lacking the gene for IL-6 (Smith et al., 2007). This rather remarkable finding that a single injection of IL-6 during gestation can have such profound impact on brain and behavior of the offspring in adulthood further emphasizes the crucial role of the immune system in brain development.

EARLY ENVIRONMENTAL CHALLENGES AND THE HYPOTHALAMIC PITUITARY ADRENAL (HPA) AXIS

The increase in cytokines does not only affect astrocytes and microglia, but is also capable of altering the major stress regulatory system: the HPA axis (Fig. 5.4). The HPA axis has its origin in the paraventricular nucleus (PVN) of the hypothalamus which lies just above the pituitary gland. When activated the so-called parvocellular cells release two neurohormones

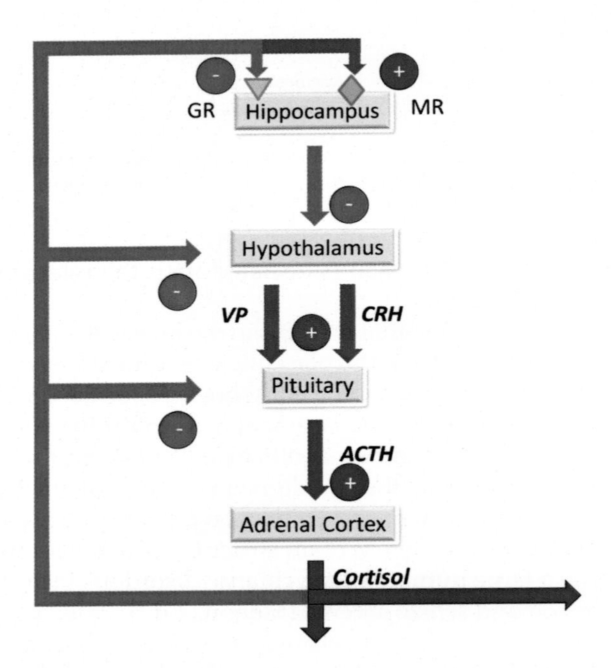

FIGURE 5.4 **The hypothalamus pituitary adrenal (HPA) axis.** The HPA axis is the most important stress system. When an organism is stressed, the paraventricular nucleus of the hypothalamus is activated and vasopressin (VP) and corticotropin releasing hormone (CRH) are being released. These neurohormones subsequently stimulate the release of adrenocorticotropic hormone (ACTH) from the pituitary gland. ACTH is released in the general bloodstream and stimulates cells in the adrenal cortex to release cortisol (corticosterone in rats and mice). These so-called glucocorticoids then prepare the body for a reaction to the stressful stimulus. Moreover these glucocorticoids negatively feedback onto the pituitary gland and the hypothalamus to reduce the release of ACTH and CRH/VP. In addition, glucocorticoids influence the hippocampus in a more complex way, leading to both activation (via the GR) and inhibition (via the mineralocorticoid receptor MR).

called corticotropin releasing hormone (CRH, sometimes referred to as corticotropin releasing factor, CRF) and vasopressin into a specific network of blood vessels in the median eminence commonly known as the hypothalamus-pituitary portal circulation. Both neurohormones are then transported to the anterior part of the pituitary gland where they synergistically stimulate endocrine cells to release another hormone, adrenocorticotropic hormone (ACTH), into the general circulation. In the final step of the HPA axis, ACTH then travels through the bloodstream to reach the adrenal glands where it binds to specific receptors (the so-called melanocortin receptor 2, MC_2), leading to the release of glucocorticoids, especially cortisol (in humans) or corticosterone (in rodents). Importantly, cortisol (and corticosterone) can influence its own release at the level of the hypothalamus (where it inhibits the release of CRH and vasopressin) and pituitary gland (where it inhibits the release of ACTH). In addition, the HPA axis is under inhibitory control of the hippocampus, which also contains receptors for cortisol. This influence is more complicated as there are two different type of corticoid receptors in the hippocampus: the so-called mineralocorticoid receptors (MR) and the glucocorticoid receptor (GR). Cortisol has a much higher affinity for the MR receptor, implying that at low concentrations, MR receptors are more or less fully occupied, while GR receptor are only occupied at high cortisol concentrations (ie, during stress). MR excite while GR inhibit hippocampal neurons leading respectively to inhibition and stimulation (disinhibition) of the PVN and the HPA axis. In other words, whether cortisol increases or decreases the HPA axis is determined by the ratio of MR/GR and the level of cortisol in the hippocampus (Struber et al., 2014). Although the HPA axis responds to different stressors, it is particularly sensitive to social threats and stressors with an unpredictable outcome.

The immune system and the HPA axis are inextricably linked, with cytokines being able to activate the HPA-axis, while glucocorticoids in turn can affect the production of cytokines. However, the effects of cytokines on the HPA axis are multi-layered and complex, but in general pro-inflammatory cytokines such as IL-1 and TNFα enhance the release of ACTH and vasopressin, and consequently of cortisol (Dejager et al., 2014), while IL-6 reduces the availability of corticosteroid binding globulins, and this enhances the free (active) concentration of cortisol. On the other hand glucocorticoids inhibit the release of pro-inflammatory cytokines, while at the same time stimulating the production of anti-inflammatory cytokines. Similar to cytokines, the HPA axis is also crucially involved in the development of the central nervous system. The development of the HPA axis is closely related to the development of the brain, and highly species-specific, as we discussed in the previous chapter. Thus for precocious species (like humans) most of the development occurs in utero, while for non-precocious animals (like rats and mice) much of the development of the HPA axis occurs in the first weeks after birth.

In rats and mice the final days of gestation are accompanied by declining levels of corticosterone, although the levels remain relatively high. This is followed by a period (from about postnatal day 4–14) where stressors that would normally elicit a corticosterone surge are virtually inactive. As a result this period has been known as the stress hyposensitive period (SHP). Only prolonged, intense stressors (such as maternal deprivation or repeated maternal separation) can elicit an HPA axis response in the SHP. Studies in humans have suggested that children also have a SHR roughly between 2 and 15 months of age. It has even been suggested that the hyposensitive period may extend from infancy throughout childhood (Lupien et al., 2009).

It is generally thought that the SHP is a protective mechanism of the organism, as corticosterone has major effects on brain development. Studies both in humans and animals have shown that early adversities not only acutely affect the HPA axis, but also have long term effects. Thus early adversities can lead to long-term changes in basal cortisol as well as in HPA axis reactivity to stressors. The exact nature of these changes is, however, still debated (Struber et al., 2014) and may, in part, depend on the timing of the adverse event. Thus, prenatal adversities and those occurring during the ages of 6–11 years are associated with higher cortisol levels, while adversities during the ages of 12–15 years are associated with lower cortisol levels (Bosch et al., 2012). Likewise, childhood adversities appear to alter the MR/GR ratio in the hippocampus resulting in an altered HPA axis feedback regulation (Fig. 5.4).

Animal studies have suggested that not only timing but also the duration of adverse life events critically influences HPA axis (re)activity. Thus, short term (15 min) separations (early handling, see Chapter 4) decrease while longer term (3–6 h) maternal separations increase the HPA axis response to stress in adulthood. One important mediating factor here is the degree of maternal care. Studies by Michael Meaney and his colleagues showed that the offspring of mothers that spent a lot of time licking and nursing the pups had relatively low HPA axis reactivity in adulthood, while the reverse was found in the offspring of mothers that spent significantly less time interacting with the pups (Liu et al., 1997). Studies in humans have also suggested that the long term effects of early adversities may be moderated by mother-child interactions.

In summary, early stressful life events in both animals and humans have been shown to lead to prolonged changes in the HPA axis, leading to chronically altered basal and stress-induced glucocorticoid levels. This has major implications for the development of the brain, as glucocorticoids are important for normal brain development and maturation. Glucocorticoids initiate maturation of nerve terminals, remodel axons, dendrites and dendritic spines and affect both neuronal and glial cell survival (Lupien et al., 2009). In this respect it is important to realize that both increases and decreases in glucocorticoids affect normal brain development, and thus slight changes in either direction may lead to long-term changes in brain structure. Most studies have focused on structural changes in the hippocampus, but stressful life events have also been found to decrease dendritic spine densities in the orbitofrontal and anterior cingulate cortex, as well as neurochemical alterations in the amygdala, especially after prenatal stressors, with stressors during adolescence having more effect on the amygdala and prefrontal cortex, again pointing to an important role of timing (Lupien et al., 2009). In addition to the role of glucocorticoids in shaping the overall structure of the brain, they also significantly affect neurogenesis, with high levels decreasing the generation, proliferation and maturation of new cells in the dentate gyrus throughout life, thus likely affecting the brain's plasticity (Schoenfeld and Cameron, 2015).

EARLY ENVIRONMENTAL CHALLENGES AND NEUROTRANSMITTER CHANGES

Most neurotransmitter systems are already expressed early in brain development, often before synaptogenesis takes place. For instance monoamines such as noradrenaline, dopamine and serotonin are expressed in the middle of gestation and surge in the last week before birth in rats. Glutamate and GABA (and especially their receptors) develop slightly differently and are often overproduced in the first period after birth (Herlenius and Lagercrantz, 2004). This development during such critical periods makes them vulnerable to early environmental challenges. Moreover, as discussed later, since many neurotransmitters (like cytokines and glucocorticoids) play important roles in shaping our brain during development, disrupting their activity may permanently alter brain morphology and connectivity.

Several environmental challenges have been found to alter the extracellular levels of neurotransmitters. Moreover, many genetic alterations lead to changes in neurotransmitter content as well. With respect to the environmental challenges, especially prenatal drug exposures such as maternal alcohol consumption and cigarette smoking are known to increase multiple neurotransmitters systems, most prominently dopamine. Likewise, prenatal exposure to valproate is known to affect the development of (among others) the serotonergic system. This effect can be quite profound. For instance, injections of valproate at gestational day 9 in the rats leads to a caudal shift in serotonergic cell bodies in (young) adulthood, most likely due to a disturbance in serotonergic neuronal differentiation and maturation (Miyazaki et al., 2005). Prenatal valproate was also found to alter the glutamate system as early as postnatal day 15 (Bristot Silvestrin et al., 2013).

However, psychological or social stressors can also alter the activity of neurotransmitters. Prenatal stress in mice leads to significant increases in brain concentrations of the serotonin transporter, especially in the striatum and hippocampus in newborn pups. Likewise, cFos immunoreactivity (an indicator of cell activity) is increased in these two areas, and highly correlated with the serotonin transporter activity (Bielas et al., 2014). Alterations in the development of the dopaminergic and noradrenergic systems have also been described after prenatal stress (Weinstock, 2007; Baier et al., 2012). Postnatal stress, in the form of maternal separation, deprivation or early handling likewise alters several neurotransmitter systems including dopamine, GABA and glutamate. These alterations are likely to be mediated in part by alterations in maternal behavior. For instance, early neonatal handling not only alters neurochemistry in pups but also significantly increases maternal behavior, as evidenced by increased arched back nursing and time spent on the nest (Stamatakis et al., 2015). As discussed by Michael Meaney and his colleagues, alterations in arched back nursing, in turn alter not only the HPA axis (as discussed earlier) but also neurotransmitter processes, such as dopamine 5-HT, oxytocin and vasopressin. Prolonged (3 h) maternal separation reduces 5-HT levels in key brain regions such as the hippocampus, amygdala and prefrontal cortex, while increasing noradrenaline levels in the hippocampus (selectively in females) and dopamine in the dorsal striatum in both genders (Matthews et al., 2001). Maternal deprivation also affects the dopaminergic system, leading, among others to increased mRNA for tyrosine hydroxylase, the rate limiting enzyme for dopamine synthesis and an increase in dopamine turnover in the prefrontal cortex, neostriatum and amygdala. In addition, alterations have been found in the glutamatergic, serotonergic and cannabinoid system (Marco et al., 2015), as well as decreases in several neuropeptide levels in the hippocampus such as neuropeptide Y and calcitonin-gene related peptide (Husum et al., 2002).

Challenges later on in life, such as isolation rearing also has long-term effects on neurotransmitter levels, especially the monoamines (Fone and Porkess, 2008), but also GABA (Matsumoto et al., 2007), while stressors around puberty are known to change vasopressin levels in the paraventricular nucleus of the hypothalamus and NMDA receptor gene expression in the hippocampus. Moreover, there is also evidence for altered dopamine functionality, as rats stressed in adolescence appear more sensitive to the hyperactivity induced by amphetamine in adulthood (McCormick and Mathews, 2007). Finally, adolescent exposure to drugs, especially cannabinoids influences the dopaminergic system as well as reducing the levels of GAD67 (the GABA synthesizing enzyme) in the prefrontal cortex (Renard et al., 2014).

There is also some evidence that early environmental challenges alter neurotransmitter levels in humans, although the evidence is much more indirect, as the neurochemistry of the brain is much less accessible than the neuroanatomy. Therefore studies have often relied on peripheral markers, such as urinary or blood levels of neurotransmitters. Thus several childhood adversities such as sexual abuse and maltreatment increase 24 h levels of urinary catecholamines, while children who experienced a motor vehicle accident had higher plasma levels of noradrenaline. Alterations in peripheral levels of serotonin and oxytocin have been described after childhood trauma (De Bellis and Zisk, 2014).

Changes in neurotransmitter levels at an early age, while the brain is still developing, can have a profound impact on the structure and function of the central nervous system, as it is increasingly recognized that many neurotransmitters also have neurotrophic effects. Thus the main excitatory neurotransmitter glutamate affects proliferation, survival, migration and differentiation of neural progenitor cells. This process is highly complex, and, time-dependent. Thus early on in development, AMPA receptors are important for proliferation and migration, while mGlu5 receptors are expressed on radial glial cells and neural progenitor cells to promote proliferation and survival. mGlu5 receptors are also particularly important for regulating the extension of radial glial processes (see Fig. 5.2) which are essential for migration. Later on in development NMDA receptors play a more prominent role (Jansson and Akerman, 2014). Similar roles in proliferation and differentiation of neural progenitor cell have also been described for the endocannabinoid system (Harkany et al., 2007).

Catecholamines like dopamine and noradrenaline also can influence neuronal cell development and function, often in a very complex manner. For instance in the cortex, dopamine can both increase (via D_2 receptors) and decrease (via D_1 receptors) neurite outgrowth, while the opposite effect (increase via D1) occurs in the neostriatum. Much attention has also focused on 5-HT, and this neurotransmitter was found to be crucially involved in several critical developmental processes, including cell division, differentiation, migration, synaptogenesis and dendritic pruning (Gaspar et al., 2003). Studies using genetically altered rats and mice, as well as those using perinatal treatment with serotonergic drugs have substantiated this claim and shown that many different serotonergic receptors, as well as the serotonin transporter are involved. Thus while 5-HT_{1A} receptor knock-out mice show deficits in dendritic maturation and neurogenesis, 5-HT_{1B} and serotonin transporter knock-out animals show deficits in axonal connections. Studies using pre- and early postnatal treatment with the SERT inhibitor fluoxetine also found developmental alterations (Olivier et al., 2011). Further emphasizing the role of 5-HT during development, it was shown that a genetic reduction of the 5-HT_{1A} receptor during the first two weeks after birth was enough to induce anxiety in adulthood, while similar reductions in adulthood were without effect (Gross et al., 2002). Likewise, the sleep deficits in adult rats with a genetically reduced SERT were prevented by pretreatment with the irreversible tryptophan hydroxylase inhibitor para-chlorophenylalanine (PCPA) or the 5-HT_{1A} receptor antagonist WAY 100,635 during the first two weeks after birth (Alexandre et al., 2006).

In summary, many environmental factors can alter the normal development of the brain resulting in impaired synaptic communication and possibly even cell death. These changes can occur through several different processes, such as changes in the immune system, the HPA axis or in neurotransmitter levels. Given the intricate relationships between these three systems, it is highly likely that the ultimate influence of environmental factors on brain functions results from an interaction between immunological, endocrinological and neuronal systems.

EARLY ENVIRONMENTAL CHALLENGES AND EPIGENETIC CHANGES

In addition to changes in the structure of the central nervous system, which is likely to cause long-term (perhaps irreversible) alterations in behavior, there is increasing evidence that early environmental challenges induce long-lasting changes in gene expression through a process generally known as epigenetics.

The term epigenetics was first introduced by Conrad Waddington as "*the branch of biology which studies the causal interactions between genes and their products which bring the phenotype into being*" (Waddington, 1942). It was subsequently redefined as "*mitotically or meiotically heritable changes in gene function that cannot be explained by changes in DNA sequence*". Although in itself a useful definition (especially in cancer research where epigenetic alterations play a very important role in tumor growth), for our purposes, this latter definition is less relevant, as brain cells are post-mitotic. Nonetheless, recent studies on the role of epigenetics in neuroscience have now also focused on transgenerational processes as we will see in subsequent chapters. A more useful definition of epigenetics in relation to gene * environment interactions is "*the study of cellular and physiological trait variations that are not caused by changes in the DNA sequence*".

The difference between genetic and epigenetic changes is illustrated in Fig. 5.5. A genetic mutation leads to a *qualitative* change in the structure of mRNA, which leads to a qualitative change in the resulting protein (although in the case of premature stopcodons and CNVs the total number of proteins may also be decreased). An epigenetic change, on the other hand leads to a change in the three-dimensional structure of DNA, leading to a *quantitative* (ie, an increase or decrease) change in mRNA and protein production. In order to understand the mechanisms underlying epigenetics, we will first need to take a closer look at the structure of chromosomes.

THE MOLECULAR STRUCTURE OF CHROMOSOMES

In Chapter 1, we already discussed the discovery of chromosomes and how it contains the DNA molecules. However, chromosomes also contain a number of proteins that are essential for the three-dimensional structure of DNA. As human DNA molecules are composed of between roughly 500,000 and 2.5 million nucleotide pairs, their average length (when fully stretched out) is between 1.7 and 8.5 cm long. Obviously, this would never fit into the nucleus of a human cell (with an average diameter of 6 μm) unless it is tightly coiled up.

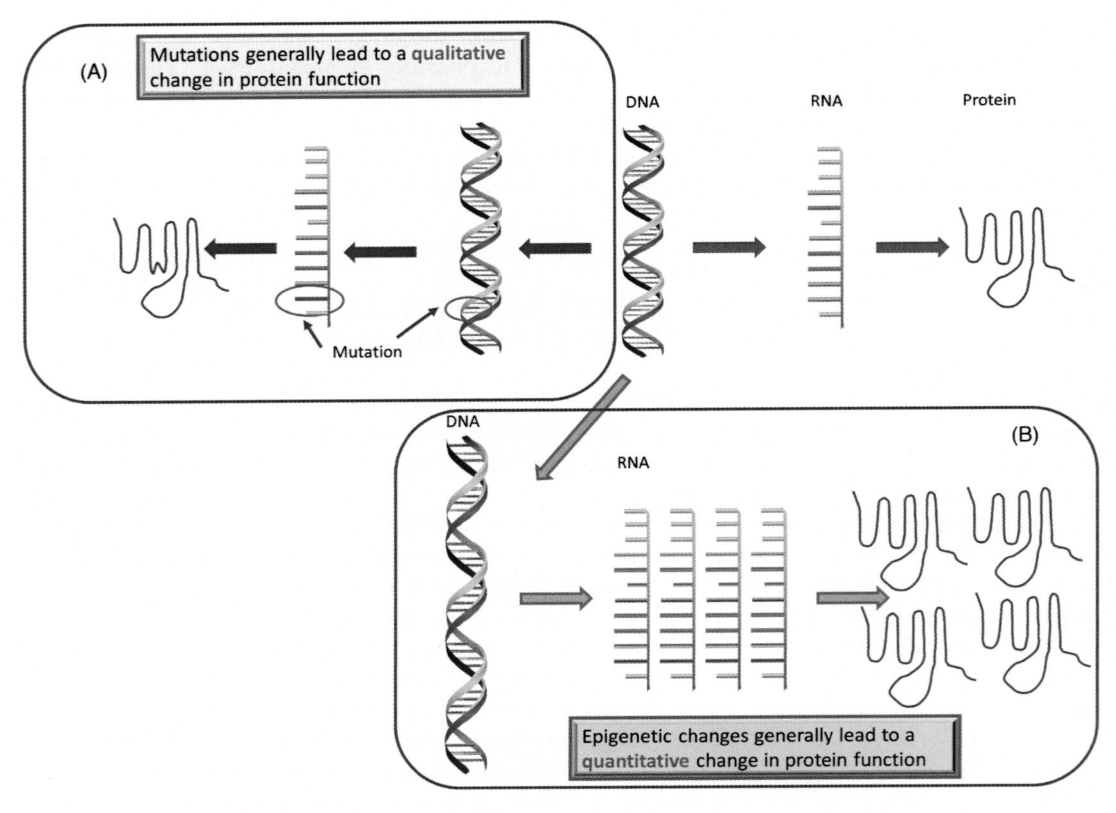

FIGURE 5.5 **Genetics and epigenetics.** (A) A genetic change (mutation) is a change in the nucleotide sequence which generally leads to a structural (*qualitative*) change in the protein function. (B) On the other hand, an epigenetic change does not alter the nucleotide sequence but changes the 3-dimensional structure of the DNA making it more or less accessible for the transcription factors, leading to a *quantitative* change in protein function.

Several different proteins are essential for this coiling process, most importantly histone and scaffolding proteins. Together with DNA these proteins make up the so-called chromatin. Several different levels of coiling can occur, depending on the specific phase within the cell cycle. At the lowest levels, DNA is wound around a cluster of 4 pairs of histones (2 copies of H_{2A}, H_{2B}, H_3 and H_4 each), forming a nucleosome. A stretch of about 146 base-pairs of double stranded DNA are coiled in 1.75 turns in each nucleosome, which is connected to the next nucleosome via a short stretch of spacer DNA. This leads to the so-called "string of beads" structure (also known as euchromatin). At the next level, these beads are tightly packed together using another histone molecule (H_1) into a so-called 30 nm fiber (also known as heterochromatin) which represents a much more tightly packed structure. During the metaphase of the cell cycles, the DNA can even be more tightly packed using specific scaffolding proteins.

From this description it is clear that histone play a prominent role in the packaging of DNA. Given that gene expression requires large molecule (such as transcription factors and

RNA polymerase, see Chapter 2) to have access to the DNA, it becomes understandable that the interaction between DNA and the histones determine the efficacy of gene expression. Thus the more open the chromatin structure is (as in the euchromatin stage) the easier transcription factors have access and the more extensive gene transcription will be, while the reverse holds true for a more closed chromatin structure (as in the heterochromatin stage).

THE HISTONE FAMILY OF PROTEINS

The histone family of proteins are a highly conserved family consisting of 5 different proteins: H_1, H_{2A}, H_{2B}, H_3, and H_4. Histone are globular proteins with short C- and N-terminal tails and are essential for stabilizing the three dimensional structure of the DNA. Unlike the other histone, the H_1 proteins is not part of the nucleosome, but rather binds to the spacer DNA in between the nucleosomes. It is particularly important for stabilizing the 30 nm chromatin fiber structure.

The H_{2A}, H_{2B}, H_3, and H_4 histones are all part of the nucleosome complex, and as indicated in Fig. 5.6 form an octameric (2 copies of each of the four histones) structure around which the double stranded DNA coils. Because of the phosphate backbone, DNA molecules are strongly negatively charged, while the histone are highly positively charged molecules, and thus binding is generally tight between them. As discussed later, modification especially on the N-terminal tail of the histones, can reduce the positive charge of the histones, leading to a reduced binding, thereby opening up the DNA and allowing transcription factors to bind more strongly and enhance gene expression.

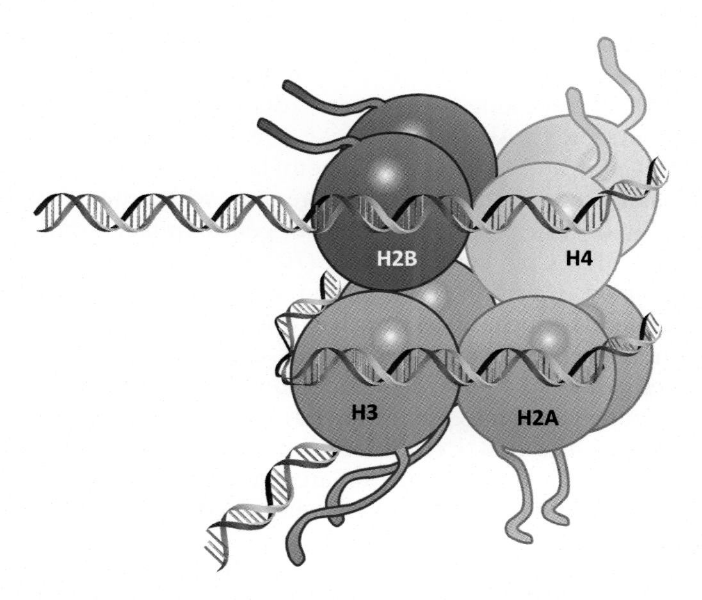

FIGURE 5.6 **The nucleosome.** The nucleosome consists of 4 pairs of histones called H_{2A}, H_{2B}, H_3, and H_4. The DNA is wrapped around this core of 8 histones.

EPIGENETIC MECHANISMS - INTRODUCTION

Epigenetic changes can occur as a result of several different mechanisms:

1. Chemical modifications at the level of the nucleotides.
2. Chemical modifications at the level of the histones.
3. Structural changes at the level of the nucleosome.

These processes involve different enzymes and can alter the accessibility of DNA for the transcriptional machinery in different ways. However, they often interact with each other. For instance chemical modifications at the level of the histones, may lead to modifications at the level of the nucleotides and vice versa. Nonetheless, these processes are very different from each other and in the following sections we will discuss them in more detail.

MODIFICATION AT THE LEVEL OF THE NUCLEOTIDES

Epigenetic changes at the level of the nucleotide include both DNA methylation and RNA interference. The first involves the covalent alteration of the DNA sequences, however, without actually changing the triplet coding or the resulting amino acid sequence. RNA interference involve the production of small RNA molecules that actively interferes with mRNA.

DNA Methylation

Actually, this term is slightly misleading, as only one nucleotide (cytosine) is methylated at the 5- position leading to 5-methylcytosine (5mC, Fig. 5.7A). Although cytosine methylation can occur at different sites within the DNA, it is most common at so-called CpG sites (ie, a two base stretch of DNA consisting of a cytosine adjacent to a guanine on the same strand), although it also occurs on CpA sites. CpG sites are symmetrical (ie, the complementary strand also contains CpG), and in most cases, methylation is symmetrical as well (Fig. 5.7B). CpG sites often cluster in so-called CpG islands (regions with unusual high CpG sequences) and are often found in or near promoter regions of genes. In fact around 70% of the human gene promoters have high CpG content. Most (70%–80%) cytosines in CpG sites are normally always methylated in vertebrates (referred to as constitutive methylation), although most CpGs in CpG islands appear unmethylated. It is important to realize that this chemical modification does not affect the base-pair rule. Thus, a methylated cytosine can still pair with a guanine. However, while spontaneous deamination of cytosine leads to uracil (which is recognized by DNA repair enzymes and removed), deamination of 5-methyl-cytosine leads to thymine which would lead to a SNP. In fact, it has been suggested that this is the reason behind the relative lack of CpG sites. The human genome has a 42% GC content, indicating that statistically $0.21 \times 0.21 = 4.41\%$ of the DNA should consist of CpG sites, while in reality it is only about 1%.

DNA methylation is mediated by enzymes from the DNA methyltransferase (DNMT) family. Several members of these family exist, and all use s-adenosyl methionine (SAM) as the methyl donor. In mammals three active members of the DNMT family have been

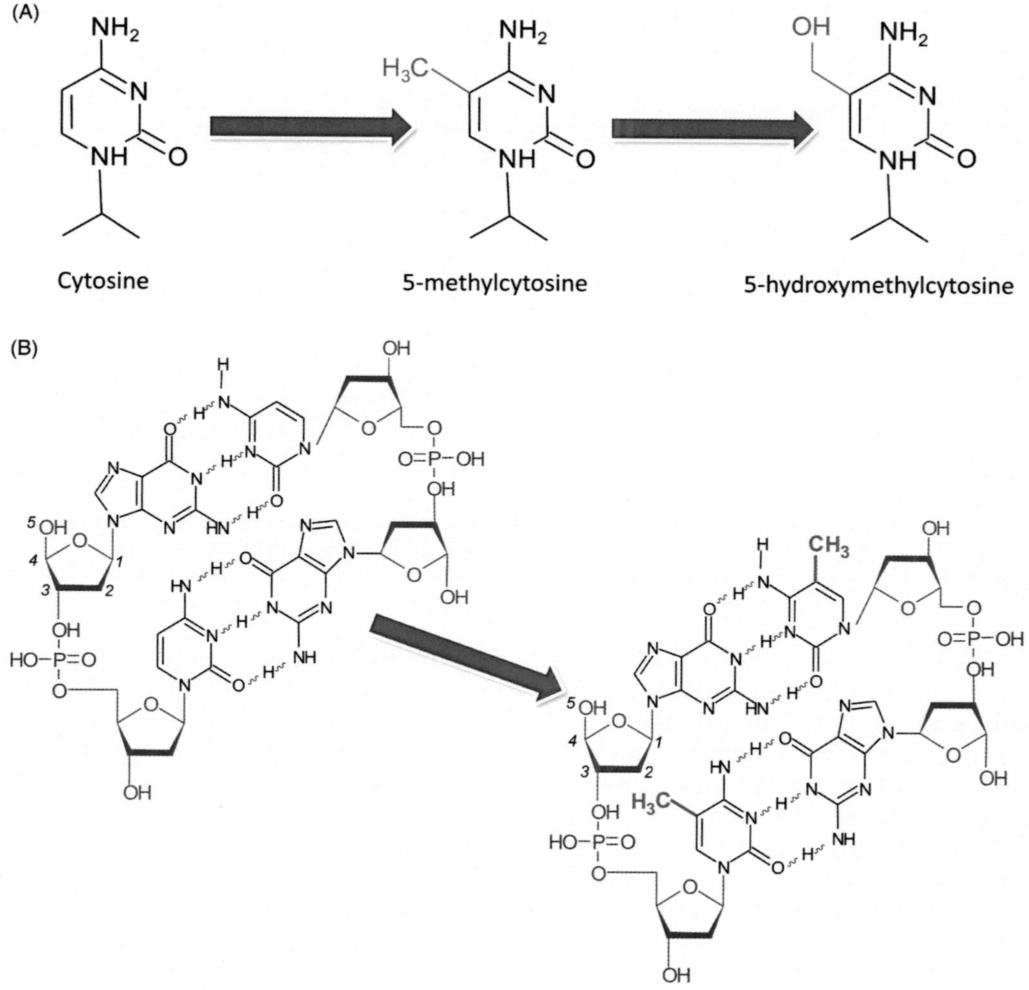

FIGURE 5.7 **DNA methylation.** (A) DNA methylation occurs only at the level of cytosine. Subsequently the 5-methylcytosine can be oxidated to 5-hydroxymethylcytosine. (B) DNA methylation often occurs on so-called CpG sites where the cytosine on both strands are usually methylated.

identified DNMT1, DNMT3a, and DNMT3b. DNMT1 is the most abundant DNA methyltransferase and is primarily involved in ensuring that the constitutive (70%–80%) methylation is faithfully conserved during DNA replication (Fig. 5.8). DNMT1 is highly expressed in dividing cells and recruited to the DNA replication sites by a complex of several proteins. Most important in this respect is UHRF1. UHRF1 binds to the original strand with the methylated cytosine and directs DNMT1 to the new non-methylated strand. The crucial importance of constitutive DNA methylation is indicated by the fact that the genetic ablation of either DNMT1 or UHRF1 induces embryonic lethality (Smith and Meissner, 2013).

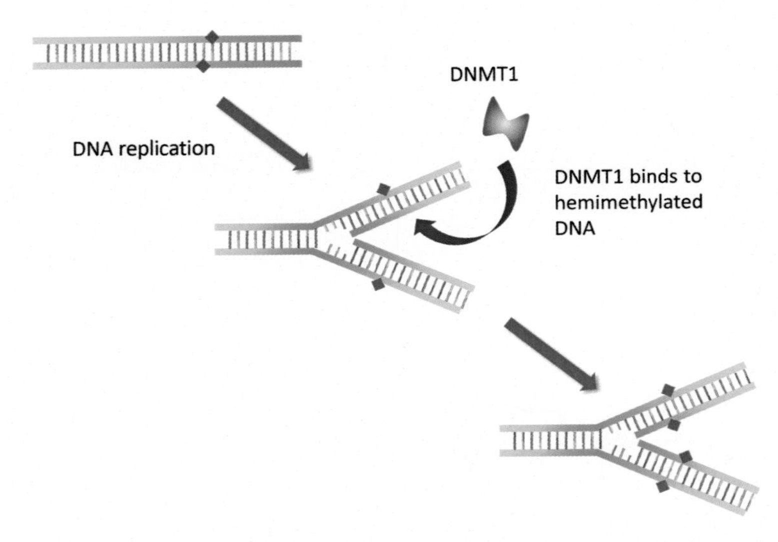

FIGURE 5.8 **Constitutive DNA methylation.** Large stretches of DNA are constitutively (constantly) methylated. This implies that when DNA is duplicated both newly synthesised strands need to be methylated again. This is accomplished by DNA methyltransferase 1 (DNMT1) which binds with high affinity to hemi-methylated CpG sites.

In contrast to this, DNMT3a and DNMT3b are involved in de novo methylation, especially during development (but also in the brain during memory formation or after environmental challenges). Although the expression profiles of DNMT3a and DNMT3b are very similar, their functions are not redundant, as evidenced by the fact that genetic deletion of either DNMT3a or DNMT3b are developmentally lethal (with DNMT3a mice dying at about 4 weeks of age, and DNMT3b mice dying before birth).

DNA methylation in general leads to a more compacted chromatin and thus repression of gene expression, although DNA promoter methylation can also occur in actively transcribed genes. The gene repression can occur through two different mechanisms, illustrated in Fig. 5.9: a direct and an indirect mechanism. First of all, the addition of the methyl groups to the two cytosine on opposite sides of the double stranded CpG sites can cause steric hindrance, thereby preventing specific transcription factors from binding to the promoter region, thus inhibiting gene transcription. The indirect mechanism is based on the binding of proteins that specifically target methylated cytosines. Two families of proteins can bind to methylated DNA: Methyl CpG binding proteins (MBD), consisting of MeCP2, MBD1, MBD2, MBD3 and MBD4 and the Kaiso protein family, consisting of Kaiso, ZBTB4 and ZBTB34 (Bogdanovic and Veenstra, 2009). Of these, most research has focused on the MBD family of proteins and especially on MeCP2 (methyl CpG binding protein 2), since mutations in this protein lead to Rett syndrome, an x-linked neurodevelopmental disorder (Amir et al., 1999).

MeCP2 has several different functions, including transcriptional repression and facilitation depending on the proteins it interacts with (Bogdanovic and Veenstra, 2009). Thus it can bind to co-repressors which has led to the suggestion that MeCP2, especially when it

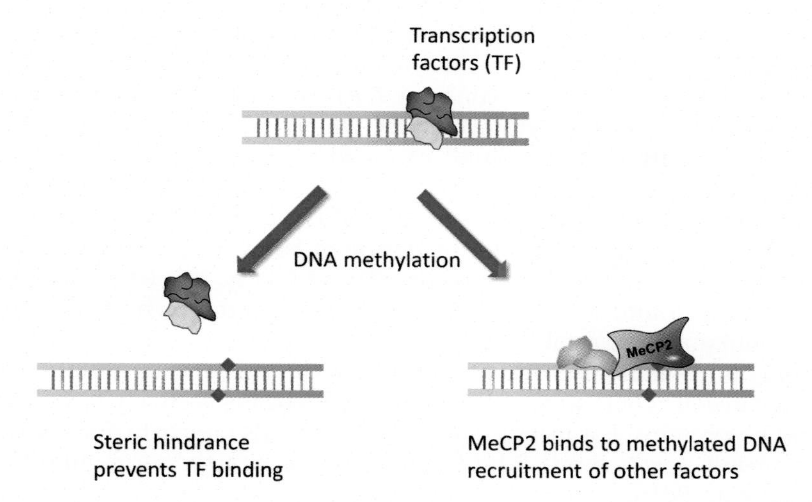

FIGURE 5.9 **The consequences of DNA methylation.** Although DNA methylation can also lead to enhanced gene transcription, it often leads to transcriptional repression. This can occur through two different mechanisms. First, the methylation itself can lead to steric hindrance, preventing the binding of transcription factors to DNA. Second, methyl CpG binding proteins such as MeCP2 (methyl CpG binding protein 2) bind selectively to methylated DNA and can subsequently recruit transcriptional repressors.

binds to DNA promoter regions is a strong transcriptional repressor. However, transcriptional profiling studies in MeCP2 mice brains have shown only subtle changes in gene expression and a study in the hypothalamus actually found both upregulated and downregulated genes in MeCp2 knock-out mice (Chahrour et al., 2008). These data were highly surprising and it was suggested that MeCP2 activates the transcription of CREB1 (c-AMP responsive element binding protein 1), which itself is a strong transcriptional activator. Whatever the underlying mechanisms are, it is now becoming clear that MeCP2 is not the global transcriptional repressor it once was thought to be. In addition to binding to transcriptional repressors, MBD can recruit enzymes involved in histone modifications, especially histone deacetylase (HDAC) enzymes, which leads to further transcriptional repression, as discussed later.

Since DNA methylation is essential for survival, as discussed earlier, and involved in cellular identity, it is important that it is both highly conserved and very stable. It was therefore long thought to be irreversible. On the one hand, this would make it an interesting candidate for long-term changes induced by external stimuli, especially since de novo methylation occurs. However, if it is to play a role in changing the brain's responsiveness after environmental challenges, it is essential that DNA methylation is reversible. Yet the idea that methyl groups can be removed from DNA has been seriously challenged for many years, especially as this process costs a large amount of energy (Szyf, 2015). As brain cells are postmitotic, demethylation can only occur enzymatically. However, recent studies have now shown that DNA methylation in the brain is dynamic and that inhibitors of DNA methylation or histone deacetylation (see below) can lead to demethylation. Moreover, it was recently shown that synchronous activation of the brain leads to wide-spread DNA demethylation (Guo et al., 2011). The question then remains what the mechanism behind this demethylation is. There has been some suggestion that DMB2 may possess demethylase activity, though that has yet to be confirmed. An alternative mechanism involves oxidation of 5mC to 5-hydroxymethylcytosine (5hmC, see Fig. 5.7A) by the so-called ten-eleven translocation (Tet) family proteins. 5hmC can lead to demethylation via several different processes, including passive demethylation (as a result of the poor binding of UHRF1 to 5hmC and subsequent lack of recruitment of DNMT1 during cell division), excision and repair from DNA (as a result of further oxidation of 5hmC to 5-formylcytosine and 5-carboxycytosine or deamination to 5-hydroxyuracil) and finally, at least in vitro, DNMT3a and DNMT3b have been found to possess dehydroxymethylating activity. Whether this also occurs in vivo remains to be established.

Interestingly 5hmC is highly concentrated in the central nervous system being about 10 fold higher than in embryonic stem cells. Moreover, 5hmC is selectively recognized by specific 5-hmC binding proteins. For instance, while MBD3 binds poorly to 5mC, it strongly binds to 5hmC. As a result, next to 5-mC, 5-hmC is now also recognized as an important epigenetic marker (Cheng et al., 2015).

RNA Interference

The elucidation of the human (and other mammalian) genomes has shown that the majority of the genome does not code for proteins. Although this part was long called "junk DNA", we now know that substantial portions of this DNA are transcribed into so-called noncoding RNA. These include transfer and ribosomal RNA as well as a series of RNAs that can interfere with gene expression (see Box 5.2 for a list of the most common non-coding RNAs). This process of RNA interference has received much attention in recent years as an alternative epigenetic mechanism at the level of the nucleotide. Most research, especially in neuroscience, has focused on microRNAs (miRNAs). These are small (22 nucleotide) RNA molecules that are originally synthesised as pri-miRNA by RNA polymerase II (Fig. 5.10). Like pre-mRNA (see Chapter 2), they usually contain a 7-methylguanoside cap on the 5′ end and a poly(A) tail on the 3′ end. However, unlike pre- mRNA, they usually take on a hairpin double stranded form. Within the nucleus, the pri-miRNA is processed by a complex consisting of the Drosha and DGCR8 proteins (in humans) to cleave off the extra parts, leading to the pre-miRNA which is then transported out of the nucleus into the cytoplasm. Here, under the influence of the protein Dicer (together with TRBP), the loop structure is cut off, leaving the mature double stranded miRNA. miRNA is subsequently incorporated in a ribonucleoprotein complex [called RNA induced silencing complex (RISC)], in which a member of the argonaute (AGO) protein family binds to the active (guide) strand of the double stranded RNA, while the non-active (sometimes referred to as passenger) strand is degraded. The guide strands then guide the RISC complex to the target mRNA and generally binds to the 3′-UTR regions. However, while the binding between the 5′ end of the miRNA and the target mRNA needs to be near perfect (referred to as the seed sequence), the binding between the 3′ end of the miRNA and the target mRNA is less stringent. As a result one miRNA can affect a large number of genes. Once the miRNA has hybridized with mRNA, the AGO mediates gene suppression, either by destabilizing the mRNA (though deadenylation, ie, the removal of the poly(A) tail, see Chapter 2) or through a variety of other processes.

BOX 5.2

THE MOST IMPORTANT NONCODING RNAS

Abbreviations	Names	Sizes (*nt)	Functions
RNA involved in protein synthesis			
mRNA	Messenger RNA	2000–5000	Codes for proteins
rRNA	Ribosomal RNA	120–5000	Components of ribosomal complex
tRNA	Transfer RNA	70–90	Connects amino acids to protein chains
RNA involved in DNA replication or modification			
snRNA	Small nuclear RNA	100–300	Splicing and other nuclear functions
snoRNA	Small nucleolar RNA	70	Involved in pre-rRNA processing
RNA involved in gene regulation			
lncRNA	Long noncoding RNA	>200	Regulator of gene transcription
miRNA	microRNA	22	Regulator of gene transcription
piRNA	Piwi associated RNA	26–32	Directs chromatin modification
siRNA	Small interfering RNA	22	Regulator of gene expression

*nt: nucleotides.

A second class of epigenetic regulator RNAs are the so-called long non-coding RNAs (lncRNAs), originally defined as longer than 200 nucleotides and lack of capability to code for more than 100 amino acids. Although this definition is rather arbitrarily, vast number of lncRNAs have been identified, some of which exceed 100,000 nucleotides. Like mRNA and miRNA, lncRNA are also transcribed by RNA polymerase II, and also often have both the 7-methylguanoside cap on the 5′- and the poly(A) tail on the 3′- end. Like miRNA, lncRNA can fold into hairpin structures, but also much more complex three dimensional structures have been described. One of the most intriguing aspects of lncRNAs is that they contain a number of different binding domains, for RNA and DNA, as well as for proteins (Mercer and Mattick, 2013), allowing for a large variety of functions. For instance the lncRNA HOTAIR guides chromatin proteins involved in histone modifications (as discussed later). Likewise, the lncRNA Xist is responsible for the epigenetic silencing of the X-chromosome.

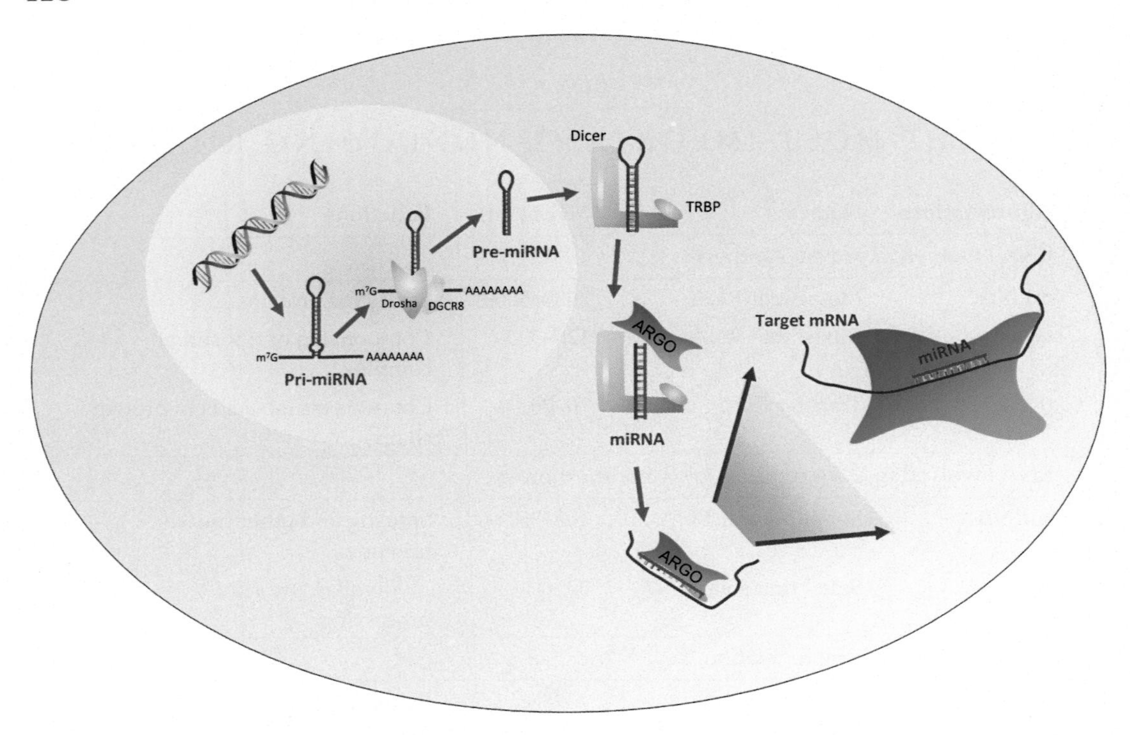

FIGURE 5.10 **The miRNA.** MicroRNAs are small RNA molecules that interfere with gene transcription by binding to specific target sequences of mRNA. Like mRNA, they are synthesized by RNA polymerase II. However, they normally form a hairpin shape. Initially they also contain a poly(A) tail and a m^7G cap, and this form is referred to as pri-miRNA. Under the influence of the enzymes Drosha and DGCR8, the tails are cut off, leaving just the hairpin structure of pre-miRNA. Pre-miRNA is transported out of the nucleus where it is altered in the mature form of (double stranded) miRNA by an enzyme complex consisting of dicer and TRBP. For miRNA to be effective, it has to form a so-called RNA induced silencing complex (RISC) in which a member of the argonaute (ARGO) protein family binds to the active strand of the miRNA (in red). The remaining strand is usually degraded. The active strand then guides the RISC complex to the RNA to allow silencing of the RNA.

CHEMICAL MODIFICATION AT THE LEVEL OF THE HISTONES

In contrast to DNA methylation, which is selective for cytosine, chemical modifications at the level of the histone are much more complex and involve a number of different modifications at different amino acids of the different histones. As discussed earlier (Fig. 5.6), histone are globular proteins with an N-terminal and C-terminal protruding tail. The majority of the histone modifications occur within the N-terminal portion of the histones. However, the chemical modifications can occur at different histones (H_{2A}, H_{2B}, H_3 or H_4), and on different amino acids. All these modifications involve the covalent binding of specific components to specific amino acids.

Histone Acetylation

Histone acetylation involves the covalent addition of an acetyl group to lysine (Fig. 5.11). Because of its $-NH_2$ group, lysine is normally a positively charged amino acid, which binds

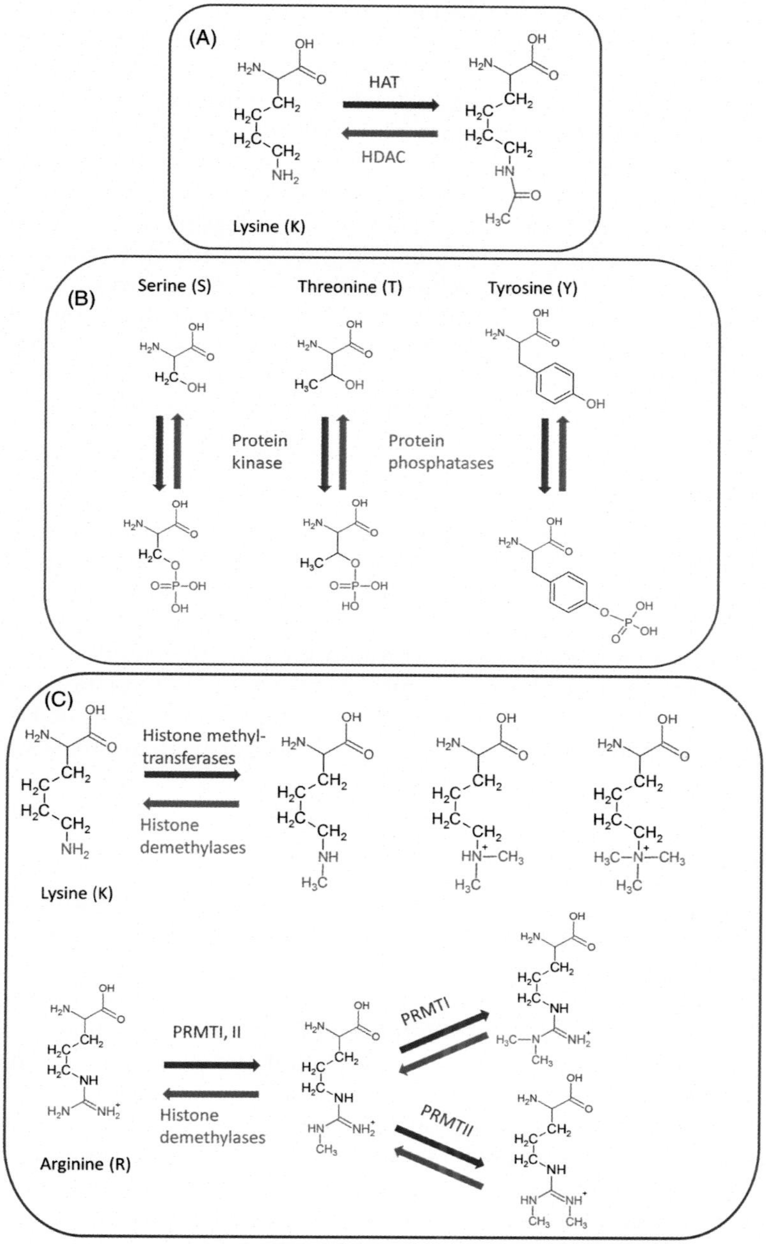

FIGURE 5.11 **Histone modification.** Chemical modifications of the histones can take a number of different form. The most important ones are: (A) Lysine acetylation; (B) Serine, threonine, and tyrosine phosphorylation, and (C) Lysine and arginine methylation, which can take the form of mono-, di-, or trimethylation.

strongly to the negatively charged DNA molecule. The addition of the acetyl group neutralizes this positive charge and hence reduces the binding between histones and DNA, leading to a more open structure which is more accessible to the transcriptional machinery. Histone acetylation therefore leads to transcriptional activation. Histone acetylation is mediated via histone acetyl transferase (HAT) enzymes which are divided in three subfamilies: GNAT, MYST, and p300/CBP. The removal of the acetyl group, on the other hand, is mediated by histone deacetylase enzymes. In mammals, four different HDAC families are known: the zinc dependent classes I, II, and IV and the NAD-dependent class III (which is also known as the sirtuin family). As an example of the interaction between the different epigenetic mechanisms, the above mentioned MeCP2, when binding to 5mC, can recruit HDAC to actively deacetylase lysine molecules, which would restore the original strong binding of positively charged lysine to DNA and reduce transcriptional activity. In addition to this interaction between histone acetylation and DNA methylation, interactions have been found between histone acetylation and histone phosphorylation and histone methylation.

Histone Phosphorylation

Histone phosphorylation can occur on serine, threonine or tyrosine (Fig. 5.11). Like acetylation, phosphorylation is generally associated with transcriptional activation. This is understandable, as the phosphate group added to the amino acids is negatively charged and thus create a repulsive force with the negatively charged phosphates of the DNA backbone. Histone phosphorylation is mediated via protein kinases (which adds phosphate groups) and protein phosphatases (which removes phosphate groups). In relation to the central nervous system, histone phosphorylation may be the most interesting of the histone modification as activation of many neurons leads to activation of protein kinases and/or protein phosphatases.

Histone Methylation

Histone methylation is more complicated than acetylation and phosphorylation, both in terms of the chemical modifications as well as the functional consequences. Methylation can occur on either lysine or arginine. However, lysine can be mono-, bi-, and trimethylated, while arginine can both be mono- or bimethylated (Fig. 5.11). Although in theory arginine can also be trimethylated, this has not (yet) been observed in vivo, however, two different bimethylated arginines occur. In contrast to acetylation and phosphorylation, the functional consequences of methylation are less easy to predict and both transcriptional repression and activation have been reported. This is probably due to the fact that different effector protein (such as transcriptional coactivators) have different affinities for mono-, bi-, and trimethylated histones. Moreover, the exact position of the methylated lysine or arginine within the gene of interest may lead to different effect. For example whereas methylation of a specific lysine on histone 3 may lead to gene repression when the histone is within the coding region, the same methylated lysine leads to enhanced gene transcription if the histone is in the promoter region (Li et al., 2007).

Histone methylation on lysine residues is mediated by histone methyltranferases (HMT, also known as PKMT, protein lysine methyltransferase), using SAM as the methyl donor, whereas PRMT (protein arginine methyltransferase) mediates arginine methylation. There are two different families of PKMT (so-called SET domain containing and non-SET domain containing). Likewise there are two different classes of PRMTs, both classes are involved in mono-methylation, while the PRMTI class is also involved in asymmetric di-methylation. The second class is also involved in symmetrical di-methylation (Fig. 5.11). Overall, though there are over 100 different histone methyltranferases enzymes known. Removal of the methyl groups from lysine and arginine residues also occurs and is mediated by demethylation and demethylination (Klose and Zhang, 2007).

Additional Histone Modifications

In addition to the three most common histone modifications just discussed, several additional covalent alterations have been described, involving the addition of much larger moieties (see also Box 5.3). Thus ubiquination involves the addition of a 76 amino acid protein to lysine residues. Although originally ubiquination was thought to occur only on lysines on H_{2A} and H_{2B} histones, recent studies have found that lysines on H_3 and H_4 can also be ubiquinated (Brunner et al., 2012). Likewise, sumoylation (small ubiquitin-like modifier proteins) involve the addition of about 100 amino acid large proteins to lysine, generally leading to transcriptional repression, both by its interaction with other repressors and because it prevents histone acetylation. In addition, ADP-ribosylation (on lysine and glutamate), glycosylation (on serine and threonine) and biotylination (on lysine) have been described. In recent years several other histone modifications have also been identified, including propionylation, butyrylination and crotonylation, all of which selectively found on lysine residues (Tweedie-Cullen et al., 2012).

BOX 5.3

THE MOST IMPORTANT HISTONE MODIFICATIONS

Modifications	Amino acids involved	Remarks
Acetylation	Lysine	Transcriptional activation
ADP-glycosylation	Lysine, glutamate	
Biotinylation	Lysine	
Crotonylation	Lysine	
Methylation	Lysine, arginine	Can be mono-, di-, or trimethylated
Phosphorylation	Serine, tyrosine, threonine	Transcriptional activation
Sumoylation	Lysine	
Ubiquination	Lysine	

The Complexity of Histone Modifications

From the earlier discussion, it is clear that histone modifications can be extremely complex and the functional consequences are difficult to predict. There are several reasons underlying this complexity:

1. There are two copies of four different histones in each nucleosome (H_{2A}, H_{2B}, H_3, H_4) each of which can be chemically modified.
2. Although most of the modifications are on the N-terminal tail of the histones, additional modifications can be found on the globular portion and the C-terminal part as well.
3. Many different residues on each of the four histones can be modified (Fig. 5.12).
4. Several of these residues can be modified by different processes. For instance the lysine in position 118 on H_{2A} ($H_{2A}K_{118}$) can be acetylated, ubiquinated, mono- and bi-methylated as well as crotonylated (see Box 5.1 and Fig. 5.12).

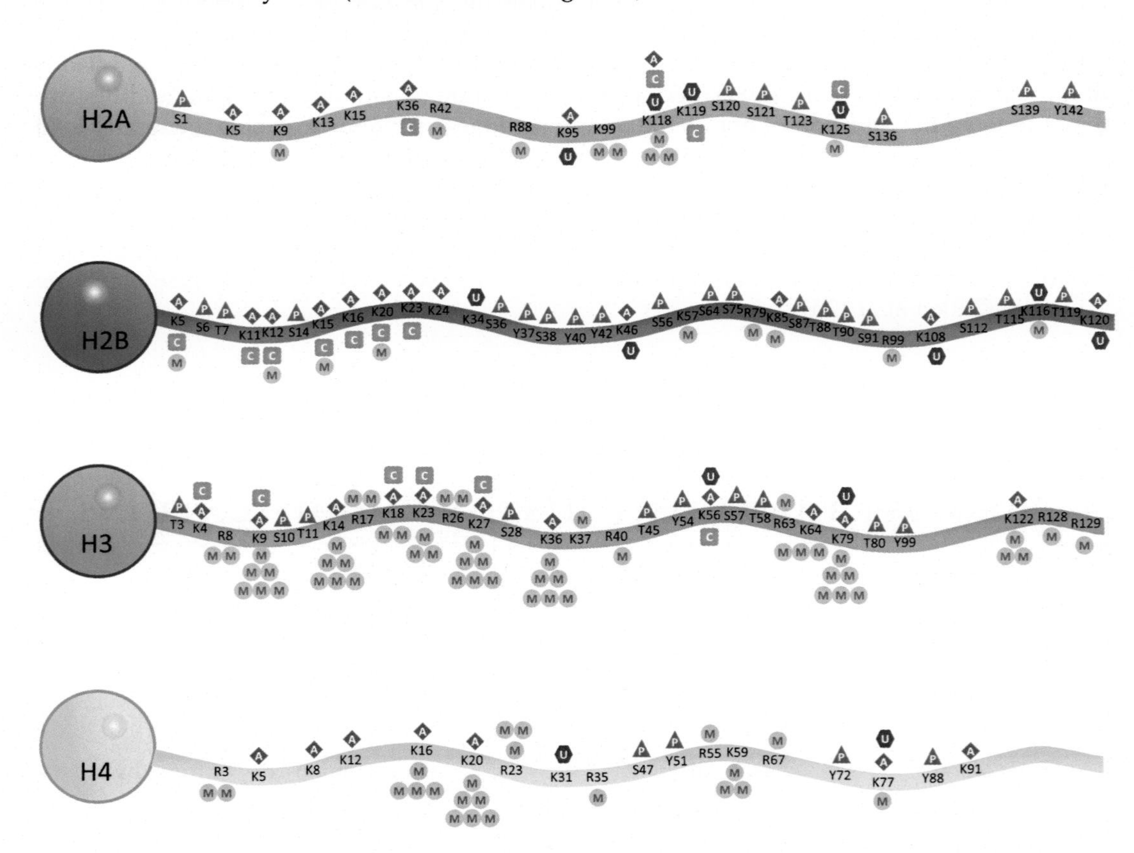

FIGURE 5.12 **Histone modification on the N-terminal regions.** One of the most complex aspects of histone modifications is that many different residues on all 4 histones can be modified. Although modification can also occur on the core protein, most modifications occur on the N-terminal tail, as indicated in this picture.

5. Several interactions have been reported between different histone modifications. For instance ubiquination of H_{2B} appear to be a prerequisite for H_3 methylation, while H_{2A} ubiquination actually inhibits H_3 methylation (Shilatifard, 2006). Likewise, a very detailed analysis of mouse brain histones showed very specific combinations of methylation and acetylation (Tweedie-Cullen et al., 2012). For instance acetylation of H_4K_5 was highly predictive of acetylation of H_4K_8, H4K12 and H_4K_{16}.
6. Multiple interactions between histone modifications and DNA methylation have been described, especially between methyl CpG binding enzymes and HDACs.
7. The effects of histone modifications on gene transcription can depend on where the position of the modified residue (and the corresponding nucleosome) is with respect to the gene. For example, H_3K_9 methylation within the coding region of a gene leads to enhanced transcription while it leads to transcriptional repression if it occurs in the promotor region (Li et al., 2007).
8. Finally, the histone modifications (as well as the chromatin modifications) are in general very regionally specific. For instance, exposure to acute stress followed by unstable housing in rats leads to an increase in DNA methylation of the *bdnf* (brain derived neurotrophic factor) gene in the dorsal CA1 and dentate gyrus, but not in the ventral CA1 and dentate gyrus (Hing et al., 2014). Likewise, in a study on post-mortem prefrontal cortical tissue of patients with schizophrenia, a significant increase in DNMT1 mRNA was found selectively for the GABAergic interneurons in layer 1 (Ruzicka et al., 2007).

STRUCTURAL CHANGES AT THE LEVEL OF THE NUCLEOSOME

The last group of epigenetic changes involves structural alterations at the level of the chromosome. Compared to modifications at the level of DNA and histones, these structural changes (also referred to as chromatic remodeling) have received much less attention, certainly in the field of neuroscience. Chromatin remodeling involves the actual movement of the nucleosome across the DNA. It is governed by large complexes of proteins known as chromatin remodeling complexes, which fall into four different large families depending on their ATPase activity. These complexes hydrolyze ATP and the resulting energy is used to disrupt the bound between the nucleosomes and DNA, allowing for several different processes to take place (Fig. 5.13) including histone variant exchanges, nucleosome sliding, nucleosome eviction or nucleosome exchange. The most interesting chromatin remodeling complex in relation to neuroscience is the BAF complex, which consists of at least 15 different subunits. The interesting aspect of BAF is that in development, the embryonic stem cell BAF complex develops into a neuronal progenitor BAF and subsequently a neuronal BAF complex (Vogel-Ciernia and Wood, 2014). Moreover, mutations in the subunits of the BAF complex have been identified in patients with autism spectrum disorder and schizophrenia. One of the component of the BAF complex (BAF53b) plays an important role in neuronal branching and synapse formation and intriguingly, BAF53b heterozygous knockout mice show significant deficits in long-term, but not short-term, memory (Vogel-Ciernia and Wood, 2014).

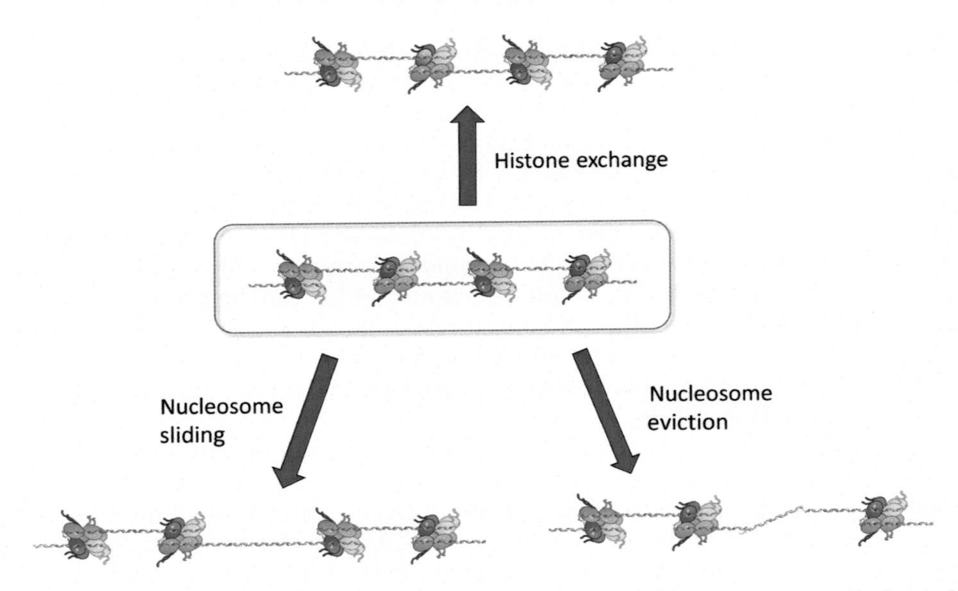

FIGURE 5.13 **Structural nucleosome modifications.** Several structural changes can occur at the level of the nucleosome, including variations in the histones composition of a nucleosome, nucleosome eviction in which an entire histone octamer is removed, or nucleosome sliding, in which the histone core slides around the DNA chain.

EARLY ENVIRONMENTAL CHALLENGES AND EPIGENETIC CHANGES

The previous sections have made it clear that epigenetic alterations can lead to increases and/or decreases in mRNA (and protein) production. Moreover, as many of the epigenetic changes may be very local and long lasting, they fulfil the prerequisites for producing long-term changes in brain activity and function. Therefore, investigations of epigenetic alterations resulting from early environmental challenges have received a lot of attention, and have now provided a wealth of data supporting this hypothesis. Several of these findings we will discuss in the next chapters, but as an illustration, we will mention several intriguing results.

The Environment and Chromatin Modification

DNA methylation is one of the best studied epigenetic changes in neuroscience, and increased DNA methylation has been observed after prenatal and early postnatal stress in rodents. As discussed earlier, prenatal stress leads to long-term changes in the HPA axis, and hence researchers have focused on DNA methylation of its components and found hypomethylation of the CRH promoter region in the hypothalamus after both prenatal restraint stress (Xu et al., 2014) and maternal deprivation (Chen et al., 2012). As a result mRNA CRH levels are higher after either manipulation. Likewise, hypomethylation of the vasopressin promoter was observed in the hypothalamus of maternally deprived mice. In addition, increases in DNA methylation were found in the promotor region of the GR, glutamate decarboxylase 1 and the α-unit of the estrogen receptor in the hippocampus and BDNF in the prefrontal cortex (Lewis and Olive, 2014;

Lutz and Turecki, 2014), underlining the complexity of epigenetic alterations. Interestingly, such alterations have not only been observed after significant stressors, but can even result from variations in normal maternal behavior. Thus it was already well known that the offspring of animals from mothers with low levels of licking, grooming/arched back nursing showed significantly stronger activation of the HPA axis accompanied by lower hippocampal GR levels, compared to the offspring of mother with high levels of maternal behaviour when exposed to a novel environment. In an influential study it was subsequently found that the offspring of mothers which nursed more frequently had decreased levels of DNA methylation of the GR promoter region and thus higher GR levels in the hippocampus (Weaver et al., 2004). Critical for the GR expression in the hippocampus is a particular CpG site to which the transcription factor nerve growth factor inducing gene A (NGFI-A) binds. However, when this CpG is methylated, binding of NGFI-A is reduced and GR expression subsequently reduced. In mothers with a high nursing level, DNA is hypomethylated, allowing NGFI-A to bind and drive GR expression.

Studies in humans have suggested that DNA methylation may also be altered by childhood adversities. Thus increased methylation of the same GR promoter region was found in post mortem hippocampal tissue of victims of suicide that were exposed to childhood maltreatment, compared to suicide victims that were not exposed to childhood maltreatment (McGowan et al., 2009). Subsequently, it was found that alterations in DNA methylation were much more widespread in suicide completers with a history of childhood maltreatment. Indeed 248 hypermethylated and 114 hypomethylated promoter regions were identified (Lutz and Turecki, 2014). In an interesting study, it was recently found that intimate partner violence during pregnancy (but not before or after pregnancy) increased DNA methylation of the GR promoter region of the offspring (but not the mothers themselves). Although this has so far only been found in the blood and needs replication in brain tissue, it does emphasize that prenatal stress can have long-term consequences for the stress responsiveness in the offspring (Radtke et al., 2011).

Up- and downregulation of miRNAs has also been reported in relation to early stressors. Maternal stress in the second half of pregnancy in rats was found to both upregulate (miR-103, miR-323, miR-98, and miR-219) and downregulate (miR145) miRNAs in the brain of the offspring immediately after birth. Since the miRNAs influence multiple other genes, the combined effect of this up- and downregulation on the offspring may be extensive (Zucchi et al., 2013). In a recent study comparing 96 subjects with borderline personality disorder that had suffered childhood maltreatment with 93 subjects with major depression disorder that had not, an association of increased miR-124 expression with childhood maltreatment was found (Prados et al., 2015). Given that one of the targets of miR-124 is the GR (NR3C1), these data are in general agreement with the evidence showing that early stressful life events alter the HPA-axis.

Several drugs, especially drugs of abuse, can also alter miRNA levels in the brain, especially miR-212 and miR-132, which are upregulated in the neostriatum 24 h after extended self-administration of cocaine. miR-212 strongly upregulates CREB, which itself is strongly linked to the rewarding properties of drugs of abuse (Kenny, 2014). Whether this upregulation is specific for cocaine or the general characteristic of drugs of abuse remains to be investigated. Voluntary alcohol intake in rats was found to lead to decreases in miR-124a in the neostriatum and miR-383 in the nucleus accumbens, while miR-206 was upregulated in the prefrontal cortex (Kenny, 2014).

The Environment and Histone Modification

Given the close interaction between DNA methylation and histone modification (as discussed earlier), it is not surprising that environmental challenges also leads to histone modifications. However, histone modifications are generally thought to be more dynamic than DNA methylation, and therefore have been less well investigated. Nonetheless, in the model of high- versus low licking/grooming arched back nursing, where high levels of nursing leads to reduced methylation of the promoter gene of the GR and of several other genes (as discussed earlier), increased levels of H_3K_{9ac} were seen on all these gene promoter regions, emphasizing the inverse relationship between DNA methylation and histone acetylation. As H_3K_{9ac} enhances gene transcription, it is often paralleled by H_3K_{4me3} (which is also an activator of gene transmission) and indeed increased levels of H_3K_{4me3} were found in the GR promoter region as well (Anacker et al., 2014).

Early postnatal stress has also been reported to increase H_3 and H_4 acetylation, the latter of which at the promoter region of the Arc and egr1, genes that are involved in neuronal plasticity (Xie et al., 2013). Likewise, postnatal stress has been linked to deceases in HDAC levels in the prefrontal cortex, hippocampus and forebrain regions (Bagot et al., 2014).

In an interesting recent paper, it was found that the male offspring of sires that were exposed to cocaine self-administration were not only less sensitive to cocaine themselves, but also had increased levels of mRNA BDNF (restricted to exon IV) in the medial prefrontal cortex. This increase was mediated by an increase in H3acetylation in the promoter region of the same exon IV transcript (Vassoler et al., 2013).

The Environment and Chromatin Remodeling

Relatively little research has so far focused on the role chromatin remodeling can play in mediating the long-term effects of environmental challenges. However, it is known to play a role in dendritic arborisation and development, and has been implicated in several neurodevelopmental disorders, such as schizophrenia and autism spectrum disorder (Vogel-Ciernia and Wood, 2014).

References

Alexandre, C., Popa, D., Fabre, V., et al., 2006. Early life blockade of 5-hydroxytryptamine 1A receptors normalizes sleep and depression-like behavior in adult knock-out mice lacking the serotonin transporter. J. Neurosci. 26, 5554–5564.

Amir, R.E., Van den Veyver, I.B., Wan, M., et al., 1999. Rett syndrome is caused by mutations in X-linked MECP2, encoding methyl-CpG-binding protein 2. Nat. Genet. 23, 185–188.

Anacker, C., O'Donnell, K.J., Meaney, M.J., 2014. Early life adversity and the epigenetic programming of hypothalamic-pituitary-adrenal function. Dialogues Clin. Neurosci. 16, 321–333.

Ashdown, H., Dumont, Y., Ng, M., et al., 2006. The role of cytokines in mediating effects of prenatal infection on the fetus: implications for schizophrenia. Mol. Psychiatry 11, 47–55.

Bagot, R.C., Labonte, B., Pena, C.J., et al., 2014. Epigenetic signaling in psychiatric disorders: stress and depression. Dialogues Clin. Neurosci. 16, 281–295.

Baharnoori, M., Brake, W.G., Srivastava, L.K., 2009. Prenatal immune challenge induces developmental changes in the morphology of pyramidal neurons of the prefrontal cortex and hippocampus in rats. Schizophrenia Res. 107, 99–109.

Baier, C.J., Katunar, M.R., Adrover, E., et al., 2012. Gestational restraint stress and the developing dopaminergic system: an overview. Neurotox. Res. 22, 16–32.

Bielas, H., Arck, P., Bruenahl, C.A., et al., 2014. Prenatal stress increases the striatal and hippocampal expression of correlating c-FOS and serotonin transporters in murine offspring. Int. J. Dev. Neurosci. 38, 30–35.

Birn, R.M., Patriat, R., Phillips, M.L., et al., 2014. Childhood maltreatment and combat posttraumatic stress differentially predict fear-related fronto-subcortical connectivity. Depress Anxiety 31, 880–892.

Bogdanovic, O., Veenstra, G.J., 2009. DNA methylation and methyl-CpG binding proteins: developmental requirements and function. Chromosoma 118, 549–565.

Bosch, N.M., Riese, H., Reijneveld, S.A., et al., 2012. Timing matters: long term effects of adversities from prenatal period up to adolescence on adolescents' cortisol stress response. The TRAILS study. Psychoneuroendocrinology 37, 1439–1447.

Bristot Silvestrin, R., Bambini-Junior, V., Galland, F., et al., 2013. Animal model of autism induced by prenatal exposure to valproate: altered glutamate metabolism in the hippocampus. Brain Res. 1495, 52–60.

Brunner, A.M., Tweedie-Cullen, R.Y., Mansuy, I.M., 2012. Epigenetic modifications of the neuroproteome. Proteomics 12, 2404–2420.

Castillo-Melendez, M., Baburamani, A.A., Cabalag, C., et al., 2013. Experimental modelling of the consequences of brief late gestation asphyxia on newborn lamb behavior and brain structure. PLoS One 8, e77377.

Chahrour, M., Jung, S.Y., Shaw, C., et al., 2008. MeCP2, a key contributor to neurological disease, activates and represses transcription. Science 320, 1224–1229.

Charil, A., Laplante, D.P., Vaillancourt, C., et al., 2010. Prenatal stress and brain development. Brain Res. Rev. 65, 56–79.

Chen, J., Evans, A.N., Liu, Y., et al., 2012. Maternal deprivation in rats is associated with corticotrophin-releasing hormone (CRH) promoter hypomethylation and enhances CRH transcriptional responses to stress in adulthood. J. Neuroendocrinol. 24, 1055–1064.

Cheng, Y., Bernstein, A., Chen, D., et al., 2015. 5-Hydroxymethylcytosine: a new player in brain disorders? Exp. Neurol. 268, 3–9.

De Bellis, M.D., Zisk, A., 2014. The biological effects of childhood trauma. Child Adolescent Psychiatric Clin. North America 23, 185.

Dejager, L., Vandevyver, S., Petta, I., et al., 2014. Dominance of the strongest: inflammatory cytokines versus glucocorticoids. Cytokine Growth Factor Rev. 25, 21–33.

Deverman, B.E., Patterson, P.H., 2009. Cytokines CNS Dev. Neuron 64, 61–78.

Edmiston, E.E., Wang, F., Mazure, C.M., et al., 2011. Corticostriatal-limbic gray matter morphology in adolescents with self-reported exposure to childhood maltreatment. Archives Pediatrics Adolescent Med. 165, 1069–1077.

Eiland, L., Romeo, R.D., 2013. Stress and the developing adolescent brain. Neuroscience 249, 162–171.

Elton, A., Tripathi, S.P., Mletzko, T., et al., 2014. Childhood maltreatment is associated with a sex-dependent functional reorganization of a brain inhibitory control network. Hum. Brain Mapping 35, 1654–1667.

Fone, K.C.F., Porkess, M.V., 2008. Behavioural and neurochemical effects of post-weaning social isolation in rodents - relevance to developmental neuropsychiatric disorders. Neurosci. Biobehav. Rev. 32, 1087–1102.

Gaspar, P., Cases, O., Maroteaux, L., 2003. The developmental role of serotonin: news from mouse molecular genetics. Nature Rev. Neurosci. 4, 1002–1012.

Gross, C., Zhuang, X., Stark, K., et al., 2002. Serotonin1A receptor acts during development to establish normal anxiety-like behavior in the adult. Nature 416, 396–400.

Guo, J.U., Ma, D.K., Mo, H., et al., 2011. Neuronal activity modifies the DNA methylation landscape in the adult brain. Nat. Neurosci. 14, 1345–1351.

Harkany, T., Guzman, M., Galve-Roperh, I., et al., 2007. The emerging functions of endocannabinoid signaling during CNS development. Trends Pharmacol. Sci. 28, 83–92.

Herlenius, E., Lagercrantz, H., 2004. Development of neurotransmitter systems during critical periods. Exp. Neurol. 190 (Suppl 1), S8–21.

Hing, B., Gardner, C., Potash, J.B., 2014. Effects of negative stressors on DNA methylation in the brain: implications for mood and anxiety disorders. Am. J. Med. Genet. B 165B, 541–554.

Husum, H., Termeer, E., Mathe, A.A., et al., 2002. Early maternal deprivation alters hippocampal levels of neuropeptide Y and calcitonin-gene related peptide in adult rats. Neuropharmacology 42, 798–806.

Jansson, L.C., Akerman, K.E., 2014. The role of glutamate and its receptors in the proliferation, migration, differentiation and survival of neural progenitor cells. J. Neural Transmission 121, 819–836.

Kenny, P.J., 2014. Epigenetics, microRNA, and addiction. Dialogues Clin. Neurosci. 16, 335–344.
Klose, R.J., Zhang, Y., 2007. Regulation of histone methylation by demethylimination and demethylation. Nature Rev. Mol. Cell Biol. 8, 307–318.
Kovelman, J.A., Scheibel, A.B., 1984. A neurohistological correlate of schizophrenia. Biol. Psychiatry 19, 1601–1621.
Lewis, C.R., Olive, M.F., 2014. Early-life stress interactions with the epigenome: potential mechanisms driving vulnerability toward psychiatric illness. Behav. Pharmacol. 25, 341–351.
Li, B., Carey, M., Workman, J.L., 2007. The role of chromatin during transcription. Cell 128, 707–719.
Liu, D., Diorio, J., Tannenbaum, B., et al., 1997. Maternal care, hippocampal glucocorticoid receptors, and hypothalamic-pituitary-adrenal responses to stress. Science 277, 1659–1662.
Lucchina, L., Depino, A.M., 2014. Altered peripheral and central inflammatory responses in a mouse model of autism. Autism Res. 7, 273–289.
Lupien, S.J., McEwen, B.S., Gunnar, M.R., et al., 2009. Effects of stress throughout the lifespan on the brain, behavior and cognition. Nature Rev. Neurosci. 10, 434–445.
Lutz, P.E., Turecki, G., 2014. DNA methylation and childhood maltreatment: from animal models to human studies. Neurosci. 264, 142–156.
Marco, E.M., Llorente, R., Lopez-Gallardo, M., et al., 2015. The maternal deprivation animal model revisited. Neurosci. Biobehavioral Rev. 51, 151–163.
Marques, A.H., O'Connor, T.G., Roth, C., et al., 2013. The influence of maternal prenatal and early childhood nutrition and maternal prenatal stress on offspring immune system development and neurodevelopmental disorders. Frontiers Neurosci. 7, 120.
Martinez-Tellez, R.I., Hernandez-Torres, E., Gamboa, C., et al., 2009. Prenatal stress alters spine density and dendritic length of nucleus accumbens and hippocampus neurons in rat offspring. Synapse 63, 794–804.
Matsumoto, K., Puia, G., Dong, E., et al., 2007. GABA(A) receptor neurotransmission dysfunction in a mouse model of social isolation-induced stress: possible insights into a non-serotonergic mechanism of action of SSRIs in mood and anxiety disorders. Stress 10, 3–12.
Matthews, K., Dalley, J.W., Matthews, C., et al., 2001. Periodic maternal separation of neonatal rats produces region- and gender-specific effects on biogenic amine content in postmortem adult brain. Synapse 40, 1–10.
McCormick, C.M., Mathews, I.Z., 2007. HPA function in adolescence: role of sex hormones in its regulation and the enduring consequences of exposure to stressors. Pharmacol. Biochem. Behav. 86, 220–233.
McGowan, P.O., Sasaki, A., D'Alessio, A.C., et al., 2009. Epigenetic regulation of the glucocorticoid receptor in human brain associates with childhood abuse. Nat Neurosci. 12, 342–348.
Menon, V., 2013. Developmental pathways to functional brain networks: emerging principles. Trends Cognitive Sci. 17, 627–640.
Mercer, T.R., Mattick, J.S., 2013. Structure and function of long noncoding RNAs in epigenetic regulation. Nat. Struct. Mol. Biol. 20, 300–307.
Miyazaki, K., Narita, N., Narita, M., 2005. Maternal administration of thalidomide or valproic acid causes abnormal serotonergic neurons in the offspring: implication for pathogenesis of autism. Int. J. Dev. Neurosci. 23, 287–297.
Monroy, E., Hernandez-Torres, E., Flores, G., 2010. Maternal separation disrupts dendritic morphology of neurons in prefrontal cortex, hippocampus, and nucleus accumbens in male rat offspring. J. Chem. Neuroanatomy 40, 93–101.
Oitzl, M.S., Workel, J.O., Fluttert, M., et al., 2000. Maternal deprivation affects behavior from youth to senescence: amplification of individual differences in spatial learning and memory in senescent Brown Norway rats. Eur. J. Neurosci. 12, 3771–3780.
Olivier, J.D.A., Blom, T., Arentsen, T., et al., 2011. The age-dependent effects of selective serotonin reuptake inhibitors in humans and rodents: a review. Progress Neuro-Psychopharmacol. Biol. Psychiatry 35, 1400–1408.
Prados, J., Stenz, L., Courtet, P., et al., 2015. Borderline personality disorder and childhood maltreatment: a genome-wide methylation analysis. Genes Brain Behav. 14, 177–188.
Radtke, K.M., Ruf, M., Gunter, H.M., et al., 2011. Transgenerational impact of intimate partner violence on methylation in the promoter of the glucocorticoid receptor. Translational Psychiatry 1, e21.
Rahdar, A., Galvan, A., 2014. The cognitive and neurobiological effects of daily stress in adolescents. Neuroimage 92, 267–273.
Renard, J., Krebs, M.O., Le Pen, G., et al., 2014. Long-term consequences of adolescent cannabinoid exposure in adult psychopathology. Frontiers Neurosci. 8, 361.

Ruzicka, W.B., Zhubi, A., Veldic, M., et al., 2007. Selective epigenetic alteration of layer I GABAergic neurons isolated from prefrontal cortex of schizophrenia patients using laser-assisted microdissection. Mol. Psychiatry 12, 385–397.

Sarkar, S., Craig, M.C., Dell'Acqua, F., et al., 2014. Prenatal stress and limbic-prefrontal white matter microstructure in children aged 6-9 years: a preliminary diffusion tensor imaging study. World J. Biol. Psychiatry 15, 346–352.

Schoenfeld, T.J., Cameron, H.A., 2015. Adult neurogenesis and mental illness. Neuropsychopharmacol. 40, 113–128.

Shilatifard, A., 2006. Chromatin modifications by methylation and ubiquitination: implications in the regulation of gene expression. Ann. Rev. Biochem. 75, 243–269.

Smith, G.N., Thornton, A.E., Lang, D.J., et al., 2014. Cortical morphology and early adverse birth events in men with first-episode psychosis. Psychol. Med., 1–13.

Smith, S.E.P., Li, J., Garbett, K., et al., 2007. Maternal immune activation alters fetal brain development through interleukin-6. J. Neurosci. 27, 10695–10702.

Smith, Z.D., Meissner, A., 2013. DNA methylation: roles in mammalian development. Nature Rev. Genetics 14, 204–220.

Stamatakis, A., Kalpachidou, T., Raftogianni, A., et al., 2015. Rat dams exposed repeatedly to a daily brief separation from the pups exhibit increased maternal behavior, decreased anxiety and altered levels of receptors for estrogens (ERalpha, ERbeta), oxytocin and serotonin (5-HT1A) in their brain. Psychoneuroendocrinology 52, 212–228.

Struber, N., Struber, D., Roth, G., 2014. Impact of early adversity on glucocorticoid regulation and later mental disorders. Neurosci. Biobehav. Rev. 38, 17–37.

Szyf, M., 2015. Epigenetics, a key for unlocking complex CNS disorders? Therapeutic implications. Eur. Neuropsychopharmacol. J. Eur. Coll. Neuropsychopharmacol. 25, 682–702.

Taylor, A.N., Chiappelli, F., Tritt, S.H., et al., 2006. Fetal alcohol syndrome, fetal, alcohol exposure and neuro-endocrine-immune interactions. Clinical Neurosci. Res. 6, 42–51.

Tweedie-Cullen, R.Y., Brunner, A.M., Grossmann, J., et al., 2012. Identification of combinatorial patterns of post-translational modifications on individual histones in the mouse brain. PLoS One 7, e36980.

van der Werff, S.J., Pannekoek, J.N., Veer, I.M., et al., 2013. Resilience to childhood maltreatment is associated with increased resting-state functional connectivity of the salience network with the lingual gyrus. Child Abuse Negl. 37, 1021–1029.

Vassoler, F.M., White, S.L., Schmidt, H.D., et al., 2013. Epigenetic inheritance of a cocaine-resistance phenotype. Nature Neurosci. 16, U42–U67.

Vogel-Ciernia, A., Wood, M.A., 2014. Neuron-specific chromatin remodeling: a missing link in epigenetic mechanisms underlying synaptic plasticity, memory, and intellectual disability disorders. Neuropharmacology 80, 18–27.

Volpe, J.J., 2012. Neonatal encephalopathy: an inadequate term for hypoxic-ischemic encephalopathy. Ann. Neurol. 72, 156–166.

Waddington, C.H., 1942. The epigenotype. Endeavour 1, 18–20.

Weaver, I.C.G., Cervoni, N., Champagne, F.A., et al., 2004. Epigenetic programming by maternal behavior. Nature Neurosci. 7, 847–854.

Weinstock, M., 2007. Gender differences in the effects of prenatal stress on brain development and behavior. Neurochem. Res. 32, 1730–1740.

Workel, J.O., Oitzl, M.S., Fluttert, M., et al., 2001. Differential and age-dependent effects of maternal deprivation on the hypothalamic-pituitary-adrenal axis of brown Norway rats from youth to senescence. J. Neuroendocrinol. 13, 569–580.

Xie, L., Korkmaz, K.S., Braun, K., et al., 2013. Early life stress-induced histone acetylations correlate with activation of the synaptic plasticity genes Arc and Egr1 in the mouse hippocampus. J. Neurochem. 125, 457–464.

Xu, L., Sun, Y., Gao, L., et al., 2014. Prenatal restraint stress is associated with demethylation of corticotrophin releasing hormone (CRH) promoter and enhances CRH transcriptional responses to stress in adolescent rats. Neurochem. Res. 39, 1193–1198.

Zucchi, F.C., Yao, Y., Ward, I.D., et al., 2013. Maternal stress induces epigenetic signatures of psychiatric and neurological diseases in the offspring. PLoS One 8, e56967.

SECTION II

GENE-ENVIRONMENT INTERACTIONS IN PSYCHIATRIC DISORDERS

CHAPTER

6

Drug Addiction

INTRODUCTION

Addictive disorders are among the most common psychiatric disorders, certainly if we include nicotine and alcohol as addictive substances. In addition to pharmacological substances, individuals can become addicted to certain situations such as gambling, internet and others. All these forms of addiction share a number of common elements, especially in relation to behavioural patterns of rigidity as we will discuss further below. However, it has yet to be shown that all addictions have a common neurobiological substrate. If we restrict ourselves to drugs, many different classes of drugs have addictive purposes, including psychostimulants, opioids, benzodiazepines and, as already mentioned, nicotine and alcohol. All these drugs have vastly different pharmacological properties. Moreover, as we will discuss below, certain genetic factors may increase the vulnerability for specific drugs of abuse, but not others. On the other hand, some genetic factors seem to be more commonly related to several different classes of drugs of abuse.

Before discussing the diagnosis of addiction, we will first briefly discuss the major classes of addictive drugs. Fig. 6.1 shows the chemical structure of several of the most commonly used addictive substances.

1. *Psychostimulants*: This class of substances induces short-term increases in mental and/or physical capabilities, leading to an increase in alertness, wakefulness, euphoria and motor activity. The most commonly used psychostimulants are amphetamine, methamphetamine, cocaine, and mephedrone. These drugs all directly enhance the extracellular levels of dopamine by influencing the dopamine transporter. This can occur through two different mechanisms: either by blocking the transporter and thus prohibiting the reuptake of dopamine into the presynaptic terminal, or by reversing the transporter which also leads to a further increase in dopamine release. Although all psychostimulants affect the extracellular concentration of dopamine, it is important to realize that they also affect noradrenaline and serotonin, albeit to a different degree. Fig. 6.2 shows the effects of the major psychostimulants on the inhibition (uptake) and reversal (release) of the catecholamine transporters. The figure clearly shows that whereas (meth)amphetamine have similar potency as inhibitors and reversers, cocaine is much more potent as a reuptake inhibitor. MDMA (3,4-methylenedioxy-methamphetamine), the active ingredient of

Gene-Environment Interactions in Psychiatry. http://dx.doi.org/10.1016/B978-0-12-801657-2.00006-9

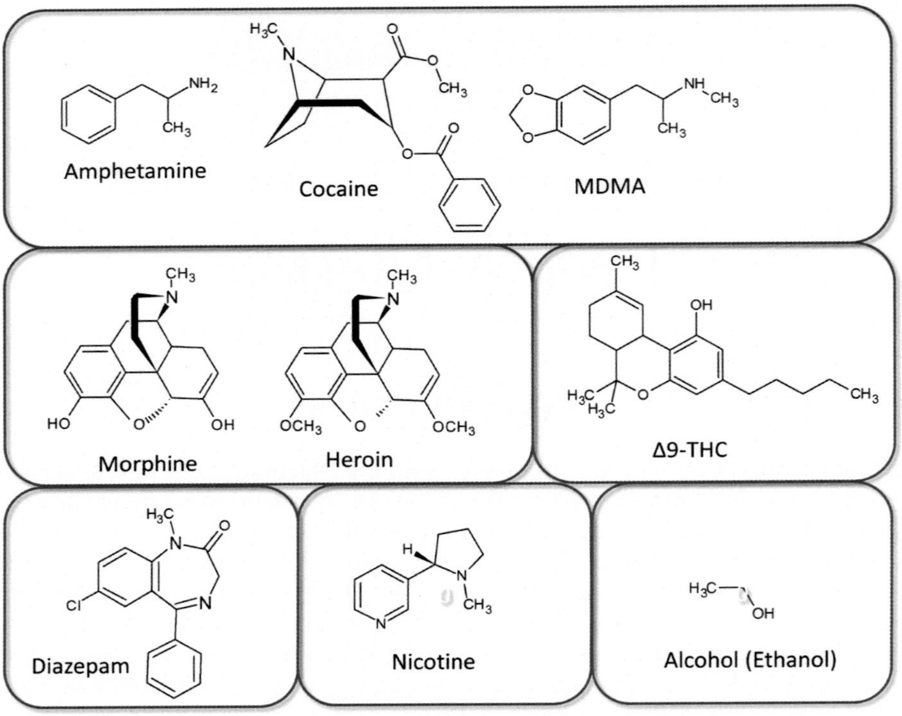

FIGURE 6.1 **Drugs of abuse.** Many drugs can lead to addictive behavior in rats and humans, belonging to a large variety of chemical classes. Based on their primary mode of action, they can be subdivided into psychostimulants (such as amphetamine, cocaine and MDMA), opiates (including morphine and heroin), cannabinoids (Δ9-THC) and benzodiazepines (diazepam). In addition, several individual compounds such as nicotine and alcohol (ethanol) can lead to addictive behavior.

ecstasy, is structurally related to (meth)amphetamine and also increases vigilance and motor activity. However, MDMA is generally not considered to be a psychostimulant as it induces several additional effects, especially an increase in empathy and feelings of closeness. For that reasons MDMA is often referred to as an entactogen. As is evident from Fig. 6.2, MDMA also differs from traditional psychostimulants in its biochemical effects, as it has a much stronger effect on serotonin than the traditional psychostimulants that have a higher affinity for dopamine.

2. *Opiates and opioids*: opiates are substances that are structurally and functionally related to morphine, while opioids share the functional effects with morphine but are structurally unrelated. Opium is the dried latex from the poppy plant (*Papaver somniferum*). It contains about 20 different chemicals, of with morphine (about 12%) is the most abundant one. Opium also contains codeine and thebaine. The opioids include fentanyl, alfentanil and methadone, as well as the endogenous opioid peptides such as the enkephalins and β-endorphin. With respect to drug addiction, the most important substance is undoubtedly heroin, which is a synthetic analogue of morphine (see Fig. 6.1). Opiates and opioids

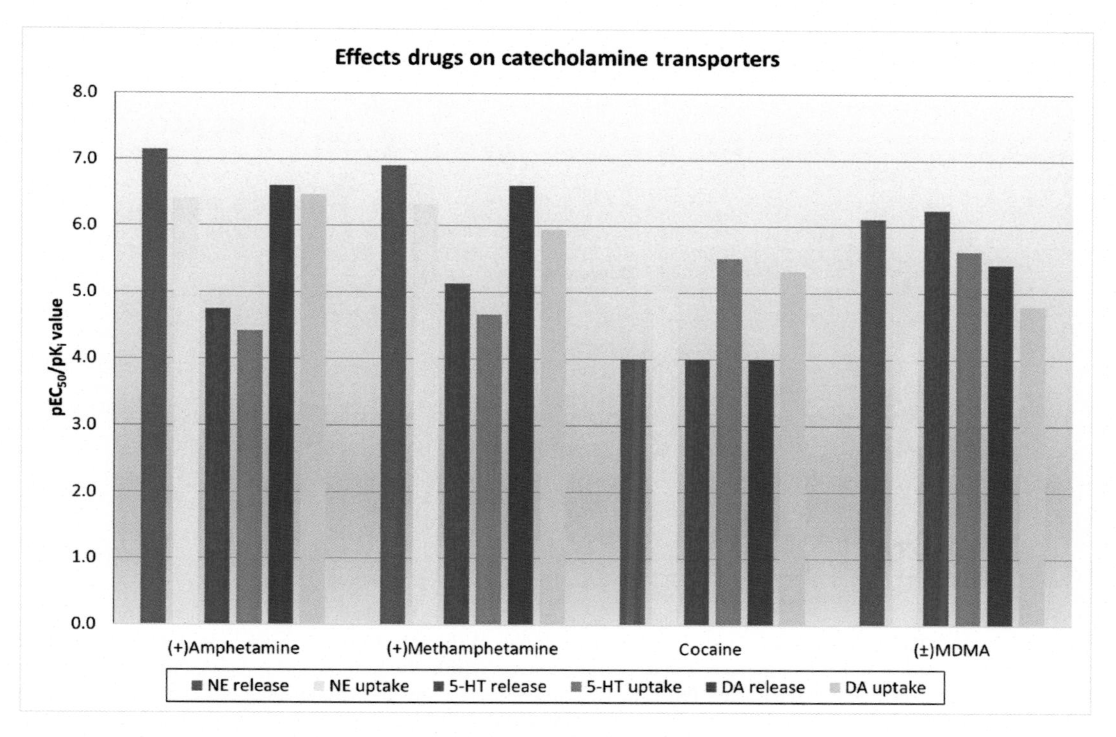

FIGURE 6.2 **Psychostimulants and monoamine transporters.** Psychostimulants primarily act on the monoamine transporters, either by blocking (uptake) or by reversing (release) the transporters. The relative efficacy is given as a negative logarithm (pK_i or pEC_{50}). This implies that a difference of 1 is equal to a factor of 10, a difference of 2 equal to 100 etc. Moreover, as the scale is logarithmic, difference of 1 represents a ten-fold difference, while a difference of 2 represent a hundred-fold difference etc.

act as direct agonist of the μ-opioid receptor, which is also stimulated by the endogenous neuropeptides met- and leu-enkephalin and β-endorphin. The μ-opioid receptors are widely expressed in the brain, and spinal cord, as well as in the peripheral nervous system, explaining the large number of symptoms these drugs can induce, including constipation, analgesia, respiratory depression, sedation and euphoria, the latter of which is likely to be of crucial importance for the addictive properties of opioids and opiates. Like the psychostimulants, opioids and opiates also increase the extracellular concentration on dopamine. However, this is due to the inhibitory effect of μ-opioid receptors on the release of GABA in the ventral tegmental area. This leads to a disinhibition of the dopaminergic cells and hence an increased release in the nucleus accumbens (see Fig. 6.3).

3. *Cannabinoids*: Cannabinoids are the active ingredients of the plant *Cannabis sativa*, a flowering plant indigenous to central and south Asia. Marihuana (or marijuana) and cannabis are interchangeably used for the preparation that is usually smoked. Hash or hashish is another product of cannabis, consisting of compressed or purified resin from the flowers. Although at least 380 different substances have been isolated from the Cannabis plants, it is the family of the cannabinoids that has received most attention,

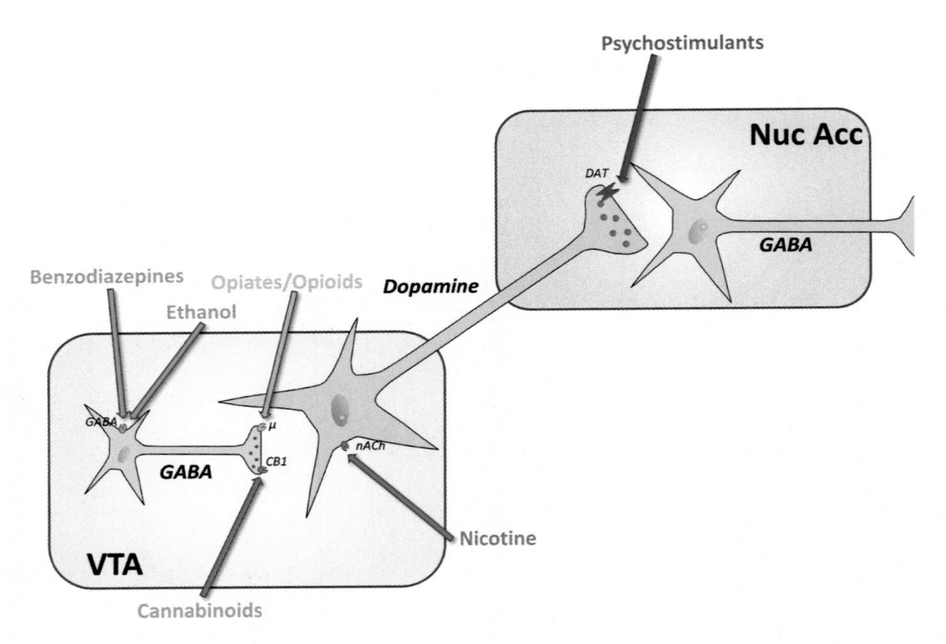

FIGURE 6.3 **The mechanism of action of drugs of abuse.** Although drugs of abuse belong to many different chemical classes, they all acutely increase the dopamine release within the nucleus accumbens (Nuc Acc). However, they do so through different mechanisms. Basically two different mechanisms of action are possible: Psychostimulants increase dopamine release by acting on the dopamine transporter in the terminal region. Since this increase in extracellular dopamine does not affect the dopaminergic cell firing, it is referred to as impulse-independent. Other drugs of abuse act more at the level of the cell bodies in the ventral tegmental area (VTA), increasing cell firing which then subsequently leads to increase extracellular levels of dopamine in the nucleus accumbens (impulse dependent release).

with tetrahydrocannabinol (THC) being the principle psychoactive substance. Cannabinoids act as agonists on the so-called cannabinoid receptors of which two different classes exist: CB_1 and CB_2, with the latter located predominantly in the immune system. The CB_1 receptors are widely distributed within the central nervous system, with high levels in the hippocampus, basal ganglia and cerebellum. The CB_1 and CB_2 receptors are normally stimulated by the so-called endocannabinoids anandamide and 2-arachidonoylglycerol. Interestingly, these endocannabinoids are highly lipophilic substances, and can therefore not be stored in presynaptic vesicles like other neurotransmitters. Instead, these endocannabinoids are synthesized on demand, when the cells are stimulated. CB_1 receptors are normally located presynaptically and primarily act by inhibiting the release of other neurotransmitters. Behaviorally, cannabinoids initially lead to a state of relaxation and euphoria combined with philosophical thinking, metacognition and introspection. However, increased anxiety and paranoia have also been reported. Like opiates and opioids, cannabinoids are thought to enhance dopamine cell firing (and subsequent release) by inhibiting the release of GABA in the ventral tegmental area (Fig. 6.3).

4. *Benzodiazepines:* Benzodiazepines belong to the category of drugs that were first used for therapeutic purposes before its addictive potential became apparent. This class of drugs

include diazepam, chlordiazepoxide and oxazepam. Benzodiazepines have powerful dampening effect on the brain, and as such are effective in the treatment of anxiety, insomnia, muscle spasms and epilepsy. Benzodiazepines act by enhancing the effects of GABA on the $GABA_A$ receptor (ie, they function as so-called positive allosteric modulators. The $GABA_A$ receptor is a tetrameric ion channel consisting of varying combinations of α-, β-, and γ-subunits with $\alpha_2\beta_2\gamma$ being the most common. While GABA itself binds to the interface between the α- and β-subunits, the benzodiazepines binds to the interface between the α- and γ-subunits. However, several different α- subunits exist and benzodiazepines only binds to some of them (most notably α_1, α_2, α_3, and α_5). The binding of benzodiazepines to the $GABA_A$ receptor changes the conformation of the receptor leading to an enhanced affinity for GABA and thus an increase in opening frequency. One important aspect of benzodiazepines is the rapid development of tolerance, leading to ever increasing dosages. As a result, benzodiazepine withdrawal can be dramatic. As with other drugs that induce tolerance, the withdrawal signs are usually the opposite of the original symptoms, and thus can include agitation, anxiety, panic attacks and convulsions. Like the previously mentioned addictive drugs, benzodiazepines can also enhance dopamine release, most likely by inhibiting GABAergic interneurons in the ventral tegmental area, thereby disinhibiting the dopaminergic cells in this region, leading to enhanced dopamine release in the nucleus accumbens.

5. *Nicotine*: Technically not a class of drugs but just a single compound, nicotine is the most important psychoactive ingredient found in species of tobacco plants. More than 70 tobacco plant species have been identified in the genus *Nicotiana*, with *Nicotiana tabacum* being the most important commercial plant. However, although nicotine is the addictive substance in tobacco smoking, many different substances have been isolated from tobacco smoke, some of which appear to enhance the effects of nicotine. Nicotine acts as an agonist on a subset of acetylcholinergic receptors, the so-called nicotinergic receptors. Like the $GABA_A$ receptor, the nicotinergic receptors are pentameric ion channels consisting of different subunits. In the central nervous system, nicotinergic receptors are either homomeric or heteromeric combinations of 12 different subunits (α_2–α_{10} and β_2–β_4) with $(\alpha_4)_3(\beta_2)_2$, $(\alpha_4)_2(\beta_2)_3$, and $(\alpha_7)_5$ being the most important receptors. Behaviorally, nicotine is a powerful parasympathicomimetic leading to a combination of both stimulant and relaxing effect. It increases arousal, alertness and enhances cognitive capacities. In addition, it can reduce body weight by decreasing appetite. However, nicotine also induces a wide range of peripheral effects such as an increase in heart rate and blood pressure. Nicotine also increases the firing rate of the dopaminergic cells within the ventral tegmental area by stimulating the nicotinergic receptors in the ventral tegmental area and therefore also increases extracellular dopamine levels within the nucleus accumbens.
6. *Ethanol*: The most used and abused substance is undoubtedly alcohol, or more accurately ethanol, the addictive substance in alcoholic beverages. In contrast to all the other addictive substances, ethanol has a very complex and to a large degree unknown pharmacology, mainly because it does not specifically bind to a single receptor or transporter. Rather it acts on multiple systems. Like the benzodiazepines, ethanol acts as a positive allosteric modulator of the $GABA_A$ receptor (mainly via an interaction with the δ-subunit). However, in addition ethanol acts as a negative allosteric modulator for several glutamate receptors (the NMDA, AMPA, and kainate receptors). Its actions on the nicotinic and

glycine receptors are more complex, being dependent on the dose, with lower doses being stimulatory and higher doses being inhibitory. As a result of this complex (dose dependent) pharmacodynamic profile, ethanol has a multitude of behavioural effects. At low doses ethanol induces euphoria, increases selfconfidence and sociability, while reducing anxiety, and shortening attention. At higher doses, sedation, ataxia, and memory impairment predominate, while even higher doses can lead to stupor and possibly coma. It is the euphoric effects of ethanol which are likely to contribute most to the initial reinforcing properties, and as with other drugs of abuse, this is likely related to an increase in dopamine release in the nucleus accumbens. Given that, especially in low doses, ethanol stimulates the $GABA_A$ receptors, similar to the benzodiazepines, the enhanced dopamine release is likely also caused by a similar disinhibition of the dopaminergic cells within the nucleus accumbens.

In addition to these major addictive substances, many others have been described, such as the hallucinogens (including drugs such as phencyclidine, ketamine, and LSD), inhalants and less well known substances such as khat and (magic) mushrooms. Although there are obvious differences between the different addictive drugs, we will focus primarily on the similarities, especially in relation to the mechanisms leading to addiction. Several recent studies from the United Kingdom, the Netherlands, and Europe have tried to evaluate the harmful effects of the different drugs of abuse. Although inherently difficult to objectively quantify, these studies asked experts to evaluate both the harm for the patients as well as the harm for others of a large number of different drugs of abuse (Nutt et al., 2010; van Amsterdam and van den Brink, 2013; van Amsterdam et al., 2015). Interestingly, between the three studies there was a strong agreement and in all studies alcohol was by far the most dangerous drug, predominantly because of its high danger for others. With respect to harm for the patients, only heroin, crack cocaine, and methamphetamine were considered more dangerous than alcohol.

DIAGNOSIS AND SYMPTOMS

Like other psychiatric disorders, drug addiction is usually defined on the presence of a number of symptoms from a checklist. One of the most often used diagnostic systems is the Diagnostic and Statistical Manual (DSM), now in its fifth edition (Apa, 2013). In earlier versions of DSM, a distinction was made between substance abuse and substance dependence (with the former being less severe than the latter), but since the distinction between these two is rather subtle, DSM-V only has one diagnosis, referred to as substance use disorders. We will however maintain the term drug abuse or drug addiction as these are still most often use in the literature. Although some researchers have suggested that there are subtle differences between these two terms, we feel that in essence they refer to the same (psychological and biological) process and thus will use them interchangeably.

Box 6.1 shows the main symptoms of substance use disorders. Although DSM-V distinguishes between different classes of drugs and have slightly different criteria for some of them, the overall list is very similar. The only exceptions are the last two criteria which we will discuss below in a bit more detail. An inspection of Box 6.1 shows that the crucial distinction between occasional use and abuse is not necessarily the quantity of drug use (although

BOX 6.1

THE DIAGNOSIS OF SUBSTANCE USE DISORDER (DSM V)

1. The substance is taken in larger amounts and/or for longer than originally intended.
2. There is a persistent desire to take the drug and/or unsuccessful attempts to control intake.
3. A great deal of time is spent in activities necessary to obtain the drug.
4. Craving or a very strong urge to take the drug is present.
5. The recurrent use of the drug results in a failure to fulfil major obligations at work, school or home.
6. Continued use of the drug in spite of having persistent negative or recurrent social or interpersonal problems.
7. Important social, occupational or recreational activities are given up or are reduced.
8. Recurrent use of the drug in situations where it is physically hazardous.
9. Continued use of the drug in spite of knowledge of having persistent or recurring physically of psychological problems likely caused by the drug.
10. The occurrence of tolerance (ie, more and more of the drug is needed to produce the same effect).
11. The occurrence of withdrawal symptoms when the drug is withheld.

it is clear that addiction is in general accompanied by a larger drug intake). It is, in fact, much more the pattern of drug intake that distinguishes the two categories of drug users. Abuse of drugs is thus accompanied by a compulsive pattern on drug seeking and taking, ignoring the obvious health-related problems, but also at the expense of work, school, and social activities. Thus rather than being able to enjoy (like) the drug, there is a craving/desperate need (wanting) of the drug. As a result the drug and everything related to its acquisition becomes the focal point of the drug abuser's life.

This switch from recreational to compulsive drug use is characteristic of all drugs of abuse, and hence criteria 1–9 holds true for all forms of addiction (and very similar criteria also apply to other forms of addiction, such as pathological gambling and internet addiction). What is different between drugs, however, are criteria 10 and 11, relating to tolerance and withdrawal. Not all drugs of abuse induce tolerance, and the degree and type of tolerance differs between different drugs. Thus, hallucinogens such as LSD and phencyclidine hardly induce any tolerance, and whereas psychostimulants induce predominantly cellular tolerance (ie, a downregulation at the cellular level), benzodiazepines and alcohol also induce a strong metabolic tolerance (ie, an upregulation of the enzymes involved in the metabolism of these drugs). In addition to tolerance, certain drugs can also induce sensitization, or reverse tolerance which is, as the term suggests the phenomenon that the same dose of a drug, upon repeated administration induces a stronger effect. Especially in animal research, psychostimulants are known to induce sensitization hyperactivity upon repeated treatment, an effect that has been linked to psychotic symptoms in humans.

Withdrawal symptoms, like tolerance is very much a function of the specific drug or drug class. In general, withdrawal symptoms are the opposite of the acute effects. Thus whereas opioids induce analgesia, hypersomnia, and respiratory depression, upon withdrawal (after chronic treatment), an individual can experience hyperalgesia, insomnia, and hyperventilation. Likewise, whereas cocaine induces strong euphoric feelings, upon withdrawal severe depression can occur. Traditionally withdrawal symptoms were subdivided into physical and psychological, with psychostimulants having more psychological withdrawal symptoms, whereas benzodiazepines and opioids were thought to have withdrawal symptoms belonging to both classes. However, as many psychological effects are accompanied by physical changes as well (such as changes in blood pressure or heart rate) such a distinction may not be that relevant.

THE EPIDEMIOLOGY OF DRUG ADDICTION

The prevalence of drug addiction differs substantially between different (classes) of drugs of abuse and depends to a certain degree on the availability of the drug. Thus, as tobacco and alcohol are much easier to obtain, nicotine addiction, and alcoholism are much more prevalent than for instance heroin or cocaine addiction. However, in addition, other factors play a role too, especially those related to social acceptance. The relatively strong reduction in nicotine addiction is to a large degree the result of a change in social attitude (as well as an increase in the price of tobacco products). Whereas in the past smoking was generally accepted, it is now largely frowned upon and in many countries in the world forbidden in public places and in the work environment.

One of the difficulties in establishing the prevalence of drug addiction is intrinsic to the definition of the disorder. It is customary, in annual drug reports such as those of the United Nations Office on Drugs and Crime (UNODC), to present data on drug use such as number of individuals who have used a specific drug at least once a year, recognizing that there is no standard definition of problematic drug use (or addiction). Fig. 6.4 shows prevalence of illicit drug use based on the 2014 world drug report (Unodc, 2014). According to the same report, the total number of people using who had used illicit drugs was estimated to be 243 million people worldwide, while the number of people with drug addiction was estimated at 27 million. This suggests that about 11% of individuals that use drugs recreationally become addicted. This number is quite similar to previous estimates. An often cited analysis by Anthony and coworkers showed that 9%–11% of cannabis smokers and 7%–12% of alcohol drinkers develop problematic patterns of consumption (Anthony et al., 2005). Interestingly, the figures are higher for cocaine (16%–17%) and especially for tobacco (30%–33%). Whether this implies that these drugs are significantly more addictive remains, however, to be investigated.

Compared to illicit drugs, tobacco, and alcohol use and addiction is a much more significant problem. In a study into the prevalence of alcohol dependence it was found that in the USA (Esser et al., 2014), although 9 out of 10 excessive drinkers did not meet the formal criteria for alcohol dependence, 29.3% of the respondent reported excessive drinking in the past month (defined as 15 or more drinks week or more than 5 during one single occasion for men, 8 respectively 4 for females). This would put the number of people alcohol dependence at about 3%, a figure quite similar to recent figures from a study in Germany (3.9%) and a

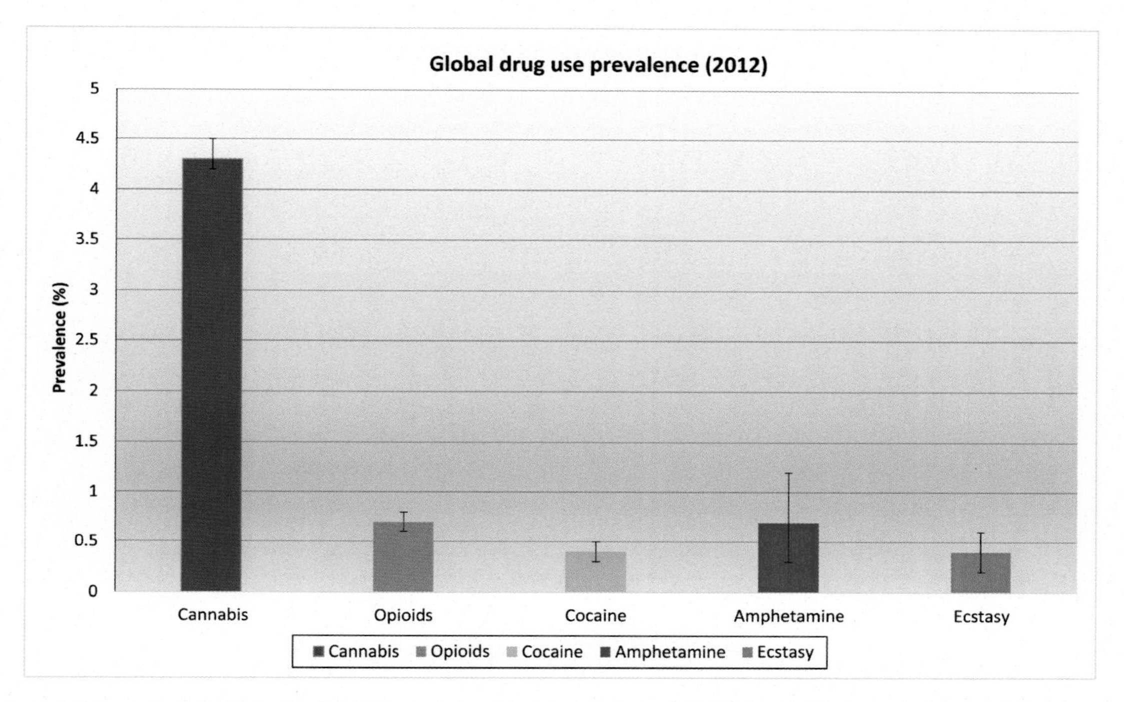

FIGURE 6.4 **Prevalence of drug use.** Given the difficulty in distinguishing between recreational and compulsive drug use, prevalence of drug use is often measured as the number of individuals that have used a drug at least once the previous 12 months. Although the number of individuals that are actually addicted to the drugs will be significantly less, the pattern will likely be similar. Thus, as evident from this picture, cannabis is the most used (and abused) illicit drug, while there are only small differences between the other drugs of abuse. The data are from 2012, obtained from the world drug report 2014 of the United Nations Office of Drug and Crime.

broad European study (3.4%; Wittchen et al., 2011; Effertz and Mann, 2013). The latter study estimated the 12 month prevalence rates for regular smoking to be about 27% for males and 20% for females.

Perhaps more important than the prevalence of addiction is the costs associated with this, both in terms of increased mortality, years of living with the disability, and overall economic costs. Here addiction are among the highest of all brain disorders, certainly when both alcohol, and tobacco are included. A recent study in Scandinavian countries found that individuals hospitalized for alcohol use disorder have, on average, a 24–28 year shorter life expectancy (Westman et al., 2015). However, drug and alcohol addiction also leads to many years of disability. The disability adjusted life years (DALY) parameter is a combination of years lost due to illness and years lived with disability. In the global burden of disease study 2010, the DALY for alcohol use disorders was estimated to be 17.6 million in 2010, and of other drug use at 20 million years (Murray et al., 2012). Together this is higher than any other psychiatric disorder with the exception of major depression and emphasizes the seriousness of drug and alcohol addiction, especially if we realize that tobacco smoking was not included. Moreover, comparing the 2010 data to the results obtained in 1990 showed a 34% increase for alcohol and a 52% increase for drug use disorders.

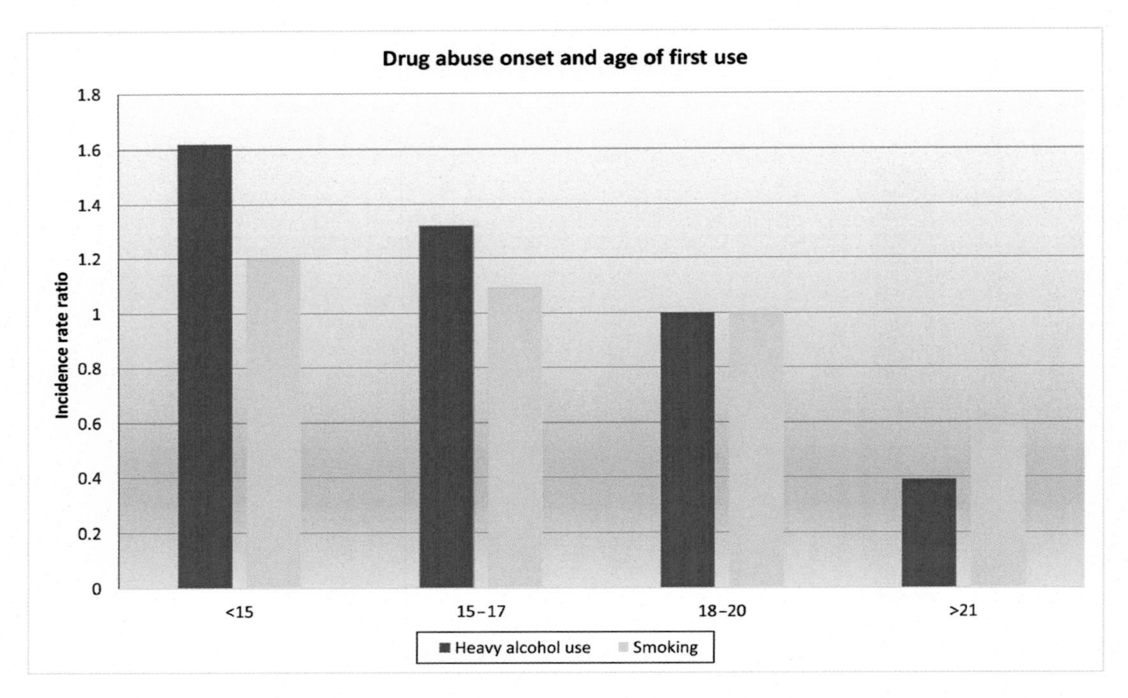

FIGURE 6.5 **Age on onset of drinking and addiction.** There is substantial evidence that the risk of drug and alcohol addiction is related to the age of first use. In a recent study it was found that the earlier children start drinking alcohol the higher the risk of heavy alcohol use later on in life is. The risk of smoking was, likewise, related to the age of first alcohol intake.

Determining the age of onset of drug, alcohol, and tobacco addiction is as difficult as determining the incidence or prevalence, if not more difficult, and therefore most studies again investigate age of first use. In a study using data from 7 international sites (including the United States, Canada, the Netherlands, Mexico, Brazil, and Germany), (first) alcohol use was at age 11 and peaked at about 18 years, after which it sharply declined again (Vega et al., 2002). The curve for cannabis was very similar, but for other drugs of abuse, the age of first use was higher and the curve broader. Age of onset is an important parameter as there is substantial evidence that the severity of drug use disorder is directly correlated with age of onset (Chen et al., 2009). For instance (see Fig. 6.5), in a recent study it was found that age of first drinking was not only related to heavier drinking but also to smoking in the last month (Liang and Chikritzhs, 2015).

THE NEUROBIOLOGY OF DRUG ADDICTION

Identifying the neurobiological substrate underlying psychiatric disorders is already very difficult (given the subtlety of most of the changes in the brain) and addiction is perhaps even more complicated, as we can only ascertain changes in the brain of patients with addiction after onset of the disorder. All drugs of abuse (including alcohol and tobacco) by definition affect brain processes, thus any alteration identified in patient with an abuse disorder can be related to the illness or the consequences of the repeated exposure to the drug. Given that all drugs of abuse enhance the extracellular release of dopamine especially in the nucleus accumbens (Di Chiara and Imperato, 1988), most of the research in drug addiction has also concentrated on dopamine, although other neurotransmitters such as serotonin, and glutamate and neuropeptides such as CRH and the endogenous opioids have also received attention. Like in rats, drugs of abuse increase dopamine release in the neostriatum and nucleus accumbens in humans (Volkow et al., 2009). These studies also showed the importance of pharmacokinetics: the faster the drug altered enters the brain (2–3 min for nicotine 4–6 for cocaine and 10–15 for methamphetamine after i.v. administration) the stronger the "high."

This dopamine increase induced by drugs of abuse such as cocaine is associated with a general inhibition of the nucleus accumbens (Kufahl et al., 2005). This deactivation correlated with the "high" experienced by cocaine abusers, while activation was correlated with craving. Given that dopamine D1 receptors lead to activation while D_2 receptors to inhibition, these data have been taken to indicate a dominance of the D_2 receptor circuitry. Whether similar effects also occur in healthy volunteers is unknown since ethical reason prohibit such studies. However, studies with methylphenidate (which like cocaine enhances the dopamine release by blocking the dopamine transporter) showed the effects on dopamine release are significantly attenuated in cocaine and alcohol abusers compared to healthy controls (Volkow et al., 2012). Although this seems paradoxical, as one would expect the effect of cocaine to increase over time, it is in line with recent studies in animals performed by Barry Everitt and coworkers in Cambridge. These studies showed that whereas acutely all drugs of abuse enhance dopamine release within the nucleus accumbens, with extended exposure, a shift to the dorsomedial and dorsolateral striatum occurs, which coincides with the shift from controlled to uncontrolled selfadministration (Willuhn et al., 2012; Everitt and Robbins, 2013).

In addition to changes in the subcortical areas, brain imaging studies have revealed decreased activity in several prefrontal areas, most notably the anterior cingulate, orbitofrontal, and dorsolateral prefrontal cortex. Impairments in the orbitofrontal (OFC) and anterior cingulate cortex (ACC) are associated with compulsivity and impulsivity, and it has been shown in animals that increased impulsivity (due to reduced striatal D_2 functioning) is predictive of compulsive cocaine use. The dorsolateral prefrontal cortex (dlPFC) plays an important role in executive functioning, and it is therefore tempting to speculate that the reduced activation within this areas in drug addicts is indicative of their loss of cognitive self-control.

One fundamental question that has yet to be answered is which of these changes in the brain might be related to the increased vulnerability to drug addiction. Studies in individuals from families with a history of alcoholism, but who were themselves not addicted, showed that these individuals had higher than normal striatal D_2 receptor availability, and associated with this a normal activation of the prefrontal areas. Thus, it was speculated that this increased D_2 receptor availability in the striatum might represent a protective factor in these individuals.

Most of the studies discussed so far have focused on the rewarding properties of drugs of abuse, and the cognitive control over these. However, enhanced negative emotions (enhanced stress, depression, and anxiety) are also an important factor in drug taking, as exemplified by the extensive comorbidity between addiction and depression, anxiety, and PTSD.

In line with this there is some evidence that drug addiction is accompanied by disturbances in the limbic system, especially the amygdala (Koob and Volkow, 2010; Jasinska et al., 2014; Zorrilla et al., 2014). In line with this, a recent study in subjects suffering from internet gaming disorder found significant deficits in amygdala grey matter, and reduced functional connectivity between the dorsolateral prefrontal cortex and the amygdala and enhanced functional connectivity between the amygdala and the insula (Ko et al., 2015). A structure that has received a lot of attention in recent years in relation to drug addiction is the lateral habenula (LH). Neuroanatomically, this structure is strategically positioned to provide a link between reward and emotion. Cells within the lateral habenula are inhibited by reward predicting stimuli, while stimuli that predicted aversive effects stimulate lateral habenular cells. The lateral habenula does not only project to the dopaminergic cells of the ventral tegmental area (VTA, that innervate the nucleus accumbens), but is also reciprocally connected to the serotonergic cells of the dorsal and ventral raphe nuclei that are crucially involved depression and anxiety (see also Chapter 7). Thus it has been postulated that the lateral habenula may be an important player in withdrawal and relapse (see Fig. 6.6).

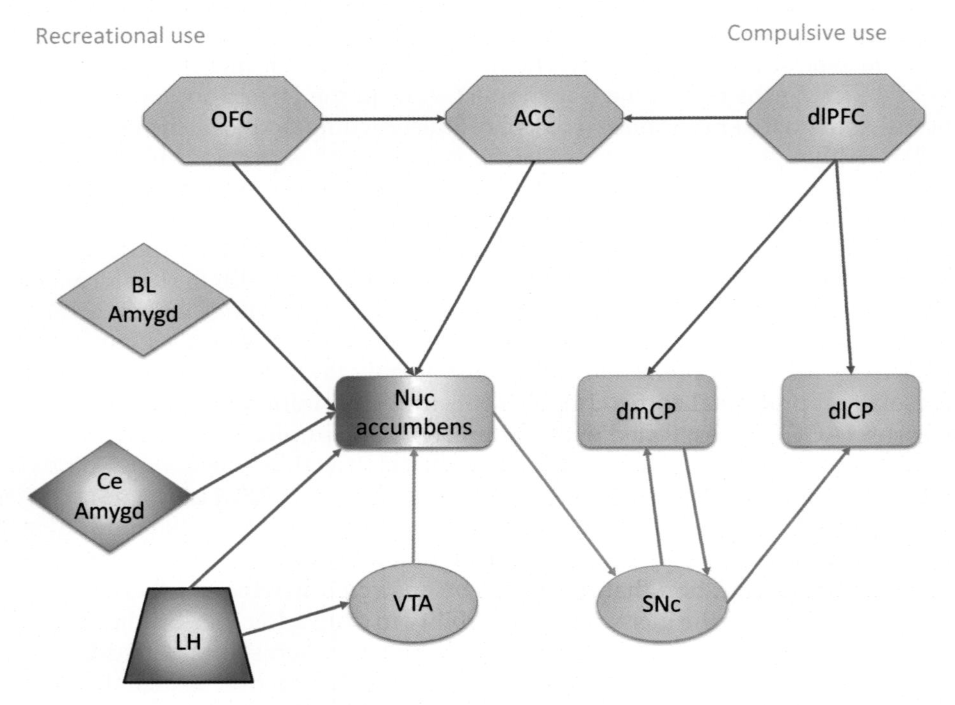

FIGURE 6.6 **The neuroanatomy of drug addiction.** Human and animal literature have identified a number of important cortical and subcortical regions related to drug addiction. These include several frontocortical regions, areas of the basal ganglia [nucleus accumbens (Nuc Accumbens) and neostriatum as well as the cell body regions of the dopaminergic and serotonergic neurons. Furthermore, increasing evidence points to a role of the lateral habenula in addictive behavior. Although most of the structures are likely to play multiple roles, the areas in orange are primarily related to the positive reinforcing effects, while the areas in green are thought to be more involved in the negative withdrawal state. Finally, the areas in blue are considered to be involved primarily in craving and reinstatement. For the other abbreviations see text.

THE PHARMACOLOGICAL TREATMENT OF DRUG ADDICTION

Effective medication for the treatment of addiction is, unfortunately, virtually absent. For some of the addictive drugs, replacement therapy has been developed, which replaces the drug of abuse with another drug of abuse, although the replacement is usually less harmful for the body and hence better tolerated. Nonetheless, most patients remain addicted. In addition, antagonist treatments have been developed for some forms of addiction, and in a few cases such as alcoholism, some alternatives are available but their effectiveness is only limited (vandenBrink, 2012).

Replacement Therapies

The best known replacement therapies have been developed for tobacco smoking. These include nicotine patches, nicotine gum, the e-cigarette, and varenicline. The nicotine patch is a transdermal patch that releases nicotine through the skin. As many of the negative symptoms [especially those related to the enhanced risk for lung (and other forms of) cancer] are due to the inhalation of harmful products through smoking (such as tar), nicotine patches are thought to be much safer. However, there are important pharmacokinetic differences between nicotine patches and smoking with the latter leading to much more rapid and higher concentrations in the blood and brain. Partly because of this, nicotine patches have only limited effect in reducing smoking. Clinical studies have shown that about 7.2% of patients in double blind studies successfully stop smoking compared to about 5.8% on placebo. Nicotine gum and lozenges or tablets are alternatives to the nicotine patch. They have the advantage that somewhat higher blood levels can be achieved but their success rate is not much higher. In addition, nicotine gum (when used in the first 12 weeks of pregnancy) has been linked to increased birth defect in an extensive study in Denmark (Morales-Suarez-Varela et al., 2006). The limitations with the various nicotine replacement strategies especially in relation to the slow release has been the reason for the introduction of the electronic cigarette (e-cigarette). The e-cigarette is an electronic nicotine delivery system often in the shape of a cigarette, which vaporizes nicotine which thus creates a much more realistic cigarette smoking simulation, without the disadvantages that smoking tobacco constitutes (including the above inhalation of tar). Unfortunately the safety of such e-cigarettes as well as its potential benefits for smoking cessation is still not completely clear (Orellana-Barrios et al., 2015).

Varenicline is a partial nicotinergic agonist of the $\alpha_4\beta_2$, $\alpha_3\beta_4$, and $\alpha_6\beta_2$ receptors and thus, compared to nicotine stimulates the receptor, albeit to a lesser degree. Thus the rationale behind its use as a nicotine replacement drug is that it is less addictive than nicotine, yet leads to enough stimulation to reduce craving (Jorenby et al., 2006). Clinical studies have confirmed that varenicline is more effective than placebo and bupropion and also more effective than other nicotine replacement therapies (Mills et al., 2009) although this latter effect was not found in a Cochrane meta-analysis (Cahill et al., 2012). Varenicline's side effect are relatively mild and include nausea and occasionally headaches, difficult in sleeping and abnormal dreams. After varenicline was launched, post marketing studies reported a number of cases of suicidal ideation and even suicidal behavior. As a result, the FDA (Food and Drug Administration in the United States of America) required varenicline to carry a so-called black box warning, even though it acknowledged that similar psychiatric events were seen in the varenicline and the placebo groups. Varenicline has also been thought to lead to increased cardiovascular events (Singh et al., 2011), but more recent reviews have found no evidence for this (Prochaska and Hilton, 2012; Mills et al., 2014). Interestingly, varenicline is a full agonist of the α_7 receptor, which has been implicated in cognition and cognitive deficits in schizophrenia. Thus the drug may have additional benefits, although if the increased risk of suicidal behavior is substantiated it may not be appropriate for the treatment of patients with schizophrenia, as they already have an increased risk of suicidal behavior.

Another well-known example of replacement therapy is methadone for the treatment of opiate (especially heroin) addiction. As with the nicotine replacement therapies, methadone is considered safer than heroin, mainly because it can be taken orally, rather than having to be injected. Moreover, methadone has a long half-life and requires only a single dose (usually 50–100 mg) per day. A Cochrane analysis of almost 2000 patients showed a significant positive effect of methadone on heroin use suppression as measured by selfreport and urine/hair analysis (Mattick et al., 2009). Buprenorphine is a partial opioid agonist (similar to varenicline) and like methadone was also found to be an effective replacement therapy (Mattick et al., 2008). Although both methadone and buprenorphine have been found to be effective as heroine replacement therapies, there is a significant proportion of patients that do not respond to either treatment. For this group of patients, heroin assisted treatment was developed and was found to be effective. However, apart from the risks associated with opioids in general (such as respiratory depression and life-threatening ventricular arrhythmias), heroin assisted therapy also has led to an intense political debate and is currently only available is a few countries in the world.

Other Therapies

Disulfiram was the first drug approved for the treatment of alcohol addiction. It works by inhibiting the enzyme aldehyde dehydrogenase which normally metabolizes acetaldehyde (which is the principle metabolite of alcohol) into acetic acid. As a consequence, disulfiram leads to high levels of acetaldehyde, even after relatively small amounts of alcohol, resulting in a number of unpleasant symptoms such as nausea, vomiting, flushing, sweating and headaches. Clinical studies have consistently found that disulfiram represents an effective intervention for the treatment of alcoholism (Jorgensen et al., 2011), although it can lead to quite serious side effects (such as epileptic seizures, cardiac complications, and neuropathic pain).

From these data it is clear that pharmacotherapy for the treatment of addiction only has limited effectiveness. Although a recent paper showed some progress has been made with respect to smoking cessation, alcoholism, opioid dependence and perhaps pathological gambling (vandenBrink, 2012), the effect sizes are relatively small, and so far very few treatment options are available for cannabis of psychostimulant addiction.

THE AETIOLOGY OF DRUG ADDICTION

There is little doubt that drug addiction, and especially alcoholism "runs in the family," and twin studies have clearly emphasized the genetic contribution to addictive behavior (see Fig. 4.6 in a previous chapter). In a recent review and meta-analysis of 12 twin and 5 adoption studies, the heritability was estimated to be about 50% and the shared environment variance about 10%, leaving another 40% for the variance determined by the unique environment (Verhulst et al., 2015). Studies on other drug addictions showed comparable findings, though there is some evidence that the influence of shared environment is stronger in younger samples and in earlier stages of drug and alcohol use (Lynskey et al., 2010).

Genetic Factors Contributing to Drug Addiction

The question which gene or genes are involved in the development of drug addiction is much more difficult to answer. Many different studies all over the world have investigated this using a variety of approaches, differing in drug of abuse investigated, type of individuals studied (ie, intake in healthy volunteers vs compulsive drug use in substance use patients) and gene(s) investigated (ie, focusing on one or a few candidate genes or genome wide association studies, GWAS). The aim of this section is not to be exhaustive and describe all the different genetic variations that have been associated with drug and alcohol addiction, but to look at the most often described genetic variations.

Table 6.1 gives a summary of the most important genes that have been implicated in substance use disorders. One of the most robust findings in nicotine dependence is the SNV rs16969968 in chromosomal region 15q25, which contains the α_5, α_3, and β_4 subunits of the

TABLE 6.1 Major Genes Implicated in Substance Use Disorder (SUD)

Genes	SUD	Functions
ADH1B, ADH1C, ADH2	Alcohol	The "G" allele of rs1229984 alters the metabolism of ethanol and is protective.
ALDH2	Alcohol	The "A" allele of rs671 blocks acetaldehyde metabolism and is protective.
ANKK1	Alcohol, Nicotine, Opioid	The "T" allele of rs1800497 reduces D2 receptor expression and enhances the risk.
CHRNA3	Nicotine	The "T" allele of rs1051730 enhances the risk.
CHRNA5	Nicotine, Opioid	The "T" allele of rs615470 leads to a reduced sensitivity of the nicotine receptor and is protective.
CHRNA5	Nicotine, Alcohol, Cocaine	The "A" allele of rs16969968 is protective for cocaine but enhances the risk for nicotine and alcohol.
CYB2A6	Nicotine	The risk allele leads to loss of function of the metabolizing enzyme.
GABRA2	Alcohol	The "G" allele of rs279858 enhances the risk.
HTR3B	Alcohol	The "A" allele of rs3758987 enhances the effects of serotonin on the 5-HT_3 receptor and increases the risk.
OPRM1	Alcohol, Opioid	The "G" allele of rs1799971 reduces the binding of endogenous opioids to the μ opioid receptor and leads to increased craving.
SLC6A4	Multiple	The short allele variant reduces the expression of the serotonin transporter by about 50% and enhances the risk of multiple addictive substances.

See the text for further explanations.

nicotine receptors. GWAS have found p values in the order of $<10^{-23}$ and 10^{-72} in combined association studies. The SNV is a nonsynonymous mutation leading to an aspartate to arginine substitution in the α5 subunit (Kendler et al., 2012). Functionally, this substitution leads to a reduced sensitivity of nicotinergic receptors for nicotinergic agonists, leading to less Ca^{2+} permeability and more rapid desensitization. Thus, it can be speculated that the reduced efficacy of the nicotinergic receptors lead to a compensatory increase in smoking. Interestingly, the same SNV has now also been linked to alcohol sensitivity (Joslyn et al., 2008), and opioid dependence (Erlich et al., 2010), but at least one report observed a protective effect on cocaine addiction (Grucza et al., 2008), further emphasizing the complexity of drug addiction.

Another example with a clear biological plausibility are genetic variations in the enzyme alcohol dehydrogenase (ADH), the rate-limiting enzyme in the metabolism of ethanol. There are seven *ADH* coding genes with the most relevant ones located on chromosome 4q22-23 (*ADH1A, ADH1B,* and *ADH1C*). Studies focusing on the *ADH1B* gene have identified three common alleles: the reference gene has an arginine at position 48 and 370 (*ADH1B*1*), while the *ADH1B*2* variant (rs 1229984, common in Asians) has a histidine in position 48 and the *ADH1B*3* variant (common in people of African descent) is characterized by a cysteine at position 180. These variations (compared to the reference allele) lead to a 40%–70% increase in metabolism causing increased levels of acetaldehyde, leading to increased flushing, nausea and headache (similar to what is observed after disulfiram treatment, see above) and hence these SNVs are protective for alcohol dependence. Similar protective effects have also been reported for specific *ADH1C* and *ADH2* alleles (Wang et al., 2012). Interestingly, the protective effects seem to depend on the genetic background. For instance, the protective effects of the ADH1C*1 allele were only seen in people of Asian or African descent. In a similar vein, two polymorphisms in the *CYP2a6* gene have been associated with nicotine addiction, which is biologically plausible as this enzyme is involved in the metabolism of nicotine (Tobacco and Genetics, 2010).

In addition to these, several neurotransmitter-related genetic polymorphisms have also been associated with drug dependence and addiction. The *GABRA2* gene, coding for the α2 subunit of the $GABA_A$ receptor, was identified in the large collaborative study on the genetics of alcoholism (COGA), with the "G" allele of rs279858 enhancing the risk for alcoholism, especially in Native Americans (Long et al., 1998). Likewise, a SNV on the α6 $GABA_A$ receptor subunit (Pro385Ser) has been associated with the sensitivity for alcohol (Goldman et al., 2005). Given GABA's specific role in the addiction circuitry (see Fig. 6.3), alterations in the $GABA_A$ receptor may well influence an individual's response to drugs of abuse such as alcohol.

Substantial research has also looked at the role of alterations in the dopaminergic pathways, given its relevance for the effects of all drugs of abuse. However, the results have so far been mixed (Gorwood et al., 2012). The most convincing evidence points to an association between the so-called TaqA1 polymorphism of the dopamine D_2 receptor and alcohol and opiate dependence (with a modest effect on nicotine dependence). Strangely enough, no convincing relation was found with stimulant dependence. The TaqA1 polymorphism is actually located about 10,000 basepairs downstream of the D_2 receptor in a gene called *ANKK1* (ankyrin repeat and kinase domain containing protein 1). However, the "T" allele of rs1800497 leads to a significant reduction in D_2 receptor expression. Additionally, some (but not all studies) have suggested associations between drug addiction and several other dopamine related genes, such as those encoding for the dopamine transporter, and the enzymes monoamine-oxidase A (MAO_A) and catechol-O-methyltransferase (COMT).

Genetic research has also focused on the role of serotonin in drug addiction. Although some of the findings could not be replicated in all studies, a recent meta-analysis identified a serotonin transporter variant as a genetic marker for addiction across cultures (Cao et al., 2013). This variation, the so-called 5-HTTLPR (5-hydroxytryptamine transporter linked promotor region) contains a 44 base-pair deletion/insertion in the promoter region of the serotonin transporter (SERT). Although it was originally thought that the short (deletion) form of the SERT was associated with a reduced expression, more recent studies have shown that there is actually an SNV in the long promoter leading to two different versions (L_A and L_G) with the (rarer) L_G also showing a reduced expression. Several other serotonin related genes have also been implicated in drug addiction, such as an SNV in the 5-HT_{3B} receptor which was predictive of alcohol dependence (Enoch et al., 2011) and an SNV in the 5-HT_{2A} receptor which was associated with heroin (Saiz et al., 2008) and cocaine (Fernandez-Castillo et al., 2013) dependence. In addition, several studies have identified variants in the μ opioid receptor gene (*OPRM1*) as a risk factor for alcohol and possibly opioid addiction (Wang et al., 2012). The most investigated SNV is rs1799971, which leads to an asparagine (A) to aspartate (G) missense mutation, leading to reduced binding (Buhler et al., 2015).

Relatively few studies have so far focused on the genetics of illicit drug use. A recent genome wide association study on opioid dependence investigated both African-Americans and European-Americans (Gelernter et al., 2014). The authors found 5 associations that met the stringent criteria ($p < 10^{-8}$), the two most interesting ones being in the potassium signaling genes *KCNC1* and *KCNG2*, although both were not found in exons but in the 3′ UTR and intron regions respectively. It is therefore at present unclear what the functional consequences of these SNVs are.

Moderation of the Effects of Genes by Environmental Factors

Twin and adoption studies have clearly shown that the heritability of drug and alcohol addiction is significantly smaller than 1. Although there appear to be some differences between different drugs of abuse, the narrow sense heritability (see Box 4.1 for a description of heritability) h^2 is roughly between 0.4 and 0.7 (Goldman et al., 2005). Moreover, the contribution of the (additive) genetic variance to addictive behavior seems to increase [at the expensive of the shared (familial) environmental influence] with age (Kendler et al., 2012), but even in adulthood, there is a significant influence of the environment (Kendler et al., 2015). Importantly, several studies have provided evidence that the effects of genes are moderated by environmental factors. Some of the most interesting findings are displayed in Table 6.2.

Several papers have investigated the influence of environmental adversities on the nicotinic receptor gene, given its involvement in drug addiction (see Table 6.1). In general these studies found a moderating effect of peer and partner smoking on the risk allele. In an interesting study it was found that peer smoking especially influences individuals that have the low risk ("AG" and "GG") variants of the rs16969968 polymorphism (Johnson et al., 2010). On the other hand, in a study investigating smoking cessation during pregnancy, it was found that individuals with the "AA" genotype were more sensitive to the presence (and absence) of a smoking partner (Chen et al., 2014).

In spite of the fact that the involvement of the 5-HTTLPR in drug and alcohol addiction is more controversial than the polymorphism in the nicotinergic receptors, several groups

TABLE 6.2 Examples of Gene Environment Interactions in Drug and Alcohol Addiction

Genes	Environments	Effects	References
ADH1B	Peer drinking	The protective effect of the "A" allele of rs1229984 is attenuated by peer drinking	Olfson et al. (2014)
ADH1B	Childhood trauma	Carriers of the "G" genotype of rs1229984 had increased severity of alcoholism in the presence of childhood trauma	Meyers et al. (2015)
CHRNA5	Peer smoking	Carriers of the "G" allele of rs16969968 showed a stronger effect of peer smoking	Johnson et al. (2010)
CHRNA5	Partner smoking	Smoking Carriers of the "A" allele of 16969968 showed less smoking reduction during pregnancy when partners smoked	Chen et al. (2014)
COMT	Childhood trauma	Alcohol dependence was highest in carriers of the "A" genotype of rs4680 that were exposed to high levels of trauma	Schellekens et al. (2013)
CRHR1	Childhood sexual abuse	Carriers of the H1 haplotype have higher alcohol consumption if exposed to childhood sexual abuse	Nelson et al. (2010)
GABRA2	Childhood trauma	Carriers of the "G" allele of rs11503014 exposed to with childhood trauma show more cocaine but not heroin or alcohol dependence	Enoch et al. (2010)
SLC6A4	Family conflict	Alcohol misuse was more prominent in carriers of the s-allele of the 5-HTTLPR exposed to higher level of family conflict	Kim et al. (2015)
SLC6A4	Residential stability	Carriers of the s-allele of the 5-HTTLPR had more substance abuse when living in an unstable environment	Windle et al. (2015)
SLC6A4	Childhood trauma	Carriers of the s-allele of the 5-HTTLPR had more alcohol use especially after childhood maltreatment	Kaufman et al. (2007)

have investigated how environmental factors impact of the 5-HTTLPR. Mostly, the findings are consistent with the idea that individuals with the lower serotonin transporter activity (the s-allele) have an increased risk of developing alcohol or drug use disorder when they are (have been) exposed to negative life events, such as childhood maltreatment, family conflict, or when they are reared in an unstable residential environment (Table 6.2).

A recurrent theme throughout this chapter has been a relative lack of data on illicit drugs use. One interesting paper investigated the interaction between childhood trauma and the *GABRA2* and drug use in African-American men (Enoch et al., 2010). The study found that exposure to childhood trauma significantly predicted all forms of drug use (heroin, cocaine and alcohol). In most haplotypes studied, severe childhood adversity increased the risk of drug addiction. The only exception was the 2111111 haplotype. Although the authors suggested that this haplotype was therefore protective against childhood adversity, it may also be due to the relatively low number of individuals with this specific haplotype. More interesting were the results obtained with the rs11502014 polymorphism in the *GABRA2* gene. This polymorphism (located in intron 2) showed a significant gene * environment interaction for cocaine dependence, but not heroin or alcohol, further emphasizing that different drugs of abuse may have different underlying etiologies.

ANIMAL MODELS FOR DRUG ADDICTION

Measuring Addiction-Like Features in Animals

One of the interesting aspects of drug of abuse is that almost invariably, rodents (and other animals) will selfadminister these drugs as well and often will go to great length to obtain access to these drugs. Thus the most often used technique for studying drug addiction in rats is the self-administration paradigm (Fig. 6.7). In this paradigm, rats are usually surgically implanted with a cannula aimed at the jugular vein allowing direct intravenous injections. After recovery, rats are placed in an operant chamber and allowed to press one of two levers. One lever (the inactive lever) has no further consequences, while pressing the other (active) lever will lead to an injection of a drug. Over the course of several days or weeks (depending on the dose and addictive strength of the drug) rats will rapidly increase pressing of the active lever. Often after the initial acquisition period, a maintenance phase follows in which the amount of work to obtain the drug is slowly increased, that is, rats have to press 2, 5, or 10 times to obtain a single injection of the drug.

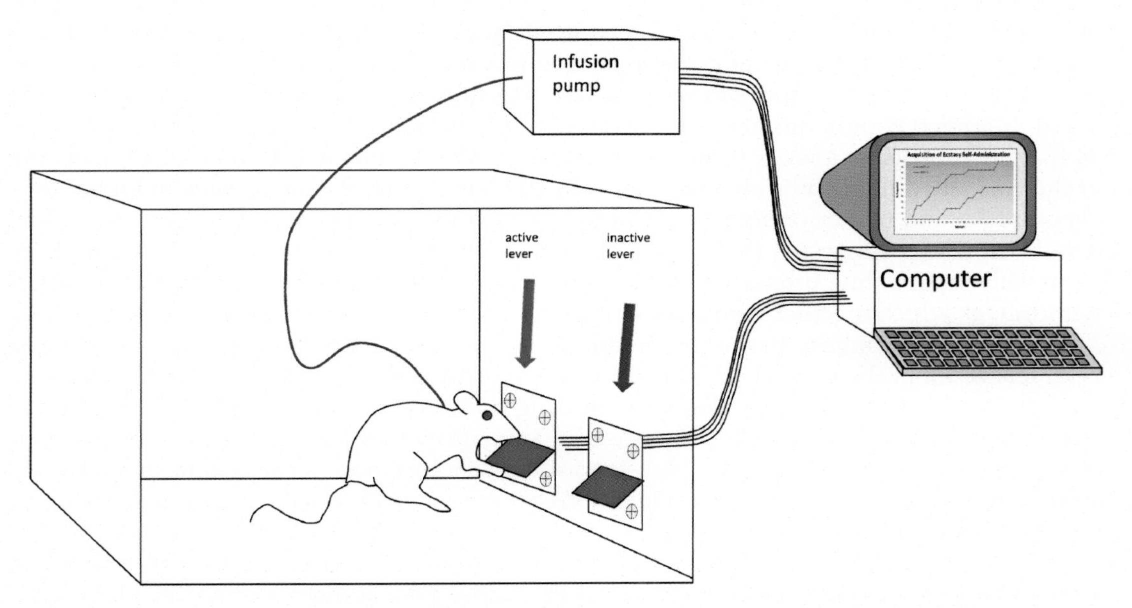

FIGURE 6.7 **The self-administration procedure.** The most often used paradigm to assess the rewarding properties of drugs of abuse in animals is the self-administration paradigm. Although different protocols exist, it often involves rats surgically implanted with a jugular vein catheter. Rats are then placed in an operant chamber and allowed to press two levers, one of which will lead to the administration of a set amount of drugs. In addition to this basic paradigm, several alternative techniques are also often used, which are described in more detail in the text.

To measure the strength of the reinforcing properties of a drug of abuse, a progressive ratio is often applied. During this paradigm, the amount of energy needed to obtain the reward is slowly increased within a single session. Thus, whereas the first injection is obtained after pressing once, the second injection requires 2 lever presses, the third 5 lever presses and so on. Often an exponential increase in lever presses is used (Richardson and Roberts, 1996). The critical dependent variable in this paradigm is the breaking point, that is, the point in time at which the animal stop responding (usually defined as not reaching the next ratio within 60 min). This is an indication of how much effort a rat is willing to invest to obtain the drug and consequently the higher the breaking point the more reinforcing the drug is (or the more reinforcing it is perceived by the animal).

Although most drug self-administration studies have been done with intravenous injections, some attempts have been made to use oral administration. For pharmacokinetics reasons (see also the discussion above about replacement therapy with tobacco patches and methadone) oral self-administration is more difficult, as there is quite a substantial time difference between the lever press and the subjective reinforcing effect of the drug. Often in these procedures rats or mice are first trained to drink or lever press for a sucrose of saccharine solution to which the drug of abuse in slowly added. This procedure has, for instance, been used to selectively breed animals based on their sensitivity for methamphetamine self-administration (Shabani et al., 2012). In this procedure mice were first trained to lever press for saccharine after which methamphetamine 20 and 40 mg/L was slowly added to the solution (fade-in phase). Once stable responding was obtained, the saccharine concentration was slowly reduced to 0 (fade out phase) after which the mice responded purely for methamphetamine. Similar strategies have been used for alcohol intake, although often just oral consumption (rather than lever presses) is used to study intake. An alternative approach to the fading-in/-out technique is the partial access technique (Simms et al., 2008). Rather than given unlimited continuous access to alcohol, in this approach rats are given access for 7 or 24 h a day three days a week (typically on Monday, Wednesday and Friday). A comparison of the two schedules (continuous vs intermittent) showed that the latter leads to higher levels of intake and higher preference. Although the reasons for this have not been studied in detail yet, it was suggested that during the intermittent schedule rats are exposed both to the positive reinforcing properties of the drug, as well as the negative reinforcement of drug withdrawal, which together would be a more powerful driver than exposure to either alone.

Although self-administration is on the surface an excellent model for drug addiction, it has been realized that there are clear limitations as well (Ahmed, 2011). These limitations are related to the problems with the definition of drug addiction as outlined above. Thus, in the standard approach (typically 2 h daily sessions) animals will self-administer a certain number of injections of a specific strength. However, if the concentration per injections is increased, animals usually reduce the number of injections per session to ensure the total amount of drug per session remains relatively constant. In other words, such rats have control over their intake. As a crucial element of drug addiction is the lack of control, this suggests that such short-term sessions may be more related to recreational use rather than addictive behavior. Alternative approaches which show more addictive/compulsive behavior have focused on either prolonging the duration (6 rather than 2 h) of each session (Ahmed, 2011) or the total number of days (3 months rather than two weeks) (Deroche-Gamonet et al., 2004) of self-administration.

An interesting approach has also been developed by Barry Everitt and coworkers in Cambridge. Recognizing that an essential part of drug addiction is compulsive seeking of the drug rather than taking, they developed a second-order schedule of selfadministration, where the animals first have to lever press to get access to a second lever which then give access to cocaine. Using the 6-h paradigm, it was shown that the rats lever pressed for cocaine even when the rats were exposed to a stimulus that was previously paired to a small electric shock (Vanderschuren and Everitt, 2004). In contrast, rats exposed to short (2 h) session of cocaine or 6 h of sucrose selfadministration significantly reduced lever pressing when the stimulus was turned on. This clearly indicates that with extended access to cocaine rats become more compulsive. Interestingly, using these extended schedules only a subpopulation of about 15%–20% of the rats become compulsive, especially when the seeking-taking schedule was modified so that on 50% of the trials pressing the seeking lever led to the cocaine taking lever and on 50% it led to a mild footshock (Pelloux et al., 2007).

Another paradigm that is often used to assess the reinforcing properties of drugs of abuse is the conditioned place preference (Tzschentke, 2007). In this paradigm (see Fig. 6.8) a drug of abuse is repeatedly paired with one compartment of a multicompartment box. After several such pairing, rats show a preference for this compartment, even in a drug-free state. The rationale behind this approach is that the positive reinforcing properties of the drug are associated

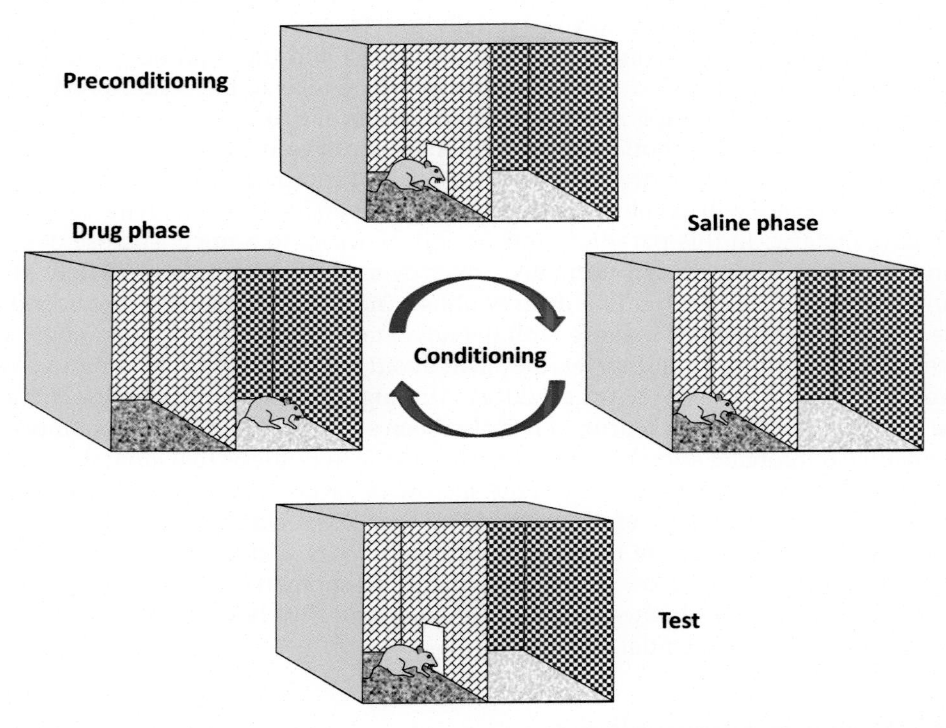

FIGURE 6.8 **The conditioned place preference procedure.** In this procedure animal are first allowed to explore a box with two (or three) compartments. In subsequent sessions animals are injected with either placebo and placed in one compartment or with the drug of abuse and placed in the other compartment (typically one session a day). After several of these pairings the animal is again allowed to freely explore the entire box. An increase in the time spent in the drug-paired compartment is an indication of the reinforcing properties of the drug. By changing the timing of the drug administration, so that it coincides with the negative feelings often seen when the drug wears off, the reverse (conditioned place aversion) can be investigated.

with the contextual cues of the box. Although the procedure appears fairly simple, many procedural aspects can complicate the experiments, such as whether the animal has an inherent preference for one compartment (the so-called biased approach) or whether a two or three compartment box is used. Nonetheless, it is an often used paradigm in drug addiction research.

A very important component of drug addiction is craving and relapse after a drug free period and several different paradigms have been developed in rodents to study this. The most often used paradigm is the reinstatement paradigm (Shaham et al., 2003). In this paradigm, animals initially learn to lever press for a drug and, after stable responding, the administration of the drug is stopped. This generally leads to a rapid decrease in responding (extinction phase). After animals have completely stopped pressing, responding can be reinstated by either giving them a priming injection of the drug of abuse, exposing them to stress, or exposing them to drug-related cues (such as a specific light stimulus that has been associated with drug administration in the past). Although there is some evidence that the neurobiological substrate underlying each of these different forms of reinstatement are different, there is also clear overlap (Shaham et al., 2003).

Finally, several different animal models have been developed to investigate the withdrawal symptoms of drugs of abuse. As these withdrawal symptoms are very drug (or drug-class) specific, the models can vary quite substantially too. However, two interesting models that have been described are the conditioned place aversion and the conditioned taste aversion (or avoidance). The conditioned place aversion is very similar to the conditioned place preference, except that the timing is slightly different. As most drugs, after their initial euphoric reaction often lead to a dysphoric response when the drug wanes off, this can induce feeling of malaise. If rats are repeatedly placed in one specific compartment during this phase, an aversion for this place will develop. Conditioned taste aversion is based on the same principle of inducing malaise. In this paradigm rats are given access to a novel sucrose of saccharin solution (a solution they would normally prefer over plain water). Immediately after this, the animal receives an injection of a drug of abuse, inducing the malaise mentioned earlier. As the rats will associate this malaise with the novel taste, the next day when given a choice between the same solution and water, they will avoid this novel taste, leading to a reduced preference for the novel taste (Lin et al., 2014). Although this model is often used as a model for the negative symptoms of a drug, it has also been suggested that it is in fact more related to the positive reinforcing effects. This theory rets on the hypothesis that animals compare the rewarding value of the drug with that of the sucrose or saccharin and, assuming the rewarding value of the drug is larger, refrain from taking the "less rewarding option"

Finally many drugs of abuse increase locomotor activity and lead to sensitization (ie, an enhanced (motor) response after chronic treatment), phenomena which have been linked to dopaminergic activity within the nucleus accumbens and thus have been thought to be a reflection of the addictive potential.

Genetic Models of Drug Addiction

Substantial animal research has been devoted to delineating the genetic basis for drug addiction, using both forward and reverse genetic approaches (see Chapter 3 for more information on the differences between these two approaches). With respect to the first, many selective breeding approaches have been employed, especially in relation to the psychoactive

TABLE 6.3 Examples of Selective Breeding Lines for the Behavioural Effects of Ethanol

Names	Species	Characteristics
UChB/UChA	Rats	Alcohol intake
AA/NAA	Rats	Alcohol intake and preference
P/NP	Rats	Alcohol intake and preference
sP/sNP	Rats	Alcohol intake and preference
HAD/LAD*	Rats	Alcohol intake and preference
HAP/LAP	Mice	Alcohol preference
HAFT/LAFT	Mice	Functional tolerance
WSP/WSR	Mice	Withdrawal induced seizures
HDID*	Mice	Alcohol intake in the dark
HOT/COLD	Mice	Alcohol induced hypothermia

**: Two different lines have been identified.*
For more details, see Crabbe, 2012; Ciccocioppo, 2013

effects of alcohol. In essence these strategies all employ the same procedure, that is, starting from within-strain individual differences in a behavioural response to ethanol and subsequently breeding high and low sensitive animals. They do differ, however, in the selection criterion, that is, which behavioural characteristic is used to identify high and low responding animals. Thus animals have been selected on the basis of differences in ethanol intake, ethanol functional tolerance, withdrawal seizure sensitivity, ethanol induced hypothermia and other specific characteristics (see Table 6.3). These selection lines have been used for studying the underlying neuronal substrate, but also to identify the underlying genetic factors. In an interesting study comparing several different selection lines, Metten and coworkers found that alcohol preference and withdrawal seem genetically inversely related (Metten et al., 1998). Thus high preferring genotypes were generally less sensitive to the withdrawal effects and vice versa. It is currently unknown what the underlying mechanisms is. Several groups have also used quantitative trait loci (QTL) analyses to identify genetic loci for alcohol intake. Using only analyses of mice derived from C57BL/6J and DBA/6J strains (which are on the top and bottom of voluntary alcohol intake), strong support was found for QTLs on mouse chromosomes 2,3, 4, and 9 (Belknap et al., 2001). Since this study several others OTLs studies have been performed, adding additional support especially to chromosome 2 and 9. Additional support for these two chromosomes was also provided by consomic strain analysis (see Chapter 3 for an explanation of consomic or chromosomal substitution strains). Substituting chromosome 2 or 9 of the (alcohol preferring) C57BL/6J mice with the same chromosome of the (alcohol nonpreferring) DBA/6J mice led to a significant reduction in alcohol drinking. Similar results, especially in relation to chromosome 2 have also been reported by others. With the reduction in costs for gene sequencing and the advent of microarrays which allow the simultaneous analysis of thousands of genes, the search is now on for the gene (or more likely genes) that underlie the difference in alcohol intake. Although this search is far from over, the syntaxin binding protein I (*Stxbp1*) gene on chromosome 2 and the sodium channel type IV (*Scn4b*) on chromosome 9 appear to be interesting candidates (Crabbe et al., 2010).

Relatively few attempts have been made to develop selection lines for other drugs of abuse, presumably because most drugs are best administered through intravenous injections, making selection of hundreds of animals an arduous and labor intensive task. As mentioned above, Tamara Phillips and coworkers using the oral saccharine fade-in-fade-out approach developed selection lines that differ significantly in methamphetamine intake (Shabani et al., 2012). In a recent genetic analysis of these mice, a QTL on chromosome 10 appeared to account for 50% of the variance between the two lines. Further microarray analyses identified altered regulation of a large number of genes, enriched in transcription factor gene. Interestingly, one of the most important genes that was found to occupy a key position in this transcription factor network was the μ opiate receptor gene (*Oprm1*).

An alternative approach to selecting animals for drug intake, is to use existing strains (like the C57BL/6J vs DBA/6J strain comparisons mentioned above) or compare selection lines that have been originally selected for other purposes. Given the prominent role of dopamine in drug addiction, rats that were originally selected for differences in the dopamine agonist apomorphine-induced stereotypy were tested for alcohol and cocaine self-administration. Not surprisingly, there were clear differences between the susceptible (APO-SUS) and unsusceptible (APO-UNSUS) rats. However, more intriguing was the finding that the differences were dependent on environmental factors, both early in life as well as directly during the test. Thus, while under baseline conditions APO-UNSUS rats consumed more alcohol and administered more cocaine than APO-SUS rats, the reverse was found after an acute challenge (van der Kam et al., 2005b; van der Kam et al., 2005a). Moreover, early maternal deprivation on postnatal day 9 significantly reduced cocaine intake in APO-UNSUS rats under nonstressed conditions, while prenatal cocaine administration increased cocaine intake (Ellenbroek et al., 2005).

So far we have described forward genetic approaches to drug addiction, but a large body of research has also made used of genetically altered mice and rats, to identify the contribution of specific genes to drug addiction. A recent review summarized the vast literature on opioid receptor and peptide gene knock out mice in drug addiction research (Charbogne et al., 2014). As self-administration is technically challenging in mice, most studies have focused on conditioned place preference as a measure of the rewarding properties of drugs of abuse. Although there are differences between individual groups and studies, most have confirmed the important role of the μ opioid receptor, not only in relation to morphine and heroin, but also for cannabinoids, cocaine and alcohol. In addition to the μ opioid receptor, the data of knock out mice also support a role for the κ-opioid receptor, while the involvement of the δ-opioid receptor is less evident.

Given the extensive evidence linking the nicotinic receptor to drug addiction (see Table 6.1), researchers have also investigated the functional consequences of genetic alterations in different nicotinic receptor (subunits) in mice on drugs of abuse. Genetic ablation of the gene for the α_5 subunit of the nicotine receptor was found to significantly increase nicotine selfadministration (Fowler et al., 2011) and conditioned place preference (Jackson et al., 2010). In addition to the α_5 gene, there are 11 additional genes encoding for neuronal nicotinergic subunits (α_2–α_{10} and β_2–β_4) and several others have also been implicated in drug addiction, such as the α_4, α_6 and β_2. The role of the α_7 subunit is more controversial. Although genetic deletion of the α_7 subunit did not reduce nicotine induced conditioned place preference (Walters et al., 2006), a recent study actually found an increase in conditioned place preference while a gain-of-function mutation in the *Chrna7* gene found led to a significant reduction (Harenza et al., 2014).

Finally, reverse genetic research has also implicated serotonergic genes in the reinforcing properties of drug addiction. For instance, serotonin transporter knock out rats showed a significant increase in cocaine self-administration and conditioned place preference (Homberg et al., 2008). An interesting species difference was observed in the effects of MDMA: whereas serotonin transporter knockout mice show a significant reduction in self-administration (Trigo et al., 2007), a significant enhancement was found in rats with a genetic ablation of the serotonin transporter (Oakly et al., 2014). It has been suggested that this difference is more related to species differences in the actions of MDMA than in the role of the serotonin transporter in drug addiction in general (Easton and Marsden, 2006). Mice lacking the 5-HT1B receptor show attenuated cocaine induced place preference while at the same time showing increased self-administration (which was interpreted as a reduction in the rewarding properties; Miszkiel et al., 2011).

In summary, quite a substantial number of studies have investigated the genetic basis of the reinforcing properties of drugs of abuse in animals. An important caveat is, however, that most studies focused on conditioned place preference and only few really looked at self-administration. Moreover, even the studies that did investigate self-administration almost invariably looked at relatively short duration. As discussed above, this generally leads to reasonably controlled drug intake, rather than compulsive drug taking. In this respect, the recently developed HDID mice (see Table 6.3) is characterized by binge drinking until intoxication (Crabbe et al., 2010) and might represent an interesting model more closely resembling the human addictive behavior profile.

Moderation of the Effects of Genes by Environmental Factors

In contrast to clinical studies, very few animal studies have looked at gene–environment interactions in relation to drug addiction. Although there is substantial evidence that early environmental challenges can affect the rewarding properties of drugs of abuse, very few studies have combined these challenges with genetic factors. Several groups have shown that early maternal separation (in the first two postnatal weeks) enhances the intake of alcohol, morphine and cocaine, although the parameters of the separation paradigm are critical (Nylander and Roman, 2013). Similar results have been found in the offspring of prenatally stressed pregnant rats, although these animals appeared not to differ in alcohol intake (Van Waes et al., 2011). Finally, several studies have shown that stress during early life can increase cocaine selfadministration in adulthood (Baarendse et al., 2014; Burke and Miczek, 2015).

One of the studies that tried to investigate how genetic factors moderate the effects of early environmental challenges was performed in Rhesus monkeys, making use of a naturally occurring insert/deletion in the serotonin transporter promoter, similar to that seen in humans (although in Rhesus monkeys it is a 21 base pair deletion/insertion). Compared to normal mother-reared monkeys, peer-reared animals show higher alcohol intake, an effect only seen in animals with the short allele of the rh5-HTTLPR (Barr, 2013). A similar moderating effect on alcohol consumption was found in monkeys with a loss-of-function mutation in the neuropeptide Y system as well as in animals with a polymorphism in the promoter region of the *CRH* gene.

Above, we already mentioned the effects of maternal deprivation on cocaine intake in APO-UNSUS rats. A similar approach investigating alcohol intake found, however, no effect in APO-SUS or APO-UNSUS rats (Sluyter et al., 2000). A recent study investigated the effects of adolescent stress on amphetamine intake in Lewis and Fischer 344 rats (Meyer and Bardo, 2015). Compared to Fischer 344, Lewis rats show increased amphetamine selfadministration under normal circumstances. However, after environmental enrichment, a significant reduction was seen only in Lewis rats, while isolation rearing lead to an increase in selfadministration which was limited to the Fischer 344 rats. Although these data are suggestive of a gene * environment interaction, it should be kept in mind that ceiling or floor effects might also play an important role here. As Lewis rats already show a high baseline intake, it might be difficult for isolation rearing to increase it further. Likewise, as Fischer 344 rats show a very low baseline levels of intake, environmental enrichment cannot reduce this any further.

In conclusion whereas the animal research overwhelmingly supports a role for genetic factors in determining the reinforcing properties of drug of abuse, the data also support a role for environmental factors. However, how these factors interact, especially in models that are more closely related to the compulsive drug intake seen in humans, is still largely unknown and requires a stronger integration between genetic and environmental models.

References

Ahmed, S.H., 2011. Escalation of Drug Use. Neuromethods 53, 267–292.

Anthony, J.C., Chen, C.Y., Storr, C.L., 2005. Drug dependence epidemiology. Clin. Neurosci. Res. 5, 55–68.

APA, 2013. Diagnostic and Statistical Manual of mental disorders, 5th edn. American Psychiatric Publishing, Arlington VA.

Baarendse, P.J., Limpens, J.H., Vanderschuren, L.J., 2014. Disrupted social development enhances the motivation for cocaine in rats. Psychopharmacol. (Berlin) 231, 1695–1704.

Barr, C.S., 2013. Non-human primate models of alcohol-related phenotypes: the influence of genetic and environmental factors. Curr. Topics Behav. Neurosci. 13, 223–249.

Belknap, J.K., Hitzemann, R., Crabbe, J.C., et al., 2001. QTL analysis and genomewide mutagenesis in mice: complementary genetic approaches to the dissection of complex traits. Behav. Genet. 31, 5–15.

Buhler, K.M., Gine, E., Echeverry-Alzate, V., et al., 2015. Common single nucleotide variants underlying drug addiction: more than a decade of research. Addict. Biol.

Burke, A.R., Miczek, K.A., 2015. Escalation of cocaine self-administration in adulthood after social defeat of adolescent rats: role of social experience and adaptive coping behavior. Psychopharmacol. (Berlin).

Cahill, K., Stead, L.F., Lancaster, T., 2012. Nicotine receptor partial agonists for smoking cessation. Cochrane Database Sys. Rev. 4, CD006103.

Cao, J., Hudziak, J.J., Li, D., 2013. Multi-cultural association of the serotonin transporter gene (SLC6A4) with substance use disorder. Neuropsychopharmacology 38, 1737–1747.

Charbogne, P., Kieffer, B.L., Befort, K., 2014. 15 years of genetic approaches in vivo for addiction research: opioid receptor and peptide gene knockout in mouse models of drug abuse. Neuropharmacology 76 (Pt B), 204–217.

Chen, C.Y., Storr, C.L., Anthony, J.C., 2009. Early-onset drug use and risk for drug dependence problems. Addict. Behav. 34, 319–322.

Chen, L.S., Baker, T.B., Piper, M.E., et al., 2014. Interplay of genetic risk (CHRNA5) and environmental risk (partner smoking) on cigarette smoking reduction. Drug Alcohol Depend 143, 36–43.

Ciccocioppo, R., 2013. Genetically selected alcohol preferring rats to model human alcoholism. Curr. Topics Behav. Neurosci. 13, 251–269.

Crabbe, J.C., 2012. Translational behavior-genetic studies of alcohol: are we there yet? Genes Brain Behav. 11, 375–386.

Crabbe, J.C., Phillips, T.J., Belknap, J.K., 2010. The complexity of alcohol drinking: studies in rodent genetic models. Behav. Genet. 40, 737–750.

Deroche-Gamonet, V., Belin, D., Piazza, P.V., 2004. Evidence for addiction-like behavior in the rat. Science 305, 1014–1017.

Di Chiara, G., Imperato, A., 1988. Drugs abused by humans preferentially increase synaptic dopamine concentrations in the mesolimbic system of freely moving rats. Proc. Natl. Acad. Sci. USA 85, 5274–5278.

Easton, N., Marsden, C.A., 2006. Ecstasy: are animal data consistent between species and can they translate to humans? J. Psychopharmacol. 20, 194–210.

Effertz, T., Mann, K., 2013. The burden and cost of disorders of the brain in Europe with the inclusion of harmful alcohol use and nicotine addiction. Eur. Neuropsychopharmacol. J. Eur. Coll. Neuropsychopharmacol. 23, 742–748.

Ellenbroek, B.A., van der Kam, E.L., van der Elst, M.C.J., et al., 2005. Individual differences in drug dependence in rats: the role of genetic factors and life events. Eur. J. Pharmacol. 526, 251–258.

Enoch, M.A., Gorodetsky, E., Hodgkinson, C., et al., 2011. Functional genetic variants that increase synaptic serotonin and 5-HT3 receptor sensitivity predict alcohol and drug dependence. Mol. Psychiatry 16, 1139–1146.

Enoch, M.A., Hodgkinson, C.A., Yuan, Q., et al., 2010. The influence of GABRA2, childhood trauma, and their interaction on alcohol, heroin, and cocaine dependence. Biol. Psychiatry 67, 20–27.

Erlich, P.M., Hoffman, S.N., Rukstalis, M., et al., 2010. Nicotinic acetylcholine receptor genes on chromosome 15q25.1 are associated with nicotine and opioid dependence severity. Hum. Genet. 128, 491–499.

Esser, M.B., Hedden, S.L., Kanny, D., et al., 2014. Prevalence of alcohol dependence among US adult drinkers, 2009-2011. Prev. Chronic. Dis. 11, E206.

Everitt, B.J., Robbins, T.W., 2013. From the ventral to the dorsal striatum: devolving views of their roles in drug addiction. Neurosci. Biobehav. Rev. 37, 1946–1954.

Fernandez-Castillo, N., Roncero, C., Grau-Lopez, L., et al., 2013. Association study of 37 genes related to serotonin and dopamine neurotransmission and neurotrophic factors in cocaine dependence. Genes Brain Behav. 12, 39–46.

Fowler, C.D., Lu, Q., Johnson, P.M., et al., 2011. Habenular alpha5 nicotinic receptor subunit signaling controls nicotine intake. Nature 471, 597–601.

Gelernter, J., Kranzler, H.R., Sherva, R., et al., 2014. Genome-wide association study of alcohol dependence: significant findings in African-and European-Americans including novel risk loci. Mol. Psychiatry 19, 41–49.

Goldman, D., Oroszi, G., Ducci, F., 2005. The genetics of addictions: uncovering the genes. Nature Rev. Genetics 6, 521–532.

Gorwood, P., Le Strat, Y., Ramoz, N., et al., 2012. Genetics of dopamine receptors and drug addiction. Hum. Genet. 131, 803–822.

Grucza, R.A., Wang, J.C., Stitzel, J.A., et al., 2008. A risk allele for nicotine dependence in CHRNA5 is a protective allele for cocaine dependence. Biol. Psychiatry 64, 922–929.

Harenza, J.L., Muldoon, P.P., De Biasi, M., et al., 2014. Genetic variation within the Chrna7 gene modulates nicotine reward-like phenotypes in mice. Genes Brain Behav. 13, 213–225.

Homberg, J.R., De Boer, S.F., Raaso¸, H.S., et al., 2008. Adaptations in pre- and postsynaptic 5-HT1A receptor function and cocaine supersensitivity in serotonin transporter knockout rats. Psychopharmacol. 200, 367–380.

Jackson, K.J., Marks, M.J., Vann, R.E., et al., 2010. Role of alpha5 nicotinic acetylcholine receptors in pharmacological and behavioral effects of nicotine in mice. J. Pharmacol. Exp. Ther. 334, 137–146.

Jasinska, A.J., Stein, E.A., Kaiser, J., et al., 2014. Factors modulating neural reactivity to drug cues in addiction: a survey of human neuroimaging studies. Neurosci. Biobehav. Rev. 38, 1–16.

Johnson, E.O., Chen, L.S., Breslau, N., et al., 2010. Peer smoking and the nicotinic receptor genes: an examination of genetic and environmental risks for nicotine dependence. Addiction 105, 2014–2022.

Jorenby, D.E., Hays, J.T., Rigotti, N.A., et al., 2006. Efficacy of varenicline, an alpha4beta2 nicotinic acetylcholine receptor partial agonist, vs placebo or sustained-release bupropion for smoking cessation: a randomized controlled trial. JAMA 296, 56–63.

Jorgensen, C.H., Pedersen, B., Tonnesen, H., 2011. The efficacy of disulfiram for the treatment of alcohol use disorder. Alcohol. Clin. Exp. Res. 35, 1749–1758.

Joslyn, G., Brush, G., Robertson, M., et al., 2008. Chromosome 15q25.1 genetic markers associated with level of response to alcohol in humans. Proc. Natl. Acad. Sci. USA 105, 20368–20373.

Kaufman, J., Yang, B.Z., Douglas-Palumberi, H., et al., 2007. Genetic and environmental predictors of early alcohol use. Biol. Psychiatry 61, 1228–1234.

Kendler, K.S., Ohlsson, H., Sundquist, J., et al., 2015. Triparental families: a new genetic-epidemiological design applied to drug abuse, alcohol use disorders, and criminal behavior in a Swedish national sample. Am. J. Psychiatry 172, 553–560.

Kendler, K.S., Chen, X., Dick, D., et al., 2012. Recent advances in the genetic epidemiology and molecular genetics of substance use disorders. Nat. Neurosci. 15, 181–189.

Kim, J., Park, A., Glatt, S.J., et al., 2015. Interaction effects between the 5-hydroxy tryptamine transporter-linked polymorphic region (5-HTTLPR) genotype and family conflict on adolescent alcohol use and misuse. Addiction 110, 289–299.

Ko, C.H., Hsieh, T.J., Wang, P.W., et al., 2015. Altered gray matter density and disrupted functional connectivity of the amygdala in adults with Internet gaming disorder. Prog. Neuropsychopharmacol. Biol. Psychiatry 57, 185–192.

Koob, G.F., Volkow, N.D., 2010. Neurocircuitry of Addiction. Neuropsychopharmacology 35, 217–238.

Kufahl, P.R., Li, Z., Risinger, R.C., et al., 2005. Neural responses to acute cocaine administration in the human brain detected by fMRI. Neuroimage 28, 904–914.

Liang, W., Chikritzhs, T., 2015. Age at first use of alcohol predicts the risk of heavy alcohol use in early adulthood: a longitudinal study in the United States. Int. J. Drug Pol. 26, 131–134.

Lin, J.Y., Arthurs, J., Reilly, S., 2014. Conditioned taste aversion, drugs of abuse and palatability. Neurosci. Biobehav. Rev. 45C, 28–45.

Long, J.C., Knowler, W.C., Hanson, R.L., et al., 1998. Evidence for genetic linkage to alcohol dependence on chromosomes 4 and 11 from an autosome-wide scan in an American Indian population. Am J Med Genet 81, 216–221.

Lynskey, M.T., Agrawal, A., Heath, A.C., 2010. Genetically informative research on adolescent substance use: methods, findings, and challenges. J Am Acad Child Adolesc Psychiatry 49, 1202–1214.

Mattick, R.P., Kimber, J., Breen, C., et al., 2008. Buprenorphine maintenance versus placebo or methadone maintenance for opioid dependence. Cochrane Database Sys. Rev. 3000, CD002207.

Mattick, R.P., Breen, C., Kimber, J., et al., 2009. Methadone maintenance therapy versus no opioid replacement therapy for opioid dependence. Cochrane Database Sys. Rev., CD002209.

Metten, P., Phillips, T.J., Crabbe, J.C., et al., 1998. High genetic susceptibility to ethanol withdrawal predicts low ethanol consumption. Mammalian Genome Official J. Int. Mammalian Genome Soc. 9, 983–990.

Meyer, A.C., Bardo, M.T., 2015. Amphetamine self-administration and dopamine function: assessment of gene x environment interactions in Lewis and Fischer 344 rats. Psychopharmacol. (Berlin) 232, 2275–2285.

Meyers, J.L., Shmulewitz, D., Wall, M.M., et al., 2015. Childhood adversity moderates the effect of ADH1B on risk for alcohol-related phenotypes in Jewish Israeli drinkers. Addict. Biol. 20, 205–214.

Mills, E.J., Wu, P., Spurden, D., et al., 2009. Efficacy of pharmacotherapies for short-term smoking abstinence: a systematic review and meta-analysis. Harm Reduction J. 6, 25.

Mills, E.J., Thorlund, K., Eapen, S., et al., 2014. Cardiovascular events associated with smoking cessation pharmacotherapies: a network meta-analysis. Circulation 129, 28–41.

Miszkiel, J., Filip, M., Przegalinski, E., 2011. Role of serotonin 5-HT1B receptors in psychostimulant addiction. Pharmacol. Rep. 63, 1310–1315.

Morales-Suarez-Varela, M.M., Bille, C., Christensen, K., et al., 2006. Smoking habits, nicotine use, and congenital malformations. Obstet. Gynecol. 107, 51–57.

Murray, C.J.L., Vos, T., Lozano, R., et al., 2012. Disability-adjusted life years (DALYs) for 291 diseases and injuries in 21 regions, 1990-2010: a systematic analysis for the Global Burden of Disease Study 2010. Lancet 380, 2197–2223.

Nelson, E.C., Agrawal, A., Pergadia, M.L., et al., 2010. H2 haplotype at chromosome 17q21.31 protects against childhood sexual abuse-associated risk for alcohol consumption and dependence. Addict. Biol. 15, 1–11.

Nutt, D.J., King, L.A., Phillips, L.D., 2010. Drug harms in the UK: a multicriteria decision analysis. Lancet 376, 1558–1565.

Nylander, I., Roman, E., 2013. Is the rodent maternal separation model a valid and effective model for studies on the early-life impact on ethanol consumption? Psychopharmacol. (Berlin) 229, 555–569.

Oakly, A.C., Brox, B.W., Schenk, S., et al., 2014. A genetic deletion of the serotonin transporter greatly enhances the reinforcing properties of MDMA in rats. Mol. Psychiatry 19, 534–535.

Olfson, E., Edenberg, H.J., Nurnberger, J., et al., 2014. An ADH1B variant and peer drinking in progression to adolescent drinking milestones: evidence of a gene-by-environment interaction. Alcohol. Clin. Experiment. Res. 38, 2541–2549.

Orellana-Barrios, M.A., Payne, D., Mulkey, Z., et al., 2015. Electronic cigarettes-a narrative review for clinicians. Am. J. Med. 128, 674–681.

Pelloux, Y., Everitt, B.J., Dickinson, A., 2007. Compulsive drug seeking by rats under punishment: effects of drug taking history. Psychopharmacol. (Berlin) 194, 127–137.

Prochaska, J.J., Hilton, J.F., 2012. Risk of cardiovascular serious adverse events associated with varenicline use for tobacco cessation: systematic review and meta-analysis. BMJ 344, e2856.

Richardson, N.R., Roberts, D.C.S., 1996. Progressive ratio schedules in drug self-administration studies in rats: a method to evaluate reinforcing efficacy. J. Neurosci. Methods 66, 1–11.
Saiz, P.A., Garcia-Portilla, M.P., Arango, C., et al., 2008. Association between heroin dependence and 5-HT2A receptor gene polymorphisms. Eur. Addict. Res. 14, 47–52.
Schellekens, A.F., Franke, B., Ellenbroek, B., et al., 2013. COMT Val158Met modulates the effect of childhood adverse experiences on the risk of alcohol dependence. Addict. Biol. 18, 344–356.
Shabani, S., Dobbs, L.K., Ford, M.M., et al., 2012. A genetic animal model of differential sensitivity to methamphetamine reinforcement. Neuropharmacology 62, 2169–2177.
Shaham, Y., Shalev, U., Lu, L., et al., 2003. The reinstatement model of drug relapse: history, methodology and major findings. Psychopharmacology 168, 3–20.
Simms, J.A., Steensland, P., Medina, B., et al., 2008. Intermittent access to 20% ethanol induces high ethanol consumption in Long-Evans and Wistar rats. Alcoholism Clin. Experiment. Res. 32, 1816–1823.
Singh, S., Loke, Y.K., Spangler, J.G., et al., 2011. Risk of serious adverse cardiovascular events associated with varenicline: a systematic review and meta-analysis. CMAJ 183, 1359–1366.
Sluyter, F., Hof, M., Ellenbroek, B.A., et al., 2000. Genetic, sex, and early environmental effects on the voluntary alcohol intake in Wistar rats. Pharmacol. Biochem. Behav. 67, 801–808.
Tobacco, Genetics, C., 2010. Genome-wide meta-analyses identify multiple loci associated with smoking behavior. Nat. Genet. 42, 441–447.
Trigo, J.M., Renoir, T., Lanfumey, L., et al., 2007. 3,4-methylenedioxymethamphetamine self-administration is abolished in serotonin transporter knockout mice. Biol. Psychiatry 62, 669–679.
Tzschentke, T.M., 2007. Measuring reward with the conditioned place preference (CPP) paradigm: update of the last decade. Addiction Biol. 12, 227–462.
UNODC, 2014. World Drug Report 2014. Edn. United Nations publications.
van Amsterdam, J., van den Brink, W., 2013. The high harm score of alcohol. Time for drug policy to be revisited? J. Psychopharmacol. 27, 248–255.
van Amsterdam, J., Nutt, D., Phillips, L., et al., 2015. European rating of drug harms. J. Psychopharmacol. 29, 655–660.
van der Kam, E.L., Ellenbroek, B.A., Cools, A.R., 2005a. Gene-environment interactions de termine the individual variability in cocaine self-administration. Neuropharmacology 48, 685–695.
van der Kam, E.L., Coolen, J.C.M., Ellenbroek, B.A., et al., 2005b. The effects of stress on alcohol consumption: mild acute and sub-chronic stressors differentially affect apomorphine susceptible and unsusceptible rats. Life Sci. 76, 1759–1770.
Van Waes, V., Enache, M., Berton, O., et al., 2011. Effect of prenatal stress on alcohol preference and sensitivity to chronic alcohol exposure in male rats. Psychopharmacol. (Berlin) 214, 197–208.
vandenBrink, W., 2012. Evidence-based pharmacological treatment of substance use disorders and pathological gambling. Curr. Drug Abuse Rev. 5, 3–31.
Vanderschuren, L.J., Everitt, B.J., 2004. Drug seeking becomes compulsive after prolonged cocaine self-administration. Science 305, 1017–1019.
Vega, W.A., Aguilar-Gaxiola, S., Andrade, L., et al., 2002. Prevalence and age of onset for drug use in seven international sites: results from the international consortium of psychiatric epidemiology. Drug Alcohol Depend. 68, 285–297.
Verhulst, B., Neale, M.C., Kendler, K.S., 2015. The heritability of alcohol use disorders: a meta-analysis of twin and adoption studies. Psychol. Med. 45, 1061–1072.
Volkow, N.D., Wang, G.J., Fowler, J.S., et al., 2012. Addiction circuitry in the human brain. Annu. Rev. Pharmacol. Toxicol. 52, 321–336.
Volkow, N.D., Fowler, J.S., Wang, G.J., et al., 2009. Imaging dopamine's role in drug abuse and addiction. Neuropharmacology 56 (Suppl 1), 3–8.
Walters, C.L., Brown, S., Changeux, J.P., et al., 2006. The beta2 but not alpha7 subunit of the nicotinic acetylcholine receptor is required for nicotine-conditioned place preference in mice. Psychopharmacol. (Berlin) 184, 339–344.
Wang, J.C., Kapoor, M., Goate, A.M., 2012. The genetics of substance dependence. Annu. Rev. Genomics Hum. Genet. 13, 241–261.
Westman, J., Wahlbeck, K., Laursen, T.M., et al., 2015. Mortality and life expectancy of people with alcohol use disorder in Denmark, Finland and Sweden. Acta Psychiatr. Scand. 131, 297–306.

Willuhn, I., Burgeno, L.M., Everitt, B.J., et al., 2012. Hierarchical recruitment of phasic dopamine signaling in the striatum during the progression of cocaine use. Proc. Natl. Acad. Sci. USA 109, 20703–20708.

Windle, M., Kogan, S.M., Lee, S., et al., 2015. Neighborhood x Serotonin Transporter Linked Polymorphic Region (5-HTTLPR) interactions for substance use from ages 10 to 24 years using a harmonized data set of African American children. Dev. Psychopathol., 1–17.

Wittchen, H.U., Jacobi, F., Rehm, J., et al., 2011. The size and burden of mental disorders and other disorders of the brain in Europe 2010. Eur. Neuropsychopharmacol. J. Eur. Coll. Neuropsychopharmacol. 21, 655–679.

Zorrilla, E.P., Logrip, M.L., Koob, G.F., 2014. Corticotropin releasing factor: a key role in the neurobiology of addiction. Front Neuroendocrinol. 35, 234–244.

CHAPTER

7

Affective Disorders

INTRODUCTION

Affective disorders, also often referred to as mood disorders is a group of psychiatric illnesses where a disturbance in mood is considered the main underlying feature. Thus, although mood disturbances occur in virtually all psychiatric disorders (most notably in ADHD, ASD, and schizophrenia), only in affective disorders is it considered the defining feature. Disturbances in mood can take the form of either elevated mood, as it occurs in mania or hypomania, or reduced (depressed) mood as it occurs in major depressive episodes. In general, two major types of affective disorders can be distinguished: (1) Major depressive disorders (MDD), mainly characterized by low mood (feelings of sadness and hopelessness); (2) bipolar disorders (BP), characterized by depressive episodes and periods of mania or hypomania. In addition, anxiety disorders, characterized by feelings of nervousness, anxiety, and fear are usually also included in the category of affective disorders. The anxiety disorder is a very broad category of disorders encompassing a number of different subtypes, including social phobia, panic disorder, obsessive compulsive disorders, and posttraumatic stress disorder (PTSD). Given this large number of different (sub)types of affective disorders and the considerable overlap in symptoms and pathology among them, we have decided to limit our discussion in this chapter to major depressive disorder (MDD) and bipolar disorder (BP).

MAJOR DEPRESSIVE DISORDER

Diagnosis and Symptoms

As discussed earlier, MDD is characterized by severe and pervasive low mood, leading to intense sadness and hopelessness. These feeling typically occur in episodes as illustrated in Fig. 7.1 and can differ both in intensity and in duration. Although the average duration of a depressive episode is usual considered to be about 6 months, they typically vary from anything between 2 and 12 months. In a recent analysis of the global burden of disease the average duration of a major depressive episode was calculated at 37.5 weeks roughly about 8–9 months (Ferrari et al., 2013). However, in contrast to schizophrenia (see Fig. 9.1), where psychotic symptoms also often occur in episodes, it is generally accepted that when symptoms subside, the patient's functioning is very similar to that before the episode started.

Gene-Environment Interactions in Psychiatry. http://dx.doi.org/10.1016/B978-0-12-801657-2.00007-0

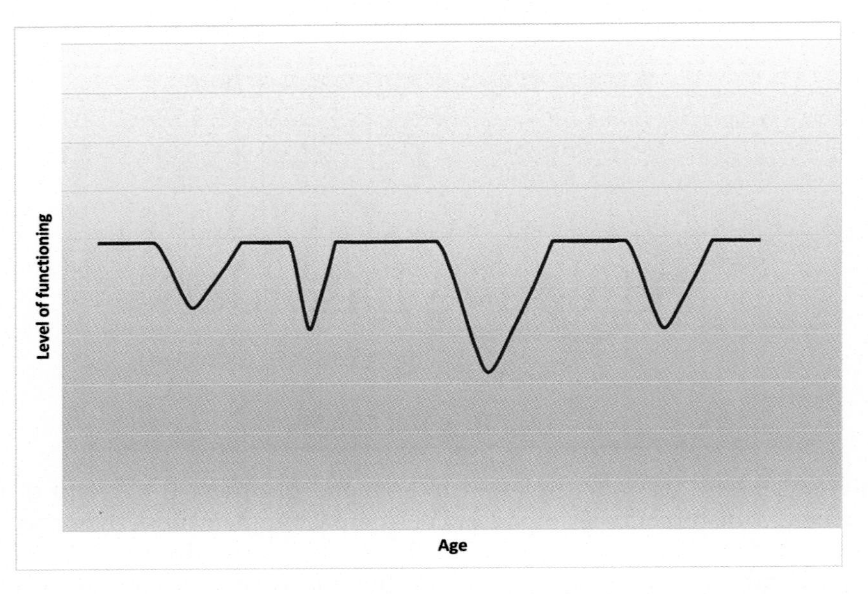

FIGURE 7.1 **The typical course of illness for MDD.** MDD is characterized by periods of depressed mood that can last between 2 and 12 months. However, in between these episodes, the level of functioning is relatively normal.

The diagnosis of MDD requires the presence of at least 5 different symptoms, one of which is either depressive mood or marked loss of interest or pleasure in most daily activities (Box 7.1). Most of the symptoms listed in Box 7.1 can occur in every human and do not inevitably lead to a diagnosis of MDD. In some sense, the distinction between unhappiness, sadness, and MDD is reminiscent of the distinction between recreational drug use and drug addiction (see Chapter 6). Thus it is the degree to which the feelings affect normal functioning that ultimately determines the diagnosis. Therefore, most of the symptoms of Box 7.1 have the additional prerequisite that they occur nearly every day, that they are present for at least 2 weeks and that they cause significant distress and affect normal functioning.

Another interesting aspect of MDD is that two patients can actually have diametrically opposite symptoms. Thus both substantial weight gain and weight loss can be indicative of MDD. Likewise, both hypersomnia and insomnia are symptoms that can occur in MDD and finally, patients can suffer from either psychomotor retardation, or psychomotor agitation. This has obvious implications for our understanding of MDD and suggests that MDD is a very heterogeneous disorder (or more likely a group of disorders). Likewise, as was touched upon in Chapter 3, these criteria make it impossible to develop an animal model that encapsulates all symptoms of MDD. Indeed, as we will discuss later, most animal models have focused on diminished pleasure (anhedonia).

Many of the symptoms of MDD can also occur in other conditions, and consequently an important aspect of the differential diagnosis is to ensure no other condition could explain the underlying symptomatology. In this respect it is important to ensure there has not been any incidence of mania or hypomania, as this would warrant a diagnosis of BP (as discussed later). This distinction is important as there is mounting evidence of substantial differences

BOX 7.1

THE DIAGNOSIS OF MDD (DSM V)

1. Five (or more) of the following symptoms have been present during the same 2-week period and represent a change from previous functioning:
   a. Depressed mood most of the day, nearly every day, as indicated by either subjective report (eg, feels sad, empty, hopeless) or observation made by others (eg, appears tearful). (Note: in children and adolescents, can be irritable mood).
   b. Markedly diminished interest or pleasure in all, or almost all, activities most of the day, nearly every day (as indicated by either subjective account or observation).
   c. Significant weight loss or weight gain (eg, a change of more than 5% of body weight in a month), or decrease or increase in appetite nearly every day.
   d. Insomnia or hypersomnia nearly every day.
   e. Psychomotor agitation or retardation nearly every day (observable by others, not merely subjective feelings of restlessness or being slowed down).
   f. Fatigue or loss of energy nearly every day.
   g. Feelings of worthlessness or excessive or inappropriate guilt (which may be delusional) nearly every day (not merely self-reproach or guilt about being sick).
   h. Diminished ability to think or concentrate, or indecisiveness, nearly every day (either by subjective account or as observed by others).
   i. Recurrent thoughts of death (not just fear of dying), recurrent suicidal ideation without a specific plan, or a suicide attempt or a specific plan for committing suicide.

Note: one of the symptoms should be either

2. The symptoms cause clinically significant distress or impairment in social, occupational, or other important areas of functioning.
3. The episode is not attributable to the physiological effects of a substance or to another medical condition.
4. The occurrence of the major depressive episode is not better explained by schizoaffective disorder, schizophrenia, schizophreniform disorder, delusional disorder, or other specified and unspecified schizophrenia spectrum and other psychotic disorders.
5. There has never been an episode of mania or hypomania.

between the depressive symptoms in MDD and BP. Similarly, loss of pleasure is an essential aspect of (the negative symptoms of) schizophrenia (see Chapter 9). Pleasure is often subdivided in two different stages: anticipatory and consummatory pleasure. The Temporal Experience of Pleasure Scale (TEPS) is an 18 item yes/no questionnaire aimed at distinguishing between these two aspects of pleasure and includes questions such as "*The smell of freshly cut grass is enjoyable to me*" (consummatory) and "*When I think about eating my favorite food, I can almost taste how good it is*" (anticipatory). Principal component analysis indeed identified two different factors assessing either anticipatory or consummatory pleasure (Gard et al., 2006). So far, only few studies in MDD have distinguished between these two aspects of pleasure. One studies found significant reduction in anticipatory but not consummatory pleasure in patients with MDD (Sherdell et al., 2012), while a more recent study found deficits in both phases of pleasure in Chinese patients with MDD (Li et al., 2015). In this latter study, both aspects of pleasure were significantly negatively correlated with the Hamilton depression test, duration of illness, as well as number of hospital admissions, suggesting that both anticipatory and consummatory pleasure is directly related to severity of the illness. Interestingly, these authors also found significant deficits in both forms of pleasure in patients with schizophrenia, but in this case only consummatory pleasure was associated with the negative symptoms.

Several scales have also been developed to assess the severity of a major depressive episode. The most often used is the Hamilton depression rating scale (HDRS, also sometimes referred to as the Ham_D), developed in 1960 (Hamilton, 1960). The original scale consisted of 17 items that were scored on a scale of 0 (absent) to 4 (very severe), thus an individual could receive a score between 0 and 68. In a subsequent version 4 additional items were added to allow for the subtyping of depression. This version is typically referred to as $HDRS_{21}$ (as opposed to the original $HDRS_{17}$). However, these 4 items do not really aim to assess the severity of the subtype. One limitation of the HDRS is that it focusses predominantly on the affective aspects of depression and does not assess symptoms such as weight gain or hypersomnia. Later adaptations such as the SIGH-SAD (Structured Interview Guide for the Hamilton Depression Rating Scale–Seasonal Affective Disorders) have included several of these atypical depressive symptoms, although, as the name indicates, the scale was originally designed for assessing seasonal affective disorder, which are distinct from MDD.

The Epidemiology of MDD

MDD is one of the most common mental illnesses, although differences exist across different countries. In one of the first international studies, Weisman and coworkers found that life-time prevalence ranged from 1.5% in Taiwan to 19% in Beirut (Weissman et al., 1996). However, it is important to emphasize that Weisman and coworkers studied the prevalence of major depressive episodes (MDE), which includes both MDD and BP. A subsequent study looking only at MDD found lifetime prevalence estimates ranging from 1% (Czech Republic) to 16.9% (United States of America). This wide variability in prevalence rates was thought to be (at least in part) attributable to differences in measurement and study design, which led to the World Mental Health (WMH) Initiative (Kessler and Bromet, 2013). Using a much more structured design, the prevalence of MDE (and other disorders) was assessed in 18 different countries. However, in spite of this more homogeneous approach, lifetime prevalence still ranged from 6.5% to 21.2%, with an average of 14.6% for high income and 11.1% for low/middle income countries (see Fig. 7.2). This study also assessed the 12 month prevalence, which was on average 5.5% in high income and 5.9% in low/middle income countries, indicating that about 1/2 to 1/3 of all patients have recurring episodes in a single year.

This large difference between countries was also apparent in the global burden of disease study (Ferrari et al., 2013). Although this study analyzed point-rather than 12-months or lifetime prevalence, the estimates varied between around 2.5% (for Asian-pacific countries including

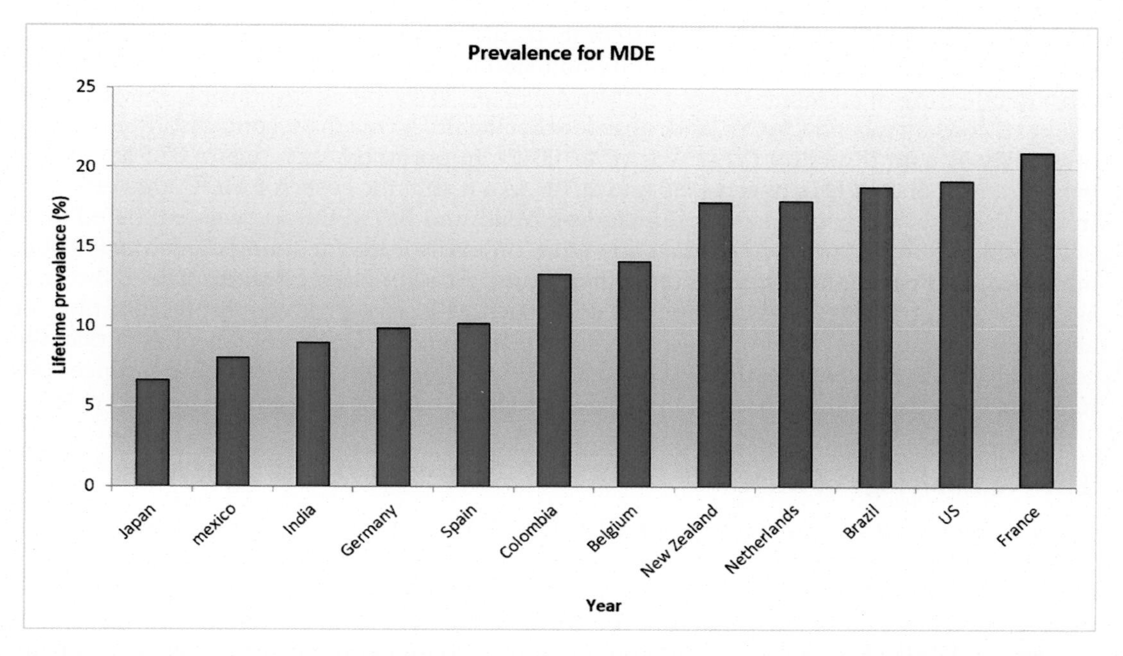

FIGURE 7.2 **The prevalence of MDD.** MDD is one of the most prevalent psychiatric disorders, with lifetime prevalences ranging from about 7% in Japan to 21% in France. *Data from Kessler and Bromet, 2013.*

Japan and Korea) to about 7.2 in North Africa and the Middle East. The study also showed that the point prevalence was relatively stable between 1990 and 2010 (both estimated at 4.1%).

Several studies have investigated the age of onset of MDD. In the already mentioned global burden of disease study, prevalence steadily increased until about 19 years of age, then peaked between 20 and 64 and declined afterwards (Ferrari et al., 2013). In the WMH survey, the age of onset was more explicitly measured and found to be less variable than the prevalence in the different countries. The average age of onset was 25.7 in high income countries (ranging from 22.7 in the United States to 30.1 in Japan) and 24.0 years in low/middle income countries (ranging from 18.8 in China to 31.9 in India)., the average age of onset is later than most other mental disorders (see Fig. 8.1).

MDD is significantly more common in females than in males. For instance, in the global burden of disease study, the point prevalences were 3.2% for males and 5.5% for females (Ferrari et al., 2013) while the odds ratio (OR) in females was 1.8 times that of males for MDE in the WMH study (Bromet et al., 2011). This last study also identified several other risk factors for MDE, all of which were consistent across most countries. Thus the OR was 3.6 for separated and 2.1 for divorced (compared to married) individuals, while the OR for living alone was 1.8 (compared to living with a spouse). Finally, compared to high income the OR for low income individuals was 1.7 (Bromet et al., 2011). Interestingly, although some studies found that the risk of MDE depends on the level of education, this was not found in the WMH study.

Overall, MDD has a high prevalence, and was estimated to affect about 187 million females and 111 million males worldwide, making it the most prevalent brain disorder after anxiety disorders (Wittchen et al., 2011). Because of its recurring nature, MDD takes a big toll on an individual, and with 63.2 million disability-adjusted life years (DALY) in 2010, it is by far the most debilitating of all mental disorder (Murray et al., 2012), being responsible for 34% of all mental disorders related DALYs, and, after low back pain, it was the second disorder in terms of year lived with disability (YLD; Vos et al., 2012). Importantly, there was a 37.5% increase in both DALY and YLD between 1990 and 2010. As a result, the costs for MDD are very high. In fact, the costs for mood disorders (including MDD and BP) in Europe was estimated to be is 113 billion Euro in 2010, higher than any other brain disorder, including dementia (103 billion Euro) and psychotic disorders (94 billion Euro). As with other brain disorders, the largest proportion of these costs are indirect costs such as loss of productivity (accounting for about 64% of all costs). Likewise, in a recent study from the United States of America, the costs of MDD in 2010 were estimated to be 211 billion dollars, an increase of 21.5% since 2005 (Greenberg et al., 2015), with again the largest proportion of the costs due to indirect costs.

The Neurobiology of MDD

Given the broad symptomatology of MDD (see Box 7.1) it is not surprising that the neurobiological substrate underlying this disorder is for the most part unknown and studies have often led to contradictory results. Most of the studies, both in humans and animals have investigated the dysregulation of emotion in patients with MDD, focusing predominantly on anhedonia and deficits in reward perception. Emotional control and regulation has traditionally been linked to the limbic system. Originally described in 1878 by the French neurologist Paul Broca, the "limbic lobe" was considered to form a border (*limbus*) around the brain stem and originally included the cingulate and temporal cortex and hippocampus. However it was not associated with emotions but rather thought to be involved in olfaction. It was not until the pioneering work of James Papez that the limbic system became associated with emotions. He also included the hypothalamus and various thalamic nuclei in his "emotional system" which has since been referred to as the "*Papez circuit*". Since then this system has been adapted to include structures such as the amygdala and to exclude other structures such as the hippocampus. Yet the association of the "*Papez circuit*" in the regulation of emotion is still largely valid and there is convincing evidence that large parts of this system are dysregulated in patients with MDD.

Nonetheless, other parts of the brain have also been implicated in depression, largely based on studies in neurological patients such as patients with Parkinson's disease and Huntington's disease. For instance, studies in individuals with Parkinson's disease have shown that 30%–50% suffer from depression and up to 60% from apathy (Aarsland et al., 2012). Likewise, depression is the most prevalent neuropsychiatric condition in Huntington's disease, and a recent survey among more than 2800 patients found that 40% were suffering from depression and more than 50% had sought help for depression in the past (Paulsen et al., 2005). Given that both neurological disorders have their main lesion within the striatal complex, this has led to the hypothesis that MDD is associated with dysfunction within the limbic-cortical-striatal-pallidal-thalamic circuitry (Ongur et al., 2003; Drevets et al., 2008). As shown in Fig. 7.3 this circuit involves the orbital and medial prefrontal cortex, amygdala, hippocampus, nucleus accumbens, mediodorsal thalamic nucleus and ventral pallidum. Further studies, especially in primates, have suggested that the orbital and medial prefrontal cortex are associated with two extended cortical networks: *the orbital prefrontal network,* which seems

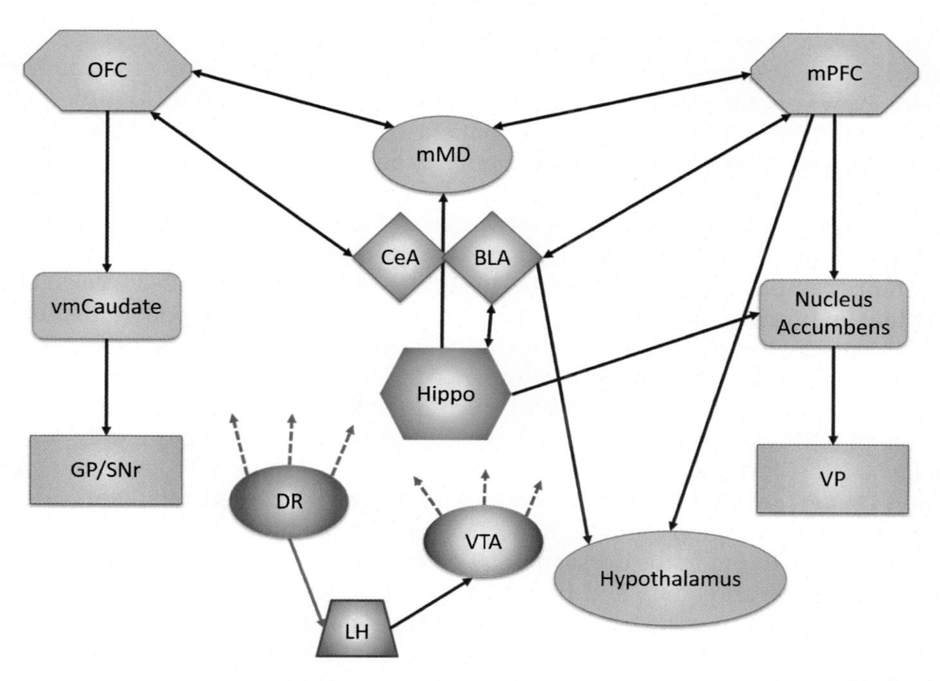

FIGURE 7.3 **The neurobiological basis of MDD.** Patients with MDD are characterised by structural and functional changes within a wide range of brain regions, encompassing cortical [medial prefrontal (mPFC) and orbitofrontal (OFC)] and basal ganglia [ventromedial (vm) caudate nucleus, nucleus accumbens, globus pallidus/substantia nigra part reticulate (GP/SNr) and Ventral Pallidum (VP)]. In addition, changes have consistently been observed in the Amygdala [both Central (CeA) and Basolateral (BLA)], the hippocampus (Hippo), and the medial part of the mediodorsal thalamic nucleus (mMD). In addition, alterations in the serotonergic system [originating from the dorsal raphe (DR)] and the dopaminergic system [originating from the ventral tegmental area (VTA)] have been found. Both these cell body regions are under the influence of the lateral habenula (LH). Because the serotonergic and dopaminergic innervation is so complex, they were deliberately omitted from this diagram.

to play an important role in sensory integration and the affective coding of stimuli (such as reward, aversion etc.), and the *medial prefrontal cortex*, which through its prominent connection to the hypothalamus seems to be involved in the visceral reactions to emotional stimuli (Saleem et al., 2008). Neuroimaging studies have corroborated the involvement of many of the structures of Fig. 7.3 in MDD (and in BP as well, as discussed later). Thus there is strong evidence for gray matter volume loss in several regions of the medial prefrontal cortex, especially the subgenual anterior cingulate cortex (sgACC) (Drevets et al., 2008). These reductions appear to exist already at first onset and occur in patients at high risk for developing MDD. Nonetheless, there seems to be a progressive loss of volume over the course of the illness (at least in patients with psychotic depression) and can be found in both sexes and in both MDD and BP patients (Drevets et al., 2008). Interestingly, there is some evidence that the volumetric reduction in sgACC, while not responsive to antidepressants (Drevets et al., 1997), is (partly) reversed by chronic lithium treatment in BP patients (Moore et al., 2009). In addition, reductions have been reported in the posterior cingulate cortex, hippocampus, and ventral striatum, including the nucleus accumbens. The findings in the hippocampus have been more varied, with some studies reporting no volume change. There is some evidence that the degree of volume loss may be related to the aetiology and/or the duration of illness, with significant volume reductions seen especially in patients exposed to childhood trauma (Chaney et al., 2014) and in more chronic patients (Sheline et al., 2003). Moreover, there is increasing evidence that the volume reduction may be specific for certain subfields within the hippocampus with the cornu ammonis and dentate gyrus being most affected (Malykhin and Coupland, 2015). There is still debate whether or not antidepressant treatment reverses this hippocampal reduction. Rodent studies have shown that virtually all antidepressant drugs increase hippocampal volume and neurogenesis after relatively short (2–3 weeks) treatment. The fact that, in humans, the effects on hippocampal volume are seen both in treated and untreated patients seems to indicate that the influence of antidepressant treatment is relatively mild, although more longitudinal studies are needed.

Volumetric alterations have also been reported for the amygdala, but the results are less obvious with both increases and decreases in volume reported. In agreement with the neuroimaging data, reduced post mortem cell counts have been reported for most of these areas (Price and Drevets, 2010). Likewise, alterations in cerebral glucose metabolism have been observed for most of the regions, although the data are more complex. Thus whereas a reduction in metabolism was found in the dorsal medial prefrontal cortex, increases have been reported in areas such as the amygdala and hippocampus (Price and Drevets, 2010). In the sgACC decreases have been reported, but corrected for the smaller volume, it has been suggested that in fact there is an increase in activity in the remaining part of this cortical area (Price and Drevets, 2010). In line with this, overall increases in metabolism are seen within the sgACC after antidepressant treatment and electroconvulsive therapy (Price and Drevets, 2010).

Although these data suggest a relationship between structural and functional data, more detailed fMRI studies have recently questioned this. Thus in a study including 50 patients and 50 controls, gray matter volume reductions, and functional abnormalities were found in different cortical regions, with neither of them correlating with the clinical symptoms (Yang et al., 2015b). A similar lack of relationship with clinical variables was found in a study investigating resting-state functional connectivity in MDD patients. Thus while the functional connectivity was reduced in the cingulate cortex and enhanced in the occipital cortex in patients with MDD, there was no correlation with symptoms (Zhang et al., 2015) as assessed with the Hamilton Depression Scale. It is, however, important to realize that patients with MDD not only suffer from depressive symptoms but also from additional symptoms, and it may be possible that the structural and/or functional abnormalities, while not directly related to the depressive symptoms may be relevant to other aspects of the disorders. For instance, in a study comparing 51 medication free MDD patients with 51 controls, significant increases in gray matter volume were found in the posterior cingulate and inferior frontal gyrus (Yang et al., 2015a), but whereas neither increases were correlated with the Hamilton scores, the increase in the inferior frontal gyrus was correlated with sustained attention deficits. Together, these data underscore the importance of the circuit shown in Fig. 7.3, especially the medial prefrontal/amygdala/hippocampal/hypothalamus connections in the regulation of MDD.

It is much more difficult to identify the accompanying biochemical alterations. One of the oldest biochemical theories of MDD is the serotonin hypothesis, based on findings from both clinical and preclinical experiments that showed that while addition of tryptophan (the precursor of serotonin) enhanced the actions of antidepressants, tryptophan depleting reversed it (Lopez-Munoz and Alamo, 2009). Determining changes in the serotonergic system in patients with MDD has proven much more difficult (Kohler et al., 2016). However, with the advent of selective radioactive probes for PET and SPECT scan, substantial progress has been made and a recent review summarizing the results found clear support for a dysfunctional serotonergic system in patients with MDD (Savitz and Drevets, 2013). Thus, in spite of differences (at least in part due to different patients, PET ligands, and medication status) reductions of about 20%–30% in 5-HT1A binding in the raphe have been reported in patients with MDD. Reductions have also been reported in the amygdala, hippocampus, and anterior cingulate cortex. Likewise, reductions in 5-HT2A binding in the cortex and 5-HT1B binding in the ventral striatum and ventral pallidum have been reported. Several reports also found reduced serotonin transporter binding within cortical and subcortical regions. Overall, there seems convincing evidence for a disturbed serotonin neurotransmission in depression.

However, in addition, there is increased attention for the dopaminergic system in depression. This is, in part, based on the role dopamine plays in the reward system (see Chapter 6). As anhedonia is closely related to abnormal perception/expression of pleasure, it has been proposed that dopamine is also involved in depression. In line with this, drugs such as reserpine, which significantly reduce dopamine neurotransmission (though it also affects noradrenaline and serotonin) can induce depressive symptoms in vulnerable individuals. In line with this, several studies in animals have shown that the dopaminergic cells within the VTA (the ventral tegmental area, which project predominantly to the nucleus accumbens and the prefrontal cortex, see Fig. 7.4) are involved in depression-like behavior. Thus phasic but not tonic stimulation induces a depressive phenotype in mice previously exposed to a learned helplessness paradigm (Chaudhury et al., 2013). Intriguingly, another study showed exactly the opposite, namely that phasic dopamine release rescued depression symptoms (Tye et al., 2013). One likely explanation for this paradox is that different dopaminergic cells within the VTA code for different stimuli. Thus it has been shown that in mice susceptible to social defeat stress (a model often used to induce a depressive phenotype, as discussed later) VTA cells projecting to the nucleus accumbens are excited while VTA cells projecting to the medial prefrontal cortex are inhibited (Chaudhury et al., 2013). Interestingly, this subdivision within the VTA is extended into the input side as well, with the lateral habenula (LH, see Fig. 7.4) projecting predominantly to the dopaminergic cells innervating the medial prefrontal cortex, while neurons in the laterodorsal tegmentum (LDT) projecting to VTA neurons innervating the nucleus accumbens. These circuits seem to encode opposite behavior with activation of the LDT increasing reward seeking and activation of the LH increasing aversion (Chaudhury et al., 2015). Given that the LH mediated aversion is suppressed by serotonin, and serotonin transmission is reduced in MDD (as discussed earlier) this may link the two major monoamine neurotransmitters (dopamine and serotonin) together. However, it should be noted that other neurotransmitter system such as noradrenaline (Chaudhury et al., 2015) and glutamate (Niciu et al., 2014) have also been implicated in the pathophysiology of MDD.

Finally, there is strong evidence implicating abnormalities within the hypothalamic-pituitary-adrenal (HPA) axis in the pathogenesis MDD. It has been well documented that stressful life events can enhance the risk for MDD, especially early in life. Likewise, there is ample evidence that early stressful life events, such as childhood physical or sexual abuse can permanently alter the HPA axis (for more details see Chapter 5). In line with that, patients with MDD have altered cortisol levels both at rest and in response to stressors. One method of

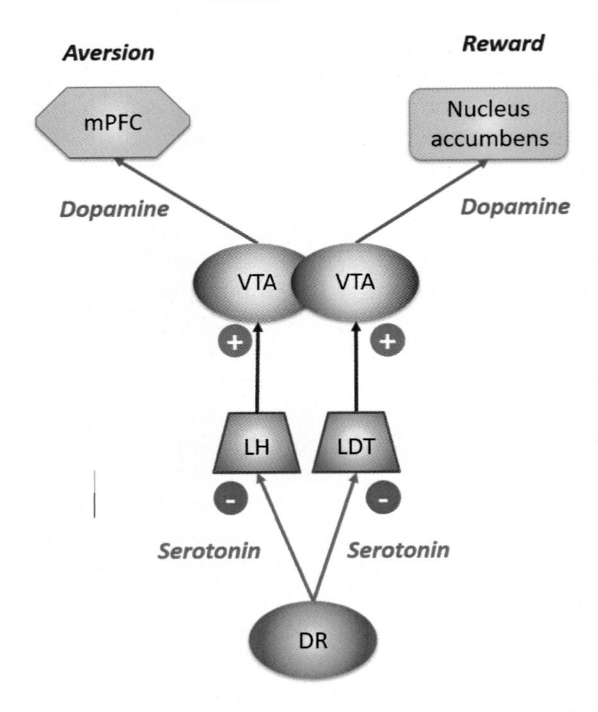

FIGURE 7.4 **The interactions between serotonin and dopamine in the pathophysiology of MDD.** Recent work using optogenetic techniques has shown that within the dopaminergic cell body region of the ventral tegmental area (VTA) two functionally different cell groups can be found. The cells projecting to the medial prefrontal cortex (mPFC) are controlled by the lateral habenula (LH) and, when stimulated, induce aversion. On the other hand the VTA cells that project to the nucleus accumbens are controlled by the laterodorsal tegmentum (LDT) and, when stimulated, induce reward seeking behavior. Both the LH and the LDT receive an inhibitory serotonergic influence from the dorsal raphe (DR), thus providing a direct interaction between serotonin and dopamine in the pathophysiology of MDD.

assessing the stress HPA-axis responsiveness is using the cortisol awakening response (CAR) as it represent a physiological response to an external stimulus. Studies in patients have suggested that the response depends on the severity of the disease with the CAR being normal in unaffected individuals, increased in patients with moderate symptoms and decreased in patients with severe levels of anhedonia (Dedovic and Ngiam, 2015). A recent, large study also reported increased CAR in both at-risk patients (a combined group including individuals with subclinical depressive symptoms and remitted patients) and patients with full-threshold MDD (Verduijn et al., 2015). These data suggest a complex interaction between stress, the HPA axis and MDD, presumably due to the complexities of the homeostatic regulation of the HPA-axis (see also Fig. 5.4). In line with an altered HPA axis system, it has been shown that a substantial number of patients with MDD have a blunted dexamethasone suppression (Frodl and O'Keane, 2013). Dexamethasone is a synthetic cortisol analogue that (as it does not penetrate the blood brain barrier) stimulates the inhibitory feedback mechanism at the level of the pituitary gland, thus leading to a reduction in circulating cortisol. As this suppression is less in patients with MDD, this suggests that the feedback inhibition is decreased. An adaptation of this tests (the DEX/CRH test which combines dexamethasone with CRH) has been shown to be somewhat more sensitive, although other psychiatric patients (such as patients with schizophrenia) and family members of MDD patients that themselves have no mental illness, also show abnormal suppression (Frodl and O'Keane, 2013).

In conclusion, much research has been done to elucidate the neurobiology of MDD and although great progress has been made, many details are still lacking, especially in relation to the integration of the currently available data. We already pointed out some of the possible relationships such as between dopamine and serotonin. Likewise, both these monoamines are crucially important in the response to stress and especially serotonin has an important influence on the HPA axis (Mahar et al., 2014). Finally chronically elevated levels of cortisol are known to downregulate the production of BDNF, which in turn reduces neurogenesis, and (at least in animals) reduces gray matter volume in several brain regions including the hippocampus where the changes are most prominent (McEwen et al., 2015).

The Pharmacological Treatment of MDD

The first effective pharmacological treatments for MDD were serendipitously discovered in the 1950s, an era often referred to as a psychopharmacological revolution, as it also saw the discovery of the first antipsychotics and anxiolytics. In 1952 studies began on a drug called iproniazid for the treatment of tuberculosis (Lopez-Munoz and Alamo, 2009) and the researchers soon found that the drug had significantly more central nervous system "side effect" than isoniazid. Indeed the psychological changes induced by iproniazid were remarkable, with an increase in vitality, social activity and authors reported "patients dancing in the halls, though there were holes in their lungs". However, as the safety profile was considered inferior to isoniazid, it was quickly abandoned for the treatment of tuberculosis. Fortunately, several researchers, of which Nathan Kline is most well-known, picked up on the "side effects" profile of the drug and showed it had in fact antidepressant effects in nontuberculosis patients as well, and by 1958 more than 400,000 patients with depression had been treated with iproniazid (in spite of the fact that it was only marketed as an anti-tuberculosis treatment). The fact that iproniazid but not isoniazid had antidepressant effects, while both had antituberculosis effects, indicated that these two drug actions must have different underlying mechanisms, and it was soon suggested that it was the inhibitory effect of iproniazid on monoamine oxidase (MAO) that was responsible for its antidepressant activity (see Fig. 7.4). MAO is an intracellular enzyme mainly located on the outer membrane of mitochondria and, as the name indicates, its main function is to metabolize monoamines through oxidative deamination both in the brain and in other parts of the body, most notably the liver. Following the success of iproniazid, a series of additional MAO-inhibitors soon followed, including tranylcypromine and phenelzine which together had a market share of about 90% in the mid-1980s. In spite of this success, MAO-inhibitors have several significant side-effects, the most dangerous of which is the so-called "tyramine" or "cheese" effect. Tyramine acts as a releaser of catecholamines, such as noradrenaline and adrenaline in the periphery (as it does not penetrate the blood brain barrier). Under normal circumstances consuming tyramine rich food (such as cheese, but also bananas) or drinks (most alcoholic beverages contain tyramine) has no significant consequences as tyramine is rapidly degraded by MAO. However, when an individual is taking MAO-inhibitors, tyramine levels can rise rapidly, leading to significant increases in (nor)adrenaline release, which in turn can lead to a strong increase in heart rate and blood pressure culminating in a (potentially fatal) hypertensive crisis. Hence, patients of MAO-inhibitors have to follow a strict diet to prevent such a crisis, especially as many MAO-inhibitors bind irreversibly.

In 1968 it was shown that there are in fact two different forms of MAO, called MAO_A and MAO_B. As the traditional MAO_i including iproniazid, tranylcypromine and phenelzine were irreversible inhibitors of both types of MAO, this offered the opportunity to develop more selective and reversible MAO inhibitors, with the potential of producing less side effects. Moclobemide was the first of the reversible selective MAO_A inhibitors, while deprenyl was found to be a selective (though irreversible) inhibitor of MAO_B. Studies have shown that deprenyl (although effective in the treatment of Parkinson's disease) has virtually no antidepressant effect, while moclobemide has (albeit to a lesser extent than the traditional nonselective irreversible MAO inhibitors). While noradrenaline and serotonin are preferentially metabolized by MAO_A, dopamine is metabolized by both forms, which suggests that the antidepressant actions seem more related to noradrenaline and serotonin.

A second major breakthrough in the treatment of MDD came with the identification of the tricyclic antidepressants. Like with iproniazid, their anti-depressive properties were discovered by a combination of serendipity and careful observation. In 1952, Delay and Deniker reported on the successful treatment of psychosis with the tricyclic phenothiazine compound chlorpromazine (see Chapter 9). Subsequently, several other compounds with similar tricyclic structures were tested. In 1956, Roland Kuhn, a psychiatrist in Münsterlingen, near Basle in Switzerland tested one of these compounds called imipramine and found it completely lacked antipsychotic efficacy, indeed it seemed to make some patients even more agitated (Brown and Rosdolsky, 2015). However, Kuhn noticed that in three patients with depressive psychosis, their depression clearly improved. Subsequent testing in another 37 patients proved imipramine's antidepressant action. Several years later additional tricyclic antidepressants (TCAs) came onto the market including amitryptyline, desipramine, and nortriptyline. It was subsequently shown that all TCAs inhibit the reuptake of noradrenaline and serotonin (Fig. 7.5), although there are clear differences in selectivity. Unfortunately, all TCAs also inhibit a number of other receptors, such as the histamine H_1 receptor, the α_1 adrenoceptor and the muscarinic acetylcholinergic receptors, leading to a variety of autonomic side effects, as well as weight gain and sedation.

In contrast to the development of the MAO-inhibitors and the TCAs, the development of the third class of antidepressants was a direct consequence of a hypothesis driven program (Lopez-Munoz and Alamo, 2009). As discussed earlier, in the mid-1960s one of the prevailing theories of depression was the serotonergic hypothesis, which suggested that reduced serotonin levels were responsible for the symptoms of depression. Hence, selective inhibitors of the serotonin transporters were developed with fluoxetine being the first to be tested (getting FDA approval in 1987). This was quickly followed by several others, including fluvoxamine, citalopram and paroxetine.

Likewise, dual serotonin/noradrenaline reuptake inhibitors (SNRIs) have been developed rationally, based on the idea that TCAs have good anti-depressant effects and that their side effects are predominantly due to the blockade of other receptors. Thus more selective SNRI's such as duloxetine, venlafaxine, and its active metabolite desvenlafaxine were developed. In addition, selective noradrenaline reuptake inhibitors such as reboxetine have been developed but are only approved in certain countries for the treatment of MDD.

Finally, several so-called atypical antidepressants have been developed which do not belong to any of these classes, such as bupropion (which could be classified as a noradrenaline/dopamine reuptake inhibitor), mirtazapine (which blocks a number of different receptors

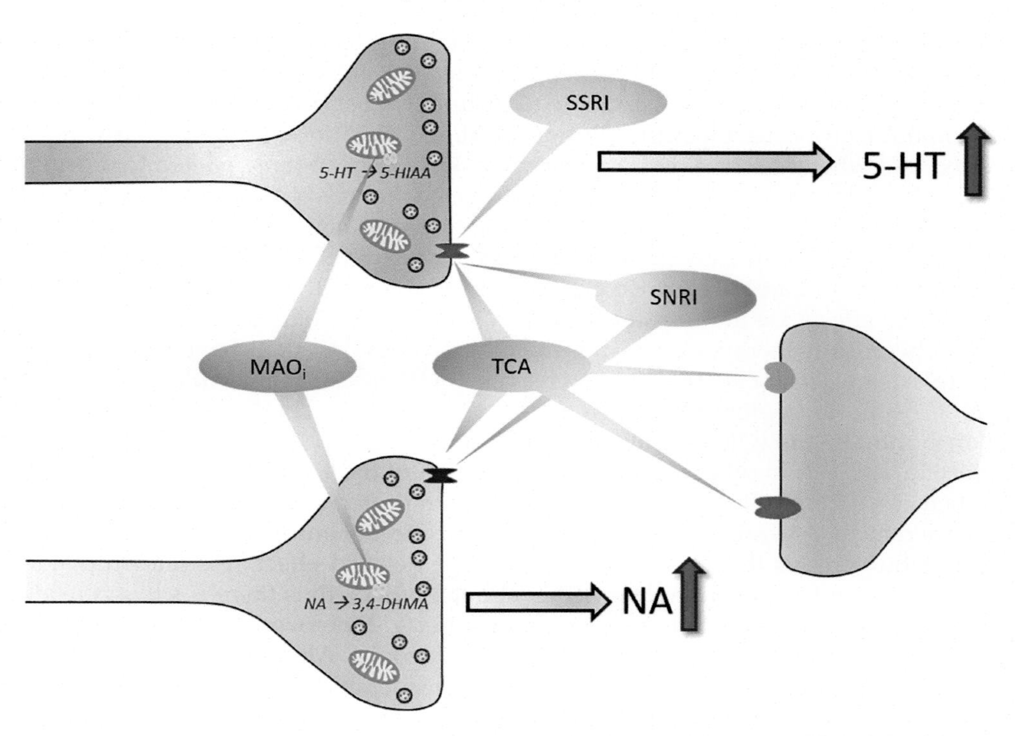

FIGURE 7.5 **The acute mechanisms of action of the most important classes of antidepressant drugs.** A large number of drugs have been found to improve depressive symptoms in patients with MDD, and all major classes acutely lead to an increase in serotonin (5-HT) and/or noradrenaline (NA). Many of these can be classified in one of 4 major groups: The monoamine oxidase inhibitors (MAO_i) block the action of the enzyme monoamine oxidase A, thereby preventing the metabolism of 5-HT into 5-hydroxyindoleacetic acid (5-HIAA) and of NA into 3,4-dihydroxymandelic acid (3,4-DHMA).Tricyclic antidepressants (TCA) block the 5-HT and NA transporter, while also affecting a variety of other postsynaptic receptors. Selective serotonin reuptake inhibitors (SSRI) on the other hand, selectively block the 5-HT transporter. Finally, serotonin, noradrenaline reuptake inhibitors (SNRI), like the TCA block both the 5-HT and NA transporter, but have much less influence on other postsynaptic receptors.

such as the $5HT_{2A}$ and $5\text{-}HT_{2C}$ receptor), and agomelatine (which blocks the $5HT_{2C}$ and stimulates the melatonergic MT_1/MT_2 receptor).

With so many different antidepressant drugs, it is difficult to discuss the risks and benefits of each of these drugs, or even the different classes of drugs. Suffice it to say that all classes of drugs show significant improvement of depressive symptoms, although the side effect profiles differ. Due to the potentially lethal tyramine-evoked hypertensive crisis, MAO-inhibitors have lost most of their attractiveness as treatments for MDD. Likewise, the presence of significant autonomic side effects limits the usefulness of TCAs. Therefore, first line of treatment for MDD is currently either an SSRI or an SNRI. Whether both types of drugs have the same beneficial effects or whether SNRIs are superior is still debated. Given that the selective noradrenaline reuptake inhibitor reboxetine has antidepressant properties (albeit inferior to SSRIs), SNRI may be superior to SSRIs. In line with this, a recent meta-analysis showed that fluoxetine was inferior to venlafaxine (Magni et al., 2013). Likewise, in a meta-analysis including 93 trials (with over 17,000 participants), SNRIs were slightly though significantly more effective than SSRIs (Papakostas et al., 2007), although whether this difference is clinically relevant is debatable. In an interesting study an attempt was made to distinguish between the roles of noradrenaline and serotonin transporter blockade (Pringle et al., 2013). Although preliminary, the results suggested that whereas serotonin transporter blockade may be more related to relieving distress, noradrenaline transporter blockade may have greater relevance for reducing anhedonia.

In spite of about 50 years of experience with antidepressants, their exact mechanism of action is still far from completely understood. As illustrated in Fig. 7.5 all antidepressants will lead to enhanced extracellular concentrations of serotonin and/or noradrenaline. However, whereas these effects are induced fairly rapidly, usually after only a few days, it takes up to 4–6 weeks before any therapeutic effect is noticeable. Several theories have been developed to explain this delay. First of all, the rapid increase in 5-HT release will stimulate 5-HT1A autoreceptors. These autoreceptors are located on the serotonergic cell bodies (and some are also located on terminals) and when stimulated inhibit serotonergic cell firing leading to a reduction in extracellular 5-HT. In other words, the stimulation of these autoreceptors counteracts the initial increase in extracellular 5-HT. However, with repeated treatment, these 5-HT_{1A} autoreceptors are rapidly downregulated allowing the full effect of antidepressants on 5-HT release to occur. Although this theory explains some of the delayed effect, it is known that 5-HT_{1A} receptor downregulation is complete within about 2 weeks, thus there is still a considerable lag between the molecular and the therapeutic effects. Moreover, some experiments have been performed combining SSRIs with a 5-HT_{1A} antagonist. Although this combination was found to be somewhat beneficial, there was still a considerable lack between the start of treatment and the onset of a therapeutic effect. One of the most compelling theories to explain the (additional) lag is illustrated in Fig. 7.6 and involves the activation of BDNF (Duman, 1998). As discussed in the previous section there is evidence that patients with MDD, presumably as a result of alterations in the HPA axis have reduced levels of BDNF, which may (at least in part) be responsible for a reduction in neurogenesis, dendritic spine arborisation and hippocampal volume. Studies have shown that chronic treatment with virtually all antidepressants (as well as electroconvulsive therapy) increase BDNF levels. This is likely due to the fact that stimulation of specific 5-HT receptors (most notable 5-HT_4, 5-HT_6 and 5-HT_7) enhances the production of cAMP. Although 5-HT_{1A} receptors inhibit the production of cAMP, given the aforementioned downregulation of these receptors upon repeated treatment, this effect will diminish over time. cAMP will subsequently activate the enzyme protein kinase A, which, as the name implies phosphorylates many different proteins. One of these is CREB (cAMP response element binding) protein. CREB is a transcription factor (see Chapter 2) which binds to specific sequences of DNA to enhance transcription of a number of different genes, one of them being BDNF. BDNF then subsequently increases dendritic arborisation, spine formation and neurogenesis. These processes obviously take time and can therefore explain the delayed therapeutic response. Although this theory is not uniformly supported, there is convincing evidence from both animal and human research that increased BDNF and enhanced hippocampal volume are prerequisites for the therapeutic actions of antidepressants (Fig. 7.6).

In addition to the side-effects and the delayed action, current antidepressant therapy suffers from relatively low response rates. Although the numbers differ from study to study, overall at least 30 to 60% of patients with MDD do not adequately response to antidepressant treatment (Colman et al., 2011; Hendrie and Pickles, 2013). Hence the need for improved

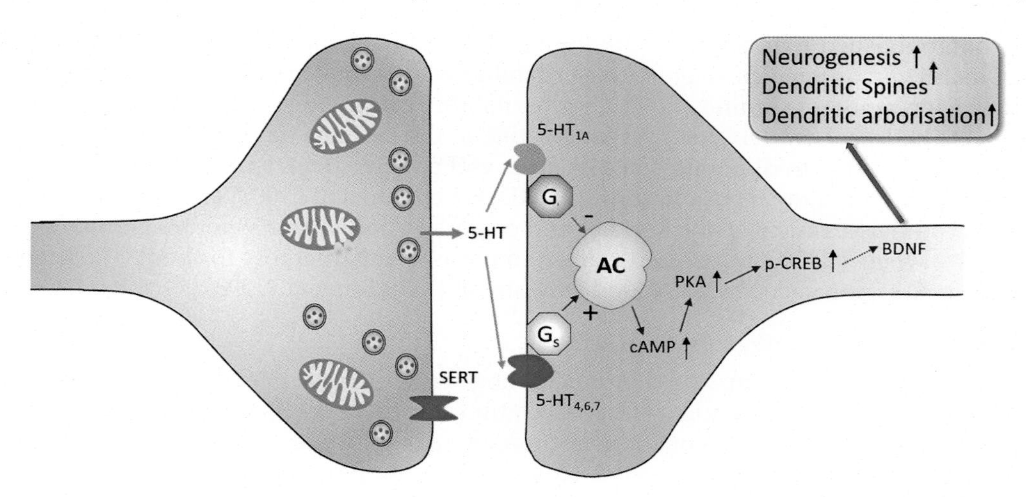

FIGURE 7.6 **The role of BDNF in the long-term effects on antidepressant drugs.** Although the mechanism by which long-term treatment with antidepressant drugs reduces the symptoms of MDD is still far from fully understood, substantial research has suggested a role for Brain Derived Neurotrophic Factor (BDNF). Ultimately almost all antidepressant drugs enhance serotonin (5-HT) neurotransmission, mostly by increasing extracellular 5-HT, which then stimulates specific receptors such as the 5-HT_{1A} and the 5-HT_4, 5-HT_6, and 5-HT_7. Especially the stimulation of the latter receptors will stimulate adenylate cyclase (AC) to produce cyclic adenosine-monophosphate (cAMP). cAMP then initiates a cascade of events, involving the activation of protein kinase A (PKA) which ultimately phosphorylates cAMP response element-binding protein (p-CREB). CREB is a transcription factor that binds to specific DNA sequences (the so called cAMP response element, CRE) to enhance the transcription of specific genes, including BDNF. BDNF subsequently induces a range of structural changes in the brain, including neurogenesis and dendritic spine growth, which are likely to underlie the therapeutic actions of antidepressant drugs. Note that the production of mRNA of BDNF takes place in the cell nucleus, but that for illustration purposes it is shown in the dendritic region here.

medication is obvious. However, for the time being, several drugs have been approved specifically for treatment-resistant depression. These include the second generation antipsychotics aripiprazole and quetiapine either alone or in combination with existing antidepressant therapy and the combination of olanzapine and fluoxetine (Kasper, 2014). Although these combinations have proven effective (Connolly and Thase, 2012; Han et al., 2013), the effect sizes are not very large, which may not be so surprising given the fact that these patients have already shown to be resistant to at least two other antidepressant drugs. How these second generation antipsychotics induce their therapeutic effect has yet to be identified conclusively. However, the fact that only some of the antipsychotics seem to be effective, suggests that their therapeutic effect is unrelated to their antipsychotics efficacy.

The Aetiology of MDD

MDD, like all other psychiatric disorders discussed in the present volume, runs in the family, and twin and adoption studies have supported the idea that MDD has a substantial heritability. In a meta-analysis of 5 family studies on the risk of MDD in first degree relatives, a remarkable degree of consistency was noticed, with odds ratios in first degree relative between 2.21 and 4.57 (Sullivan et al., 2000). In the same analysis, heritability was estimated on the basis of 6 published twin studies. Although some studies found differences in the heritability of males versus females, overall the heritability was estimated to be 37%. However, other studies have reported significantly higher heritability rates. This is likely to be due (at least in part) to different models being used to estimate the nongenetic component (see Box 4.1) but may also be due to differences in diagnosis and patient characteristics. Thus, heritability seems higher with more recurrent forms of MDD (Kendler et al., 2007) and seems to increase with repeated assessment. Thus Kendler and coworkers found that whereas heritability was about 40% when MDD was assessed using a single assessment, it rose to about 70% when the diagnosis was confirmed with a second assessment (Kendler et al., 1993).

Genetic Factors Contributing to MDD

Many studies have been aimed at elucidating the molecular genetic factors involved in MDD, using linkage, and genome wide association studies (GWAS) as well as candidate gene approaches. As discussed in a previous section, there is evidence linking MDD to alterations in monoaminergic pathways, the HPA-axis and neurotrophic factors. Hence, many of the candidate gene studies have focused on genes related to these systems (see Table 7.1).

With respect to the monoaminergic system, most attention has focused on serotonergic neurotransmission, and an association has been reported with several components, although not all were strong enough to be confirmed in a meta-analysis (Lopez-Leon et al., 2008). As discussed in previous chapters, one of the most important determinants of serotonergic neurotransmission is the serotonin transporter and arguably the most investigated genetic marker in psychiatry is the 5-HTTLPR (serotonin transporter linked polymorphic region) which consists of several different variants, the most important of which are the s- and l-allele which codes for a short and long repeat variant of the promoter with the s-allele leading to significantly lower levels of serotonin transporter than the l-allele. Several studies have suggested that the s-allele is associated with MDD, although others studies have failed to confirm this. However, in a meta-analysis, the association was confirmed (Lopez-Leon et al., 2008), although the effect is relatively small (odds ratio of about 1.1 for each s-allele). Nonetheless, given the frequency of the s-allele (up to 40% of Caucasians and about 70%–80% of Asians carry at least one copy of the s-allele) on a population level, the effects of the s-allele on MDD may be more important than its odds ratio suggests. Interestingly, in addition to the s-allele being associated with MDD, the l-allele is associated with response to SSRI efficacy (at least in Caucasians). Given that the l-allele leads to higher levels of serotonin transporter, this association is not unexpected (Porcelli et al., 2012).

In addition, MDD has been associated with several other components of the serotonergic system, most notably the synthesizing enzyme tryptophan hydroxylase, and the serotonergic receptors 5-HT1A and 5-HT2A. With respect to the *HTR1A* gene, two variants have been studied. The first, rs6295 is located in the promoter region of the gene, and the risk (G) allele is known to influence transcription. In a recent meta-analysis involving over 3,000 MDD patients and 4,000 controls, a significant association was found between MDD and rs6295 (Kishi et al., 2013). The second variant rs878567 also affects the binding of 5-HT$_{1A}$ receptors. In a study in healthy volunteers, individuals with a "C" allele had significantly higher binding potential than individuals homozygous for the "T" allele. Studies in patients with MDD have found a significant association between rs876567 and MDD, which was confirmed in a meta-analysis with over 1,000 patients and 1,700 controls (Kishi et al., 2013). The most studied

TABLE 7.1 Major Genetic Factors Implicated in MDD

Genes/Regions	Remarks
Single nucleotide variants	
CRHR1	Corticotropin release hormone receptor 1: association between the "T-A-T-G-G" haplotype of rs7209436 and rs173365 and the diagnosis of MDD
CRHR2	Corticotropin release hormone receptor 2: association between the "C" allele of rs3779250 and diagnosis of MDD
DBH	Dopamine-β-hydroxylase: association between a 19 base pair deletion in the promoter region and diagnosis of MDD
FKBP5	FK506 Binding Protein 5: association between several variants and diagnosis of MDD.
HTR1A	5-HT1A receptor: association between the "G" allele of rs6295 and the "T" allele of rs878567 and diagnosis of MDD
MAOA	Monoamine oxidase A: association between the VNTR in the promoter region and the "T" allele of rs1137070 and diagnosis of MDD.
PCLO	Piccolo: association between the "C" allele of rs2522833 and the diagnosis of MDD
SLC6A3	Dopamine transporter: association between a 40 base pair VNTR in the 3′-UTR and diagnosis of MDD.
SLC6A4	Serotonin transporter: association between the "s" allele of the 5-HTTLPR and diagnosis of MDD
Copy number variations	
4q28.3	Deletions have been associated with MDD
5q35.1	Duplication was found in 5 unrelated patients with MDD
7p21.3	Microdeletions have been associated with MDD
9q23	Deletions have been associated with MDD
15q11.2	Duplications have been associated with MDD
15q26.3	Microduplication have been associated with MDD
16p11.2	Both deletions and duplications have been associated with MDD
17q21.31	Duplications have been associated with MDD
18p11.31	Microdeletions have been associated with MDD

Please see text for further explanations.

variant in the *HTR2A* gene is rs6311. Although a recent meta-analysis showed no significant association between MDD and the risk (G) allele, there was a strong tendency and with relatively small number of patients (1491 vs 2937 controls) the sensitivity analysis showed the meta-analysis to be unstable (Jin et al., 2013). However, even with larger number of patients and controls, no significiant association with MDD was found with this variant or rs6313 (Tan et al., 2014; Zhao et al., 2014).

Another monoamine related gene that has been associated with MDD is the metabolizing enzyme *MAOA*. Several different variations have been investigated, including a VNTR in the promoter region, a CA repeat microsatellite in intron 2 and rs6313, which is a synonymous mutation, although there is some evidence that the "G" allele leads to higher MAO_A activity. Several studies have shown that the VNTR in the promoter region is associated with MDD, which was supported by a meta-analysis (Fan et al., 2010), while the association with the other variant was not reliably proven. Recently another variant (rs1137070) showed significant associations with major depressive disorder with an odds ratio of 1.3 for the "T" allele (Liu et al., 2015).

MAO_A is not only important for the breakdown of serotonin, but also of noradrenaline and (to a lesser extent) dopamine. Hence, studies have also looked at other genes within these monoamines. Whereas some studies have reported an association between MDD and rs2242446 in *SLC6A2* (coding for the noradrenaline transporter), which is located in the 5′ flanking promoter region and could influence transcription, two meta-analyses failed to find convincing support (Lopez-Leon et al., 2008; Zhou et al., 2014). On the other hand, association was found with a 19 base-pair deletion in the promoter region of the *DBH* gene which codes for dopamine-β-hydroxylase, the enzyme that converts dopamine to noradrenaline (Zhou et al., 2015).

Related to dopamine neurotransmission, a 40 base pair VNTR within the 3′ untranslated region of the dopamine transporter (*SLC6A3*) was also convincingly associated with MDD (Lopez-Leon et al., 2008). In addition, an association was found between rs2399496 of the *DRD3* gene and MDD, with the "AA" genotype significantly enhancing the risk for MDD as well as for MDD with nicotine dependence comorbidity (Korhonen et al., 2014). Finally, several studies have identified an association between MDD and the *PCLO* gene. *PCLO* encodes for the protein piccolo, which is involved in the presynaptic regulation of monoamine release. Several studies have reported an association between MDD and rs2522833 with the "C" allele conferring an increased risk (Hek et al., 2010; Minelli et al., 2012). Interestingly, this genotype is also associated with lower activation in the insula and cingulate cortex during an emotional task (Woudstra et al., 2013), and stronger alterations in the HPA axis (higher baseline cortisol levels, and more blunted dexamethasone/CRH suppression (Schuhmacher et al., 2011).

With respect to the HPA-axis, several different genes have been investigated. One of the most important component of the HPA axis is corticotropin releasing hormone (CRH) which is known to interact with two different receptors, CRH_1 and CRH_2, coded by the genes *CRHR1* and *CRHR2*. Significant associations have been found between variants in both receptor genes and MDD. For instance in a study from Japan, the SNP rs110402 and rs242924 in the *CRHR1* and the rs3779250 in the *CRHR2* were associated with MDD. Moreover, the T-A-T-G-G haplotype consisting of rs7209436 and rs173365 in *CRHR1* was positively associated with MDD (Ishitobi et al., 2012). This last haplotype was also found to be associated with MDD in a study in Spain (Ching-Lopez et al., 2015). Downstream from CRH, cortisol also interacts with two receptors, the glucocorticoid (GR) and the mineralocorticoid receptor (MR). Although variants in these receptors have not been convincingly associated to MDD, some papers have found an association with the *FKBP5* gene. The gene codes for a co-chaperone protein essential in GR signaling. FKBP5 binds to heat-shock protein 90 (HSP-90) which is important in the translocation of the cortisol/GR complex to the cell nucleus where it can interfere with gene transcription (Zannas et al., 2016). In fact, FKBP5 inhibits this translocation, and it needs to be exchanged by FKBP4 before translocation can take place. Hence genetic alterations within the *FKBP5* gene can significant alter the effects of cortisol. Several different variants have been found associated with MDD (Szczepankiewicz et al., 2014). Interestingly, the rs1360780 SNP within the *FBKP5* has also been associated with the response to antidepressant drugs (Lekman et al., 2008; Horstmann et al., 2010; Stamm et al., 2016).

Finally, several genes related to neurodevelopment have been studied, most extensively the genes for BDNF and its receptor (*BDNF* and *NTRK2*). However, in spite of the clear role of BDNF in the pathophysiology of MDD (as discussed earlier), there is no hard evidence for a genetic association (Gyekis et al., 2013). Moreover, although there is some evidence that depressed patients homozygous for the "A" allele of rs11140714 of *NTRK2* have significantly reduced volumes of specific brain regions involved in emotional regulation (Murphy et al., 2012) an association between this, or any other variant in the NTRK2 gene and MDD has yet to be conclusively proven.

GWAS, in contrast to candidate gene studies, are hypothesis free and aim to investigate a very large number of genes and gene variants. As a result, large numbers of patients are required, especially given that most genetic factors have only a small effect on the total MDD phenotype (Odds ratios are generally around 1.2–1.4). The first GWAS study in depressed patients was part of the Genetic Association Information Network (GAIN) and involved 1,738 MDD cases and 1,802 controls (Sullivan et al., 2009b). Although over 435,000 single nucleotide variants were analyzed, none reached the GWAS level of significance ($p < 5 \times 10^{-8}$). However, among the 200 most significant variants were 11 within the *PCLO* gene, which was already mentioned earlier. Moreover, although a replication in an independent sample did not lead to significant effects, when the two samples were combined, the significance of rs2522833 was very close to the GWAS significance (6.4×10^{-8}). A GWAS study focused on early onset depression (GenRED, Genetic of Recurrent Early onset Depression) identified a location on 18q22.1 which was close to GWAS significance (Shi et al., 2011). A GWAS study from the STAR*D (Sequenced Treatment Alternatives to Relieve Depression) again failed to find a significant signal (Shyn et al., 2011). However, by combining the GAIN, GenRED, and STAR*D three suggestive associations were identfied: *ATP6V1B2* (which codes for a subunit of a vacuolar protein pump ATPase), *SP4* (which codes for a brain specific Sp4 zinc finger transcription factor) and GRM7 (which encodes for the $mGluR_7$ receptor). In a meta-analysis of two European cohorts, *GRM7* again was the best performing gene, although it failed to reach statistical significance (Muglia et al., 2010). The paucity of statistically significant results from GWAS studies has been suggested to be due to a lack of power. However, when a mega-analysis was done on 9,240 cases of MDD and 9,519 controls, again no single variant reached genome-wide significance (Hindorff et al., 2009). Since then several different GWAS studies have been performed. In a recent review, fifteen GWAS studies were reviewed and in only one study a single GWAS significant association was found (Dunn et al., 2015): a variant in rs1545843 of the *SLC6A15* gene, a neutral amino acid transporter. None of the other fourteen studies found any significant association.

In addition to single nucleotide variants and other relatively small variations, there is also evidence for an association between more structural chromosomal alterations and MDD. Several GWAS studies have been performed in this respect. In a study from Germany with 604 MDD patients and 1,643 controls four different chromosomal regions were identified in patients: microdeletions in 7p21.3 and 18p11.32, microduplications in 15q26.3 and both deletions and multiplications in 16p11.2 (Degenhardt et al., 2012). In another study including over 1,600 patients and more than 4,000 controls, a specific duplication was found in 5q35.1 in five unrelated cases of MDD (Glessner et al., 2010). This duplication encompassed (among others) the *SLIT3* gene which plays an important role in axon guidance. In a recent study involving treatment-resistant patients with MDD there was some evidence for increased duplication events (especially smaller ones in the 100–200 kb range) most notable in 15q11.2 and 17q21.31, in addition to deletions of 4q28.3 and 9q23 (O'Dushlaine et al., 2014).

Moderation of the Effects of Genes by Environmental Factors

Arguably one of the most influential papers in the field of gene–environment interactions in MDD (and probably in psychiatry in general) was the paper by Caspi and his coworkers published in 2003 on the interaction between stressful life events and the 5-HTTLPR (Caspi et al., 2003). In this paper, based on the Dunedin Multidisciplinary Health and Development Study, the authors investigated 847 Caucasian individuals at age 26 and stratified them into three genotypes based on the deletion/insertion in the promoter region of the *SLC6A4* gene into "ss," "sl," and "ll." In addition, both childhood adversities and stressful life events in the last five years (between 21 and 26) were assessed. Overall, the study found that individuals with the "ss" genotype were much more sensitive to the influence of stressful life events in the past 5 years than individuals with the "ll" genotype, with the "sl" genotype showing an intermediate response. This increased sensitivity was seen with respect to self-reported depression, probability of a major depressive episode, probability of suicide attempt and informant reports of depression. A similar gene–environment effect was also seen with respect to childhood maltreatment, with the "ss" genotype again being much more sensitive to "probable" and "severe" maltreatment than individuals with the "ll" genotype. This paper sparked a flurry of research aimed at replicating these important findings, with many groups replicating these findings and others failing. In a very detailed review 81 studies (including 55,269 participants) were identified and compared (Sharpley et al., 2014) and overall the meta-analysis supported the hypothesis that individuals with the "ss" genotype were more sensitive to the detrimental effects of stressful life events, irrespective of whether these were assessed objectively or through self-report (although the effects were slightly stronger with the objective assessment). Likewise, the interaction was seen with both childhood and adult stressful life events, with the effects of childhood adversity being stronger than later life stressful events. Finally, the interactive effects were seen independent of the research design (longitudinal, case-control, or cross-sectional design). Thus the majority of studies seems to be supportive of the original studies of Caspi and coworkers, which begs the question about the underlying mechanisms. In this respect a recent paper found that while individuals with the "s" allele have lower levels of mRNA (in blood cells), this effect is potentiated by both prenatal maternal stress and childhood trauma (Wankerl et al., 2014). Interestingly, this effect was not mediated via an increase in DNA methylation in CpG sites in the promoter region of the *SLC6A4* gene, which is an epigenetic mechanism generally leading to a suppression in gene expression (see Chapter 5). In addition to the interaction between the 5-HTTLPR genotype and stressful life events, several other papers investigated the relationship between the serotonergic system and stress. For instance, carriers of the "C" (rs878567) allele of the *HTR1A* gene exposed to childhood physical abuse are more likely diagnosed with MDD than individuals with the "TT" genotype (Brezo et al., 2010). These authors found similar interactions between childhood physical abuse and the "AA" allele of rs3794808 of *SLC6A4* and several single nucleotide variants in the *HTR2A* gene (see Table 7.2). It was also found that carriers of the "T" allele of rs6313 within the *HTR2A* gene were more likely to develop

TABLE 7.2 Examples of Gene–Environment Interactions in MDD

Genes	Environments	Effects	References
BDNF	Early life stressors	Carriers of the "A" allele of rs 6265 exposed to early life stress have smaller amygdala, hippocampus and lateral prefrontal cortex and more likely to have MDD	Gatt et al. (2009)
BDNF	Stressful life events	Carriers of the "G" allele of rs6265 are more susceptible to stressful life events	Kim et al. (2007); Chen et al. (2012)
CRHR1	Early and later life stressful events	Carrier of the "A" allele of rs110402 exposed to both early and late stressful life events have more symptoms of depression	Starr et al. (2014)
CRN1	CPA	Carriers of the "G" allele of rs 1049353 exposed to CPA showed more self-reported anhedonia	Agrawal et al. (2012)
DRD3	Early life stressors	Carriers of the "A" allele of rs6280 exposed to early life stressors are more likely to have MDD	Henderson et al. (2000)
FKBP5	CPA	Carrier of the "TT" genotype of rs1360780 exposed to CPA are more likely to have MDD	Appel et al. (2011)
HTR1A	CPA	Carrier of the "C" allele of rs878567 exposed to CPA are more likely to have MDD	Brezo et al. (2010)
HTR2A	CPA	Carrier of the "A" allele of rs6561333 or the "GG" genotype of rs1885884 or the "C" allele of rs7997012 exposed to CPA have higher incidence of suicide attempts	Brezo et al. (2010)
HTR2A	Urbanity	Carriers of the "T" allele of rs6313 living in remote rural areas are more likely to have MDD	Jokela et al. (2007a)
MAOA	Childhood maltreatment	Carrier of <3.5 repeats of uVNTR exposed to maltreatments are more likely to have MDD	Cicchetti et al. (2007)
MR	Childhood maltreatment	Female carriers of the GA haplotype of rs5522 and rs2070951 exposed to childhood maltreatment are more likely to have MDD	Vinkers et al. (2015)
SLC6A2	Stressful Life Events	Carriers of the "C" allele of rs2242446 exposed to stressful life events are more likely to have MDD	Sun et al. (2008)
SLC6A4	CPA	Female carrier of the "ss" genotype exposed to CPA are more likely to have MDD	Caspi et al. (2003)
SLC6A4	Hurricane and social support	Carriers of the "ss" genotype exposed to hurricanes and low social support are more likely to have MDD	Kilpatrick et al. (2007)
SLC6A4	CPA	Carriers of the "AA" genotype of rs3794808 exposed to CPA are more likely to have MDD	Brezo et al. (2010)
TPH1	Social support	Carriers of the "A" allele of rs1800532 and of rs1799913 exposed to low levels of social support have more depressive symptoms	Jokela et al. (2007b)

Please see text for further explanations.
CPA, Childhood Physical Abuse.

MDD when they were reared in a remote rural area (Jokela et al., 2007a). Finally, an interaction between the *TPH1* gene and social support was found, showing that carriers with the "AA" genotype of rs1800532 and rs1799913 who were exposed to low levels of social support showed more depressive symptoms than individuals with the "GG" genotype, with the "AG" genotype showing an intermediate sensitivity (Jokela et al., 2007b). Overall, these data support the hypothesis that genetic modifications within the serotonergic system predisposes these individuals to the detrimental effects of stressful life events.

Given the significant influence of stress on the aetiology of depression, its moderation by HPA-axis related genes has also received considerable attention. In particular, the *FKBP5* gene has received substantial interest in this respect. Studies have shown that stressful life events interact with several different variants within the *FKBP5* gene (see Table 7.2) to enhance the risk of MDD. Moreover, the presence of severe traumatic life events in combination with the "CC" genotype of rs3800373 or the "AA" genotype of rs4713916 led to an earlier age of onset of MDD (Zimmermann et al., 2011). The influence of childhood physical abuse on the stress-induced cortisol response is moderated by rs1360780 so that carriers of the "CC" genotype have a blunted cortisol response when exposed to childhood adversity (Buchmann et al., 2014). Interestingly, the interaction between this genotype and childhood maltreatment also leads to changes in connectivity, as evidenced by higher mean diffusion rates and lower fractional anisotropy in several key brain regions such as the insula and the inferior frontal gyrus (Tozzi et al., 2016) in MDD patients with the "T" allele exposed to childhood maltreatment compared to patients with the "CC" genotype. Several other interactions between stressful life events and genetic variants with the HPA-axis have also been studied in relation to MDD. Vinkers and coworkers showed that carriers of the haplotype composed of rs5522 and rs2070951 within the *MR* gene interacts with childhood maltreatment in a complex sex-dependent manner (Vinkers et al., 2015). Thus whereas in both a clinical and population sample the interaction between the haplotype and childhood maltreatment was significant, this was due to the fact that female carriers of the "GA" haplotype displayed increased vulnerability (especially in the population-based sample), while male "CG" carriers showed increased resistance (especially in the clinical sample). These data emphasize the importance of including both males and female in the analysis and that differences may be observed depending on the severity of the symptoms (ie, whether depressive symptoms are studied in the general population or within a clinical setting). Finally, an interaction was found between both early and later life stressful events and the *CRHR1* gene, which codes for the corticotropin releasing hormone receptor 1 such that carriers of the "A" allele of rs110402 showed more depressive symptoms when exposed to stressful life events (Starr et al., 2014).

Several studies have also investigated the interaction between stress and BDNF, in line with the (presumed) role of BDNF in the pathophysiology of MDD and the influence of stress on BDNF levels (as discussed earlier). In line with this, several papers have found that carriers of the "G" allele of rs6265 (coding for valine at protein position 66) exposed to stressful life events have a higher incidence of depressive symptoms than carriers of the "AA" genotype (coding for methionine at position 66) (Kim et al., 2007; Chen et al., 2012). On the other hand, carriers of the "A" allele of rs6265 exposed to early stressful life events were found to have smaller hippocampus, amygdala, and lateral prefrontal cortex (Gatt et al., 2009). In addition, these carriers exposed to early stressful life events also showed reduced accuracy in a working memory task, increased resting heart rate, and heart rate response to several events (startle response, executive, and working memory task). Moreover, it leads to a reduced positive bias for happy faces in a working memory task (Vrijsen et al., 2014).

It is at present difficult to explain these different findings although the difference in timing of the stressor (early vs later life) may be an important factor. Another possible explanation is provided by several recent papers showing that the interaction between BDNF and stressful life events is mediated via an interaction with the 5-HTTLPR genotype. Thus several studies have found that childhood maltreatment is especially predictive of MDD in individuals that carry both the "s" allele of the 5-HTTLPR and the "G" allele of r6265 of the *BDNF* gene (Kaufman et al., 2006; Comasco et al., 2013; Ignacio et al., 2014).

Overall, many gene–environment studies have been conducted in relation to MDD, and there is strong evidence that stressful life events (both during childhood and later in life) enhances the risk of MDD, and that this effect is significantly stronger in individuals with genetic alterations in the serotonergic system, the HPA axis or BDNF. Moreover, the first studies have now been published investigating the neurobiological underpinnings of this interaction including animal studies as discussed later.

Animal models for MDD

Measuring MDD-Like Features in Animals

Modelling MDD in animals has proven particularly challenging. As with most other psychiatric disturbances we discuss in this book, there is considerable heterogeneity within the symptoms and thus in the patients, and likely in the aetiology and neurobiology as well. However, in addition to this, as discussed before and illustrated in Box 7.1, MDD patients can have diametrically opposite symptoms thus making it impossible to model the entire spectrum of MDD symptomatology in one model. As a result, most animal models have focused on the most salient symptoms of MDD.

One of the most often used test to assess a "depressive phenotype" in rodents is the forced swim test (FST). Originally developed as a screening test (see Chapter 3) for detecting antidepressant activity (Porsolt et al., 1991), it has since been repurposed as a model to detect "depressive symptoms" (see Fig. 7.7). In its simplest form the FST consists of two phases. In the first phase an animal is immersed in a relatively small cylinder of water with the water being deep enough to force the animals to swim (ie, the tail is not allowed to be able to touch the bottom). This phase is typically 10–15 min long. More importantly, in phase II (typically 24 h after phase I), the animals are placed back for an additional 5 min and the resulting behavior is analyzed. In this phase animals typically show floating behavior (in which the animals inflate their lungs and remain as immobile as possible). Since several (though not all) antidepressants decrease this immobility, it has been inferred that an increase in immobility is a sign of depression. Although this sounds like a logical assumption, it is important to recall the difference between a simulation model and a screening test (see for more details Chapter 3). Thus a screening test often uses a parameter that is not necessarily related to the disease itself. As an example, another screening test that was used for detecting antidepressants was the mouse-killing test, in which a mouse was placed together with a rat to induce aggressive behavior (Eisenstein et al., 1982). Although for obvious ethical reasons this test is no longer used, like the FST, several antidepressants significantly reduced mouse-killing behavior. Yet it seems unreasonable to assume that an increase in mouse-killing behavior is a

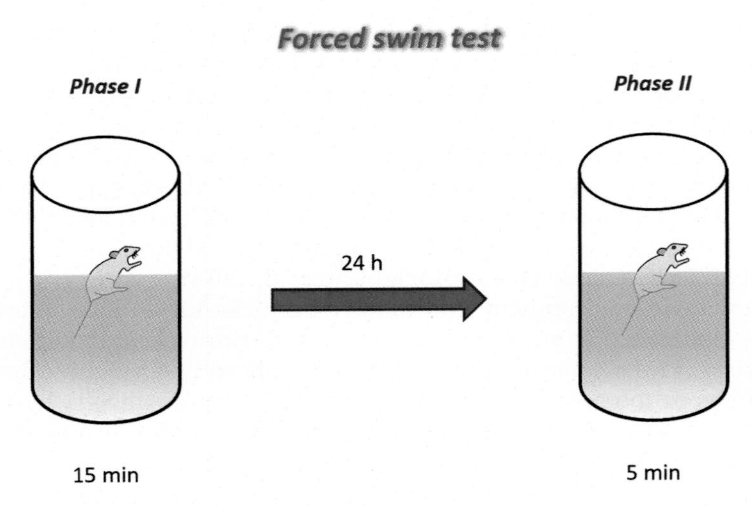

FIGURE 7.7 **The forced swim test.** Originally designed as a test for detecting antidepressant drugs, the forced swim test is still widely used as a test for identifying a "depression-like" phenotype in mice and rats. See text for a more detailed discussion about the relevance of the forced swim test for depression.

sign of depression. In the FST, it has been theorised that the increased immobility is a sign of "learned helplessness," which in turn is thought to be related to depression. However, there is not much convincing evidence for that. In fact, from an ethological point of view, inflating your lungs and remaining as immobile as possible seems the most efficient strategy to survive, especially as 24 h before, the animal was saved from the water after some time. Thus the validity of the FST (and the related tail suspension test which is often used in mice) is questionable. It is, nonetheless, still often used.

In relation to anhedonia, one of the most often used paradigms is the sucrose preference test (Willner, 1984). In this test, animals are typically given free access to two bottles, one filled with normal tap water, and the other with a moderate (usually 5%) sucrose solution. Normal animals will show a strong preference for the sweet sucrose solution, and a reduction in this preference is considered to be an indication of a reduced perception of pleasure. In some paradigms the sucrose intake is combined with lever presses, thus allowing to assess the effort an animal is willing to invest in order to obtain the sucrose. Related to this is the intracranial self-stimulation paradigm (D'Souza and Markou, 2010). It has long been known that animals will electrically stimulate themselves when they are implanted with an electrode in some specific brain regions, such as the medial forebrain bundle, and this paradigm has been extensively used to map the "reward system" in the brain. It has been suggested that, conversely, it may also be used to assess reduced pleasure. Typically, this is done by using a graded electrical current and determining the threshold for self-stimulation. The theory behind this is that an increase in threshold would indicate that the animals need a stronger stimulation in order to feel the reward/pleasure (ie, they exhibit anhedonia).

As mentioned earlier, there is evidence that pleasure (and hence anhedonia) has different aspects, such as anticipatory and consummatory pleasure and there is some data that patients with MDD suffer more from deficits in anticipatory than consummatory pleasure, and thus in addition to test such as the sucrose preference test, paradigms for anticipatory pleasure should also be included in models for depression. Several tests have been developed such as the anticipatory locomotor activity paradigm (Barbano and Cador, 2006), and the anticipatory vocalizations paradigm (Buck et al., 2014). In both paradigms, animals are food restricted. In the anticipatory locomotor paradigm, animals are usually habituated to an open field for several days before the experimental phase begins, so as to reduce the baseline locomotor (and rearing) activity. In the experimental phase, the animals are again placed in the open field but now after a set time, food is placed in the open field. Over the course of several days, locomotor activity and especially rearing) will increase rapidly in anticipation of the food being introduced. Likewise in the anticipatory vocalizations paradigm, animals are placed in a box and after a set time a stimulus is given (usually a light or a sound), followed after a fixed interval (2–3 min) by the presentation of food. However, rather than assessing motor activity, ultrasonic vocalizations are recorded. As discussed more fully in chapter 8, rats can vocalize in three distinct frequency ranges, with the range of 50 kHz and higher specific for positive affect. Hence a reduction in anticipatory pleasure should be accompanied by a reduction in 50 kHz calls.

In addition to these tests, many other aspects of MDD (such as alterations in sleep, HPA-axis activity and food intake) are also included from time to time in animal models. In the next section we will discuss several of the most commonly used genetic models for MDD.

Genetic Models for MDD

Over the years, numerous forward and reverse genetic animal models have been proposed for MDD, in addition to models using environmental manipulations such as maternal separation and chronic mild stress (Harro, 2013). Among the forward genetic models, the Flinders Sensitive (FSL) and Resistant (FRL) lines are probably the most often used (Wegener et al., 2012). These rats were originally selected on the basis of their response to the acetylcholinesterase inhibitor diisopropyl fluorophosphate (DFP). Although the aim was to develop a strain resistant to DFP, the selection actually led to the development of a strain that became progressively more sensitive (FSL) while the FRL line remained similarly sensitive to the Sprague Dawley rats from which they were originally selected. Further analysis showed that the FSL differed from the FRL primarily in an enhanced sensitivity of the muscarinic receptors. Although the serotonergic theories of depression were still quite dominant, some researchers also suggested that an overactive muscarinic cholinergic system may be related to (aspects of) depression and it was therefore suggested that the FSL rat may represent an animal model for depression (Overstreet and Wegener, 2013). Indeed over 25 years of research with this model, numerous depression-like phenotypes have been identified such as reduced bodyweight and locomotor activity, increased rapid eye movement (REM) sleep and cognitive deficits, in addition to an increased immobility in the forced swim test (Overstreet, 1993). However, no deficit was found in the sucrose preference test (Rea et al., 2014). On the other hand there is some evidence of a deficit in reward, as conditioned place preference in response to mother pup interaction is reduced in FSL rats (Lavi-Avnon et al., 2008). In agreement with this the release of dopamine in the accumbens was significantly reduced during mother–pup interaction in FSL rat. Moreover, the FSL rat seems to have a good predictive validity with respect to various classes of antidepressant drugs (Overstreet and Wegener, 2013).

In addition to the FSL, several other forward genetic models have been used in the study of MDD, such as the Fawn Hooded (FH) rats and several strains that were selectively bred for behavior in the forced swim test (Weiss et al., 1998) or the learned helplessness test (Vollmayr and Henn, 2001). The FH rats were originally selected from a selective breeding program for hypertension but were subsequently found to show increased immobility in the forced swim test. In addition, these rats show increased alcohol consumption (Overstreet et al., 2007). Intriguingly genetic research found that only one specific substrains of FH rats show both increased immobility and increased alcohol consumption indicating a different genetic basis for these two behaviors. The FH rats also shows depression-like alterations in the HPA axis, such as increased basal levels of corticosterone (Aulakh et al., 1993) and have alterations in the serotonergic system, including a blunted corticosterone response (Aulakh et al., 1993) and hypothermic response (Aulakh et al., 1994). These data are compatible with reduced serotonin receptor binding seen in patients with MDD. In addition, alteration in neuropeptide Y have been reported in depression-relevant brain regions such as the hippocampus (Mathe et al., 1998). However, as with the FSL rats, FH rats do not show signs of anhedonia in the sucrose preference test (Hall et al., 1998).

Given the biochemical and genetic association between MDD and serotonin, it is not surprising that many reverse genetic models targeting components of the serotonergic system have been investigated for (anti)depression-like phenotypes. For instance the serotonin transporter knock-out rats and mice show several typical depressive-like alterations such as increased immobility in the forced swim test and reduced sucrose preference (Olivier et al., 2008). However, the effects in mice were found to be dependent on the background strain, as SERT knockout mice on a 129/S background showed increased immobility (Holmes et al., 2002), while mice with a C57BL/6 background did not show increased immobility (Wellman et al., 2007) nor altered sucrose preference (Kalueff et al., 2006). On the other hand most of the other genetic alterations within the serotonin system result in an "antidepressant"-like phenotype rather than a depression-like phenotype. Thus, 5-HT_{1A} (Ramboz et al., 1998), 5-HT_{1B} (Mayorga et al., 2001) or 5-HT_7 (Guscott et al., 2005) receptor knock-out mice do not show changes in immobility in the forced swim test and/or the tail suspension or other aspects of anhedonia.

Genetic alterations in other monoaminergic (especially the noradrenergic) systems have also been investigated in relation to MDD but again, few of them lead to depression-like phenotypes (Cryan and Mombereau, 2004). However, the phenotype depends strongly on the receptor subtype that is altered. Thus, while the α_{2A} adrenoceptor knockout mice (Schramm et al., 2001) shows increase immobility in the forced swim test, as well as a disturbed diurnal sleep-wake cycle (Lahdesmaki et al., 2002), genetic deletions of the α_{2C} adrenoceptor lead to decreased immobility (Sallinen et al., 1999). A similar opposite effect has been found in mice overexpressing selective α1 adrenoceptor subtypes, with α_{1A} adrenoceptor overexpressing mice having decreased and α1B adrenoceptor overexpressing mice having increased immobility in the forced swim test (Barkus, 2013). Finally, in relation to noradrenaline, the noradrenaline transporter knockout mice seem to present a more antidepressant like phenotype as measured in the forced swim and tail suspension test. Unfortunately very few studies have been done in relation to anhedonia, and when performed, genetic alterations in the noradrenergic system generally do not affect sucrose preference. In this respect it is important to once more emphasize that MDD patients primarily suffer from deficits in anticipatory pleasure, rather than consummatory pleasure. Thus the inability to find differences in sucrose consumption may be more related to the nature of the test (measuring consummatory pleasure) rather than to the relevance of the genetic alteration for depression. Recently, the behavioural profile of a mouse with a mutation in the *PCLO* gene was investigated based on the rs2522833 variant associated with MDD. Unfortunately the animal did not show any alterations in either anxiety of depression like behaviors as measured in the forced swim test (Giniatullina et al., 2015).

In line with the clinical studies emphasizing alterations in the HPA-axis, several groups have also studied mice with a genetically altered HPA-axis with most studies focusing on CRH and its receptors. As with noradrenaline, opposing effects have been found with regard to the CRH receptors, with genetic ablation of CRH2 receptor leading to increased anxiety and immobility (Bale et al., 2000; Bale and Vale, 2003) while ablation of the CRH1 receptor leads to decreased anxiety (Timpl et al., 1998).

Overall then, although many different genetic animal models have been evaluated for a "depression"-like phenotype, few consistent findings have been reported. Moreover most of the studies have used the forced swim test, with few models showing anhedonia-like deficits. Moreover, very few models have been investigated with respect to anticipatory versus consummatory pleasure. Fortunately, there have been several studies showing that when specific genetic models are exposed to environmental challenges, the depressive-like phenotype may change, in line with the overwhelming evidence from clinical studies. We will discuss these interaction studies next.

Moderation of the Effects of Genes by Environmental Factors

Studies in the FSL rats showed that mothers from this line spent significantly less time nurturing (licking and grooming) compared to FRL rats (Lavi-Avnon et al., 2005). Cross-fostering of FSL to FRL mothers significantly diminished the immobility in the forced swim test, although the reverse was not true (ie, FRL pups reared by FSL mothers), thus indicating a gene–environment interaction (Friedman et al., 2006). As discussed earlier, FSL rats do not show an anhedonic phenotype (as measured using the sucrose preference test). However, when both strains were exposed to chronic mild stress the FSL rats showed a significantly stronger decrease in sucrose preference compared to the FRL rats (Pucilowski et al., 1993). Likewise, FSL rats were found to be more susceptible to early life stressors (see Table 7.3). Thus, when exposed to maternal separation (3h daily between postnatal day 2 and 14), FSL show a more severe exacerbation of the depressive phenotype, in terms of weight loss and immobility in the forced swim test (El Khoury et al., 2006). Moreover this same treatment led to significantly reduced levels of neuropeptide Y in the hippocampus and increases in neurotensin in the nucleus accumbens specifically in FSL rats (Jimenez-Vasquez et al., 2001; Wortwein et al., 2006). Several other FSL-specific molecular changes have also been reported (Wegener et al., 2012) some of which may have relevance for the depressive phenotype in FSL rats.

The impact of environmental challenges has also been studied in animals with a genetically compromised serotonin transporter (SERT). These studies have typically (although not exclusively) been done with animal heterozygous for the null mutation, as these animals usually have about a 50% reduction in transporters, similar to humans with the s-allele. Studies in mice have shown that heterozygous SERT knockout but not wild type mice, when

TABLE 7.3 Examples of Gene–Environment Interactions in Animal Models for MDD

Genes/strains	Environments	Effects	References
Forward genetics			
FSL	Maternal care	FSL exposed to low level maternal care show stronger immobility in FST	Friedman et al. (2006)
FSL	CMS	FSL exposed to CMS show decreased sucrose consumption	Pucilowski et al. (1993)
FSL	MS	FSL rats exposed to MS show increased weight loss and immobility in FST	El Khoury et al. (2006)
Reverse genetics			
SERT	Maternal care	SERT heterozygous mice exposed to low level maternal care have reduced latency to immobility in TST	Carola et al. (2011)
GR	Stress	GR overexpressing mice exposed to stress show increased learned helplessness and corticosterone release	Ridder et al. (2005)
BDNF	Maternal care	Heterozygous BDNF knock-outs mice exposed to low level maternal care have a more anxious phenotype	Carola and Gross (2010)
BDNF	Chronic adolescent corticosterone	Heterozygous BDNF knock-outs rats exposed to chronic corticosterone show decreased fear extinction	Gururajan et al. (2015)

See text for further explanations.
CMS, chronic mild stress; FST, forced swim test; MS, maternal separation; TST, tail suspension test.

exposed to low levels of maternal care show a more severe phenotype in the tail suspension test, as evidenced by a decrease in the latency to immobility (Carola et al., 2011). However, the interactive effects were stronger in models of anxiety. This same pattern was also seen when heterozygous or homozygous SERT knockout mice were exposed to adult life stressors (Carola and Gross, 2012). Studies in monkeys have further underlined the interaction between genetic variations in the serotonin transporter and early environmental factors. Rhesus monkeys, like humans, have a 21 basepair polymorphism in the serotonin transporter (sh5-HTTLPR) analogous (though not identical) to the 5-HTTLPR in humans. It has been shown that monkeys with the short allele when raised with peers rather than the mother show an exaggerated HPA axis response upon social separation. Moreover, they show abnormal play behavior and increased aggression (Barr et al., 2003). Although these monkeys have not been evaluated in more traditional models of depression, they do show enhanced alcohol intake, a frequent comorbidity of patients with MDD. In a similar vein, several groups have looked at BDNF heterozygous knock out animal (as homozygous animals are not viable) and studied the effects of stress or chronic corticosterone treatment (see Table 7.3) and although several gene–environment interactions have been found, these were more related to anxiety and cognition than to depression. Thus there is an urgent need for more research in this area, not only in relation to BDNF but also in relation to the genes that have consistently been associated with MDD (see Tables 7.1 and 7.2) such as the *SLC6A4*, the *PCLO* and the *FKBP5* gene.

BIPOLAR DISORDER (BP)

Diagnosis and Symptoms

Although the ancient Greeks already proposed a relationship between melancholia and mania, it was in the mid-19th century (1854) when two French psychiatrists (Jean-Pierre Falret and Jules Baillarger) independent from each other described patients that showed both manic and depressive symptoms and referred to the disorder as *folie circulaire* (circular insanity) and *folie à double forme* (dual-form insanity) respectively. The famous German psychiatrist Emil Kraepelin in the beginning of the last century then coined the term "manic-depressive psychosis" to distinguish it from the "dementia praecox" to emphasize that manic-depressive psychosis has a more episodic course and less severe outcome. It was not until the 1980s that the disorder was renamed bipolar disorder (BP) which was thought to be less stigmatizing that manic-depressive psychosis or illness.

Three different forms of BP are generally recognized on the basis of the severity of the symptoms: BP-I, BP-II and cyclothymic disorders. This is schematically illustrated in Fig. 7.8.

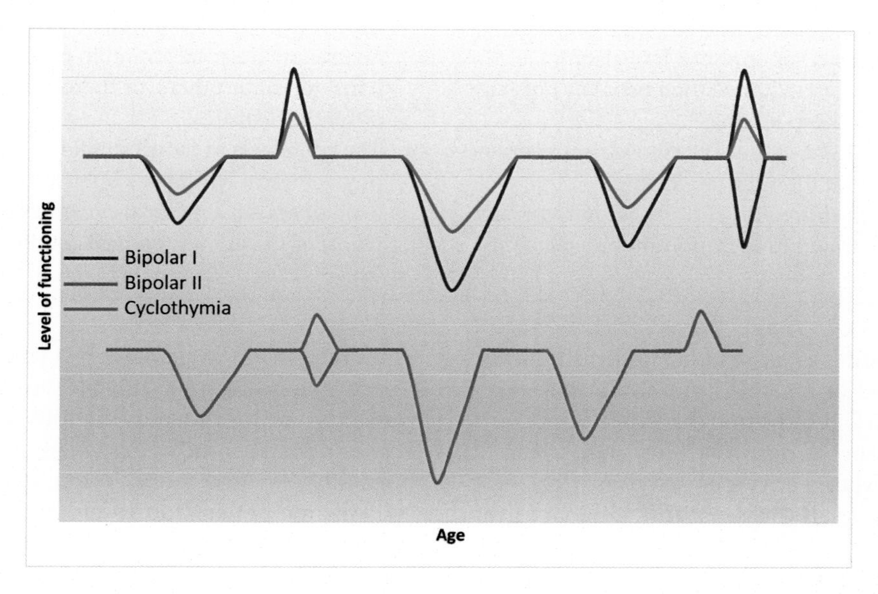

FIGURE 7.8 **The typical course of illness for BP.** BP is characterized by the presence of (at least one) episode of elevated mood (mania in BP-I and hypomania in BP-II). Most patients however, cycle through multiple period of mania and with the duration of manic episodes being shorter than the periods of depressive mood. In addition, so called mixed episodes can occur in which both manic and depressive symptoms can occur within a single episode. Whereas the depressive episodes are more or less equally severe in BPI and BP-II, BP-II is characterised by less severe manic symptoms. In addition cyclothymia is characterised by periods of elevated and depressed mood, but the intensity of both is less than seen in patients with BP-I.

BOX 7.2

THE DIAGNOSIS OF A MANIC EPISODE (DSM V)

1. A distinct period of abnormally and persistently elevated, expansive, or irritable mood and abnormally and persistently increased goal-directed activity or energy, lasting at least 1 week and present most of the day, nearly every day (or any duration if hospitalization is necessary).
2. During the period of mood disturbance and increased energy or activity, three (or more) of the following symptoms (four if the mood is only irritable) are present to a significant degree and represent a noticeable change from usual behavior:
   a. Inflated self-esteem or grandiosity.
   b. Decreased need for sleep (eg, feels rested after only 3 h of sleep).
   c. More talkative than usual or pressure to keep talking.
   d. Flight of ideas or subjective experience that thoughts are racing.
   e. Distractibility (ie, attention too easily drawn to unimportant or irrelevant external stimuli), as reported or observed.
   f. Increase in goal-directed activity (either socially, at work or school, or sexually), or psychomotor agitation (ie, purposeless non-goal-directed activity).
   g. Excessive involvement in activities that have a high potential for painful consequences (eg, engaging in unrestrained buying sprees, sexual indiscretions, or foolish business investments).
3. The mood disturbance is sufficiently severe to cause marked impairment in social or occupational functioning or to necessitate hospitalization to prevent harm to self or others, or there are psychotic features.
4. The episode is not attributable to the physiological effects of a substance (eg, a drug of abuse, a medication, other treatment) or to another medical condition.

BP I disorder is associated with full manic periods (blue curve), whereas BP II (red curve) is associated only with hypomanic periods, while both generally have full-blown depressive episodes. Cyclothymic disorder (green curve) represents a milder form with both hypomanic and less severe depressive episodes. The DSM-V criteria for a manic episode are displayed in Box 7.2. Manic episodes are characterized by abnormally elevated mood, often accompanied by inflated self-esteem and feelings of grandiosity. This may even lead to delusions of grandeur. In most cases patients with BP-I have excessive energy, sleeping little and very talkative. Important for the diagnosis of a manic episode is the severity of the symptoms. Not only should they be noticeable by both the patients and his/her environment, but it also has to significantly impair functioning. This, in essence, is the defining difference between a manic and a hypomanic episode. Although the later should also be clearly noticeable and distinguishable from the "normal" behavioural pattern of the patient, but it should not be severe enough to cause a marked impairment in normal functioning. Moreover, whereas a manic episode has to last for at least 1 week, a hypomanic episode can last for only a few days.

The vast majority of patients with BP also suffer from depressive episodes during which symptoms identical to those described in Box 7.1 can occur. However, for the diagnosis of BP-I such episodes are not essential. An important complication in the diagnosis of BP is the fact that many different type of drugs can induce symptoms of mania and thus a critical analysis of past drug use is essential for the diagnosis.

Finally, cyclothymic disorders are defined as the presence of numerous periods with hypomanic symptoms and numerous periods with depressive symptoms for at least 2 years in adults. However, the symptoms should not classify as a hypomanic or a major depressive episodes and hence cyclothymic disorder is generally regarded as a much milder form of BP.

In addition to manic and depressive episodes, many patients with BP also suffer from so-called "mixed episodes" (or mixed states) in which manic and depressive symptoms occurs within a single episode. As Fig. 7.8 further illustrated, in between manic and depressive (or mixed) episodes overall mood is greatly improved. However, during significantly parts of this "recovery" state patients show subsyndromal symptoms. In a 20 year follow-up study with 157 BP- I patients, it was found that patients were symptomatic slightly less than half of the time (47%). However, during the symptomatic period, only during 12% of the time were syndromal manic or depressive symptoms present, with the rest of the time spent with subsyndromal or minor symptoms (Judd et al., 2002). During the symptomatic phase, patients spends more time showing depressive signs (68%), as compared to (hypo)manic signs (20%) or to mixed states (12%). A comparable analysis was done in BP-II patients, with several interesting differences (Judd et al., 2003). Thus BP-II patients were symptomatic slightly more than half of the time (54%). However, contrary to BP-I, BP-II patients spend the vast majority of time showing depressive symptoms (94%) with only 2% spend showing hypomanic symptoms and the remaining 4% of the time in mixed states. As a result, BP-II patients showed significantly fewer changes in weekly symptoms than BP-I patients. Thus these studies clearly show that the course of symptoms is different between BP-I and BP-II and that the BP-II is not a "milder form" of BP-I. indeed studies suggest that the stability of the diagnosis of either BP-I or BP II is relatively high and patients rarely switch.

Several different rating scales have been developed to measure the severity of the manic phase and, consequently, the effects of the different treatments. However, probably the most used scale is the Young Mania Rating Scale [YMRS, with young not referring to the age of the patients but to the author of the scale: (Young et al., 1978)]. The YMRS consists of 11 items that are scored on a scale of 0 (absent) to 4 (severe). Given that cooperation of the patient is not always guaranteed, four items which can be objectively assessed by the examiner (irritability, speech (rate and amount) content of conversation and disruptive-aggressive behavior) are scored on a scale from 0 to 8 and thus weigh particularly high in the overall score. The total score thus varies between 0 and 60, with a score below 20 considered to be minimally severe and above 38 as severe. Studies have shown the scale to have good interrater reliability (Furukawa, 2010).

The Epidemiology of BP

In a historical overview Alvarez Aria and coworkers gave a description of BP before the age of drug therapy and concluded that many of the early descriptions about the natural course of BP are still valid these days (Alvarez Ariza et al., 2009). Thus BP typically starts with a depressive episode. However, when the first episode is a manic episode (in about 20% of the cases) the rate of recurrence is higher. Manic episodes are mostly shorter than depressive episodes but recovery between episodes is usually good (see also Fig. 7.8). Moreover, elderly patients often have mixed states and delusional symptoms and their episodes last longer. However, an investigation of all BP patients in the STEP-BD (Systematic Treatment Enhancement Program for Bipolar Disorder) did not find significant age-dependent differences in either depressive or manic symptoms (Al Jurdi et al., 2012). However, this study divided patients in only two age groups (between 20 and 59 and >60 years) which may have obscured more subtle differences.

BP typically develops shortly after puberty with an average age of onset between 17 and 23 years of age (Mitchell et al., 2011), with onset often slightly earlier in individuals with a family history of BP. In this respect it has been suggested that there are distinct subgroups of BP patients with different age of onset. Thus Bellivier and coworkers distinguished three groups in their sample of 368 patients: approximately 28% of patients had an early (17 years) age of onset, about 50% an intermediate (25) and the remaining 22% a late (40) age of onset (Bellivier et al., 2003). A similar distribution was found in a subsequent larger sample (Hamshere et al., 2009). Several studies have investigated clinical risk factors for the development of BP and a recent systematic review of the literature showed that especially (early onset) anxiety disorders and ADHD are among the most important risk factors (Faedda et al., 2014).

Once BP has fully developed, patients often show a fairly consistent pattern of cycling, and individuals can be subdivided into depressive-mania-interval (DMI) and mania-depressive interval (MDI) patients based on which transition predominantly occurs before recovery. Clinically, there is some evidence that MDI patients respond better to lithium while DMI patients respond better to antidepressants (Mitchell et al., 2011). A special phenomenon in BP is the occurring of "rapid cycling." This is generally defined as having four or more distinct episodes of hypomania, depression or mixed episodes within a single year. Cross-sectional studies have shown that between 23% and 45% of BP patients show rapid cycling (Mitchell et al., 2009). However, it appears that rapid cycling is only seen during a relatively short period of time for most patients. For instance in the large STEP-BD study mentioned earlier, while at entry 34% of patients were classified as rapid cyclers, only 5% of patients were still rapidly cycling upon follow-up after 12 months (Schneck et al., 2008).

Given the chronic character of BP, 6 and 12 month point prevalence are comparable and have been estimated to be around 0.6%–0.8% (Merikangas et al., 2007; Ferrari et al., 2011), with BP-I slightly more prevalent than BP-II. When analyzing the complete BP spectrum, the aggregate lifetime prevalence rates were 0.4, 0.4, and 1.4 for BP-I, BP-II and subthreshold BP (Merikangas et al., 2011). There is some evidence that rates of BP are different for different countries, although world-wide studies show some discrepancies. Thus whereas one study found the highest rates in the USA and the lowest in India and Bulgaria (Merikangas et al., 2011), another study found the highest rates in the Australasia and Eastern Asian region and the lowest in Asia pacific and Eastern Europe (Ferrari et al., 2011). An important question in relation to the epidemiology of BP is the presence or absence of a sex bias. There have been some reports that, in contrast to MDD, BP is more prevalent in males than in females. However, the most recent findings seem to suggest that there is no real bias, with a mean (6 or 12 month) prevalence of 0.63% in male and 0.97% in females (Ferrari et al., 2011).

Although the prevalence of BP is thus much lower than for MDD, it still causes significant burden to the patient, his or her family and friends and society as a whole. In fact it has been considered one of the ten leading causes of disability in the world. In the global burden of disease study 2010, the daily disability adjusted life years for bipolar disorder was estimated to be 12.8 million years, an increase of 41% over 1990 (Murray et al., 2012). Estimates regarding the financial burden suggests that the costs of BP (BP-I and BP-II combined) may be as high as $151 billion (Dilsaver, 2011), including $120 billion of indirect costs (such as costs associated with suicide, lost income and caregivers). Thus, although substantial differences exist between different estimates of overall costs (especially in relation to indirect costs), it is clear that BP puts a significant economic burden on society (Kleine-Budde et al., 2014).

The Neurobiology of BP

The symptom similarity between the depressive symptoms in MDD and BP explains the significant overlap in the neuropathology and neurobiology between these two affective disorders. Overall, the functional studies suggest that BP is associated with limbic hyper- and cortical hypoactivity (Kupferschmidt and Zakzanis, 2011). Similar to MDD, patients with BP have significantly reduced volume of the subgenual anterior cingulate cortex (sgACC) and there is some evidence that this can be normalized by chronic lithium treatment (Moore et al., 2009). In line with these structural findings, significant alterations in chemical composition have also been reported in post mortem studies such as reductions in marker for both glutamate and GABAergic neurons (Savitz et al., 2014), as well as significantly reduced activation (Kupferschmidt and Zakzanis, 2011). GABAergic neuron density is also reduced in other part of the ACC, such as the pregenual and the supragenual ACC (Savitz et al., 2014). Interestingly, while a reduction in gray matter volume has also been reported for the orbital frontal cortex, this area seems to show an increase in GABAergic neuron density. In addition, whereas this area is usually hypoactive in MDD it shows hyperactivity in BP patients. Given its role in impulsivity (see also Chapter 8) it is not surprising that an overactivity of the OFC is seen during the manic phase of BP (Chen et al., 2011a). Reductions in hippocampal volume have been reported, although the effect sizes seem relatively small and are confounded by the fact that lithium treatment increases the hippocampal size (Otten and Meeter, 2015). As with the ACC, reduction in GABAergic cells within the hippocampus have been described in BP patients, especially in the stratum oriens and pyramidale of CA2/3 (Benes, 2011). In line with studies in MDD (see Fig. 7.3), structural abnormalities within the amygdala have been reported, with BP patients, that were at least 2 months medication free, showing reduced amygdala volume, while people on active medication (lithium or valproate) had greater volume compared to controls, suggesting a neurotrophic action of these drugs. Concomitant with this reduced volume, decreased neuronal size and reduced cell numbers within specific subregions of the amygdala have been reported in post mortem tissue from patients with BP (Berretta et al., 2007).

In addition to these shared features, several abnormalities in brain structure and function have been found in BP that seem unique compared to MDD patients. However, very few studies have directly compared MDD and BP patients. One of the areas of interest is the basal ganglia, where there is a general overactivity in both the striatum and the globus pallidus in BP especially in mania (Malhi et al., 2004; Caligiuri et al., 2006). On the other hand, in

MDD, there is convincing support for a decreased activation of the striatum (Marchand and Yurgelun-Todd, 2010) especially in relation to a decreased subjective experience of pleasure (Osuch et al., 2009). A similarly divergent pattern has been found for the ventrolateral prefrontal cortex, where overactivity has been reported in BP while underactivity is most often seen in patients with MDD. The overactivity is most prominently seen in the right hemisphere (Houenou et al., 2011) and is most robustly found during mania (Chen et al., 2011a).

Investigations of the neurotransmitters involved in BP show a similar pattern as the volumetric and functional studies, showing both similarities and differences with MDD. Thus, similar to MDD, patients with BP have reduced binding of 5-HT_2 receptors in several cortical areas (Yatham et al., 2010) and 5-HT_{1A} in the dorsal raphe (Drevets et al., 1999), although an increase in 5-HT_{1A} binding has also been reported (Sullivan et al., 2009a). On the other hand, while reduced SERT binding has repeatedly been found throughout the brain in MDD, several reports have found increased SERT binding in the thalamus, prefrontal cortex and striatum in BP, while reductions were only found in the dorsal raphe (Walderhaug et al., 2011).

Within the dopaminergic system both increases and decreases have been reported in patients with BP, likely associated with manic and depressive episodes respectively. Thus studies have shown a decrease in prefrontal D1 receptor density, and increased striatal D2 receptor densities (Cousins et al., 2009). In addition, during mania the putamen is significantly enlarged and blood flow through the basal ganglia is also increased (Cousins et al., 2009). On the other hand, amphetamine administered to BP patients does not lead to enhanced dopamine release in the striatal (Anand et al., 2000), suggesting change at postsynaptic sites rather than an overall increase in dopamine release. This contrasts with schizophrenia, where predominantly presynaptic changes in dopamine neurotransmission have been reported (see Chapter 9). Finally there is substantial evidence of dysfunctional HPA axis activity in BP. A recent meta-analysis of 41 studies found significantly elevated basal and dexamethasone induced cortisol levels (Belvederi Murri et al., 2016). These levels were strongly and positively associated with the manic phase and negatively associated with antipsychotic use. Likewise, BP is associated with increased levels of ACTH but not CRF.

The Pharmacological Treatment of BP

Several different drugs have been approved for the treatment of BP (see Table 7.4) with several drugs only approved for specific phases of the illness. Thus, while quetiapine is approved for the treatment of acute mania and depression as well as for maintenance treatment, chlorpromazine is only approved for the treatment of mania, and lamotrigine should only be used in maintenance treatment.

The first drug to be approved for the treatment of BP was lithium. Although John Cade from Melbourne is often regarded as the first person to use lithium in the treatment of mania, it had already been used in 1870 in Philadelphia and in 1894 in Denmark (Shorter, 2009). However, it was Cade's work in 1949 on 10 patients and subsequently Noack and Trautner in 1951 reporting on 100 patients that renewed the interest in lithium as a treatment for mania. The breakthrough came with the work of Shou and Stromgren from Aarhus (Denmark) who used a randomized placebo control to show lithium's effectiveness. Subsequent studies confirmed the effectiveness in the treatment of mania (Gershon et al., 2009). However, it is less effective in the treatment of the depressive episodes. An often used parameter to indicate the

TABLE 7.4 Drugs Approved by the FDA for the Treatment of the Different Phases of BP

Drug names	Drug types	Years of approval
Drugs approved for acute bipolar depressive episode		
Olanzapine/Fluoxetine	Second generation antipsychotic/SSRI	2003
Quetiapine	Second generation antipsychotic	2006
Lurasidone[a]	Second generation antipsychotic	2013
Drugs approved for acute bipolar manic episode		
Lithium	Mood stabilizer	1970
Chlorpromazine	First generation antipsychotic	1973
Divalproex	Mood stabilizer	1994
Olanzapine	Second generation antipsychotic	2000
Risperidone	Second generation antipsychotic	2003
Quetiapine	Second generation antipsychotic	2004
Ziprasidone	Second generation antipsychotic	2004
Aripiprazole	Second generation antipsychotic	2004
Carbamazepine	Mood stabilizer	2004
Asenapine	Second generation antipsychotic	2009
Cariprazine	Second generation antipsychotic	2015
Drugs approved for bipolar maintenance treatment		
Lithium	Mood stabilizer	1974
Lamotrigine	Mood stabilizer	2003
Olanzapine	Second generation antipsychotic	2004
Aripiprazole	Second generation antipsychotic	2005
Quetiapine	Second generation antipsychotic	2008
Risperidone	Second generation antipsychotic	2009
Ziprasidone[b]	Second generation antipsychotic	2009

[a] *Lurasidone is approved both as monotherapy as well as in combination with lithium or valproate.*
[b] *only approved in in combination with lithium or valproate.*

relative effectiveness of bipolar treatments in the two phases is the so-called polarity index (PI). A PI above 1 indicates a stronger effect on mania, while a polarity index below 1 is indicative of a stronger antidepressant effect (Popovic et al., 2012). As can be seen in Table 7.5, the PI for lithium is 1.39 indicating that it is indeed less effective for the treatment of bipolar depression. As a result, it is neither indicated nor approved for this condition. However, it is useful for the maintenance treatment and there is substantial support for the effectiveness of lithium as a prophylactic drug and hence it is still generally used as a first drug of choice (Gershon et al., 2009; Grof and Muller-Oerlinghausen, 2009). The exact working mechanism of lithium, however, is still far from understood. As mentioned before, lithium has neurotrophic effects, some of which may be related to symptom improvement. In line with this,

TABLE 7.5 Polarity index of BP pharmacological treatment
(Popovic et al., 2012; Carvalho et al., 2015). The index for quetiapine is for a combined preparation with either lithium or valproate

Drug name	Polarity Index
Lithium	1.4
Lamotrigine	0.4
Valproate	0.5
Olanzapine	3.9
Olanzapine plus lithium/valproate	0.5
Aripiprazole	10.4
Aripiprazole plus lithium/valproate	4.2
Risperidone	9.1
Quetiapine	1.4
Quetiapine plus lithium/valproate	0.8
Ziprasidone plus lithium/valproate	3.9

the effectiveness of lithium is moderated by the BDNF genotype (Rybakowski, 2014). On the other hand, lithium also affects the dopamine neurotransmission, leading to an attenuation in D2 receptor signaling (Cousins et al., 2009).

In addition to lithium several antipsychotics are also indicated for the management of acute mania, including both first- and second-generation drugs. Given the similarities between psychotic symptoms and mania, the usefulness of antipsychotics is not really surprising and it is suggested that these drugs work via a blockade of the dopamine D_2 receptor (see also Chapter 9). Most of these antipsychotic drugs (especially the second generation drugs) are also effective in the maintenance phase of the treatment. When analyzing the antipsychotics, several interesting differences emerge. As indicated in Table 7.5, most of the antipsychotics are more effective against the manic phase with PIs in the order of 3.9 and higher. The only exception appears to be quetiapine which (at least in combination with lithium or valproate) has a PI below 1.0. In line with this only quetiapine has been approved as a monotherapy for the treatment of bipolar depression. Finally, valproate has been approved for the treatment of acute mania in the USA (although carbamazepine is also approved in other countries). Valproate was originally developed for the treatment of epilepsy and its effectiveness was believed to be due to an increase in GABAergic neurotransmission. However, the pharmacology of valproate is highly complex (Rosenberg, 2007; Ranger and Ellenbroek, 2016), affecting epigenetic mechanisms (valproate acts as a histone deacetylase inhibitor, see Chapter 5), oxidative stress processes and several second messenger systems. In addition, it affects dopamine neurotransmission and together with its neurotrophic effects, these effects may be related to the clinical effectiveness in mania. In summary, there are numerous options for the treatment of acute mania, and a recent review has suggested that while both second generation antipsychotics and mood stabilizers by themselves are effective, the combined treatment may have a stronger effect (Glue and Herbison, 2015). Using a network meta-analysis, the authors showed that the odds ratios for a therapeutic effect increased from 2.2 (for antipsychotics alone) and 2.1 (for mood stabilizers alone) to 3.3 for the combined treatment. Interestingly, the study also suggested that antipsychotics were more effective than mood stabilizers. Although this was less evident from the odds ratios, it was clear from a probability of best treatment analysis. Thus, in relation to the overall scores (using the YMRS), antipsychotics had a probability of 14% of being the best treatment, compared to 0% for mood stabilizers (in the remaining 86% of the studies the combined treatment was superior). Interestingly, a similar conclusion was recently drawn for the long-term treatment study of BP, with combination therapy usually more effective than mood stabilizers alone, although at the same time the side-effects increased (Buoli et al., 2014).

In contrast to the therapeutic possibilities for the treatment of acute mania, far fewer drugs are available for the treatment of bipolar depression (see Table 7.4). Intriguingly, none of the antidepressant drugs are indicated for the treatment of bipolar depression. The main reason for this is that, while these drugs may be effective in the reduction of depressive symptoms they have an increasing risk of inducing (hypo)mania or rapid cycling (Pacchiarotti et al., 2013). Thus it is of vital important to assess the presence (current or in the past) of a (hypo)manic episode. Until 2003 no treatment was approved by the FDA for the treatment of bipolar depression. In that year, the combination of olanzapine and fluoxetine was approved, followed by quetiapine in 2006 (and the extended release form in 2008) and lurasidone (alone or in combination with either lithium or valproate) in 2013.

In line with these guidelines, studies have shown that whereas olanzapine was effective in short term bipolar depression, the combination with fluoxetine was superior especially with respect to onset of action and remission rate (Tohen et al., 2003; McIntyre et al., 2013). The olanzapine/fluoxetine combination was shown to be similarly effective to treatment with quetiapine monotherapy as well as with lurasidone (both monotherapy and adjunctive to valproate or lithium). In all cases, response rates (defined as a >50% reduction on the Montgomery-Asberg Depression Rating Scale, MADRS) were between 52 and 60% (Citrome, 2014). Some differences, however, occurred with respect to the side effects. Here, lurasidone (both as mono- and combined therapy) was superior compared to the olanzapine/fluoxetine combination (which led to significant weight gain and diarrhea) and quetiapine (which led to significant sleepiness and dry mouth). Overall, there is convincing evidence that some second generation antipsychotic drugs are effective in the treatment of bipolar depression. However, there are substantial differences between drugs and there is no evidence for a "class" effect (De Fruyt et al., 2012). Thus while the effect ratio was 4.7 for quetiapine it was close to 3 for olanzapine and 0.19 for aripiprazole. Why specific antipsychotics such as quetiapine and lurasidone are effective in bipolar depression while others are not is not clear. One potential mechanisms of action is blockade of $5\text{-}HT_7$ receptors. Both lurasidone and quetiapine have a strong $5\text{-}HT_7$ receptor affinity, and animal research has suggested that $5\text{-}HT_7$ antagonists may have antidepressant activity (Gellynck et al., 2013; Naumenko et al., 2014). Moreover, the recently approved antidepressant vortioxetine is, among others, a potent $5\text{-}HT_7$ antagonist.

The Aetiology of BP

There is broad consensus, based on a multitude of family and adoption studies that BP, like virtually all psychiatric disorders has a substantial genetic component. Familial aggregation

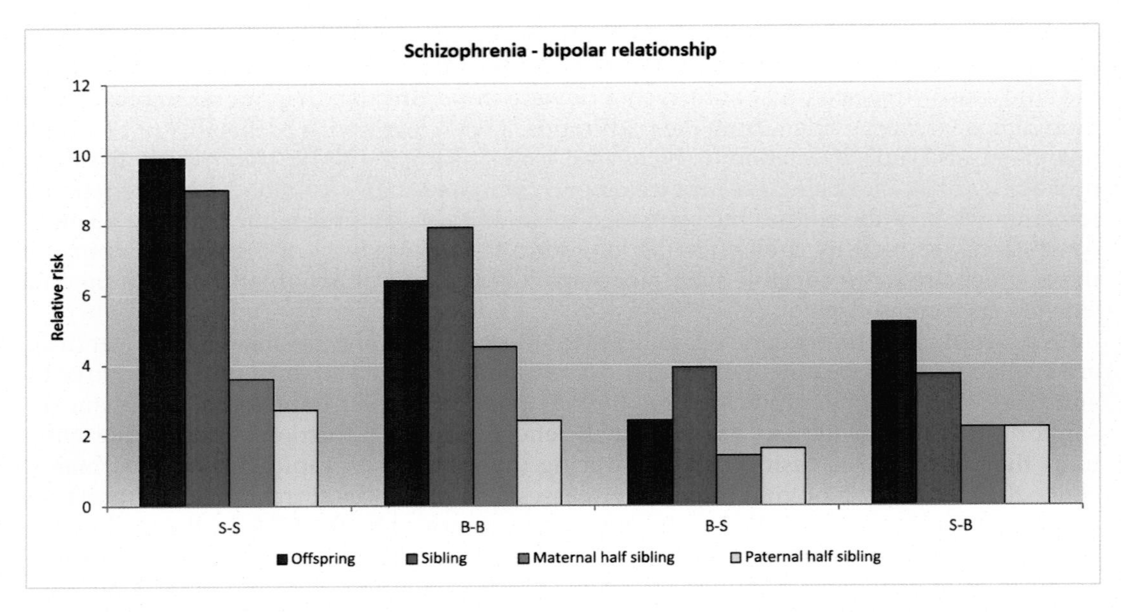

FIGURE 7.9 **The genetic relationship between BP and schizophrenia.** Family studies have shown substantial overlap between the genetic risk for BP and schizophrenia (Lichtenstein et al., 2009; Guxens et al., 2016). The S-S groups shows the risk of family members developing schizophrenia when the proband (the first diagnosed family member) has schizophrenia and B-B shows the risk of family members developing BP when the proband has BP. The B-S group, on the other hand shows the risk of the family members developing schizophrenia when the proband has BP, while the S-B group the risk of the family members developing BP when the proband has schizophrenia. From this picture it is clear that the same disorder risk (S-S and B-B) is the highest, yet there is still a substantial risk for the offspring developing the other disorder (B-S and S-B).

has been demonstrated by well-controlled systematic studies (Smoller and Finn, 2003), and although heritability estimates vary from study to study it has been suggested that BP is one of the most heritable of all major psychiatric disorder. Thus, using the Finnish national hospital discharge register, monozygotic twin concordance rates for BP-I were estimated between 43% (narrow definition) and 75% (broad definition, including schizoaffective disorder and BP-II). The concordance rate for dizygotic twins was between 5.6% and 10.3%. Model fitting of these data suggested that a heritability of 93% (Kieseppa et al., 2004), although other studies have reported somewhat lower heritability estimates (Wendland and McMahon, 2011). In an interesting study, Lichtenstein and coworkers using two Swedish national registries, investigated the genetic relationship between schizophrenia and BP (Lichtenstein et al., 2009; Guxens et al., 2016). Their results are illustrated in Fig. 7.9 and show several interesting features. First of all, the risk for first degree relatives (parent-offspring and siblings) is much larger than for half-siblings. Secondly, the risk for the same diagnosis (schizophrenia–schizophrenia or BP–BP) is higher than the risk for cross diagnosis. Nonetheless, the risk for schizophrenia is around 2–3 times higher in first degree relatives with BP and vice versa than the risk in the general population. A further analysis of these data led to the conclusion that the overall heritability for BP is around 59%, 40% of which seems to be shared genetic factors with schizophrenia. The remaining 41% consists of 3% shared childhood environmental effect and 38% unique environmental effects. This study also investigated the risk of BP in adopted

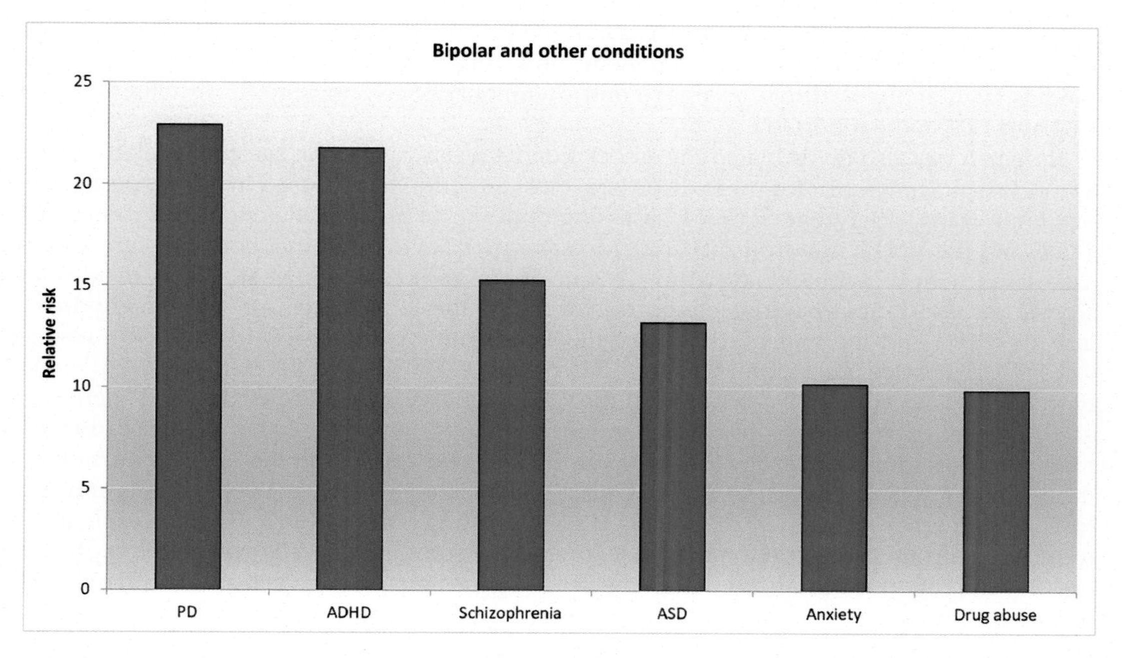

FIGURE 7.10 **The relationship between BP and other disorders.** In an extensive family study from Sweden the relative risk of children from parents with BP was studied. As the figure indicates the risk of developing personality disorder (PD), attentional deficit hyperactivity disorders (ADHD), schizophrenia, autism spectrum disorder (ASD), anxiety disorder or drug abuse are all enhanced when one of the parents suffers from BP.

away children and found that the risk was slightly, but nonsignificantly smaller (4.3 vs 6.4) compared to nonadopted away children. In a follow-up study the genetic relationship of BP was extended to other psychiatric conditions as well (Song et al., 2015), leading to relative risk ratios of between 9.7 for drug abuse and 22.9 for personality disorders (see Fig. 7.10).

Genetic Factors Contributing to BP

The clear evidence for a substantial genetic component has sparked wide-spread investigations to identify the underlying genes. Unfortunately, but in line with the other psychiatric disorders described in this book, these studies have not led to clear and reproducibe genetic markers for BP. Rather, again in line with most other disorder described in this book, the general consensus is that there are no common variants of major effect size for BP. Nonetheless, several genes have been identified in candidate gene and/or genome wide association studies and the most important ones will be discussed here.

With respect to candidate genes, attention has focused among others on genes that have also been implicated in MDD, such as *BDNF*, *MAOA* and *SLC6A4*. With respect to the *BDNF* gene most attention has been directed to rs6262, which leads to a Val to Met substitution at position 66. At least 22 different studies have been published so far and the results seem inconclusive. Thus, while a first meta-analysis found no significant association (Kanazawa et al., 2007), a subsequent metaanalysis did find a small but significant association with an Odds Ratio (OR) of 1.13 (Seifuddin et al., 2012). The most recent metaanalysis, however, seems to support the initial lack of association (Gonzalez-Castro et al., 2015). Thus, analyzing 22 studies (involving over 9,000 patients and 7,000 controls) and using several different models such as the additive, recessive and dominant model, nonsignificant ORs (between 1.02 and 1.07) were reported.

Research has also investigated the association between variants in serotonin-related genes and BP. As we have seen for other disorders, most attention has focused in this respect on the serotonin transporter gene *SLC6A4*, assessing both the 44 basepair deletion/insertion 5-HTTLPR and the VNTR in intron 2 (STin2). Metaanalyses have found that both are significantly associated with BP with ORs of 1.1 to 1.2 (Cho et al., 2005; Lasky-Su et al., 2005). In addition significant associations have been reported for the synthesizing enzyme tryptophan hydroxylase 1 (*TPH1*) and the metabolizing enzyme Monoamine oxidase A (*MAOA*). With respect to *TPH1*, a positive association was found between rs1800532 and BP, with the "AA" genotype having an OR of 1.37 compared to the "AC" genotype and 1.41 vs the "CC" genotype according to a metaanalysis of 7 independent studies (Chen et al., 2008). Likewise, a recent meta-analyses found an association between BP and two variants in the *TPH2* gene, notably the rs4760280 (OR 0.62) and rs7954758 (OR 1.32) in an analysis of over 1,000 patients and controls in both cases (Gao et al., 2016). This is an interesting observation as *TPH2* is exclusively found in neurons and is predominantly expressed in serotonergic cells within the raphe nuclei. Several different variations within the *MAOA* gene have been investigated with only two reaching significance in meta-analyses, namely the CA repeat in intro 2, where both the a5 and the a6 allele enhance the OR (Furlong et al., 1999; Fan et al., 2010) and the often studied rs6323 where the T allele is associated with an enhanced risk for BP (Fan et al., 2010). Another gene involved in the metabolic pathway of 5-HT is *ASMT*, which codes for acetylserotonin-O-methyltransferase, the last step in the transformation of 5-HT to melatonin. The "G" allele of either rs4446909 or rs5989681 have been found to increase the risk for BP. A recent meta-analysis estimated the odds ratio for both cases to be about 1.2–1.3 (Etain et al., 2012). Interestingly, studies investigating the association between genetic variants in the serotonin receptor genes and BP have found protective variants for both the *HRT1A* (rs6295) and the *HRT3B* (rs1176744) gene (Kishi et al., 2011; Hammer et al., 2012).

Given the proposed (or theorized) role of dopamine in BP, candidate gene studies have also focused on elements within this neurotransmitter system, but mostly with negative findings. Thus meta-analyses were unable to find positive associations between BP and the *DRD2*, *DRD4* and the *SLC6A3* gene (Gatt et al., 2015). However, significant associations have been found for the *DRD4* gene. More specifically a significant association was found for the 2 repeats of the 48 basepair VNTR in exon 3 and BP (Lopez Leon et al., 2005). In addition the "A" allele of rs4680 of the *COMT* gene has been associated with BP (Zhang et al., 2009).

Studies investigating other neurotransmitter systems have also yielded positive associations. For instance several allelic variants within the *DAOA* gene have been positively associated with BP (Detera-Wadleigh and McMahon, 2006). *DAOA* codes for D-amino acid oxidase activator, a protein that activates the enzyme D-amino oxidase. This enzyme metabolizes D-serine, which is an essential co-activator of the NMDA receptor, particularly in the forebrain. Likewise, *GRIN3* codes for the 3A subunit of the NMDA receptor and associations between the "A" allele of rs4743473 and BP have been reported (Cichon et al., 2011).

Several other genes that seem to have more organizational functions have also been positively associated with BP. These include *ANK3*, *BRD1*, *MTHFR* and *ZNF618* (see Table 7.6).

TABLE 7.6 Major Genetic Factors Implicated in BP

Genes/regions	Remarks
Single nucleotide variants	
ANK3	Ankyrin 3: associations between the "C" allele of rs9804190, the "T" allele of rs10994336 and the "G" allele of rs1938526 and the diagnosis of BP
ASMT	Acetylserotonin O-methyltransferase: associations between the "G" allele of rs4446906 or of rs5989681 and the diagnosis of BP
BRD1	Bromodomain containing 1: associations between several variants and the occurrence of BP
BRE	Brain and reproductive organ-expressed: association between the "T" allele of rs6547829 and the diagnosis of BP
CACNA1C	The α1C subunit of the Cav1.2 channel: several associations, especially with variants in intron3 have been associated with BP, both in candidate gene and in GWAS studies.
COMT	Catechol-O-methyltransferase: association between the "A" allele of rs4680 and the diagnosis of BP
DAOA	D-amino acid oxidase activator: associations between the "C" allele of rs1935058 and the "A" allele of rs1935062 and the diagnosis of BP
DRD4	Dopamine D4 receptor: associations between the 2 repeats of the 48 bp VNTR and the diagnosis of BP
GRIN3A	Glutamate ionotropic receptor NMDA type subunit 3A: association between the "A" allele of rs4743473 and the diagnosis of BP
HTR1A	5-HT1A receptor: the "G" allele of rs6295 is protective for BP
HTR3B	5-HT3 receptor subunit B: the "A" allele of rs1176744 is protective for BP
MAD1L1	MAD1 mitotic arrest deficient-like 1: associations between the "T" allele of rs11764590, rs10278591 and rs3996329 and the diagnosis of BP
MAOA	Monoamine oxidase A: association between the a5 and a6 alleles of the CA repeat in intron2 and the diagnosis of BP
MTHFR	Methylenetetrahydrofolate reductase: association between the "C" allele of rs1801131 and the diagnosis of BP
SLC6A4	Serotonin transporter: association between the "s" allele of the 5-HTTLPR and diagnosis of MDD
TPH1	Tryptophan hydroxylase 1: association between the "AA" genotype of rs1800532 and the diagnosis of BP
ZNF618	Zinc Finger Protein 618: association between the "T" allele of rs7023951 and the diagnosis of BP
Copy number variations	
1q21.1	Duplications have been associated with BP
3q29	Deletions have been associated with BP
16p11.2	Duplications have been associated with BP

Please see text for further explanations.

ANK3 is a member of the ankyrin family, consisting of proteins that are involved in cell motility, activation, proliferation and cell-cell contact. ANK3 was originally found in neuronal axons, especially in the nodes of Ranvier. At least three different variants have been associated with BP (rs9804190, rs10994336, and rs1928526), which was confirmed in several meta-analyses (Schulze et al., 2009; Takata et al., 2011). *BRD1* and *ZNF618* both code for DNA binding proteins that are thought to be involved in gene expression. BRD1 interact both with DNA and histone tails and stimulates acetylation of H_3 and H_4 histones, thus leading to transcriptional activation (see Chapter 5 for more details on histone acetylation). ZNF618, on the other hand codes for a protein belonging to the zinc finger protein family that are known to bind to DNA and regulate gene transcription. Several different single nucleotide variants of *BRD1* have been associated with BP (Nyegaard et al., 2010). Interestingly some of variants in the neighboring gene, another zinc finger binding protein (*ZBDE4*) were also associated with BP (Nyegaard et al., 2010). In relation to *ZNF618*, a single variant (the "T" allele of rs7023951 was positively associated with BP (Cichon et al., 2011).

In addition to candidate gene studies, several genome wide association studies (GWAS) have been performed in BP patients varying in size from several hundreds to over 9,500 patients (Muhleisen et al., 2014; Acikel et al., 2015). In 2011 the Psychiatric GWAS Consortium published the first meta-analysis with 11,974 cases and 51,792 controls (Psychiatric, 2011), followed by several additional meta-analyses. These analyses are large enough to identify genome wide associations, and about 15 genes have now been found below the threshold ($p < 5 \times 10^{-8}$) for genome wide significance, five of which have been replicated at least once. These include *ANK3* (discussed earlier) and *CACNA1C*. *CACNA1C* codes for the α_{1C} subunit of the L-type calcium channel Cav1.2 (Bhat et al., 2012). This channel plays an important role in the depolarization-induced Ca^{2+} influx related to neurotransmitter release in several important brain regions, including the mesolimbic dopaminergic pathway (Bhat et al., 2012). Several different associations have been identified with BP, most of these located in intron 3 (rs1006737, rs4765913, rs4765914, and rs1024582). Especially the "A" allele of rs4765913 has shown strong associations with p values between 1.5×10^{-8} and 9.8×10^{-10} (Psychiatric, 2011; Green et al., 2013). These variants were recently replicated and several additional variants were found, including one within the promotor region of *CACNA1C* in a study including over 1500 patients (Fiorentino et al., 2014). In addition, three other genes *TRANK1*, *ODZ4* and *NCAN* have been repeatedly identified as risk genes in GWAS studies with p values 5×10^{-8} and 2.4×10^{-11} (Goes, 2016).

Finally, several structural variations have been found to be more common in BP. These copy number variations (CNVs) include duplications of 1q21.1 or 16p11.2 and deletions of 3q29. Although the occurrence of these CNVs are very rare (between about 0.1 and 0.01%), they have relatively high ORs (between 2.6 for 1q21.1 and 17.3 for 3q11.2) (Green et al., 2016). A recent study also found significantly more de novo CNVs (defined as deletions or duplications > 10 kilobases) in BP patients.

Moderation of the Effects of Genes by Environmental Factors

Several studies have investigated whether specific environmental factors enhance the risk of BP and, if so, whether this effect depends on the genetic background of the patients. A recent study compared the brains of BP patients with both healthy controls and co-twins using structural MRI (Bootsman et al., 2015). By including both concordant and discordant mono- and dizygotic twins, the authors were able to parse out the relative contribution of genes and environment. Although the authors did not specifically look at the interaction between these two factors, they found that while baseline subcortical brain volumes were predominantly determined by genetic factors (h^2 in the range of 0.64 (thalamus) to 0.85 (hippocampus)), volume changes over time were much more dependent on environmental factors (with h^2 being between 0.00 (hippocampus) and 0.11 (nucleus accumbens)). The only exception was the caudate nucleus which, with an h^2 of 0.43, had a moderate genetic heritability).

Some studies have more specifically focused on the interaction between individual genes and environmental factors in the development of BP. For instance, two studies investigated the influence of environmental effects on individuals with the rs6265 variants of the *BDNF* gene. In one study the influence of family cohesion on individuals with the "A" allele was studied, given the well replicated finding that families with BP patients often show high levels of dysfunction (Zeni et al., 2016). Comparing 20 children and adolescents with BP with 22 healthy controls, the hippocampal volumes were measured using structural MRI. Although no main effect of genotype was found, a significant interaction was found. Specifically individuals with the "A" allele that were living in families with low cohesion had significantly smaller hippocampal volumes. In a similar study it was found that carriers of the "A" allele exposed to a dysfunctional family environment have significantly higher anxiety scores (Park et al., 2015). Several other studies have also investigated the effects of environmental stressors on the rs6265 single nucleotide variant of the *BDNF* gene, generally confirming the increased vulnerability of carriers of the "A" allele (see also Table 7.7).

Studies have also investigated the effects of early traumatic experiences in individuals with the "s" and "l" alleles of the 5-HTTLPR of *SLC6A4*. In line with studies in MDD (see Table 7.2), individuals with one or two copies of the "s" allele are more sensitive to childhood traumatic experiences, leading to an earlier age of onset (Etain et al., 2015) and more psychotic symptoms (De Pradier et al., 2010). In the latter study, the interaction showed a more complex pattern. Thus, by themselves both the "s" allele and cannabis use were significantly associated with psychotic symptoms. In the presence of childhood sexual abuse, individuals with the "s" allele showed more psychotic symptoms (a direct interactions) as well as more cannabis use (an indirect interaction).

Overall, although most studies have relied on relatively small numbers of patients and controls (and often included schizophrenic or MDD patients as well) there is clear (at least preliminary) evidence that specific genes (most notably *BDNF* and *SLC4A6*) interact with childhood and familial stressors in the development of BP.

Animal models for BP

Measuring BP-Like Features in Animals

Although depressive episodes and symptoms are more prevalent in BP, it is the manic episode that is essential for the diagnosis (see Box 7.2). Thus most animal models for BP have focused on modelling the manic aspects of BP. Obviously, as is the case with most psychiatric disorders many of the characteristic symptoms (such as inflated self-esteem, grandiosity or increased pressure to talk) of BP are impossible to assess in animals. However, several of the other characteristics mentioned in Box 7.2 can be operationalized in rats and mice (Einat, 2011; Young et al., 2011). Figure 7.11 shows a number of different paradigms to

TABLE 7.7 Examples of Gene–Environment Interactions in BP

Genes	Environments	Effects	References
BDNF	Family cohesion	Carriers of the "A" allele of rs 6265 reared in unstable families have smaller hippocampus volumes and more anxiety	Zeni et al. (2016); Park et al. (2015)
BDNF	Stressful life events	Carriers of the "A" allele of rs6265 exposed to stressful life events have more severe depressive symptoms	Hosang et al. (2010)
BDNF	Early life stress	Carriers of the "A" allele of rs6265 exposed to early life stressors have more severe BP and earlier age of onset	Miller et al. (2013)
BDNF	Childhood sexual abuse	Carriers of the "A" allele of rs6265 exposed to childhood sexual abuse have significantly smaller right hippocampal volume, and larger right and left lateral ventricular size.	Aas et al. (2013)
BDNF	Childhood emotional abuse	Carriers of the "A" allele of rs6265 exposed to childhood emotional abuse have verbal fluency deficit	Aas et al. (2013)
COMT	Childhood trauma	Carriers of the "G" allele of rs4680 exposed to childhood trauma have higher schizotypy symptoms	Savitz et al. (2010)
SLC6A4	Childhood trauma	Carriers of the "ss" genotype of the 5-HTTLPR exposed to childhood trauma had significantly earlier age of onset	Etain et al. (2015)
SLC6A4	Childhood sexual abuse	Carriers of the "s" allele of the 5-HTTLPR exposed to childhood sexual abuse show more psychotic symptoms	De Pradier et al. (2010)

Please see text for further explanations.

specifically assess mania-related behaviors. Hyperactivity in the open field is generally considered an animal analogue of psychomotor agitation and is often measured in locomotor boxes, using either photocell beam interruptions or video tracking software (see Chapter 8 for more details). Although this test is also used for measuring hyperactivity in relation to ADHD, studies in humans have shown that the hyperactivity pattern seen in patients with BP (Perry et al., 2009) is different from that seen in patients with ADHD (Young et al., 2007) and schizophrenia (Perry et al., 2009). These difference were visible using a more complex open field design than the standard open field. By including several objects in the room and using sophisticated movement analysis (Young et al., 2007), a more detailed analysis of motor behavior could be obtained. Interestingly, the human version of this behavioural pattern monitor (BPM) was actually based on the rodent version (Geyer et al., 1986), thus representing an interesting example of "back-translation".

Another characteristic of BP is reduced sleep, which can be assessed in rodents using EEG recordings. In addition, BP patients, during a manic episode consistently show signs of excessive behaviors, manifest, among others, as increased sexual behavior and aggressiveness and often associated with increased risk taking behavior. Increased sexual behavior can be

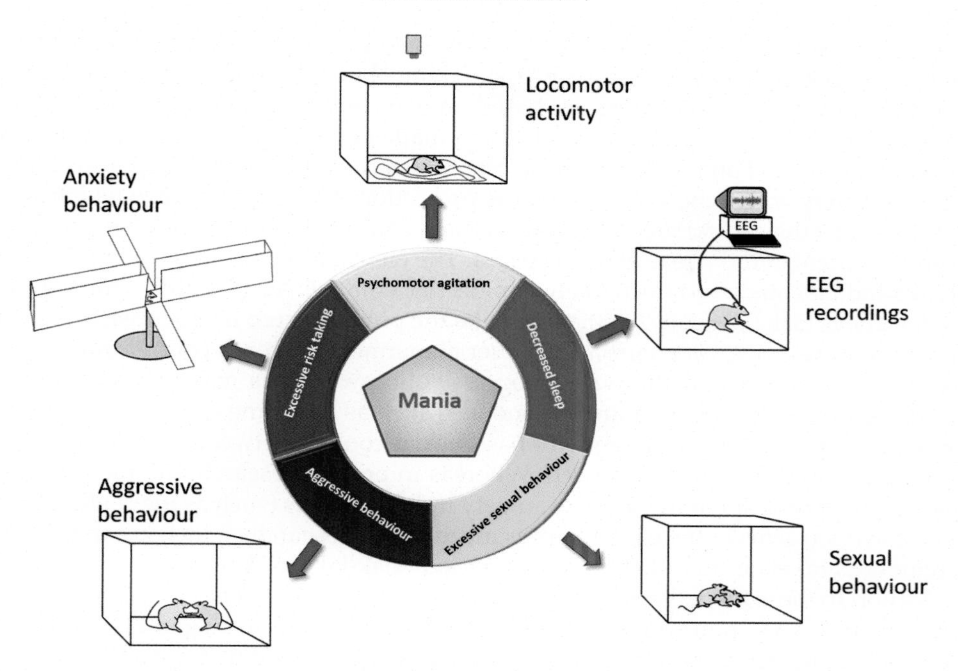

FIGURE 7.11 **Modelling mania in animals.** Animal models for BP have focused on the presence of mania-like symptoms, as they are defining features of the disorder. Several different aspects of mania-like behavior can be assessed in animals, including psychomotor agitation (using locomotor analysis), decreased sleep (EEG recordings of sleep/wake cycle), excessive sexual and aggressive behavior and increased risk-taking behavior (using anxiety-like tests such as the elevated plus maze).

studied in rodents as well, especially in males. Male rodent sexual behavior has characteristic structures, each bout starting with (usually multiple) mountings and intromissions, finishing with ejaculation. Increases in this behavior and reductions in the latency of mounting are usually taken as an indication of increased sexual drive. As sexual behavior is predominantly driven by male behavior, increased sexual drive in females is much less easily assessed. The most prominent sexual behavior displayed by females is lordosis (during which the females curve their body in a U shaped form) which is usually seen as an invitation to the male to initiate sexual interaction.

One of the most frequently used paradigms for assessing aggressive behavior is the resident-intruder test, based on territorial behavior. In this test a resident rat is placed in an experimental box for a period of time, often with a female to allow the animal to establish a territory. After this, during the actual experiment, the female rat is removed and a novel rat (the intruder) is introduced. As rats and especially mice display strong territorial behavior, this usually leads to aggressive behavior of the resident males generally leading to submissive behavior in the intruder males. Although this set-up can reliably induce aggressive behavior, it is important to realize that this is a relatively "natural" form of aggressive behavior aimed at survival and the distinction between "normal" and "pathological" aggression is not always easy to make.

Finally, several different techniques have been used to assess increased risk-taking behavior. One of the classical way of assessing "risk-taking" behavior is through the use of anxiety models. Many of the anxiety models are based on two conflicting tendencies. For example, in the classical open field test a distinction is often made between the center and the periphery of the open field, based on the idea that the center is perceived as more dangerous for rodents than the periphery where the walls offer more protection. Thus the degree to which an animal will venture into the central area is based on the balance between the tendency to explore a novel environment and the fear of open spaces. The elevated plus maze (see Fig. 7.11) and the closely related elevated zero mice likely exploit this conflict. A somewhat similar task is the successive alleys task (Deacon, 2013), which like the plus and zero maze consists of four compartment, they are, however connected in a serial manner, with each compartment becoming progressively narrower and the walls progressively lower, thus making the animals more and more exposed. Several other anxiety models, such as the home cage emergence task and the novelty suppressed feeding, equally based on a conflict between two tendencies, have been routinely used in anxiety research. Whereas in anxiety research, the focus is generally on a reduction of time spent in the more anxiety inducing areas (such as the center of the open field and the open arms in the elevated plus maze), in BP research an increase in time spent in these regions is taken as an indication of increased "risk-taking" behavior.

In addition to these anxiety-related paradigms, more cognitive tasks have been used to assess "risk-taking" and "poor-decision" making in BP, such as the Iowa gambling task (IGT). In this task, originally developed for humans, rodents are presented with (usually) two options: a large reward/large risk and a small reward/small risk options (van den Bos et al., 2014). In the long run, the small reward/small risk options is generally the most rewarding one, and therefore an increase in the large reward/large risk (and in the long run less advantageous) choice is considered a model for mania-related "risk-taking" behavior.

Patients with BP also suffer from cognitive symptoms. Although not essential for the diagnosis, deficits in prepulse inhibition, executive function, impulsivity and vigilance are often seen in patients with BP (Young et al., 2011), and many of these aspects can be modelled in rodents. As many of these cognitive deficits are also prominent in schizophrenia, they will be more specifically discussed in chapter 9.

Genetic Models of BP

Several genetic models have been proposed for modelling the manic phase of BP in animals. Using a forward-genetic approach, Flaisher-Grinberg and coworkers investigated sucrose consumption in four different mouse strains assumed to model BP-related goal-directed (reward-seeking) behavior (Flaisher-Grinberg et al., 2009). The authors found that Black Swiss mice showed the greatest preference for a sweet sucrose solution, and, more importantly, that this preference was reduced by repeated injections of either lithium or valproate, while the preference of other strains was not affected. In addition, Black Swiss mice were found to show increased aggressive and risk-taking behavior, as well as an increase in goal-directed behavior (Flaisher-Grinberg and Einat, 2010; Kara et al., 2015).

Among the reverse genetic models for BP, most attention has focused on models reducing the dopamine transporter (DAT) activity. Mice with a complete lack of the DAT were found to exhibit a strong hyperactivity in a novel open field (Giros et al., 1996), which was reduced by valproate, providing a link to BP (Ralph-Williams et al., 2003). However, as the hyperactivity was also reduced by psychostimulants, the DAT knock-out (KO) mice is perhaps better suited as an animal model for ADHD (see Chapter 8). Moreover, a more detailed analysis of the hyperactivity of the DAT KO mice shows no increase in specific exploratory behavior (such as rearing, nose poking etc.) which are characteristic for BP patients. Interestingly, subsequent work using a genetic knock-down (KD) model of the DAT (reducing levels by about 90%) not only led to hyperactivity, but also to enhanced exploration. Thus, using the BPM, Young and coworkers showed that such mice exhibit a very similar type of hyperactivity as BP patients, characterized by increased nose-pokes and rearing (exploration) and, at the same time, moving in a much straighter line (Perry et al., 2009; Young et al., 2010). Intriguingly, the DAT KD mice also show increased risk-taking behavior in the IGT and reduced vigilance (van Enkhuizen et al., 2015). Thus, these data suggest that the DAT KD mouse represents an interesting model for the manic phase of BP.

Two additional models are based on genes related to the circadian rhythm. Mice with a mutation in the *clock* gene (so called *ClockΔ19* or *mClock*) show several similarities with BP, including hyperactivity and hyperexploration [although the behavioural pattern is not as similar to BP as that of the DAT KD mice (van Enkhuizen et al., 2013)], decreased sleep and increased risk-taking behavior and sucrose preference (Chen et al., 2011b). Interestingly, a viral vector mediated restoration of the *clock* gene in the dopaminergic cell body regions of the ventral tegmental area rescued the phenotype, emphasizing the importance of the clock receptors on the dopaminergic cells bodies (Mukherjee et al., 2010). In line with this, and consistent with the neurobiology of mania (as discussed above), a recent study showed that deletion of the nuclear receptor for clock genes (REV-Erbα) in the midbrain leads to mania-like symptoms in mice and to a hyperdopaminergic state (Chung et al., 2014). One of the most intriguing genetic model for BP is the D-box binding protein (*DBP*) KO mice. Like the *clock* gene *DBP* is involved in the regulation of the circadian cycle. The interesting aspect of this knock-out model is that while the animals exhibit a depressive like phenotype at baseline, after stress (such as handling or single housing) the mice exhibit a mania-like phenotype with hyperactivity and increased hedonic activity (Le-Niculescu et al., 2008). So, not only does this point to a gene–environmental interaction, it is also one of very few models for BP that encompasses both the depressive and manic phase of BP.

Two additional genetic models that have been associated with mania are the *GRIK2* and *ERK1* knock-out mice (Chen et al., 2011b). The *GRIK2* gene codes for a subunit of the ionotropic glutamatergic kainate glutamate 6 receptors, and the mouse model is actually based on the more traditional pharmacological model for BP of cocaine sensitization (Post, 1992) following repeated administration. Since this sensitization is partly reversed by lithium, it has been suggested to represent an animal model for BP. One of the changes in the brain after repeated (but not acute) administration of cocaine is a downregulation of mRNA levels of the kainate receptor. In line with this, mice with a genetic deletion of this receptor show hyperactivity on multiple tests, increased goal directed behavior, increase saccharin preference, increased female urine sniffing and increased risk-taking on the elevated pus maze and reduced sleep (Shaltiel et al., 2008; Malkesman et al., 2010). In line with the mania phenotype, several of these aspects (such as increased risk taking and hyperactivity), were reversed by chronic lithium treatment.

The rationale behind the *ERK1* KO mice stems from the fact that both lithium and valproate enhance ERK1 signaling (Engel et al., 2009). *ERK1* codes for a protein called extracellular signal-regulated kinase 1, which, together with the closely related ERK2, are protein serine/threonine kinases that form part of a cascade of intracellular processes involved in a large variety of processes including cell adhesion, cell cycle progression, cell migration, cell survival, differentiation, metabolism, proliferation, and transcription (Roskoski, 2012). *ERK1* KO mice show a number of the mania-related symptoms also seen in the previous models, such as hyperactivity, increased goal-directed behavior and increased risk-taking behavior in the elevated plus maze (Engel et al., 2009). However, while treatment with valproate and olanzapine reverses several aspect of the phenotype, lithium is ineffective, thus limiting the predictive validity.

Moderation of the Effects of Genes by Environmental Factors

Very few attempts have been made to investigate whether the effect of specific genetic alterations are influenced by environmental factors. We already discussed the DBP KO mice which show a depressive-like phenotype under normal conditions but a more mania-like phenotype when stressed. In another study, the effects of chronic stress on the phenotype of Ank3 KO mice was investigated (Leussis et al., 2013). As homozygous Ank3 KO mice show severe early-onset ataxia that interferes with normal motor activity, only heterozygous KO mice were studied. Compared to WT mice, these mice showed increased risk-taking behavior in the elevated plus maze, the light-dark box and the novelty suppressed feeding test. Likewise, these mice had a significantly greater sucrose preference. However, prepulse inhibition and locomotor activity were normal. Interestingly, after chronic stress (6 weeks single housing) virtually all these behavior reversed. Thus in the elevated plus maze and the light-dark box, the heterozygous Ank KO mice showed an anxious phenotype, while sucrose preference was not different. Behavior in the forced swim test, on the other hand shows an increase immobility, taken as indicative of a depressive phenotype. Thus, while the DBP KO mice switch from a depressive to a manic phenotype after stress, the reverse seems to occur in Ank3 KO mice.

Unfortunately, there appear to be very few additional animal models in which genetic and environmental factors are combined, in spite of the fact that, especially early stressful life events can significantly moderate the effects of specific genetic variants in humans (see Table 7.7). It is therefore important that additional research in this area is undertaken, especially as the few studies that have looked at interactions seem to represent models in which aspects of both mania and depression can be observed.

References

Aarsland, D., Pahlhagen, S., Ballard, C.G., et al., 2012. Depression in Parkinson disease-epidemiology, mechanisms and management. Nat. Rev. Neurol. 8, 35–47.

Aas, M., Haukvik, U.K., Djurovic, S., et al., 2013. BDNF val66met modulates the association between childhood trauma, cognitive and brain abnormalities in psychoses. Prog. Neuro-Psychopharmacol. Biol. Psychiatr. 46, 181–188.

Acikel, C., Son, Y.A., Celik, C., et al., 2015. Evaluation of whole genome association study data in bipolar disorders: potential novel snps and genes. Klinik Psikofarmakoloji Bulteni Bull. Clin. Psychopharmacol. 25, 12–18.

Agrawal, A., Nelson, E.C., Littlefield, A.K., et al., 2012. Cannabinoid receptor genotype moderation of the effects of childhood physical abuse on anhedonia and depression. Arch. Gen. Psychiatr. 69, 732–740.

Al Jurdi, R.K., Nguyen, Q.X., Petersen, N.J., et al., 2012. Acute bipolar I affective episode presentation across life span. J. Geriatr. Psychiatr. Neurol. 25, 6–14.

Alvarez Ariza, M., Mateos Alvarez, R., Berrios, G.E., 2009. A review of the natural course of bipolar disorders (manic-depressive psychosis) in the pre-drug era: review of studies prior to 1950. J. Affect. Disord. 115, 293–301.

Anand, A., Verhoeff, P., Seneca, N., et al., 2000. Brain SPECT imaging of amphetamine-induced dopamine release in euthymic bipolar disorder patients. Am. J. Psychiatr. 157, 1108–1114.

Appel, K., Schwahn, C., Mahler, J., et al., 2011. Moderation of adult depression by a polymorphism in the FKBP5 gene and childhood physical abuse in the general population. Neuropsychopharmacology 36, 1982–1991.

Aulakh, C.S., Hill, J.L., Murphy, D.L., 1993. Attenuation of hypercortisolemia in fawn-hooded rats by antidepressant drugs. Eur. J. Pharmacol. 240, 85–88.

Aulakh, C.S., Tolliver, T., Wozniak, K.M., et al., 1994. Functional and biochemical evidence for altered serotonergic function in the fawn-hooded rat strain. Pharmacol. Biochem. Behav. 49, 615–620.

Bale, T.L., Vale, W.W., 2003. Increased depression-like behaviors in corticotropin-releasing factor receptor-2-deficient mice: sexually dichotomous responses. J. Neurosci. 23, 5295–5301.

Bale, T.L., Contarino, A., Smith, G.W., et al., 2000. Mice deficient for corticotropin-releasing hormone receptor-2 display anxiety-like behaviour and are hypersensitive to stress. Nat. Genet. 24, 410–414.

Barbano, M.F., Cador, M., 2006. Differential regulation of the consummatory, motivational and anticipatory aspects of feeding behavior by dopaminergic and opioidergic drugs. Neuropsychopharmacology 31, 1371–1381.

Barkus, C., 2013. Genetic mouse models of depression. Curr. Topics Behav. Neurosci. 14, 55–78.

Barr, C.S., Newman, T.K., Becker, M.L., et al., 2003. The utility of the non-human primate model for studying gene by environment interactions in behavioral research. Genes Brain Behav. 2, 336–340.

Bellivier, F., Golmard, J.L., Rietschel, M., et al., 2003. Age at onset in bipolar I affective disorder: further evidence for three subgroups. Am. J. Psychiatr. 160, 999–1001.

Belvederi Murri, M., Prestia, D., Mondelli, V., et al., 2016. The HPA axis in bipolar disorder: systematic review and meta-analysis. Psychoneuroendocrinology 63, 327–342.

Benes, F.M., 2011. The neurobiology of bipolar disorder: from circuits to cells to molecular regulation. Curr. Topics Behav. Neurosci. 5, 127–138.

Berretta, S., Pantazopoulos, H., Lange, N., 2007. Neuron numbers and volume of the amygdala in subjects diagnosed with bipolar disorder or schizophrenia. Biol. Psychiatr. 62, 884–893.

Bhat, S., Dao, D.T., Terrillion, C.E., et al., 2012. CACNA1C (Cav1.2) in the pathophysiology of psychiatric disease. Prog. Neurobiol. 99, 1–14.

Bootsman, F., Brouwer, R.M., Kemner, S.M., et al., 2015. Contribution of genes and unique environment to cross-sectional and longitudinal measures of subcortical volumes in bipolar disorder. Eur. Neuropsychopharmacol. 25, 2197–2209.

Brezo, J., Bureau, A., Merette, C., et al., 2010. Differences and similarities in the serotonergic diathesis for suicide attempts and mood disorders: a 22-year longitudinal gene-environment study. Mol. Psychiatr. 15, 831–843.

Bromet, E., Andrade, L.H., Hwang, I., et al., 2011. Cross-national epidemiology of DSM-IV major depressive episode. BMC Med. 9.

Brown, W.A., Rosdolsky, M., 2015. The clinical discovery of imipramine. Am. J. Psychiatr. 172, 426–429.

Buchmann, A.F., Holz, N., Boecker, R., et al., 2014. Moderating role of FKBP5 genotype in the impact of childhood adversity on cortisol stress response during adulthood. Eur. Neuropsychopharmacol. 24, 837–845.

Buck, C.L., Vendruscolo, L.F., Koob, G.F., et al., 2014. Dopamine D1 and mu-opioid receptor antagonism blocks anticipatory 50 kHz ultrasonic vocalizations induced by palatable food cues in Wistar rats. Psychopharmacology (Berlin) 231, 929–937.

Buoli, M., Serati, M., Altamura, A.C., 2014. Is the combination of a mood stabilizer plus an antipsychotic more effective than mono-therapies in long-term treatment of bipolar disorder? A systematic review. J. Affect. Disord. 152, 12–18.

Caligiuri, M.P., Brown, G.G., Meloy, M.J., et al., 2006. Striatopallidal regulation of affect in bipolar disorder. J. Affect. Disord. 91, 235–242.

Carola, V., Gross, C., 2010. BDNF moderates early environmental risk factors for anxiety in mouse. Genes Brain Behav. 9, 379–389.

Carola, V., Gross, C., 2012. Mouse models of the 5-HTTLPR x stress risk factor for depression. Curr. Topics Behav. Neurosci. 12, 59–72.

Carola, V., Pascucci, T., Puglisi-Allegra, S., et al., 2011. Effect of the interaction between the serotonin transporter gene and maternal environment on developing mouse brain. Behav. Brain Res. 217, 188–194.

Carvalho, A.F., Quevedo, J., McIntyre, R.S., et al., 2015. Treatment implications of predominant polarity and the polarity index: a comprehensive review. Int. J. Neuropsychopharmacol., 18.
Caspi, A., Sugden, K., Moffitt, T.E., et al., 2003. Influence of life stress on depression: moderation by a polymorphism in the 5-HTT gene. Science 301, 386–389.
Chaney, A., Carballedo, A., Amico, F., et al., 2014. Effect of childhood maltreatment on brain structure in adult patients with major depressive disorder and healthy participants. JPN 39, 50–59.
Chaudhury, D., Liu, H., Han, M.H., 2015. Neuronal correlates of depression. CMLS 72, 4825–4848.
Chaudhury, D., Walsh, J.J., Friedman, A.K., et al., 2013. Rapid regulation of depression-related behaviours by control of midbrain dopamine neurons. Nature 493, 532–536.
Chen, C., Glatt, S.J., Tsuang, M.T., 2008. The tryptophan hydroxylase gene influences risk for bipolar disorder but not major depressive disorder: results of meta-analyses. Bipolar Disord. 10, 816–821.
Chen, C.H., Suckling, J., Lennox, B.R., et al., 2011a. A quantitative meta-analysis of fMRI studies in bipolar disorder. Bipolar Disord. 13, 1–15.
Chen, G., Henter, I.D., Manji, H.K., 2011b. Partial rodent genetic models for bipolar disorder. Curr. Topics Behav. Neurosci. 5, 89–106.
Chen, J., Li, X., McGue, M., 2012. Interacting effect of BDNF Val66Met polymorphism and stressful life events on adolescent depression. Genes Brain Behav. 11, 958–965.
Ching-Lopez, A., Cervilla, J., Rivera, M., et al., 2015. Epidemiological support for genetic variability at hypothalamic-pituitary-adrenal axis and serotonergic system as risk factors for major depression. Neuropsychiatr. Dis. Treat. 11, 2743–2754.
Cho, H.J., Meira-Lima, I., Cordeiro, Q., et al., 2005. Population-based and family-based studies on the serotonin transporter gene polymorphisms and bipolar disorder: a systematic review and meta-analysis. Mol. Psychiatr. 10, 771–781.
Chung, S., Lee, E.J., Yun, S., et al., 2014. Impact of circadian nuclear receptor REV-ERBalpha on midbrain dopamine production and mood regulation. Cell 157, 858–868.
Cicchetti, D., Rogosch, F.A., Sturge-Apple, M.L., 2007. Interactions of child maltreatment and serotonin transporter and monoamine oxidase A polymorphisms: depressive symptornatology among adolescents from low socioeconomic status backgrounds. Dev. Psychopathol. 19, 1161–1180.
Cichon, S., Muhleisen, T.W., Degenhardt, F.A., et al., 2011. Genome-wide association study identifies genetic variation in neurocan as a susceptibility factor for bipolar disorder. Am. J. Hum. Genet. 88, 372–381.
Citrome, L., 2014. Treatment of bipolar depression: making sensible decisions. CNS Spectr. 19 (Suppl 1), 4–11, quiz 11-13, 12.
Colman, I., Naicker, K., Zeng, Y.Y., et al., 2011. Predictors of long-term prognosis of depression. Can. Med. Assoc. J. 183, 1969–1976.
Comasco, E., Aslund, C., Oreland, L., et al., 2013. Three-way interaction effect of 5-HTTLPR, BDNF Val66Met, and childhood adversity on depression: a replication study. Eur. Neuropsychopharmacol.: J. Eur. Coll. Neuropsychopharmacol. 23, 1300–1306.
Connolly, K.R., Thase, M.E., 2012. Emerging drugs for major depressive disorder. Expert Opin. Emerg. Drugs 17, 105–126.
Cousins, D.A., Butts, K., Young, A.H., 2009. The role of dopamine in bipolar disorder. Bipolar Disord. 11, 787–806.
Cryan, J.F., Mombereau, C., 2004. In search of a depressed mouse: utility of models for studying depression-related behavior in genetically modified mice. Mol. Psychiatr. 9, 326–357.
D'Souza, M.S., Markou, A., 2010. Neural substrates of psychostimulant withdrawal-induced anhedonia. Curr. Topics Behav. Neurosci. 3, 119–178.
De Fruyt, J., Deschepper, E., Audenaert, K., et al., 2012. Second generation antipsychotics in the treatment of bipolar depression: a systematic review and meta-analysis. J. Psychopharmacol. 26, 603–617.
De Pradier, M., Gorwood, P., Beaufils, B., et al., 2010. Influence of the serotonin transporter gene polymorphism, cannabis and childhood sexual abuse on phenotype of bipolar disorder: a preliminary study. Eur. Psychiatr. J. Assoc. Eur. Psychiatr. 25, 323–327.
Deacon, R.M., 2013. The successive alleys test of anxiety in mice and rats. JoVE.
Dedovic, K., Ngiam, J., 2015. The cortisol awakening response and major depression: examining the evidence. Neuropsychiatr. Dis. Treat. 11, 1181–1189.
Degenhardt, F., Priebe, L., Herms, S., et al., 2012. Association between copy number variants in 16p11.2 and major depressive disorder in a German case-control sample. Am. J. Med. Genet. B 159B, 263–273.

Detera-Wadleigh, S.D., McMahon, F.J., 2006. G72/G30 in schizophrenia and bipolar disorder: review and meta-analysis. Biol. Psychiatr. 60, 106–114.

Dilsaver, S.C., 2011. An estimate of the minimum economic burden of bipolar I and II disorders in the United States: 2009. J. Affect. Disord. 129, 79–83.

Drevets, W.C., Price, J.L., Furey, M.L., 2008. Brain structural and functional abnormalities in mood disorders: implications for neurocircuitry models of depression. Brain Struct. Funct. 213, 93–118.

Drevets, W.C., Price, J.L., Simpson, Jr., J.R., et al., 1997. Subgenual prefrontal cortex abnormalities in mood disorders. Nature 386, 824–827.

Drevets, W.C., Frank, E., Price, J.C., et al., 1999. PET imaging of serotonin 1A receptor binding in depression. Biol. Psychiatr. 46, 1375–1387.

Duman, R.S., 1998. Novel therapeutic approaches beyond the serotonin receptor. Biol. Psychiatr. 44, 324–335.

Dunn, E.C., Brown, R.C., Dai, Y., et al., 2015. Genetic determinants of depression: recent findings and future directions. Harvard Rev. Psychiatr. 23, 1–18.

Einat, H., 2011. Strategies for the development of animal models for bipolar disorder: new opportunities and new challenges. Curr. Topics Behav. Neurosci. 5, 69–87.

Eisenstein, N., Iorio, L.C., Clody, D.E., 1982. Role of serotonin in the blockade of muricidal behavior by tricyclic antidepressants. Pharmacol. Biochem. Behav. 17, 847–849.

El Khoury, A., Gruber, S.H.M., Mork, A., et al., 2006. Adult life behavioral consequences of early maternal separation are alleviated by escitalopram treatment in a rat model of depression. Prog. Neuro-Psychopharmacol. Biol. Psychiatr. 30, 535–540.

Engel, S.R., Creson, T.K., Hao, Y., et al., 2009. The extracellular signal-regulated kinase pathway contributes to the control of behavioral excitement. Mol. Psychiatr. 14, 448–461.

Etain, B., Lajnef, M., Henrion, A., et al., 2015. Interaction between SLC6A4 promoter variants and childhood trauma on the age at onset of bipolar disorders. Sci. Rep. 5, 16301.

Etain, B., Dumaine, A., Bellivier, F., et al., 2012. Genetic and functional abnormalities of the melatonin biosynthesis pathway in patients with bipolar disorder. Hum. Mol. Genet. 21, 4030–4037.

Faedda, G.L., Serra, G., Marangoni, C., et al., 2014. Clinical risk factors for bipolar disorders: a systematic review of prospective studies. J. Affect. Disord. 168, 314–321.

Fan, M., Liu, B., Jiang, T., et al., 2010. Meta-analysis of the association between the monoamine oxidase-A gene and mood disorders. Psychiatr. Genet. 20, 1–7.

Ferrari, A.J., Baxter, A.J., Whiteford, H.A., 2011. A systematic review of the global distribution and availability of prevalence data for bipolar disorder. J. Affect. Disord. 134, 1–13.

Ferrari, A.J., Charlson, F.J., Norman, R.E., et al., 2013. Burden of depressive disorders by country, sex, age, and year: findings from the global burden of disease study 2010. Plos Med., 10.

Fiorentino, A., O'Brien, N.L., Locke, D.P., et al., 2014. Analysis of ANK3 and CACNA1C variants identified in bipolar disorder whole genome sequence data. Bipolar Disord. 16, 583–591.

Flaisher-Grinberg, S., Einat, H., 2010. Strain-specific battery of tests for domains of mania: effects of valproate, lithium and imipramine. Front. Psychiatr. 1, 10.

Flaisher-Grinberg, S., Overgaard, S., Einat, H., 2009. Attenuation of high sweet solution preference by mood stabilizers: a possible mouse model for the increased reward-seeking domain of mania. J. Neurosci. Methods 177, 44–50.

Friedman, E., Berman, M., Overstreet, D., 2006. Swim test immobility in a genetic rat model of depression is modified by maternal environment: a cross-foster study. Dev. Psychobiol. 48, 169–177.

Frodl, T., O'Keane, V., 2013. How does the brain deal with cumulative stress? A review with focus on developmental stress, HPA axis function and hippocampal structure in humans. Neurobiol. Dis. 52, 24–37.

Furlong, R.A., Ho, L., Rubinsztein, J.S., et al., 1999. Analysis of the monoamine oxidase A (MAOA) gene in bipolar affective disorder by association studies, meta-analyses, and sequencing of the promoter. Am. J. Med. Genet. 88, 398–406.

Furukawa, T.A., 2010. Assessment of mood: guides for clinicians. J. Psychosom. Res. 68, 581–589.

Gao, J., Jia, M., Qiao, D., et al., 2016. TPH2 gene polymorphisms and bipolar disorder: a meta-analysis. Am. J. Med. Genet. B 171, 145–152.

Gard, D.E., Gard, M.G., Kring, A.M., et al., 2006. Anticipatory and consummatory components of the experience of pleasure: a scale development study. J. Res. Personality 40, 1086–1102.

Gatt, J.M., Burton, K.L.O., Williams, L.M., et al., 2015. Specific and common genes implicated across major mental disorders: a review of meta-analysis studies. J. Psychiatr. Res. 60, 1–13.

Gatt, J.M., Nemeroff, C.B., Dobson-Stone, C., et al., 2009. Interactions between BDNF Val66Met polymorphism and early life stress predict brain and arousal pathways to syndromal depression and anxiety. Mol. Psychiatr. 14, 681–695.

Gellynck, E., Heyninck, K., Andressen, K.W., et al., 2013. The serotonin 5-HT7 receptors: two decades of research. Exp. Brain Res. 230, 555–568.

Gershon, S., Chengappa, K.N., Malhi, G.S., 2009. Lithium specificity in bipolar illness: a classic agent for the classic disorder. Bipolar Disord. 11 (Suppl 2), 34–44.

Geyer, M.A., Russo, P.V., Masten, V.L., 1986. Multivariate assessment of locomotor behavior: pharmacological and behavioral analyses. Pharmacol. Biochem. Behav. 25, 277–288.

Giniatullina, A., Maroteaux, G., Geerts, C.J., et al., 2015. Functional characterization of the PCLO p.Ser4814Ala variant associated with major depressive disorder reveals cellular but not behavioral differences. Neuroscience 300, 518–538.

Giros, B., Jaber, M., Jones, S.R., et al., 1996. Hyperlocomotion and indifference to cocaine and amphetamine in mice lacking the dopamine transporter. Nature 379, 606–612.

Glessner, J.T., Wang, K., Sleiman, P.M., et al., 2010. Duplication of the SLIT3 locus on 5q35.1 predisposes to major depressive disorder. PLoS One 5, e15463.

Glue, P., Herbison, P., 2015. Comparative efficacy and acceptability of combined antipsychotics and mood stabilizers versus individual drug classes for acute mania: network meta-analysis. Australian New Zealand J. Psychiatr. 49, 1215–1220.

Goes, F.S., 2016. Genetics of bipolar disorder: recent update and future directions. Psychiatr. Clin. North America 39, 139–155.

Gonzalez-Castro, T.B., Nicolini, H., Lanzagorta, N., et al., 2015. The role of brain-derived neurotrophic factor (BDNF) Val66Met genetic polymorphism in bipolar disorder: a case-control study, comorbidities, and meta-analysis of 16,786 subjects. Bipolar Disord.ers 17, 27–38.

Green, E.K., Hamshere, M., Forty, L., et al., 2013. Replication of bipolar disorder susceptibility alleles and identification of two novel genome-wide significant associations in a new bipolar disorder case-control sample. Mol. Psychiatr. 18, 1302–1307.

Green, E.K., Rees, E., Walters, J.T., et al., 2016. Copy number variation in bipolar disorder. Mol. Psychiatr. 21, 89–93.

Greenberg, P.E., Fournier, A.A., Sisitsky, T., et al., 2015. The economic burden of adults with major depressive disorder in the united states (2005 and 2010). J. Clin. Psychiatr. 76, 155–U115.

Grof, P., Muller-Oerlinghausen, B., 2009. A critical appraisal of lithium's efficacy and effectiveness: the last 60 years. Bipolar Disord. 11 (Suppl 2), 10–19.

Gururajan, A., Hill, R.A., van den Buuse, M., 2015. Brain-derived neurotrophic factor heterozygous mutant rats show selective cognitive changes and vulnerability to chronic corticosterone treatment. Neuroscience 284, 297–310.

Guscott, M., Bristow, L.J., Hadingham, K., et al., 2005. Genetic knockout and pharmacological blockade studies of the 5-HT7 receptor suggest therapeutic potential in depression. Neuropharmacology 48, 492–502.

Guxens, M., Ghassabian, A., Gong, T., et al., 2016. Air pollution exposure during pregnancy and childhood autistic traits in four European population-based cohort studies: the ESCAPE project. Environ. Health Perspect. 124, 133–140.

Gyekis, J.P., Yu, W., Dong, S., et al., 2013. No association of genetic variants in BDNF with major depression: a meta- and gene-based analysis. Am. J. Med. Genet. B 162B, 61–70.

Hall, F.S., Huang, S., Fong, G.W., et al., 1998. Effects of isolation-rearing on voluntary consumption of ethanol, sucrose and saccharin solutions in Fawn hooded and Wistar rats. Psychopharmacology 139, 216.

Hamilton, M., 1960. A rating scale for depression. J. Neurol. Neurosurg. Psychiatr. 23, 56–62.

Hammer, C., Cichon, S., Muhleisen, T.W., et al., 2012. Replication of functional serotonin receptor type 3A and B variants in bipolar affective disorder: a European multicenter study. Trans. Psychiatr. 2, e103.

Hamshere, M.L., Gordon-Smith, K., Forty, L., et al., 2009. Age-at-onset in bipolar-I disorder: mixture analysis of 1369 cases identifies three distinct clinical sub-groups. J. Affect. Disord. 116, 23–29.

Han, C., Wang, S.M., Kato, M., et al., 2013. Second-generation antipsychotics in the treatment of major depressive disorder: current evidence. Expert Rev. Neurotherapeutic. 13, 851–870.

Harro, J., 2013. Animal models of depression vulnerability. Curr. Topics Behav. Neurosci. 14, 29–54.

Hek, K., Mulder, C.L., Luijendijk, H.J., et al., 2010. The PCLO gene and depressive disorders: replication in a population-based study. Hum. Mol. Genet. 19, 731–734.

Henderson, A.S., Korten, A.E., Jorm, A.F., et al., 2000. COMT and DRD3 polymorphisms, environmental exposures, and personality traits related to common mental disorders. Am. J. Med. Genet. 96, 102–107.

Hendrie, C., Pickles, A., 2013. The failure of the antidepressant drug discovery process is systemic. J. Psychopharmacol. 27, 407–416.

Hindorff, L.A., Sethupathy, P., Junkins, H.A., et al., 2009. Potential etiologic and functional implications of genome-wide association loci for human diseases and traits. Proc. Natl. Acad. Sci. USA 106, 9362–9367.

Holmes, A., Yang, R.J., Murphy, D.L., et al., 2002. Evaluation of antidepressant-related behavioral responses in mice lacking the serotonin transporter. Neuropsychopharmacology 27, 914–923.

Horstmann, S., Lucae, S., Menke, A., et al., 2010. Polymorphisms in GRIK4, HTR2A, and FKBP5 show interactive effects in predicting remission to antidepressant treatment. Neuropsychopharmacology 35, 727–740.

Hosang, G.M., Uher, R., Keers, R., et al., 2010. Stressful life events and the brain-derived neurotrophic factor gene in bipolar disorder. J. Affect. Disord. 125, 345–349.

Houenou, J., Frommberger, J., Carde, S., et al., 2011. Neuroimaging-based markers of bipolar disorder: evidence from two meta-analyses. J. Affect. Disord. 132, 344–355.

Ignacio, Z.M., Reus, G.Z., Abelaira, H.M., et al., 2014. Epigenetic and epistatic interactions between serotonin transporter and brain-derived neurotrophic factor genetic polymorphism: insights in depression. Neuroscience 275, 455–468.

Ishitobi, Y., Nakayama, S., Yamaguchi, K., et al., 2012. Association of CRHR1 and CRHR2 with major depressive disorder and panic disorder in a Japanese population. Am. J. Med. Genet. B 159B, 429–436.

Jimenez-Vasquez, P.A., Mathe, A.A., Thomas, J.D., et al., 2001. Early maternal separation alters neuropeptide Y concentrations in selected brain regions in adult rats. Dev. Brain Res. 131, 149–152.

Jin, C., Xu, W., Yuan, J., et al., 2013. Meta-analysis of association between the -1438A/G (rs6311) polymorphism of the serotonin 2A receptor gene and major depressive disorder. Neurol. Res. 35, 7–14.

Jokela, M., Lehtimaki, T., Keltikangas-Jarvinen, L., 2007a. The influence of urban/rural residency on depressive symptoms is moderated by the serotonin receptor 2A gene. Am. J. Medical Genetic. Part B 144B, 918–922.

Jokela, M., Raikkonen, K., Lehtimaki, T., et al., 2007b. Tryptophan hydroxylase 1 gene (TPH1) moderates the influence of social support on depressive symptoms in adults. J. Affect. Disord. 100, 191–197.

Judd, L.L., Akiskal, H.S., Schettler, P.J., et al., 2002. The long-term natural history of the weekly symptomatic status of bipolar I disorder. Arch. Gen. Psychiatr. 59, 530–537.

Judd, L.L., Akiskal, H.S., Schettler, P.J., et al., 2003. A prospective investigation of the natural history of the long-term weekly symptomatic status of bipolar II disorder. Arch. Gen. Psychiatr. 60, 261–269.

Kalueff, A.V., Gallagher, P.S., Murphy, D.L., 2006. Are serotonin transporter knockout mice 'depressed'?: hypoactivity but no anhedonia. Neuroreport 17, 1347–1351.

Kanazawa, T., Glatt, S.J., Kia-Keating, B., et al., 2007. Meta-analysis reveals no association of the Val66Met polymorphism of brain-derived neurotrophic factor with either schizophrenia or bipolar disorder. Psychiatr. Genet. 17, 165–170.

Kara, N.Z., Flaisher-Grinberg, S., Einat, H., 2015. Partial effects of the AMPAkine CX717 in a strain specific battery of tests for manic-like behavior in black Swiss mice. Pharmacol. Rep. 67, 928–933.

Kasper, S., 2014. Treatment-resistant depression: a challenge for future research. Acta Neuropsychiatr. 26, 131–133.

Kaufman, J., Yang, B.Z., Douglas-Palumberi, H., et al., 2006. Brain-derived neurotrophic factor-5-HTTLPR gene interactions and environmental modifiers of depression in children. Biol. Psychiatr. 59, 673–680.

Kendler, K.S., Gatz, M., Gardner, C.O., et al., 2007. Clinical indices of familial depression in the Swedish Twin Registry. Acta Psychiatr. Scandinavica 115, 214–220.

Kendler, K.S., Neale, M.C., Kessler, R.C., et al., 1993. The lifetime history of major depression in women - reliability of diagnosis and heritability. Archiv. General Psychiatr. 50, 863–870.

Kessler, R.C., Bromet, E.J., 2013. The epidemiology of depression across cultures. Annu. Rev. Publ. Health 34, 119–138.

Kieseppa, T., Partonen, T., Haukka, J., et al., 2004. High concordance of bipolar I disorder in a nationwide sample of twins. Am. J. Psychiatr. 161, 1814–1821.

Kilpatrick, D.G., Koenen, K.C., Ruggiero, K.J., et al., 2007. The serotonin transporter genotype and social support and moderation of posttraumatic stress disorder and depression in hurricane-exposed adults. Am. J. Psychiatr. 164, 1693–1699.

Kim, J.M., Stewart, R., Kim, S.W., et al., 2007. Interactions between life stressors and susceptibility genes (5-HTTLPR and BDNF) on depression in Korean elders. Biol. Psychiatr. 62, 423–428.

Kishi, T., Yoshimura, R., Fukuo, Y., et al., 2013. The serotonin 1A receptor gene confer susceptibility to mood disorders: results from an extended meta-analysis of patients with major depression and bipolar disorder. Eur. Arch. Psychiatry Clin. Neurosci. 263, 105–118.

Kishi, T., Okochi, T., Tsunoka, T., et al., 2011. Serotonin 1A receptor gene, schizophrenia and bipolar disorder: an association study and meta-analysis. Psychiatry Res. 185, 20–26.

Kleine-Budde, K., Touil, E., Moock, J., et al., 2014. Cost of illness for bipolar disorder: a systematic review of the economic burden. Bipolar Disord. 16, 337–353.

Kohler, S., Cierpinsky, K., Kronenberg, G., et al., 2016. The serotonergic system in the neurobiology of depression: relevance for novel antidepressants. J. Psychopharmacol. 30, 13–22.

Korhonen, T., Loukola, A., Wedenoja, J., et al., 2014. Role of nicotine dependence in the association between the dopamine receptor gene DRD3 and major depressive disorder. PLoS One 9, e98199.

Kupferschmidt, D.A., Zakzanis, K.K., 2011. Toward a functional neuroanatomical signature of bipolar disorder: quantitative evidence from the neuroimaging literature. Psychiatr. Res. 193, 71–79.

Lahdesmaki, J., Sallinen, J., MacDonald, E., et al., 2002. Behavioral and neurochemical characterization of alpha(2A)-adrenergic receptor knockout mice. Neuroscience 113, 289–299.

Lasky-Su, J.A., Faraone, S.V., Glatt, S.J., et al., 2005. Meta-analysis of the association between two polymorphisms in the serotonin transporter gene and affective disorders. Am. J. Med. Genet. B 133B, 110–115.

Lavi-Avnon, Y., Yadid, G., Overstreet, D.H., et al., 2005. Abnormal patterns of maternal behavior in a genetic animal model of depression. Physiol. Behav. 84, 607–615.

Lavi-Avnon, Y., Weller, A., Finberg, J.P., et al., 2008. The reward system and maternal behavior in an animal model of depression: a microdialysis study. Psychopharmacology (Berlin) 196, 281–291.

Le-Niculescu, H., McFarland, M.J., Ogden, C.A., et al., 2008. Phenomic, convergent functional genomic, and biomarker studies in a stress-reactive genetic animal model of bipolar disorder and co-morbid alcoholism. Am. J. Med. Genet B 147B, 134–166.

Lekman, M., Laje, G., Charney, D., et al., 2008. The FKBP5-gene in depression and treatment response--an association study in the sequenced treatment alternatives to relieve depression (STAR*D) cohort. Biol. Psychiatr. 63, 1103–1110.

Leussis, M.P., Berry-Scott, E.M., Saito, M., et al., 2013. The ANK3 bipolar disorder gene regulates psychiatric-related behaviors that are modulated by lithium and stress. Biol. Psychiatr. 73, 683–690.

Li, Y.H., Mou, X.D., Jiang, W.H., et al., 2015. A comparative study of anhedonia components between major depression and schizophrenia in Chinese populations. Ann. General Psychiatr. 14.

Lichtenstein, P., Yip, B.H., Bjork, C., et al., 2009. Common genetic determinants of schizophrenia and bipolar disorder in Swedish families: a population-based study. Lancet 373, 234–239.

Liu, Z., Huang, L., Luo, X.J., et al., 2015. MAOA variants and genetic susceptibility to major psychiatric disorders. Mol. Neurobiol.

Lopez-Leon, S., Janssens, A.C.J.W., Ladd, A.M.G.Z., et al., 2008. Meta-analyses of genetic studies on major depressive disorder. Mol. Psychiatr. 13, 772–785.

Lopez-Munoz, F., Alamo, C., 2009. Monoaminergic neurotransmission: the history of the discovery of antidepressants from 1950s until today. Curr. Pharmaceutic. Des. 15, 1563–1586.

Lopez Leon, S., Croes, E.A., Sayed-Tabatabaei, F.A., et al., 2005. The dopamine D4 receptor gene 48-base-pair-repeat polymorphism and mood disorders: a meta-analysis. Biol. Psychiatr. 57, 999–1003.

Magni, L.R., Purgato, M., Gastaldon, C., et al., 2013. Fluoxetine versus other types of pharmacotherapy for depression. Cochrane Database Sys. Rev.

Mahar, I., Bambico, F.R., Mechawar, N., et al., 2014. Stress, serotonin, and hippocampal neurogenesis in relation to depression and antidepressant effects. Neurosci. Biobehav. Rev. 38, 173–192.

Malhi, G.S., Lagopoulos, J., Sachdev, P., et al., 2004. Cognitive generation of affect in hypomania: an fMRI study. Bipolar Disord. 6, 271–285.

Malkesman, O., Scattoni, M.L., Paredes, D., et al., 2010. The female urine sniffing test: a novel approach for assessing reward-seeking behavior in rodents. Biol. Psychiatr. 67, 864–871.

Malykhin, N.V., Coupland, N.J., 2015. Hippocampal neuroplasticity in major depressive disorder. Neuroscience 309, 200–213.

Marchand, W.R., Yurgelun-Todd, D., 2010. Striatal structure and function in mood disorders: a comprehensive review. Bipolar Disord. 12, 764–785.

Mathe, A.A., Jimenez, P.A., Theodorsson, E., et al., 1998. Neuropeptide Y, Neurokinin A and neurotensin in brain regions of Fawn Hooded "depressed", Wistar and Sprague Dawley rats. Effects of electroconvulsive stimuli. Prog. Neuropsychopharmacol. Biol. Psychiatr. 22, 529–546.

Mayorga, A.J., Dalvi, A., Page, M.E., et al., 2001. Antidepressant-like behavioral effects in 5-hydroxytryptamine(1A) and 5-hydroxytryptamine(1B) receptor mutant mice. J. Pharmacol. Exp. Ther. 298, 1101–1107.

McEwen, B.S., Gray, J.D., Nasca, C., 2015. 60 years of neuroendocrinology redefining neuroendocrinology: stress, sex and cognitive and emotional regulation. J. Endocrinol. 226, T67–T83.

McIntyre, R.S., Cha, D.S., Kim, R.D., et al., 2013. A review of FDA-approved treatment options in bipolar depression. CNS Spectr. 18 (Suppl 1), 4–20, quiz 21.

Merikangas, K.R., Akiskal, H.S., Angst, J., et al., 2007. Lifetime and 12-month prevalence of bipolar spectrum disorder in the National Comorbidity Survey replication. Arch. Gen. Psychiatr. 64, 543–552.

Merikangas, K.R., Jin, R., He, J.P., et al., 2011. Prevalence and correlates of bipolar spectrum disorder in the world mental health survey initiative. Arch. Gen. Psychiatr. 68, 241–251.

Miller, S., Hallmayer, J., Wang, P.W., et al., 2013. Brain-derived neurotrophic factor val66met genotype and early life stress effects upon bipolar course. J. Psychiatr. Res. 47, 252–258.

Minelli, A., Scassellati, C., Cloninger, C.R., et al., 2012. PCLO gene: its role in vulnerability to major depressive disorder. J. Affect. Disord. 139, 250–255.

Mitchell, P.B., Hadzi-Pavlovic, D., Loo, C.K., 2011. Course and outcome of bipolar disorder. Curr. Topics Behav. Neurosci. 5, 1–18.

Mitchell, P.B., Johnston, A.K., Corry, J., et al., 2009. Characteristics of bipolar disorder in an Australian specialist outpatient clinic: comparison across large datasets. Australian New Zealand J. Psychiatr. 43, 109–117.

Moore, G.J., Cortese, B.M., Glitz, D.A., et al., 2009. A longitudinal study of the effects of lithium treatment on prefrontal and subgenual prefrontal gray matter volume in treatment-responsive bipolar disorder patients. J. Clin. Psychiatr. 70, 699–705.

Muglia, P., Tozzi, F., Galwey, N.W., et al., 2010. Genome-wide association study of recurrent major depressive disorder in two European case-control cohorts. Mol. Psychiatr. 15, 589–601.

Muhleisen, T.W., Leber, M., Schulze, T.G., et al., 2014. Genome-wide association study reveals two new risk loci for bipolar disorder. Nature Comm., 5.

Mukherjee, S., Coque, L., Cao, J.L., et al., 2010. Knockdown of clock in the ventral tegmental area through RNA interference results in a mixed state of mania and depression-like behavior. Biol. Psychiatr. 68, 503–511.

Murphy, M.L., Carballedo, A., Fagan, A.J., et al., 2012. Neurotrophic tyrosine kinase polymorphism impacts white matter connections in patients with major depressive disorder. Biol. Psychiatr. 72, 663–670.

Murray, C.J.L., Vos, T., Lozano, R., et al., 2012. Disability-adjusted life years (DALYs) for 291 diseases and injuries in 21 regions, 1990-2010: a systematic analysis for the Global Burden of Disease Study 2010. Lancet 380, 2197–2223.

Naumenko, V.S., Popova, N.K., Lacivita, E., et al., 2014. Interplay between serotonin 5-HT1A and 5-HT7 receptors in depressive disorders. CNS Neurosci. Ther. 20, 582–590.

Niciu, M.J., Ionescu, D.F., Richards, E.M., et al., 2014. Glutamate and its receptors in the pathophysiology and treatment of major depressive disorder. J. Neural Transm. (Vienna) 121, 907–924.

Nyegaard, M., Severinsen, J.E., Als, T.D., et al., 2010. Support of association between BRD1 and both schizophrenia and bipolar affective disorder. Am. J. Med. Genet. B 153B, 582–591.

O'Dushlaine, C., Ripke, S., Ruderfer, D.M., et al., 2014. Rare copy number variation in treatment-resistant major depressive disorder. Biol. Psychiatr. 76, 536–541.

Olivier, J.D.A., Van Der Hart, M.G.C., Van Swelm, R.P.L., et al., 2008. A study in male and female 5-HT transporter knockout rats: an animal model for anxiety and depression disorders. Neuroscience 152, 573–584.

Ongur, D., Ferry, A.T., Price, J.L., 2003. Architectonic subdivision of the human orbital and medial prefrontal cortex. J. Comparative Neurol. 460, 425–449.

Osuch, E.A., Bluhm, R.L., Williamson, P.C., et al., 2009. Brain activation to favorite music in healthy controls and depressed patients. Neuroreport 20, 1204–1208.

Otten, M., Meeter, M., 2015. Hippocampal structure and function in individuals with bipolar disorder: a systematic review. J. Affect. Disord. 174, 113–125.

Overstreet, D.H., 1993. The Flinders sensitive line rats: a genetic animal model of depression. Neurosci. Biobehav. Rev 17, 51–68.

Overstreet, D.H., Wegener, G., 2013. The flinders sensitive line rat model of depression--25 years and still producing. Pharmacol. Rev. 65, 143–155.

Overstreet, D.H., Rezvani, A.H., Djouma, E., et al., 2007. Depressive-like behavior and high alcohol drinking co-occur in the FH/WJD rat but appear to be under independent genetic control. Neurosci. Biobehav. Rev. 31, 103–114.

Pacchiarotti, I., Bond, D.J., Baldessarini, R.J., et al., 2013. The International Society for Bipolar Disorders (ISBD) task force report on antidepressant use in bipolar disorders. Am. J. Psychiatr. 170, 1249–1262.

Papakostas, G.I., Thase, M.E., Fava, M., et al., 2007. Are antidepressant drugs that combine serotonergic and noradrenergic mechanisms of action more effective than the selective serotonin reuptake inhibitors in treating major depressive disorder? A meta-analysis of studies of newer agents. Biol. Psychiatr. 62, 1217–1227.

Park, M.H., Chang, K.D., Hallmayer, J., et al., 2015. Preliminary study of anxiety symptoms, family dysfunction, and the brain-derived neurotrophic factor (BDNF) Val66Met genotype in offspring of parents with bipolar disorder. J. Psychiatr. Res. 61, 81–88.

Paulsen, J.S., Nehl, C., Hoth, K.F., et al., 2005. Depression and stages of Huntington's disease. J. Neuropsychiatr. Clin. Neurosci. 17, 496–502.

Perry, W., Minassian, A., Paulus, M.P., et al., 2009. A reverse-translational study of dysfunctional exploration in psychiatric disorders: from mice to men. Arch. Gen. Psychiatr. 66, 1072–1080.

Popovic, D., Reinares, M., Goikolea, J.M., et al., 2012. Polarity index of pharmacological agents used for maintenance treatment of bipolar disorder. Eur. Neuropsychopharmacol. J. Eur. Coll. Neuropsychopharmacol. 22, 339–346.

Porcelli, S., Fabbri, C., Serretti, A., 2012. Meta-analysis of serotonin transporter gene promoter polymorphism (5-HTTLPR) association with antidepressant efficacy. Eur. Neuropsychopharmacol. J. Eur. Coll. Neuropsychopharmacol. 22, 239–258.

Porsolt, R.D., Lenegre, A., McArthur, R.A., 1991. Pharmacological models of depression. In: Olivier, B., Mos, J., Slangen, JL., (Eds.), Adv. Phar. Sci., edn. Basel Birkhauser, pp 137–159.

Post, R.M., 1992. Transduction of psychosocial stress into the neurobiology of recurrent affective-disorder. Am. J. Psychiatr. 149, 999–1010.

Price, J.L., Drevets, W.C., 2010. Neurocircuitry of mood disorders. Neuropsychopharmacology 35, 192–216.

Pringle, A., McCabe, C., Cowen, P.J., et al., 2013. Antidepressant treatment and emotional processing: can we dissociate the roles of serotonin and noradrenaline? J. Psychopharmacol. 27, 719–731.

Psychiatric, G.C.B.D.W.G., 2011. Large-scale genome-wide association analysis of bipolar disorder identifies a new susceptibility locus near ODZ4. Nat. Genet. 43, 977–983.

Pucilowski, O., Overstreet, D.H., Rezvani, A.H., et al., 1993. Chronic mild stress-induced anhedonia: greater effect in a genetic rat model of depression. Physiol. Behav. 54, 1215–1220.

Ralph-Williams, R.J., Paulus, M.P., Zhuang, X.X., et al., 2003. Valproate attenuates hyperactive and perseverative behaviors in mutant mice with a dysregulated dopamine system. Biol. Psychiatr. 53, 352–359.

Ramboz, S., Oosting, R., Amara, D.A., et al., 1998. Serotonin receptor 1A knockout: an animal model of anxiety-related disorder. Proc. Natl. Acad. Sci. USA 95, 14476–14481.

Ranger, P., Ellenbroek, B.A., 2016. Perinatal influences of valproate on brain and behaviour: an animal model for autism. Curr. Topic. Behav. Neurosci, [e- pub ahead of print].

Rea, E., Rummel, J., Schmidt, T.T., et al., 2014. Anti-anhedonic effect of deep brain stimulation of the prefrontal cortex and the dopaminergic reward system in a genetic rat model of depression: an intracranial self-stimulation paradigm study. Brain Stimulation 7, 21–28.

Ridder, S., Chourbaji, S., Hellweg, R., et al., 2005. Mice with genetically altered glucocorticoid receptor expression show altered sensitivity for stress-induced depressive reactions. J. Neurosci. 25, 6243–6250.

Rosenberg, G., 2007. The mechanisms of action of valproate in neuropsychiatric disorders: can we see the forest for the trees? CMLS 64, 2090–2103.

Roskoski, Jr., R., 2012. ERK1/2 MAP kinases: structure, function, and regulation. Pharmacol. Res. 66, 105–143.

Rybakowski, J.K., 2014. Response to lithium in bipolar disorder: clinical and genetic findings. ACS Chem. Neurosci. 5, 413–421.

Saleem, K.S., Kondo, H., Price, J.L., 2008. Complementary circuits connecting the orbital and medial prefrontal networks with the temporal, insular, and opercular cortex in the macaque monkey. J. Comparative Neurol. 506, 659–693.

Sallinen, J., Haapalinna, A., MacDonald, E., et al., 1999. Genetic alteration of the alpha2-adrenoceptor subtype c in mice affects the development of behavioral despair and stress-induced increases in plasma corticosterone levels. Mol. Psychiatr. 4, 443–452.

Savitz, J., van der Merwe, L., Newman, T.K., et al., 2010. Catechol-o-methyltransferase genotype and childhood trauma may interact to impact schizotypal personality traits. Behav. Genet. 40, 415–423.

Savitz, J.B., Drevets, W.C., 2013. Neuroreceptor imaging in depression. Neurobiol. Dis. 52, 49–65.

Savitz, J.B., Price, J.L., Drevets, W.C., 2014. Neuropathological and neuromorphometric abnormalities in bipolar disorder: view from the medial prefrontal cortical network. Neurosci. Biobehav. Rev. 42, 132–147.

Schneck, C.D., Miklowitz, D.J., Miyahara, S., et al., 2008. The prospective course of rapid-cycling bipolar disorder: findings from the STEP-BD. Am. J. Psychiatr. 165, 370–377, quiz 410.

Schramm, N.L., McDonald, M.P., Limbird, L.E., 2001. The alpha(2a)-adrenergic receptor plays a protective role in mouse behavioral models of depression and anxiety. J. Neurosci. 21, 4875–4882.

Schuhmacher, A., Mossner, R., Hofels, S., et al., 2011. PCLO rs2522833 modulates HPA system response to antidepressant treatment in major depressive disorder. Int. J. Neuropsychopharmacol. 14, 237–245.

Schulze, T.G., Detera-Wadleigh, S.D., Akula, N., et al., 2009. Two variants in Ankyrin 3 (ANK3) are independent genetic risk factors for bipolar disorder. Mol. Psychiatr. 14, 487–491.

Seifuddin, F., Mahon, P.B., Judy, J., et al., 2012. Meta-analysis of genetic association studies on bipolar disorder. Am. J. Med. Genet. B 159B, 508–518.

Shaltiel, G., Maeng, S., Malkesman, O., et al., 2008. Evidence for the involvement of the kainate receptor subunit GluR6 (GRIK2) in mediating behavioral displays related to behavioral symptoms of mania. Mol. Psychiatr. 13, 858–872.

Sharpley, C.F., Palanisamy, S.K.A., Glyde, N.S., et al., 2014. An update on the interaction between the serotonin transporter promoter variant (5-HTIIPR), stress and depression, plus an exploration of non-confirming findings. Behav. Brain Res. 273, 89–105.

Sheline, Y.I., Gado, M.H., Kraemer, H.C., 2003. Untreated depression and hippocampal volume loss. Am. J. Psychiatr. 160, 1516–1518.

Sherdell, L., Waugh, C.E., Gotlib, I.H., 2012. Anticipatory pleasure predicts motivation for reward in major depression. J. Abnormal Psychol. 121, 51–60.

Shi, J., Potash, J.B., Knowles, J.A., et al., 2011. Genome-wide association study of recurrent early-onset major depressive disorder. Mol. Psychiatr. 16, 193–201.

Shorter, E., 2009. The history of lithium therapy. Bipolar Disord. 11 (Suppl 2), 4–9.

Shyn, S.I., Shi, J., Kraft, J.B., et al., 2011. Novel loci for major depression identified by genome-wide association study of sequenced treatment alternatives to relieve depression and meta-analysis of three studies. Mol. Psychiatr. 16, 202–215.

Smoller, J.W., Finn, C.T., 2003. Family, twin, and adoption studies of bipolar disorder. Am. J. Med. Genet. C Semin Med. Genet. 123C, 48–58.

Song, J., Bergen, S.E., Kuja-Halkola, R., et al., 2015. Bipolar disorder and its relation to major psychiatric disorders: a family-based study in the Swedish population. Bipolar Disord. 17, 184–193.

Stamm, T.J., Rampp, C., Wiethoff, K., et al., 2016. The FKBP5 polymorphism rs1360780 influences the effect of an algorithm-based antidepressant treatment and is associated with remission in patients with major depression. J. Psychopharmacol. 30, 40–47.

Starr, L.R., Hammen, C., Conway, C.C., et al., 2014. Sensitizing effect of early adversity on depressive reactions to later proximal stress: moderation by polymorphisms in serotonin transporter and corticotropin releasing hormone receptor genes in a 20-year longitudinal study. Dev. Psychopathol. 26, 1241–1254.

Sullivan, G.M., Ogden, R.T., Oquendo, M.A., et al., 2009a. Positron emission tomography quantification of serotonin-1A receptor binding in medication-free bipolar depression. Biol. Psychiatr. 66, 223–230.

Sullivan, P.F., Neale, M.C., Kendler, K.S., 2000. Genetic epidemiology of major depression: review and meta-analysis. Am. J. Psychiatr. 157, 1552–1562.

Sullivan, P.F., de Geus, E.J., Willemsen, G., et al., 2009b. Genome-wide association for major depressive disorder: a possible role for the presynaptic protein piccolo. Mol. Psychiatr. 14, 359–375.

Sun, N., Xu, Y., Wang, Y., et al., 2008. The combined effect of norepinephrine transporter gene and negative life events in major depression of Chinese Han population. J. Neural Transm. (Vienna) 115, 1681–1686.

Szczepankiewicz, A., Leszczynska-Rodziewicz, A., Pawlak, J., et al., 2014. FKBP5 polymorphism is associated with major depression but not with bipolar disorder. J. Affect. Disord. 164, 33–37.

Takata, A., Kim, S.H., Ozaki, N., et al., 2011. Association of ANK3 with bipolar disorder confirmed in East Asia. Am. J. Med. Genet B 156B, 312–315.

Tan, J., Chen, S., Su, L., et al., 2014. Association of the T102C polymorphism in the HTR2A gene with major depressive disorder, bipolar disorder, and schizophrenia. Am. J. Med. Genet B 165B, 438–455.

Timpl, P., Spanagel, R., Sillaber, I., et al., 1998. Impaired stress response and reduced anxiety in mice lacking a functional corticotropin-releasing hormone receptor 1. Nat. Genet. 19, 162–166.
Tohen, M., Vieta, E., Calabrese, J., et al., 2003. Efficacy of olanzapine and olanzapine-fluoxetine combination in the treatment of bipolar I depression. Arch. Gen. Psychiatr. 60, 1079–1088.
Tozzi, L., Carballedo, A., Wetterling, F., et al., 2016. Single-nucleotide polymorphism of the FKBP5 gene and childhood maltreatment as predictors of structural changes in brain areas involved in emotional processing in depression. Neuropsychopharmacology 41, 487–497.
Tye, K.M., Mirzabekov, J.J., Warden, M.R., et al., 2013. Dopamine neurons modulate neural encoding and expression of depression-related behaviour. Nature 493, 537–541.
van den Bos, R., Koot, S., de Visser, L., 2014. A rodent version of the Iowa Gambling Task: 7 years of progress. Front. Psychol. 5.
van Enkhuizen, J., Minassian, A., Young, J.W., 2013. Further evidence for ClockDelta19 mice as a model for bipolar disorder mania using cross-species tests of exploration and sensorimotor gating. Behav. Brain Res. 249, 44–54.
van Enkhuizen, J., Geyer, M.A., Minassian, A., et al., 2015. Investigating the underlying mechanisms of aberrant behaviors in bipolar disorder from patients to models: rodent and human studies. Neurosci. Biobehav. Rev. 58, 4–18.
Veletza, S., Samakouri, M., Emmanouil, G., et al., 2009. Psychological vulnerability differences in students--carriers or not of the serotonin transporter promoter allele S: effect of adverse experiences. Synapse 63, 193–200.
Verduijn, J., Milaneschi, Y., Schoevers, R.A., et al., 2015. Pathophysiology of major depressive disorder: mechanisms involved in etiology are not associated with clinical progression. Trans. Psychiatr. 5, e649.
Vinkers, C.H., Joels, M., Milaneschi, Y., et al., 2015. Mineralocorticoid receptor haplotypes sex-dependently moderate depression susceptibility following childhood maltreatment. Psychoneuroendocrinology 54, 90–102.
Vollmayr, B., Henn, F.A., 2001. Learned helplessness in the rat: improvements in validity and reliability. Brain Res. Brain Res. Protoc. 8, 1–7.
Vos, T., Flaxman, A.D., Naghavi, M., et al., 2012. Years lived with disability (YLDs) for 1160 sequelae of 289 diseases and injuries 1990-2010: a systematic analysis for the Global Burden of Disease Study 2010. Lancet 380, 2163–2196.
Vrijsen, J.N., van Oostrom, I., Arias-Vasquez, A., et al., 2014. Association between genes, stressful childhood events and processing bias in depression vulnerable individuals. Genes Brain Behav. 13, 508–516.
Walderhaug, E., Varga, M., Pedro, M.S., et al., 2011. The role of the aminergic systems in the pathophysiology of bipolar disorder. Curr. Topics Behav. Neurosci. 5, 107–126.
Wankerl, M., Miller, R., Kirschbaum, C., et al., 2014. Effects of genetic and early environmental risk factors for depression on serotonin transporter expression and methylation profiles. Translational psychiatry 4.
Wegener, G., Mathe, A.A., Neumann, I.D., 2012. Selectively bred rodents as models of depression and anxiety. Curr. Topics Behav. Neurosci. 12, 139–187.
Weiss, J.M., Cierpial, M.A., West, C.H., 1998. Selective breeding of rats for high and low motor activity in a swim test: toward a new animal model of depression. Pharmacol. Biochem. Behav. 61, 49–66.
Weissman, M.M., Bland, R.C., Canino, G.J., et al., 1996. Cross-national epidemiology of major depression and bipolar disorder. Jama-J. Am. Med. Assoc. 276, 293–299.
Wellman, C.L., Izquierdo, A., Garrett, J.E., et al., 2007. Impaired stress-coping and fear extinction and abnormal corticolimbic morphology in serotonin transporter knock-out mice. J. Neurosci. 27, 684–691.
Wendland, J.R., McMahon, F.J., 2011. Genetics of bipolar disorder. Curr. Topics Behav. Neurosci. 5, 19–30.
Willner, P., 1984. The validity of animals models of depression. Psychopharmacology 83, 1–16.
Wittchen, H.U., Jacobi, F., Rehm, J., et al., 2011. The size and burden of mental disorders and other disorders of the brain in Europe 2010. Eur. Neuropsychopharmacol. J. Eur. Coll. Neuropsychopharmacol. 21, 655–679.
Wortwein, G., Husum, H., Andersson, W., et al., 2006. Effects of maternal separation on neuropetide Y and calcitonin gene-related peptide in "depressed" Flinders Sensitive Line rats: a study of gene-environment interactions. Prog. Neuro-Psychopharmacol. Biol. Psychiatr. 30, 684–693.
Woudstra, S., van Tol, M.J., Bochdanovits, Z., et al., 2013. Modulatory effects of the piccolo genotype on emotional memory in health and depression. PLoS One 8, e61494.
Yang, X., Ma, X., Huang, B., et al., 2015a. Gray matter volume abnormalities were associated with sustained attention in unmedicated major depression. Compr. Psychiatr. 63, 71–79.
Yang, X., Ma, X., Li, M., et al., 2015b. Anatomical and functional brain abnormalities in unmedicated major depressive disorder. Neuropsychiatr. Dis. Treat. 11, 2415–2423.

in the absence of any obvious distraction).

iv. Often does not follow through on instructions and fails to finish schoolwork, chores, or duties in the workplace (eg, starts tasks but quickly loses focus and is easily sidetracked).

v. Often has difficulty organizing tasks and activities (eg, difficulty managing sequential tasks; difficulty keeping materials and belongings in order; messy, disorganized work; has poor time management; fails to meet deadlines).

vi. Often avoids, dislikes, or is reluctant to engage in tasks that require sustained mental effort (eg, schoolwork or homework; for older adolescents and adults, preparing reports, completing forms, reviewing lengthy papers).

vii. Often loses things necessary for tasks or activities (eg, school materials, pencils, books, tools, wallets, keys, paperwork, eyeglasses, mobile telephones).

viii. Is often easily distracted by extraneous stimuli (for older adolescents and adults, may include unrelated thoughts).

ix. Is often forgetful in daily activities (eg, doing chores, running errands; for older adolescents and adults, returning calls, paying bills, keeping appointments).

b. *Hyperactivity and impulsivity*: Six (or more) of the following symptoms have persisted for at least 6 months to a degree that is inconsistent with developmental level and that negatively impacts directly on social and academic/occupational activities: Note: The symptoms are not solely a manifestation of oppositional behaviour, defiance, hostility, or a failure to understand tasks or instructions. For older adolescents and adults (age 17 and older), at least five symptoms are required

i. Often fidgets with or taps hands or feet or squirms in seat.

ii. Often leaves seat in situations when remaining seated is expected (eg, leaves his or her place in the classroom, in the office or other workplace, or in other situations that require remaining in place).

iii. Often runs about or climbs in situations where it is inappropriate. (Note: In adolescents or adults, may be limited to feeling restless.)

iv. Often unable to play or engage in leisure activities quietly.

v. Is often "on the go," acting as if "driven by a motor" (eg, is unable to be or uncomfortable being still for extended time, as in restaurants, meetings; may be experienced by others as being restless or difficult to keep up with).

vi. Often talks excessively.

vii. Often blurts out an answer before a question has been completed (eg, completes people's sentences; cannot wait for turn in conversation).

viii. Often has difficulty waiting his or her turn (eg, while waiting in line).

ix. Often interrupts or intrudes on others (eg, butts into conversations, games, or activities; may start using other people's things without asking or receiving permission; for adolescents and adults, may intrude into or take over what others are doing).

at the predictive validity of the two classes of symptoms for high school graduates found clear differences: whereas 40% of individuals with high levels of hyperactivity (measured at age 6 to 12) graduated from high school, only 29% of children with high inattention scores did (Pingault et al., 2011). In a follow-up study, the authors showed that graduation failure is particularly high in children with a rising pattern of inattentiveness, rather than a stable or fluctuating pattern (Pingault et al., 2014). High inattention during childhood was also associated with higher levels of nicotine use, while high levels of oppositional behaviour was associated with higher levels of nicotine, cannabis and cocaine use (Pingault et al., 2013). Together these data indicate that the subtype with predominantly inattention symptoms has a more negative outcome than the subtype with predominantly hyperactivity symptoms. In addition to the symptoms, several other characteristics are important for a full diagnosis of ADHD. Similar to the diagnosis of other diseases, the symptoms of ADHD must be significantly severe to affect normal functioning, other diagnoses should be excluded and some of the symptoms should be present before the age of 12. This last criterion has changed from the previous version of DSM (DSM-IV-TR) where symptoms were required to be present before the age of 7. This change is in line with studies that show that only about 50% of patients with ADHD recall symptoms before the age of 7, while 95% recall symptoms before the age of 12 (Kessler et al., 2005).

Another change in the diagnosis between DSM-IV-TR and DSM-V is the addition of several items related to ADHD in later life. For instance criterion a.v. (poor time management; fails to meet deadlines), a.vi. (avoid preparing reports, completing forms, reviewing lengthy papers) and a.vii. (loosing wallets, keys, paperwork) specifically address issues that are more related to adolescents or adults. This is in line with the findings that a significant percentage of children with ADHD continue to show signs in adolescence and adulthood.

Although the symptoms do not differ fundamentally between children, adolescents and adults with ADHD, the consequences differ for the different age categories. Whereas ADHD in children is often associated with poor school performance, suspension and expulsion, in adolescents it can lead to poor peer or family relations, conduct problems, delinquency and experimentation with illicit drugs. In adulthood, ADHD can lead to driving accidents, difficulties in social relationships, including marriage and problems at work, but also increased criminal behaviour, alcoholism and an increased risk of suicidality.

One important aspect of ADHD is the high comorbidity with other psychiatric disorders. In a national comorbidity survey, the odds ratios of most major childhood psychiatric disorders was significantly elevated in patients with ADHD at age 15 (Fig. 8.2). Interestingly, the odds ratios were much closer to 1 in adults with ADHD (Kessler et al., 2005). The only exception was obsessive compulsive disorder, which in adult ADHD patients had an odds ratio of 5.7. However, as the authors point out, this comorbidity was the rarest and the estimate rather unstable.

The Epidemiology of ADHD

In a large national comorbidity survey from the USA performed in the period between Feb. 2001 and Apr. 2003, the life-time prevalence of ADHD was estimated to be 8.1% (Kessler et al., 2005). Moreover the study showed that the age of onset was relatively early. In line with figure 8.1, ADHD starts at a relatively early age with 99% of all cases diagnosed before the age of 16 (Fig. 8.3). Several other impulse-control disorders, such as oppositional-defiant disorder and conduct disorder had similar lifetime prevalence (8.5% resp. 9.5%) and a similar age of

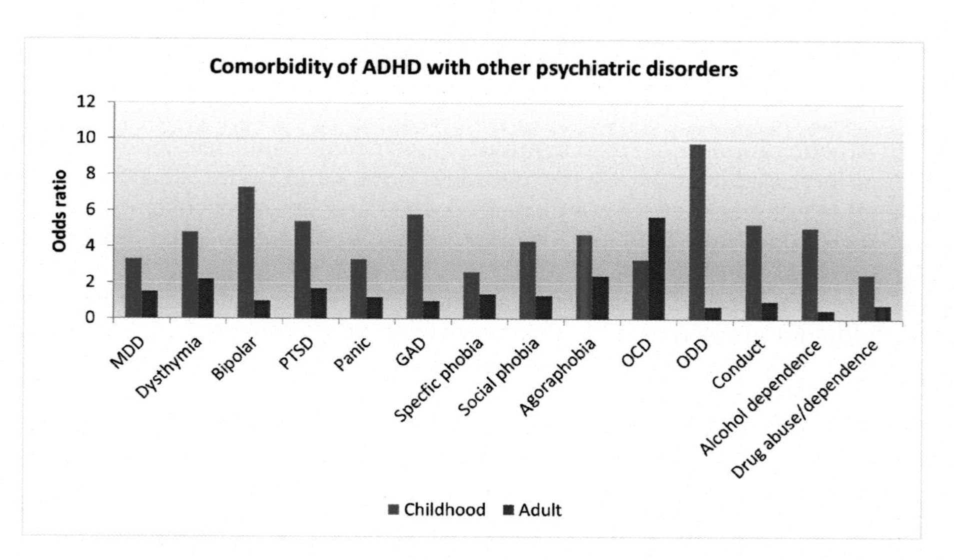

FIGURE 8.2 **The comorbidity risk of ADHD.** Patients with ADHD are known to suffer from a wide range of additional symptoms, especially mood disturbances (major depressive disorder, bipolar disorders and various anxiety disorders) and oppositional defiant disorder. These comorbidities are especially prominent during childhood, and tend to be much less significant in adult ADHD cases, although these patients have other comorbidities (especially alcoholism and drug misuse). Abbreviations: MDD: major depressive disorder; PTSD: posttraumatic stress disorder; GAD: generalized anxiety disorder; OCD: obsessive compulsive disorder; ODD: positional defiant disorder.

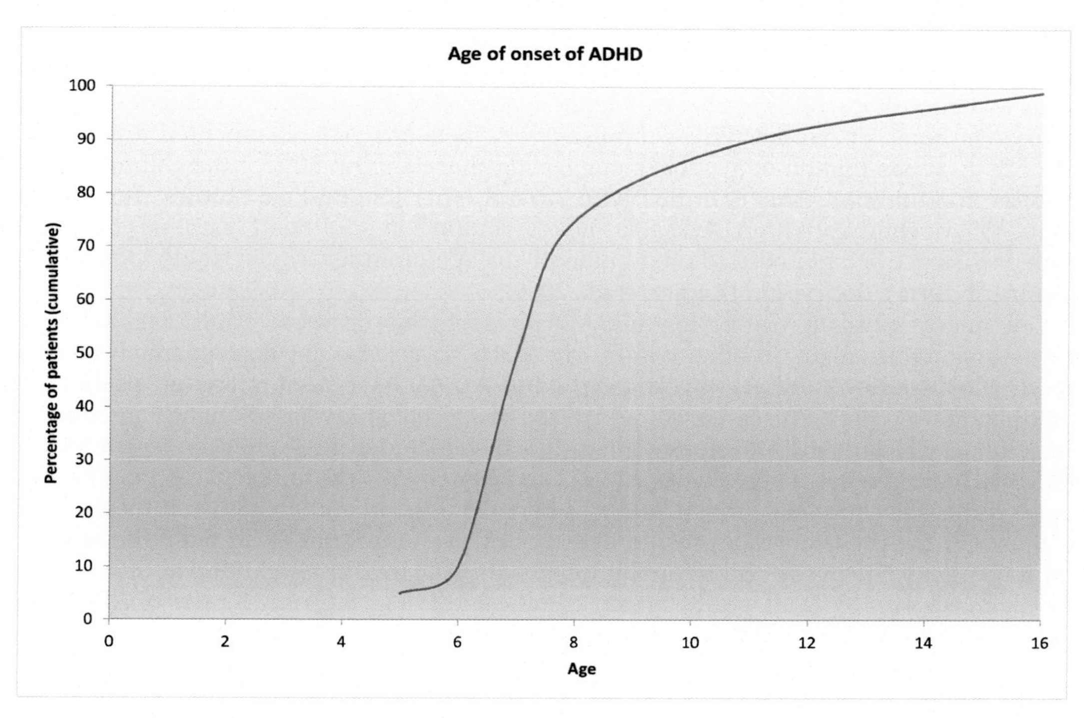

FIGURE 8.3 **The age on onset of ADHD.** Based on a large studies in the USA between 2001 and 2003 (Kessler et al., 2005), the first children receive the diagnosis of ADHD around six years of age and by age eleven about 90% of all cased are diagnosed.

onset. It has been suggested that the prevalence differs between different regions, being higher in North America and Europe than in middle/low income countries, and that the incidence of ADHD has increased over the last few decades (Timimi and Taylor, 2004; Singh, 2008). Two USA national telephone surveys conducted on 2003 and 2007, for instance, found ADHD prevalence increasing from 7.8% to 9.5%, a trend that was also observed in Canada and the UK (Polanczyk et al., 2014).

However, several detailed meta-analyses indicate that geographical differences are mainly due to methodological discrepancies (Polanczyk et al., 2007; Willcutt, 2012). A recent analysis of 30 years of ADHD research confirmed that geographical location was not associated with differences in ADHD prevalence. Interestingly, this study also showed that year of study did not influence ADHD prevalence, thus contradicting the earlier mentioned trend of an increase in ADHD prevalence (Polanczyk et al., 2014). The authors propose that the studies that found an increase in prevalence over time may be flawed as they rely on administrative data, or on diagnoses reported by physicians or parents. As their meta-analysis, on the other hand, are based on standard diagnostic criteria (such as DSM-III, DSM-IV or DSM-IV-TR), they represent more reliable prevalence rates.

One important aspect of the epidemiology of ADHD is the sex ratio. Most studies have indicated that ADHD is more prevalent in boys (males) than in females. For instance in the earlier mentioned study by Polanczyk et al. (2007), the prevalence in boys was 2.45 higher than that in females. Similar higher risks for boys were also reported by others (Kessler et al., 2005). Although the higher incidence in boys may be indicative of a higher liability to develop ADHD, it may also result from a decreased recognition in girls. As boys often show more overtly disruptive behaviour, ADHD in girls may go unnoticed. In addition, there is some evidence that while girls have a higher incidence of the inattentive subtype boys show more symptoms of hyperactivity/impulsivity. A meta-analysis documented a prevalence of 2.5% in adults (Simon et al., 2009), suggesting that in about 50% of the children ADHD persists in adulthood. This is in line with several other longitudinal studies showing that about 65% of children with ADHD also show symptoms in adulthood (Faraone et al., 2006), with persistence being predicted predominantly by symptom severity in childhood as well as treatment during this period (Kessler et al., 2006).

One important issue relating to adult ADHD is whether or not all adult cases are neurodevelopmental in origin. In other words, can ADHD cases also develop separately in adulthood? Most studies that have investigated adult ADHD have been follow-up studies from childhood onset cases. This issue was addressed in a recent study based on the Dunedin Multidisciplinary Health and Development Study which included 1,037 children (encompassing 93% of all children born) between April 1972 and March 1973 (Moffitt et al., 2015). The study showed that the prevalence for ADHD was 6% in childhood and about 3% at the age of 38, in line with the prevalences discussed earlier. However, intriguingly, in the follow-forward approach, only 5% of the children diagnosed with ADHD during childhood also received the diagnosis of ADHD at age 38. In fact, as indicated in Fig. 8.4, only 3 cases of ADHD that were diagnosed during childhood retained the diagnosis in adulthood. Thus the majority of adult cases were not diagnosed as ADHD during childhood. As these adult onset cases also did not show any neuropsychological deficits as children, this raises the possibility that adult onset ADHD is fundamentally different from childhood onset, and may not have a neurodevelopmental component. On the other hand, a recent longitudinal study from China reported that 46.4% of all children diagnosed with ADHD retained their diagnosis in adulthood (Gao

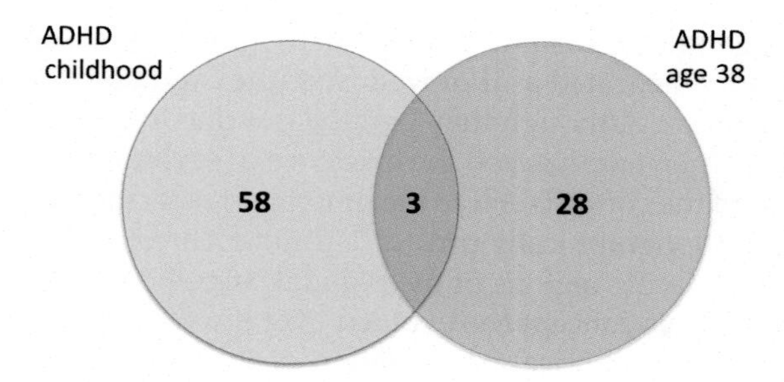

FIGURE 8.4 **The persistence of ADHD into adulthood.** Although ADHD occurs both in childhood and adulthood, it is still unclear how many children actually continue to have symptoms in adulthood and how many adults have had ADHD symptoms in childhood. In a recent study from Dunedin it was suggested that only 3 out of 61 children with ADHD had persistent symptoms at age 38. *Adapted from Moffitt et al., 2015.*

et al., 2015). Apart from potential genetic differences between the two studies, it is important to realize that in the latter study the follow up was done much earlier (age 18–24) than in the Dunedin study. Additionally, it is important to point out that a substantial proportion of children with ADHD develop bipolar disorder later in life (Faedda et al., 2014). The study from China also looked at predictors for persistence of symptoms, identifying lower IQ at baseline, earlier age of onset and presence of generalized anxiety disorder at baseline as the most reliable indicators of persistence of ADHD in adulthood.

ADHD represents a considerable burden for families and society. For instance studies have shown that mothers of preschool–age children with ADHD have significantly more depressive symptoms and report a feeling of poor parental competence (Byrne et al., 1998). Likewise, ADHD in adulthood leads to more traffic accidents and license suspensions (Barkley and Cox, 2007) and to higher divorce rates (Barkley and Fischer, 2010). Calculations of the total burden, in terms of disability or costs for ADHD is difficult as most reports combine childhood disorders. For instance in the 2010 study from Europe, the total costs for childhood disorders was calculated as 21.3 billion euro (Gustavsson et al., 2011). In the global burden of disease study from 2010, the disability-adjusted life years for ADHD was calculated as 491,000 years, an increase of 15% over 1990. Although this is significantly less than diseases such as addiction or schizophrenia, it nevertheless represents a considerable problem. This is also evident from calculations of the total costs of ADHD. A study published in 2005, calculated the total costs in the US at $31.6 billion dollars, subdivided into $1.6 billion for treatment, $12.1 billion for other healthcare costs for patients and another $14.2 billion for healthcare costs for the family of patients with ADHD. The remaining $3.7 billion was for work loss costs of adults with ADHD and their family members.

The Neurobiology of ADHD

Substantial progress has been made in delineating the neurobiological substrate underlying ADHD, with information obtained from imaging studies, pharmacological interventions, animal models and basic neuroscience research. Most of the imaging studies have focused on the fronto-striato-cerebellar loops. Conceptually, this research is based on work performed by Alexander et al. (1986) who were among the first to provide a detailed organization of the basal ganglia. Until their pioneering work using electrical microstimulation in monkeys, the general idea was that the basal ganglia acted as a "funnel" with information from different cortical areas becoming more and more integrated as it moved from the caudate-putamen via the pallido/nigral complex to the thalamus. Although they pointed out that in their model a certain degree of funneling is still present, they emphasized the existence of several different parallel loops, starting from different areas of the cortex (such as the motor cortex, oculomotor cortex or dorsolateral prefrontal cortex) which topographically project to different areas of the caudate putamen complex and from there, via specific regions of the pallidal/nigral complex, to separate subregions of the thalamus. Although this concept has evolved over the years, it is still generally considered to accurately represent the anatomical organization of the fronto-striatal network. As Fig. 8.5 illustrates, and Alexander and coworkers emphasized, this structural network also underlies the functional organization of the basal ganglia and likely extend beyond this to include other networks as well, such as the cortico-thalamo-cerebellar network. Indeed, as was reiterated in a recent review (Arnsten and Rubia, 2012) there is evidence that disturbances in three different networks exist in ADHD (Fig. 8.5), mediating affect, cognition and motor functioning.

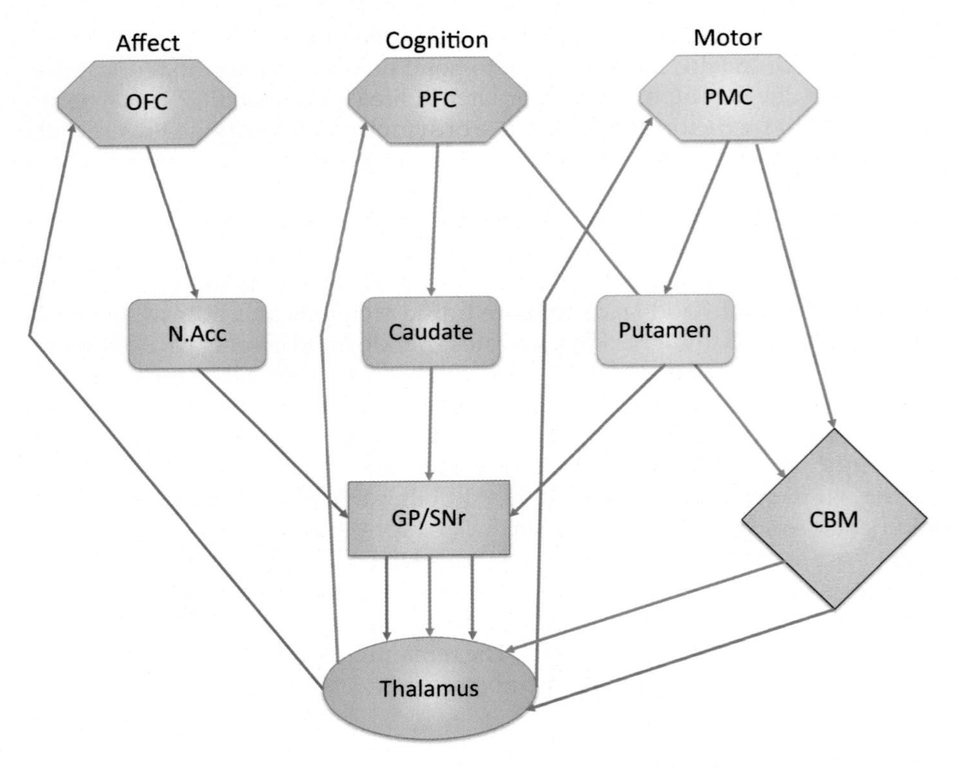

FIGURE 8.5 **The neurobiology of ADHD.** Based on a large number of studies, there is ample evidence for the involvement of both cortical and subcortical structures in the neuropathology of ADHD. Although these structures and networks can never be assigned a single function, there is evidence that the orbitofrontal cortex (OFC) and the nucleus accumbens (N. Acc) are predominantly involved in disturbances in affect, while the prefrontal cortex (PFC) and caudate are more likely to play a role in cognitive deficits. Finally the premotor cortex (PMC) and the putamen are more likely involved in motor dysfunction. Other abbreviations: GP/SNr: Globus Pallidus/Substantia Nigra pars reticulata; CBM: Cerebellum.

Given that ADHD is primarily characterized by cognitive deficits (inattention and impulsivity), it is not surprising to find structural and functional deficits within the prefrontal cortex [especially the dorsolateral and inferior prefrontal cortex (Cubillo et al., 2012)]. Structural changes, such as reduced volume of the caudate nucleus, especially in the right hemisphere (Nakao et al., 2011) are prominent especially in pre-pubertal ADHD patients (Carrey et al., 2012). Interestingly there is evidence that this reduction is reversible as (partial) normalization was found with treatment (Nakao et al., 2011). In line with the structural findings, disturbed functional connections within this network have been reported, including a hypo-activation of the prefrontal-parietal network, relevant for executive functioning (Cortese et al., 2012). Likewise, using event related functional MRI (fMRI), patients with ADHD were found to have reduced caudate activation during a cognitive switching task and an attentional task (Rubia et al., 2011).

The second major behavioural hallmark of ADHD is hyperactivity, which, according to Fig. 8.5, is primarily mediated via a network including the premotor cortex, putamen and cerebellum. Structural and functional changes similar to those of the cognitive network have been identified in patients with ADHD in all these regions. Thus significantly smaller putamen have repeatedly been observed (Nakao et al., 2011) as well as a reduction in specific cerebellar structures, such as the inferior vermis, although this has not been found consistently (Cortese, 2012).

Whereas most MRI studies have concentrated on grey matter, there is also ample evidence for white matter abnormalities. In fact effect sizes for white matter (0.30–0.64) are actually larger than for grey matter (0.27–0.35; Castellanos et al., 2002). White matter abnormalities (ie, abnormalities in fiber bundles) have repeatedly been reported in ADHD patients, especially in the inferior longitudinal fasciculus (connecting the temporal lobe with the cerebellum) and the occipitofrontal fasciculus (connecting occipital and frontal lobes). A recent review analyzed the literature with respect to a relatively new imaging technique, called DTI (Diffusion Tensor Imaging), which enables a more detailed analysis of white matter integrity (Alexander, 2005). This meta-analysis further confirmed the importance of the fronto-striato-cerebellar circuitry, with significant alterations in white matter in the basal ganglia and the cerebellum (van Ewijk et al., 2012).

These structural MRI studies support the idea that the circuitry displayed in Fig. 8.5 plays a role in the symptoms of ADHD. Moreover, given the evidence for structural alterations in the connectivity within this network, this suggests that the functional connectivity is also affected. In line with this, several papers have shown hypoperfusion in the orbitofrontal cortex, the putamen and the cerebellum (Kim et al., 2010). fMRI studies have strengthened the hypothesis of functional dysconnectivity in the brains of patients with ADHD. During normal resting state a large and robustly replicated network of brain regions associated with task-irrelevant activity has been identified, the so-called Default Mode Network (DMN), including the posterior cingulate, medial prefrontal and medial, lateral and inferior parietal cortex (Schilbach et al., 2008). Studies in patients with ADHD showed clear disturbances in the normal DMN network activity (Konrad and Eickhoff, 2010). However, the findings are rather inconsistent, with some reporting higher resting activity in (lower level) sensory cortical areas (Tian et al., 2006), while others using similar methods found reduced connectivity between the anterior cingulate cortex and various parts of the DMN, as well as within the DMN itself (Castellanos et al., 2008). In addition to changes in resting brain functional connectivity, altered activity patterns have also been reported during cognitive tasks. Thus several studies found reduced connectivity within the prefronto-striato-parieto-cerebellar network during a working memory task (Wolf et al., 2009), or a continuous performance task (Vloet et al., 2010). In an interesting study, a dissociation was found between two cognitive tasks, with reduced fronto-parietal coupling during an interference task, and reduced fronto-cerebellar connectivity during a time discrimination task.

Together, both structural and functional brain imaging studies have clearly supported the theory of a dysfunctional fronto-striato-cerebellar circuitry and the question remains which neurotransmitters are responsible for this altered connectivity. Most research in this area has focused on the catecholamines, especially dopamine and noradrenaline. Both these neurotransmitters are highly expressed in the prefrontal cortex, and dopamine is also highly concentrated in the nucleus accumbens and the caudate and putamen. Moreover, both neurotransmitters play a more diffuse neuromodulatory role in the central nervous system. Much of the evidence pointing to a role of dopamine and noradrenaline in ADHD comes from animal research, as neurochemical imaging still remains more complicated than structural or functional imaging. However, the exact deficits in dopaminergic and noradrenergic neurotransmission accompanying ADHD are still far from fully understood. For instance, a recent meta-analysis examined the (peripheral) alterations in catecholaminergic markers and found significantly reduced levels of MHPG (3-methoxy-4-hydroxyophenylethylene glycol, a metabolite of noradrenaline) in patients with ADHD. Yet improvement after psychostimulant treatment was associated with further decreases in MHPG. The same meta-analysis also found elevated urinary levels of noradrenaline, and reduced platelet activity of monoamine oxidase (Scassellati et al., 2012), the latter being associated with increased inattention (Shekim et al., 1986) and impulsivity (Coccini et al., 2009). In contrast to the MHPG, the increased noradrenaline and decreased MAO levels were normalized by psychostimulant treatment.

Although these data point to a role of the noradrenergic system in ADHD, it is unclear how peripheral levels are related to changes within the central nervous system. In order to investigate this, PET (positron emission tomography) or SPECT (single photon emission computer tomography) studies are required. However, few of these studies have been done, and most of these were done on adults with ADHD, focusing primarily on the dopaminergic system rather than the noradrenergic system (Weyandt et al., 2013).

Most of these studies investigated the density of dopamine transporters and a meta-analysis of 9 studies found a small (14%) but significant increase in dopamine transporter binding in the striatum of ADHD patients (Fusar-Poli et al., 2012). However, as the study clearly showed, there was widespread heterogeneity among the studies, with 6 finding a significant increase, 2 a significant decrease and 2 finding no effect. As many of the psychostimulant treatment for ADHD are acting of the dopaminergic and/or noradrenergic system, it is important to take this into account. Indeed as Fusar-Poli and coworkers point out, the observed increased in transporter binding may be secondary to prolonged psychostimulant treatment rather than being part of the ADHD pathophysiology. Thus it seems likely that ADHD is actually associated with significantly lower striatal dopamine transporter levels which are increased after chronic treatment. This idea is supported by the fact that the two studies that found significantly lower transporter concentrations (Hesse et al., 2009; Volkow et al., 2009) were both performed in drug naïve patients. Studies using [^{18}F]DOPA or [^{11}C] DOPA further support the idea of reduced dopamine transmission resulting from a lower

TABLE 8.1 Stimulant and Nonstimulant Drugs for the Treatment of ADHD

Drug	Mechanisms of action
STIMULANT DRUGS	
Amphetamine	Competitive substrate for the dopamine and noradrenaline transporters. High doses reverse the flux of these transporters
Methylphenidate	Blocks dopamine and noradrenaline transporters
NONSTIMULANT DRUGS	
Atomoxetine	Blocks noradrenaline transporters
Clonidine	α_2 adrenoceptor agonist
Guanfacine	α_2 adrenoceptor agonist

than normal dopamine synthesis (Forssberg et al., 2006; Ludolph et al., 2008). Finally, D_2/D_3 receptor concentrations (measured using [^{11}C]raclopride) were significantly reduced in the striatum of adult ADHD patients (Volkow et al., 2007). In line with the reduced presynaptic dopamine synthesis and transporter capacity, the increase in dopamine transmission induced by methylphenidate in the caudate nucleus is smaller in patients compared to control (Volkow et al., 2007). Interestingly, this effect was not limited to the caudate nucleus, but was also found in the left amygdala and the hippocampus (although dopamine concentrations are very low in this latter brain area). Imaging studies investigating changes in the dopaminergic system in prefrontal cortex have so far been rare. This is mostly due to practical limitations, such as low levels of dopamine transporters. However, as we will discuss later, the prefrontal cortex and especially the catecholaminergic system plays an important role in the therapeutic effects of anti-ADHD drugs.

The Pharmacological Treatment of ADHD

For the treatment of ADHD, several drugs have been approved in most countries in the world, generally subdivided into two classes: stimulant and nonstimulant drugs (Table 8.1).

Stimulant Drugs

The first drug to be approved by the FDA for ADHD was the psychostimulant methylphenidate. Like other psychostimulants (Chapter 6), methylphenidate blocks the catecholamine transporters. As shown in Fig. 8.6, methylphenidate has the highest affinity for the dopamine transporter, with negligible binding to the serotonin transporter. Despite the much higher affinity for the dopamine transporter, at clinically relevant doses, methylphenidate seems to bind to the noradrenaline transporter in the prefrontal cortex (Hannestad et al., 2010), presumably because of the much higher concentrations of this transporter compared to the dopamine transporter in this brain region. In fact, it has been shown that extracellular dopamine is taken up by the noradrenaline transporter in the prefrontal cortex (and hippocampus, as discussed earlier; Moron et al., 2002). As a result, methylphenidate increases both extracellular dopamine and noradrenaline in this brain region. Methylphenidate is both available in immediate release and extended release form (in which methylphenidate is attached to an inert

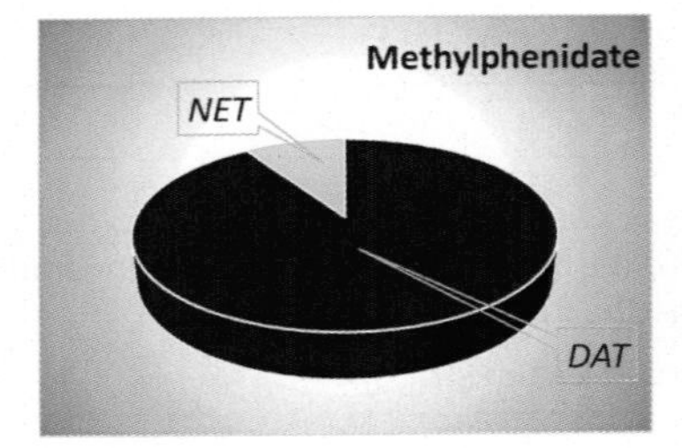

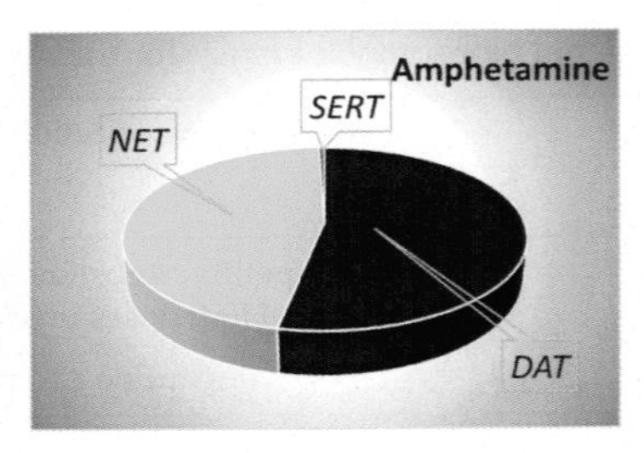

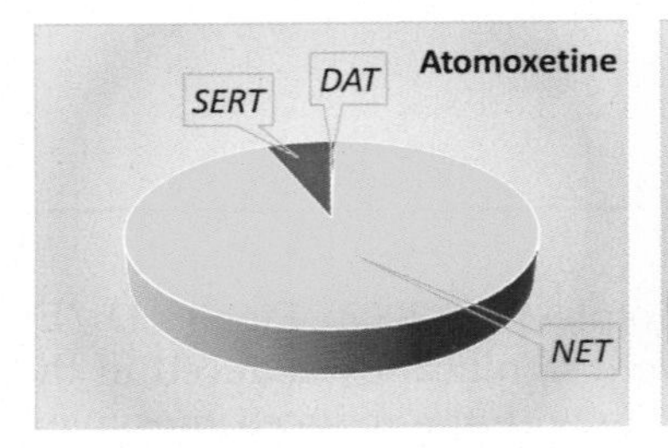

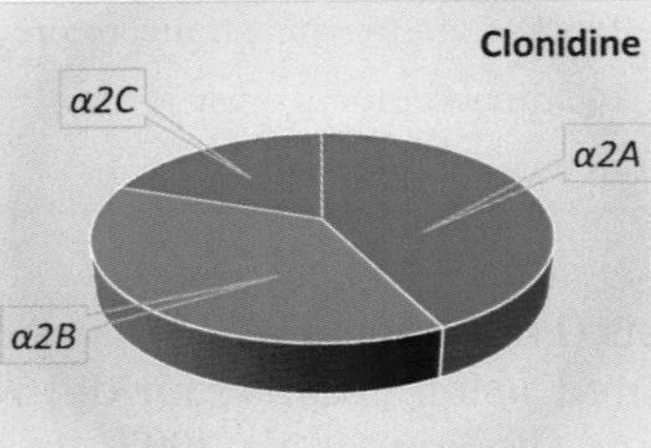

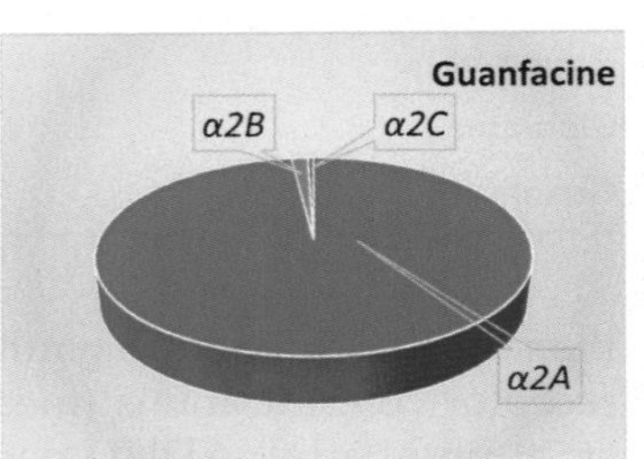

FIGURE 8.6 **Pie charts showing the relative affinities of therapeutic drugs for the different monoamine transporters or the α_2-adrenoceptors.** Top row shows the psychostimulant drugs, bottom row the nonstimulant drugs. It is important, however, to realize that these are based on in vitro affinities and that they do not take into account differences in pharmacokinetics of these drugs.

matrix, leading to a much slower release and prolonged action). A meta-analysis of 13 studies comparing immediate and extended release found that neither form of methylphenidate affected all aspects of ADHD (Punja et al., 2013). Intriguingly, the three studies that used parent ratings to assess hyperactivity and impulsivity found the extended release superior to the immediate release, while three studies that used teacher reports assessing only hyperactivity, found the opposite effect, with the immediate release showing a stronger effect. Side effects were similar for both preparations with anorexia, irritability, insomnia and gastro-intestinal symptoms being most often reported.

Like methylphenidate, amphetamine has been approved for the treatment of ADHD, with several different forms currently available, including normal amphetamine (which contains both the *d*- and *l*- enantiomer of amphetamine), extended release amphetamine, and a prodrug form lisdexamfetamine mesylate. This preparation only contains the *d*-enantiomer of amphetamine conjugated to the amino acid lysine, which requires cleavage by peptides in the bloodstream, leading to a sustained release of active *d*-amphetamine. Like methylphenidate, amphetamine affects the dopamine and noradrenaline transporters. However, whereas methylphenidate blocks the transporter, amphetamine is a competitive substrate. Moreover, at high doses amphetamine reverses the direction of the transporter, and thus leads to an active release of dopamine, noradrenaline and (to a lesser extent) serotonin.

Although methylphenidate is often the preferred first choice treatment for ADHD, a meta-analysis comparing immediate release amphetamine and methylphenidate in over 100 patients actually found a small but significant advantage of amphetamine (Faraone et al., 2002), while the side-effects profile was similar. The two main concerns with both methylphenidate and amphetamine is the relatively high rate of nonresponders (up to 35%–40% of patients) and the misuse potential, especially in adolescent children (Chapter 6). Although it has been suggested that extended release preparations have a lower abuse potential (Faraone and Upadhyaya, 2007), all amphetamine and methylphenidate formulations have a DEA schedule II classification.

One of the intriguing aspects of stimulant treatment of ADHD is the fact that these drugs reduce hyperactivity, while they are normally known (and classified) as activity enhancing. Although it was long thought that this hyperactivity reducing effect was specific for ADHD patients, and thus may be related to the specific ADHD neurobiology, it is now well accepted that it is a specific dose-related phenomenon that also occurs in healthy volunteers. While higher doses induce hyperactivity, (much) lower doses of both methylphenidate and amphetamine actually reduce hyperactivity. Although there are different explanations for this phenomenon, which would go beyond the aim of this paragraph, differences in cortical vs. subcortical dopamine transmission may be partly responsible for this phenomenon. This is studied in much more detail in animals. In a recent paper it was shown that while low doses of methylphenidate (0.5 mg/kg) enhanced extracellular dopamine concentrations in the prefrontal cortex, higher doses (4 mg/kg) are needed to enhance dopamine release within the nucleus accumbens (Spencer et al., 2015). It has been known for quite some time that there is a reciprocal relationship between cortical and subcortical dopamine, and thus the increase in prefrontal dopamine release would lead to a reduction in accumbal dopamine transmission which in turn would lead to reduced locomotor activity. Moreover, this same dose of 0.5 mg/kg increase working memory performance, while higher doses (2 mg/kg) inhibit performance, and even higher doses (4 mg/kg) induce hyperactivity (Spencer et al., 2015). In this respect it is important to realize that very few dopamine transporters are present in the prefrontal cortex, and most of the dopamine is taken up by the noradrenaline transporter. Given the differential influence of methylphenidate and amphetamine on both transporters (Fig. 8.6) these drugs will have a different influence on the concentration of dopamine within the prefrontal cortex and therefore may have different behavioural effects as well.

Nonstimulant Drugs

At present, three different nonstimulant drugs have been approved for the treatment of ADHD: atomoxetine, clonidine and guanfacine (approved in the USA, not in most European countries). Like amphetamine and methylphenidate, atomoxetine blocks monoamine transporters. However, it has a much higher affinity for the noradrenaline transporter (Fig. 8.6) and in therapeutically relevant doses it blocks neither the dopamine nor the serotonin transporter. Nonetheless, atomoxetine has been known to increase both extracellular noradrenaline and dopamine within the prefrontal cortex. The reason for this is, as mentioned before, given the high concentration of noradrenaline transporters within the prefrontal cortex (and the virtual absence of dopamine transporters) dopamine is taken up by the noradrenergic transporters. In fact dopamine has a higher affinity for the noradrenaline transporter than noradrenaline (Rothman and Baumann, 2003). However, in subcortical areas the situation is much different, and therefore within the nucleus accumbens or the striatum, atomoxetine does not increase the extracellular dopamine concentration. Consequently, it is neither a psychostimulant, nor does it have addictive properties.

Clinically, atomoxetine is an effective drug in the treatment of ADHD. However, two large meta-analyses showed that while atomoxetine was similar to immediate release methylphenidate, it was less effective compared to extended release amphetamine and methylphenidate (Hanwella et al., 2011; Stuhec et al., 2015); with both studies again suggesting that amphetamine (and lisdexamphetamine) were the most effective drugs. However, there is also evidence that atomoxetine's effects increase with prolonged treatment and appears to last longer (Clemow and Bushe, 2015). The main side effects of atomoxetine are nausea, dry mouth and anorexia. In addition, atomoxetine has been associated with suicidal ideation in adolescence and as a consequence, it carries a FDA designated "black box" warning.

In contrast to drugs for the treatment of ADHD discussed so far, clonidine and guanfacine do not affect monoamine transporters. Rather they stimulate noradrenergic α_2 receptors. Traditionally, α_2 receptors were thought to be exclusively presynaptically located and stimulation decreased noradrenaline release. As a consequence both drugs reduce blood pressure (and were originally approved for that purpose). However, it has since been shown that in several brain regions, α_2 receptors are located postsynaptically. Thus, both drugs are thought to influence the symptoms of ADHD via stimulation of postsynaptic α_2 receptors on pyramidal neurons within the prefrontal cortex (Sallee, 2010). There are three different subtypes of α_2 receptors (α_{2A}, α_{2B}, and α_{2C}) and within the prefrontal cortex it is predominantly the α_{2A} receptors which are located on pyramidal cells. As shown in Fig. 8.6, whereas clonidine influences all three α_2 receptors, guanfacine has a much greater selectivity for the α_{2A} receptors. As a result, one would expect less side effects of guanfacine compared to clonidine.

Like the psychostimulants, both clonidine and guanfacine are available in immediate and extended release forms. In addition both drugs are approved as stand-alone medication and as add-on to existing psychostimulant treatment. Both drugs are effective in the treatment of ADHD, with little difference between them as evidenced in meta-analysis studies (see Fig. 8.7; Hirota et al., 2014; Ruggiero et al., 2014). Both drugs also show added benefit when combined

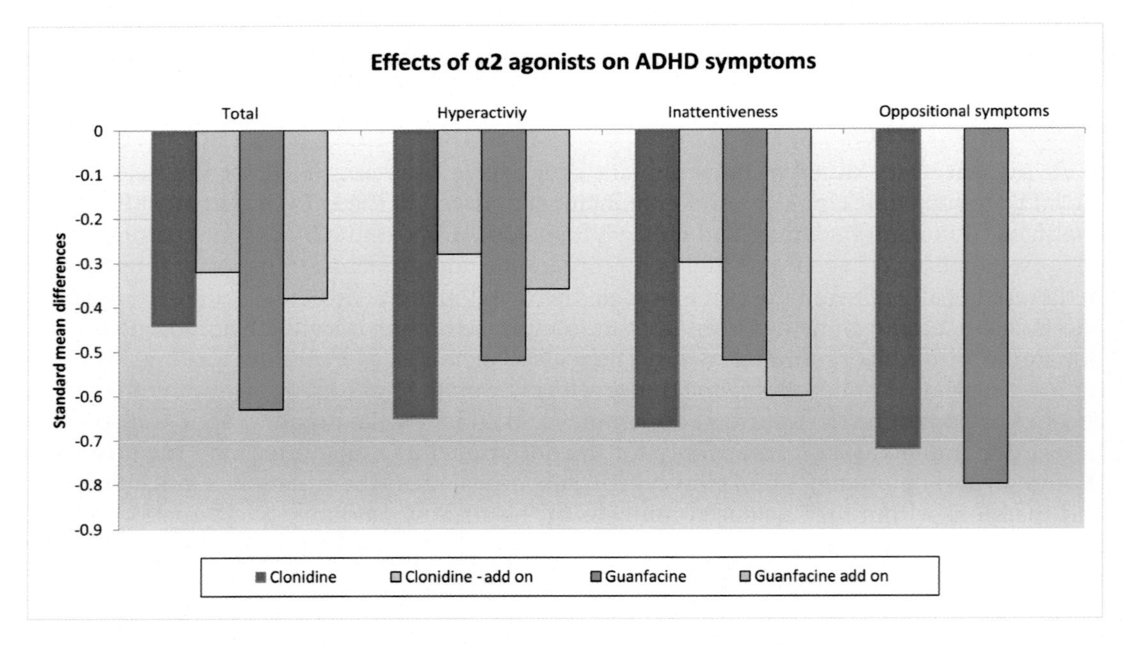

FIGURE 8.7 **The therapeutic effects of α_2 adrenoceptor agonists.** Although guanfacine has a much higher selectivity for the α_{2A} receptor, therapeutically clonidine and guanfacine are similar both as stand-alone drugs and as add-ons to other therapies. *Data from Hirota et al., 2014.*

with other anti-ADHD medication. The most important side-effects of both α_2 agonists are fatigue, sedation and somnolence. In addition, clonidine induced significant hypotension and bradycardia while guanfacine led to QTc prolongation (Hirota et al., 2014).

The finding that psychostimulants (that predominantly increase dopamine transmission in the prefrontal cortex) and nonstimulants (that predominantly increase noradrenaline transmission in the prefrontal cortex) both improve the symptoms of ADHD, suggests an interaction between both neurotransmitters within this critical brain region. Based on clinical as well as animal data, a model has been proposed that incorporates all effective treatments (Fig. 8.8) (Berridge and Arnsten, 2015). According to this model noradrenaline acts by stimulating the postsynaptic α_{2A} receptor on dendritic spines of pyramidal cells in the prefrontal cortex. This leads to an inhibition of cAMP production and strengthening of the NMDA signaling. On the other hand, stimulation of D_1 receptors by dopamine on other spines increased the production of cAMP, leading to a weakening of the NMDA signaling. Thus it is proposed that α_{2A} receptor stimulation increases the "signal" while D_1 receptor stimulation reduces the "noise" (Berridge and Arnsten, 2015). As a result, the top-down control of the prefrontal cortex on

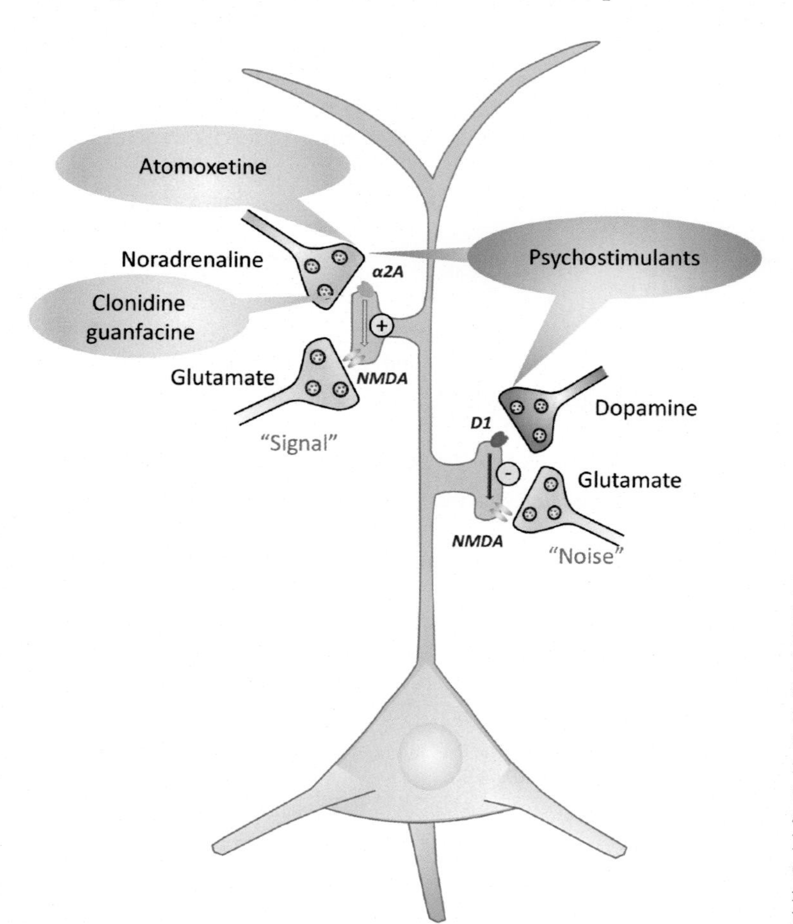

FIGURE 8.8 **A proposed mechanism for the therapeutic effects of drugs to treat ADHD.** Based on (Berridge and Arnsten, 2015), both stimulant and non-stimulant drugs are thought to influence the signal to noise ratio, by increasing the signal (through stimulation of the noradrenaline neurotransmission, left) or by decreasing the noise (through stimulation of the dopamine neurotransmission, right). For more details see text.

cognition, impulsivity and motor control is strengthened. As attractive at this model is, it should be mentioned that there are also alternative theories, focusing more on the effects of anti-ADHD medication on subcortical structures.

The Aetiology of ADHD

Twin and adoption studies have underscored the importance of genes in the development of ADHD, with the heritability estimations ranging from 0.6 to 0.9 (Brikell et al., 2015). As with other disorders, the advent of ever simpler and cheaper molecular genetic techniques has led to an intensified search for the genes involved in ADHD, using both specific candidate gene studies as well as more global genome wide association studies. Unfortunately, in line with most psychiatric disorders, this search has led to a plethora of genetic markers, each of which contributes only a small risk to the overall occurrence of ADHD.

Genetic Factors Contributing to ADHD

Given the extensive involvement of the catecholamine system in the neurobiology and treatment of ADHD, it is not surprising that many candidate gene studies focused on dopamine and noradrenaline (Table 8.2). One of the first candidates that was investigated was the dopamine transporter (*SLC6A3*), as this is a major target for the psychostimulants amphetamine and methylphenidate and, as mentioned earlier, significantly reduced levels are found

TABLE 8.2 Major Single Nucleotide Variants and Copy Number Variations Implicated in ADHD, See Text for Further Explanations.

Gene/regions	Remarks
SINGLE NUCLEOTIDE VARIANTS	
BDNF	Brain derived neurotrophic factor: Association found with the "A" allele of rs6265
DRD4	Dopamine D_4 receptor: Association with VNTR (7 repeats) in exon3 and rs1800955 (the "T" allele) in the promoter region
DRD5	Dopamine D_5 receptor: Association with the 148 base pair repeat upstream of the DRD5
HTR1B	Serotonin 5-HT_{1B} receptor: Association with the "G" allele of the rs6296 polymorphism of the 5-HT1B receptor
SLC6A3	Dopamine transporter: association with dinucleotide repeat (148 bp) on 5′-flank and the "G" allele of rs27027
SLC6A4	Serotonin transporter. Most convincing evidence is for an association with the 3′UTR (both rs27027 ("G" allele) and VNTR (10 repeats)). Association also found with a VNTR (6 repeats) in intron 8
SNAP25	SNAP25: Association with several different SNVs within the gene
TACR1	Neurokinin 1 receptor: association with the "A" allele of rs3771833, the "C" allele of rs3771829 and the "G" allele of rs17011370
COPY NUMBER VARIATIONS	
15q13.3	2 Mb deletion
22q11.2	1.2–3 Mb deletion

in drug-naïve patients with ADHD. The most widely studied polymorphism in the dopamine transporter is a variable number of tandem repeats (VNTR, see Chapter 2) in the 3′ untranslated region (3′-UTR) on the transporter that is 40 base pairs in length. The most common alleles are the 9 (440 base pairs) and 10 (480 base pairs) repeats. Using SPECT it has been shown that individuals with the 10 repeat have lower transporter density (Jacobsen et al., 2000), making it an obvious risk allele for ADHD. Indeed the first study investigating the VNTR using a sample of 57 children with ADHD found a significant association with the 10 repeat allele (Cook et al., 1995). Since then, over 30 studies have investigated this association and, although both confirmations and replication failures have been reported, a meta-analysis found a small (Odds Ratio, $OR = 1.1$) but statistically significant effect (Gizer et al., 2009). In addition to the VNTR, another single nucleotide variant (SNV) has been identified in the 3′-UTR, namely rs27027, which the "G" allele, like the VNTR have a small ($OR = 1.2$) but significant association with ADHD (Gizer et al., 2009). While the same meta-analysis failed to find a significant association with two other SNVs, a positive (small $OR = 1.2$) association was also seen with a VNTR in intron 8, where the 6 repeats was associated with ADHD. Interestingly, the meta-analysis showed significant heterogeneity in all these associations, which could be due to several different factors, such as the additional effects of other genetic or environmental factors (as discussed later) or alternatively it could be due to the fact that the dopamine transporter is only related to a subset of ADHD pathology and symptoms. In this respect it has been suggested that the dopamine transporter has a stronger association with the hyperactivity/impulsivity features of ADHD than with the attention deficit aspects (Waldman et al., 1998). Overall, 19 different genetic variations on the dopamine transporter gene have been studied in relation to ADHD, with the majority (43 out of 70 studies) finding a significant association (Li et al., 2014).

Although (slightly) less extensively studied than the dopamine transporter, several other genetic alterations in dopamine neurotransmission have also been linked to ADHD, most notably a 48 basepair VNTR in exon 3 of the dopamine D_4 receptor (*DRD4*). Several different alleles have been identified, with the 2, 4 and 7 repeats being the most common. The interest for the D_4 receptor in relation to ADHD was sparked by findings which linked this receptor with the novelty seeking personality trait (Ebstein et al., 1996), which in turn is often associated with impulsivity. At least 25 studies have looked at the association between the 7 repeat VNTR and a meta-analysis confirmed a positive linkage with an OR of 1.3. A positive association was also found with another genetic alteration, an SNV (rs1800955) in the promoter region of the D_4 receptor ($OR = 1.2$) (Gizer et al., 2009). This is an interesting association, as the risk allele (T) leads to a 40% reduction in promoter activity (Okuyama et al., 1999). Again, the meta-analysis showed significant heterogeneity between the different studies, which could be explained by studies which have linked the D_4 receptor to attentional deficits rather than hyperactivity or impulsivity (Lasky-Su et al., 2008). Several additional genetic variations in the D_4 receptor have also been investigated and according to a recent meta-analysis, evidence linking the D_4 receptor to ADHD is the strongest among all genes so far investigated with 49 out of 67 studies finding a significant association (Li et al., 2014).

In addition, genetic variations in other receptor such as the D_1, D_2 and D_3 receptors have been found, but these could not be substantiated in a meta-analysis. However, one specific genetic variation in the D_5 receptor (*DRD5*) was strongly associated with ADHD, namely a dinucleotide repeat in the 5′-flank of the receptor. This is a highly polymorphic region about

18 kilobases upstream from the D_5 receptor gene for which at least 12 different alleles (ranging from 134 to 156 basepairs) are known to exist (Sherrington et al., 1993). A recent meta-analysis confirmed that the 148 basepair repeat is more often found in patients with ADHD (Gizer et al., 2009). However, the relevance of this dinucleotide repeat is as yet unclear, as it seems unlikely that it actually altered the D_5 receptor structure or density. Nonetheless, 14 out of 22 studies have found a positive association between genetic variations in the D_5 receptor and ADHD (Li et al., 2014).

In contrast to the convincing association of several genes within the dopaminergic system, studies investigating the noradrenergic system have generally failed to find significant associations with ADHD. Thus, the already mentioned meta-analysis did not find significant associations between ADHD and variations in the genes for dopamine-β-hydroxylase (*DBH*), monoamine oxidase A (*MAOA*), the noradrenaline transporter (*SLC6A2*) or the α_{2A} adrenoceptor (*ADRA2A*; Gizer et al., 2009). However, more recent data suggest a strong association between *SLC6A2* and ADHD, with at least 10 out of 16 association studies showing a significant effect.

Several genes that influence serotonergic transmission on the other hand, were positively coupled to ADHD, especially the serotonin transporter (*SLC6A4*) and the 5-HT_{1B} receptor gene (*HTR1B*). With respect to the first, arguably the most studied genetic variation is a polymorphic region within the promoter region of the serotonin transporter, generally referred to as 5-HTTLPR (5-HT transporter linked polymorphic region). In its simplest version, the 5-HTTLPR consist of a 44 base pair insertion/deletion, leading to a long (l) and short (s) alleles (Lesch et al., 1996). Although several additional variations have been identified with the region, the s-allele has been characterized as leading to approximately 50% lower density of transporters. Whereas the s-allele has been linked to several major psychiatric disorders, such as depression (Chapter 7), anxiety and drug addiction (Chapter 6), it appears to be the l-allele which is moderately associated with ADHD (Gizer et al., 2009). Although several other common variants within the serotonin transporter have also been investigated, such as a 17 base pair repeat in intron2 (STin2), which acts as a transcriptional regulator, and a single nucleotide variant in the 3′-UTR, none seems to be consistently associated with ADHD. Overall, the number of positive and negative studies regarding the linkage between the serotonin transporter and ADHD are in balance with 13 studies each (Li et al., 2014). With respect to the 5-HT_{1B} receptor, a single nucleotide variant (rs6296) has been identified in the single exon of the *HTR1B* gene, leading to a G → C transition. The meta-analysis confirmed a significantly increased risk for ADHD ($OR = 1.1$) for individuals with the "G"-allele (Gizer et al., 2009). However, a more recent meta-analysis was less positive about the association between the *HTR1B* gene and ADHD (Li et al., 2014), with only 5 out of 14 studies showing a significant linkage.

An interesting candidate gene that was recently identified in ADHD patients is the *TACR1* gene. This gene codes for the human Substance-P preferring neurokinin receptor (NK_{1R}) and several variants have been associated with ADHD, most significantly the "A" allele of rs3771883 and the "C" allele of rs3771829 (Yan et al., 2010). This association is interesting not just because it was replicated in an independent cohort, but also because it is an example of a "reverse translational" process. In other words, theinvestigation of these genetic variants in patients was based upon the finding that NK1 knock-out mice showed an ADHD phenotype (as discussed later). Moreover, these same variants have been associated with bipolar disorder and alcoholism (Sharp et al., 2014), both of which are well-known comorbidities for (adult) ADHD (as discussed earlier).

Finally, two candidate genes that has been extensively investigated in relation to ADHD are *BDNF* (brain derived neurotrophic factor) and *SNAP25* (Synaptosomal-associated protein 25). Given the earlier summarized literature showing widespread dysconnectivity in the brains of patients with ADHD, proteins involved in brain development and neurotransmitter release are obvious candidates. Moreover, BDNF is functionally linked to serotonin and dopamine, making it an obvious candidate. The most studied genetic variation in the *BDNF* gene is a missense SNV (rs6265) leading to valine → methionine substitution, with the methionine variation leading to significantly reduced BDNF activity. Studies have found a positive association between this variant of the *BDNF* gene and ADHD although overall the number of positive and negative findings balance each other out (Li et al., 2014). Like BDNF, SNAP-25 plays an important role in neurodevelopment and neuroplasticity and is particularly involved in axonal growth and synaptic plasticity. Moreover, it plays an essential role in docking and fusion of synaptic vesicles (Sollner et al., 1993). In line with the neurodevelopmental hypothesis of ADHD, many studies (16/21) have found an association between genetic variations within the *SNAP25* gene and ADHD (Li et al., 2014), making it one of the most reliably identified genes, though again the overall odds ratio is small.

Candidate gene studies, while based on a clear hypothesis are obviously limited by our restricted knowledge of the neurobiology and neuropathology of ADHD. Genetic linkage studies, on the other hand, are hypothesis-free and thus allow us to identify novel gene targets. However, such studies often identify genetic markers, which do not necessarily alter gene transcription or protein function and thus their relevance for ADHD (or indeed any other disorder) is not immediately clear. Moreover, the candidate gene studies, as we have seen earlier, have only found genes with limited effects (ie, ORs of 1.1 to 1.3). For genetic linkage studies to be successful, it has been estimated that the identified genes should account for at least 10% of the genetic variation (Risch and Merikangas, 1996). The more recently developed genome wide association studies allow for the detection of genes with such small effects, by the sheer power of their analysis, usually involving the analysis of between 100,000 and over 1,000,000 genetic variations in individual subjects (Neale and Purcell, 2008). Although theoretically highly successful, such an approach requires large numbers (several 1,000s to 10,000) of subjects (Burton et al., 2009). As a result, such studies have only recently begun to appear in the literature, as a direct consequence of reduced costs for genotyping and the formation of large research consortia. In the field of ADHD several different genome wide association studies have been performed, however, the results have so far been quite disappointing. Not only were most of these studies unable to confirm any of the candidate genes discussed earlier, but in fact, no single gene could be found that surpassed the threshold for significance (Franke et al., 2009). As mentioned earlier, it has been suggested that the different symptom domains of ADHD may have different aetiology and recent attempts have therefore been undertaken to combine genome wide associations with behavioural characteristics such as endophenotypes. In a recent study, 479 adult ADHD patients were evaluated on a continuous performance test, in which individuals are shown a sequence of letters on the screen and they are required to press the space-bar when any letter except "X" is presented. The study measured errors of commission (ie, the space-bar is pressed when the letter "X" was presented) and several other measures of reaction time. Again, no single genetic variation was found to reach the threshold of significance, with the highest associations reaching 7.9×10^{-7} (Alemany et al., 2015). However, in line with the idea that different aspects of ADHD might be mediated via different etiologies, different subsets of genes were identified. Moreover, using gene set enrichment analysis, aimed at identifying genes that are part of a common functional pathway, several interesting avenues were identified.

One limitation of (traditional) genome wide association studies and candidate gene studies is that they focus on relatively simple genetic variations, such as single nucleotide variants or slightly longer variations (such as the variable number of tandem repeats in the *DRD4* gene). However, as discussed in Chapter 2, there are multiple additional genetic variations of which copy number variations (CNVs) have received increasing attention in psychiatry in recent years. Also, within the field of ADHD, several studies have focused on CNVs. For instance Elia and coworkers performed a genome wide CNV study comparing 4,105 healthy children with 1,013 children with ADHD (Elia et al., 2012). In addition, they compared their results in multiple independent cohorts of up to 2,493 cases and 9,222 controls. Overall, this study detected 222 CNVs in patients with ADHD. Interestingly, the authors found evidence of alterations in the metabotropic glutamate receptor gene network as CNVs involving these receptors were identified in several different cohorts. Furthermore, duplication of 15q13.3 and a deletion of 22q11.2 have been associated with ADHD, although a recent detailed analysis found clear differences between patients with the 22q11.2 deletion syndrome and ADHD (Niarchou et al., 2015). Finally, a recent study, focusing more on the structure of the CNVs, rather than the location, found an excess of small (from 100 to 300 kb) insertions in the genome of patients with ADHD (Ramos-Quiroga et al., 2014), a phenomenon which has also been observed in several other psychiatric disorders such as schizophrenia (Chapter 9).

Moderation of the Effects of Genes by Environmental Factors

The earlier analysis underscores the complexity of the genetic basis of ADHD and suggests that in addition to genes, other factors must also play a role. Indeed, several studies have emphasized the role environmental factors play in moderating the effects of genes. The first gene–environment study in ADHD investigated the interaction between maternal smoking and the dopamine transporter (Kahn et al., 2003). The authors found a significant interaction between prenatal nicotine exposure and the 3′-UTR VNTR of the dopamine transporter such that homozygous carriers of the 10 repeat allele that were exposed to maternal smoking showed more hyperactivity/impulsivity and oppositional symptoms. Interestingly, there was no such interaction with respect to inattention, further emphasizing the heterogeneity of ADHD symptoms. This interaction was also found in another study, though only in boys (Becker et al., 2008). On the other hand several studies subsequently failed to replicate this interaction (Altink et al., 2008; Kieling et al., 2013). Interestingly, in another study, while the authors again did not find an interaction between the dopamine transporter genotype and maternal smoking, a significant interaction was found with maternal alcohol (Brookes et al., 2006). Thus, the interactions between the dopamine transporter and maternal smoking in relation to ADHD remains unclear, possibly due to the heterogeneity of ADHD. Another obvious caveat with these studies in humans is that maternal smoking and alcohol use may be secondary to other maternal problems (such as stress or economic hardship). A recent study, focusing on externalizing behaviour (encompassing anti-social impulsive and substance abuse disorder) found a significant interaction between the 10 repeat VNTR of the dopamine transporter and maternal smoking in boys (O'Brien et al., 2013). Genetic alterations within the dopamine transporter have also been studied in relation to adverse life events with a significant interaction being observed between the 10 repeat VNTR and adverse life events both in relation to attentional problems as well as to hyperactivity/impulsivity symptoms (Laucht et al., 2007).

Gene–environment interactions have also been investigated in relation to the *DRD4* gene, especially in relation to the VNTR in exon 3. However, several studies failed to find a significant interaction between the *DRD4* gene and childhood adversities, prenatal smoking and prenatal alcohol exposure (Nigg et al., 2010). On the other hand an interesting interaction was found between season and birth and DRD4 genotype, such that the risk of developing hyperkinetic conduct disorder symptoms was reduced in winter-born children with the 7-repeat VNTR, while it was increased in summer (Seeger et al., 2004).

Several studies have investigated the relationship between environmental risk factors and the serotonin transporter gene (*SLC6A4*). In a group of delinquent adults, childhood ADHD was associated with the l-allele of the 5-HTTLPR as well as with increased childhood adversity (Retz et al., 2008). However, the most interesting finding was a 5-HTTLPR * childhood adversity interaction, such that individuals with the low risk 5-HTTLPR genotype (ss- and sl-) were more likely to develop childhood ADHD when having been exposed to adverse childhood events (Fig. 8.9), an interaction that was also found for ADHD that persisted into adulthood. Similar interactions between stressful events and the 5-HTTLPR on ADHD have

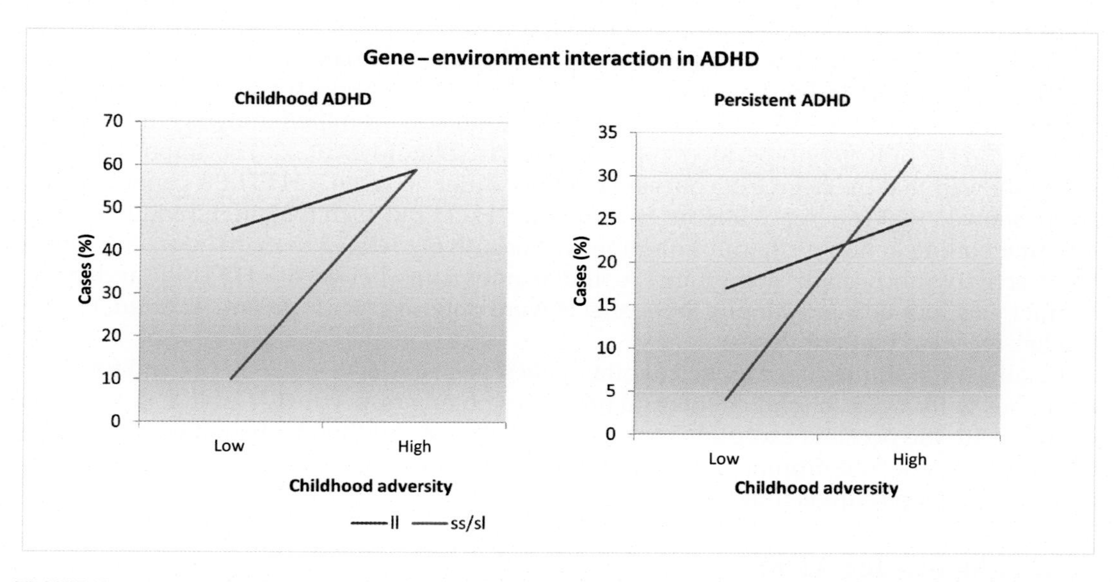

FIGURE 8.9 **An example of a gene–environment interaction.** In this study, childhood adversity and the serotonin transporter gene were found to interact, both in determining childhood ADHD, as well as persistent ADHD (Retz et al., 2008). The results showed that individuals with the "ll" allele had significantly higher risk of developing ADHD in the absence of adversity, this difference disappeared in the presence of childhood adversity (as the individuals with the "s" allele were much more sensitive to adversities). Interestingly, in cases with persistent ADHD symptoms, there was a complete cross-over with the "ll" cases more sensitive to ADHD in the absence of adversities, but the "s" allele carriers more sensitive in the presence of childhood adversities.

TABLE 8.3 Examples of Genetic–Environmental Interactions in ADHD

Gene	Environment	Effect	References
ANKK1	Familial adversities	Carriers of the TaqI A2 allele had increased risk when mothers were divorced	Waldman (2007)
BDNF	Socioeconomic status (SES)	Carriers of the "A" allele of rs1013442 or the "T" allele of rs6265 or "C" or the "A" allele of rs 1387144 exposed to lower SES environments showed more symptoms of inattention	Lasky-Su et al. (2007)
CHRNA4	Prenatal smoking	Carriers of the "C" allele of rs1044396 that were exposed to prenatal smoking had more ADHD symptoms	Todd et al. (2007)
DRD4	Winter birth	Carriers of the 7 repeat allele had increased risk when born in summer	Seeger et al. (2004)
MAOA	Parenting behaviour	Carriers of 4 repeat allele have higher risk when father's emotional distance is perceived as more positive	Vanyukov et al. (2007)
SLC6A3	Prenatal smoking	Carriers of the 10 repeat allele of the 3′-UTR exposed to prenatal smoking had higher levels of hyperactivity/impulsivity	Kahn et al. (2003)
SLC6A4	Familial adversities	Carriers of the s-allele of the 5-HTTLPR exposed to familial conflict or incoherence have increased inattention	Elmore et al. (2015)

since been reported in several studies of both adults (Muller et al., 2008) and adolescents (van der Meer et al., 2014), with increased exposure to stress enhancing the risk or severity of ADHD symptoms in individuals carrying the s-allele but not homozygous l-allele carriers. A recent study looking specifically at cohesion and conflict within the family again confirmed that the 5-HTTLPR genotype moderates the effects (Elmore et al., 2015). Importantly, the study showed that the degree of cohesion does not differ between 5-HTTLPR genotypes. The study showed that coherence interacted with the 5-HTTLPR, such that, in individuals with a low functioning genotype, family cohesion was negatively related to inattention, but not for hyperactivity/impulsivity symptoms. A similar interaction between 5-HTTLPR and family conflict was found. Interestingly, these effects were only seen with parent-reported, but not teacher-reported inattentiveness.

Finally, some studies have looked at other genes, such as *BDNF* or *CHRNA4* and found interactions with socioeconomic status and prenatal smoking respectively (Table 8.3). However, overall, with the possible exception of the 5-HTTLPR and childhood adversities, the number of studies of gene * environment interactions in ADHD have been limited and there is an urgent need for replications (Nigg et al., 2010).

Animal Models for ADHD

Measuring ADHD-Like Features in Animals

Given that one of the main symptoms of ADHD is hyperactivity, which can easily be assessed in animals, it is not surprising that most of the early animal research in ADHD focused on measuring locomotor activity. Although there are many different ways in which locomotor activity can be measured, the two most basic techniques involve either infrared beam interruption or video tracking (Fig. 8.10). In the first technique, animals are placed in an open field that is equipped with a series of infrared beams. Interruption of the beam (or beams) indicates the position of the animal in the open field. The software generally allows to distinguish between single beam interruption (often referred to as stereotyped or small movements) and the subsequent interruption of several beams, indicative of locomotor activity. In most cases a second row of beams is mounted higher to detect vertical (rearing) movements. Although this technique is perfectly capable of assessing total locomotor activity and distance moved, it does not allow for a more detailed analysis of locomotion.

The video-tracking technique, was designed to analyze locomotor patterns. The essence of the technique is that a video-camera is mounted above the open field and the software allows to distinguish between the animal and the background (usually by contrast; ie, the animal is white and the open field is black). By reducing the animal to a single point, the animal's movements can be recorded in great detail, thus allowing for an analysis of movement patterns as well. However, as traditional video tracking software only records a single point per rat, it is incapable of measuring rearing activity. More modern methods can distinguish multiple body points per animal (such as the tip of the nose and the base of the tail, in addition to the middle point of the animal), allowing for even greater detail. By analyzing the distance between the three body points it is even possible to get an indication of rearing, although it is not as reliable as the detection using infrared beams. Finally, it is worth mentioning that in addition to infrared beam interruptions and video-tracking, visual observation of behaviour is important as well. As sophisticated as video-tracking analysis can be, it cannot compete with a human observer, who will be able to identify other (perhaps competing) behaviors. For instance the NK_{1R}-/- mouse (as discussed later) was identified as an animal model for ADHD on the basis of visual inspection, while computer software had initially been unable to detect signs of hyperactivity.

However, as it has become clear in the preceding section, hyperactivity is just one of the symptoms of ADHD, with impulsivity and inattention being equally important. Several

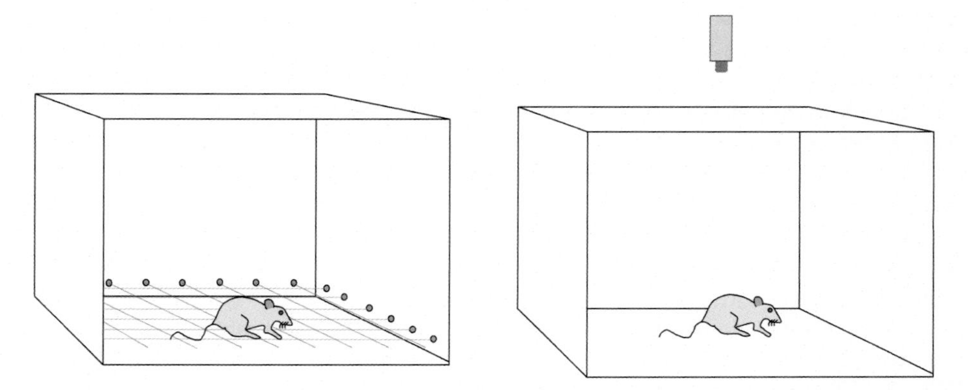

FIGURE 8.10 **Measuring locomotor activity in rodents.** Traditionally, locomotor activity is measured in boxes equipped with infrared beams (see left). Through interruption of the beams, the position of the animal can be determined and thus locomotor activity (and fine movements) can be detected. Using a second row of beams, rearing activity can also be assessed (not shown). More modern video-tracking equipment (right) allow for a much more detailed analysis of locomotor activity including an analysis of the motor pattern.

different paradigms have been developed to assess impulsive behaviors, including the 5-choice serial reaction time task (5-CSRTT), the stop-signal task (SST) and the delay discounting task (DDT). The 5-CSRTT requires an animal to nose-poke into one of five holes after a brief visual stimulus has been turned on and off again (eg, a light). In addition to measuring correct and incorrect responses and omissions, the test detects premature responding, that is, animals making a response while the cue light is still on. Premature responding is generally seen as an indicator of impulsivity. It is, however, important to realize that there are several different aspects of impulsivity, often subdivided into cognitive impulsivity (sometimes referred to as impulsive choice) and motor impulsivity (impulsive action such as premature responding). The SST task measures aspects of impulsive action, although there is evidence that the 5-CSRTT and the SST assess different aspects of impulsivity. While the 5-CSRTT measures the ability to correctly initiate a movement, the SST assesses the ability to inhibit an already initiated response. In this paradigm animals generally are required to make two responses (such as two lever presses) in quick succession to obtain a food reward (the go-trials). However, in a set number of trials (say 20%) a tone is given in between the two responses and the animal is required to withhold the second response in order to obtain a reward. The DDT also requires animals to withhold a pre-potent response. However, rather than inhibiting a response all together, in the DDT, animals are given a choice, an immediate small reward or a delayed larger reward. When the delays are relatively small (in the order of 10 to 20 seconds) the majority of normal, untreated animals will prefer to wait for the larger reward. An increase in (choice) impulsivity would lead to preference for the faster smaller rewards.

Genetic Models for ADHD

Arguably one of the most studied and oldest (genetic) rodent models in ADHD is the spontaneous hypertensive rat (SHR). As the name suggests, these rats were originally identified on the basis of their increased blood pressure and subsequently bred for this characteristic. Interestingly they were also studied, for many years for their high stress responsivity, whereas stress is said to calm ADHD patients. However, behavioural analysis showed that in addition, these rats showed hyperactivity. By selectively breeding the SHR and Wistar Kyoto (WKY; serving as controls), it was proven possible to differentiate between hypertension and hyperactivity, leading to a strain that it hyperactive but not hypertensive (Sagvolden et al., 1992; Sagvolden, 2000). Subsequent behavioural analysis showed that these animals also have deficits in impulse control and (sustained) attention, although they appear normal in the 5-CSRTT (Dommett, 2014). In addition, while some anti-ADHD drugs such as amphetamine and guanfacine ameliorate the behavioural deficits in these animals (Sagvolden, 2000), other drugs such as atomoxetine did not (Turner et al., 2013), thus questioning the predictive validity.

The *coloboma* mouse was originally developed using irradiation techniques and subsequent investigation showed a deletion of about 2 cM encompassing about 20 genes including the *SNAP25* gene. The *coloboma* mouse shows spontaneous hyperactivity, which can be reversed by amphetamine, though not by methylphenidate or by guanfacine (in moderate nonsedative doses). Intriguingly, transgenic replacement of SNAP-25 ameliorated the hyperactivity, suggesting that it is indeed the deletion of *SNAP25* which is responsible for the resulting hyperactivity. However, the *coloboma* mouse also shows other ADHD-like features including increased impulsivity in delayed discounting tasks and latent inhibition indicative of a deficit in attention (Bruno et al., 2007). Although this makes the *coloboma* mouse an interesting model for ADHD, a recent study found severe deficits in motor coordination and balance suggesting that the model may be more related to ataxia (Gunn et al., 2011). This, combined with the limited predictive validity, seriously questions the validity of the coloboma mouse as an animal model for ADHD, in spite of the quite strong genetic association between SNAP-25 and ADHD (as discussed earlier).

Among the reverse transgenic models, those affecting the dopamine transporters have received considerable attention. A complete deletion of the dopamine transporter leads to extremely high levels of hyperactivity, which can be reversed by psychostimulants. Although these animals show learning and memory deficits, these may be secondary to the extreme hyperactivity. The animals also show deficits in impulse control (Leo and Gainetdinov, 2013). In addition to the full dopamine transporter knock-out (DAT-KO) mice, several other transgenic models of the dopamine transporter have been developed, including a dopamine transporter knockdown (DAT-KD, in which the dopamine transporter levels are reduced to about 10% of the wild type mice) and a mouse with a triple-point mutation in the cocaine binding site of the dopamine transporter (DAT-CI). Both these models show hyperactivity that is reduced by prior stimulant treatment. However, whether they also show other ADHD-like symptoms such as inattentiveness or enhanced impulsivity remains to be investigated (Leo and Gainetdinov, 2013). Transgenic mouse models for the different dopamine receptors have also been investigated but none of these show ADHD-like abnormalities. The dopamine D4 knock-out mouse, for instance, shows hypolocomotion and normal impulse control (Helms et al., 2008).

In addition several other genetically altered mouse models have been described (Table 8.4), including the neurokinin 1 receptor knock-out mouse (NK-1 KO), the casein kinase1δ (CK1δ)

TABLE 8.4 Genetic Animal Models for ADHD

Models	ADHD-like features	Effects of drugs
FORWARD GENETIC MODELS		
SHR	Hyperactivity, deficits in impulsivity and cognition	Reversed by amphetamine and guanfacine, not by methylphenidate
REVERSE GENETIC MODELS		
coloboma	Hyperactivity, deficits in latent inhibition and impulsivity	Reversed by amphetamine but not methylphenidate
DAT-KO	Hyperactivity, deficits in impulsivity and cognition	Reversed by stimulants
DAT-KD	Hyperactivity	Reversed by stimulants
DAT-CI	Hyperactivity	Reversed by stimulants
NK1-KO	Hyperactivity, inattentiveness and deficits in impulsivity	Reversed by stimulants, atomoxetine and guanfacine
CK1δ	Hyperactivity	Reversed by stimulants
nAChR-β2 KO	Hyperactivity and deficits in impulsivity	

See text for more details.

over-expression mouse and the knock-out of the β_2 subunit of the nicotine receptor. The NK-1 knock-out mouse has been extensively studied both in relation to locomotor activity as well as in relation to other ADHD-like aspects, and has been shown to exhibit hyperactivity, inattentiveness, impulsivity and increased perseverance (difficulty in switching to a new, correct answer) on the 5-CSRTT (Yan et al., 2009; Porter et al., 2015; Porter et al., 2016). Moreover, this model seems to have excellent predictive validity, as both stimulant (methylphenidate) and nonstimulant (atomoxetine, guanfacine) treatments appear to reduce the behavioural deficits (Pillidge et al., 2014b; Pillidge et al., 2014a). Although the construct validity of this model may appear less obvious at first sight, there is evidence that the genetic deletion of the NK_1 receptor in mice leads to alterations in prefrontal and striatal dopamine transmission (Yan et al., 2010). Moreover, several recent reports have now found an association between ADHD and the human NK_1 receptor gene *TACR1* (Yan et al., 2010; Sharp et al., 2014).

Most recently, a double mutant mouse lacking two important proteins involved in actin dynamics (Actin Depolymerizing Factor and n-cofilin) was shown to exhibit hyperactivity, combined with impaired impulsivity (although this was assessed in the elevated plus maze and needs to be replicated in more accepted models of impulsivity such as the 5-CSRTT or DDT) and cognitive deficits (Zimmermann et al., 2015). Given that the hyperactivity and the deficits in impulsivity were reduced by methylphenidate, this model seems to exhibit both face and predictive validity. Interestingly, the double knock-out leads to a disturbance in the morphology of excitatory synapses in the dorsal striatum, accompanied by large increase in glutamate release.

Moderation of the Effects of Genes by Environmental Factors

In line with the relative lack of gene–environment interaction studies in patients with ADHD, few studies have investigated such interactions in animal models. A few studies have been performed with the SHR rat. Pamplona and coworkers, rather than studying a risk factor that could exacerbate the ADHD-like pathology, investigated to what extent environmental enrichment could reduce the pathology (Pamplona et al., 2009). Unfortunately, few effects of environmental enrichment were found. Moreover, as the authors did not perform a two-way ANOVA with enrichment and strain as independent variables, it is difficult to verify whether the SHR were more or less susceptible to the effects of enrichment. On the other hand, a study using maternal separation (3 h per day from postnatal day 2–14) found that while control (WKY) rats showed an increase in anxiety like behaviour, this was not seen in the SHR animals. Interestingly, in the SHR (but not the WKY) rats maternal separation led to a significant reduction in dopamine clearance, suggesting a genotype-dependent alteration in dopamine transporter function after maternal separation.

Focusing on the NK1R knock-out mice, Porter and colleagues studied the role of breeding strategy on the phenotypical expression. When breeding genetically altered animals, two principle breeding strategies are most often used. In the first, simplest breeding strategy, the offspring of inbred wildtype males and wildtype females are compared to the offspring of inbred homozygous males and females. The advantage of this strategy is that no genotyping needs to be done and no heterozygous animals are created (which often are not needed). The disadvantage is that the homozygous animals are reared by homozygous mothers, whereas the wildtype animals are reared by wildtype mothers. If there are (subtle) differences in maternal behaviour this may interact with the genotype. Moreover, in the long run this breeding strategy may lead to genetic drift [ie, additional (spontaneous) genetic alterations may be introduced in one but not the other line]. In order to prevent this, the second breeding strategy involves breeding heterozygous males and females, thus ensuring that all genotypes are present and they are all reared by the same mother (although it is still theoretically possible that the mother shows a different behaviour towards animals with a different genotype). The disadvantage of this strategy is that all offspring needs to be genotyped and that on average 50% of the offspring are heterozygous, which are often not used. In order to assess the role of maternal (or littermate) behaviour, Porter and coworkers did a head-to-head comparison of the behaviour of homozygous NK_1 knockout derived from the two different breeding strategies but otherwise all of the same genetic background (Porter et al., 2015). Whereas animals from both breeding strategies did not differ in hyperactivity during the dark phase or perseverative behaviour, only NK_1 knock-out mice from the hom/hom breeding strategy showed increased impulsiveness, while those derived from the het/het strategy were not, indicating that impulsivity depends on an interaction between a lack of functional NK_{1R} and breeding environment.

Although not specifically focusing on ADHD, Papaleo and coworkers studied the interaction between the COMT genotype and mild stress on impulsivity and cognition in mice (Papaleo et al., 2012). The authors found that a mild stressor led to a significantly higher impulsivity in COMT knock-out mice compared to controls as assessed with the 5-CSRTT.

Unfortunately, studies which have attempted to mimic some of the environmental risk factors discussed earlier (Table 8.3), such as prenatal nicotine or alcohol exposure or early postnatal stress exposure in one of the genetic models described in Table 8.4 are so far lacking. Given the dearth of replicated findings in human studies, more animal research would be of great importance to better understand the role of genes and environment in the development of ADHD.

AUTISM SPECTRUM DISORDERS

Diagnosis and Symptoms

With an average age of onset between 1 and 3 years, autism spectrum disorder (ASD) is the earliest psychiatric disorder discussed in this book (Fig. 8.1). According to the DSM-V system, ASD is characterized by significant deficits in social communication and interaction as well as the presence of repetitive and restricted behaviors and/or interests (Box 8.2). These symptoms usually occur early during development and should be so severe as to impact on the patient's overall functioning. One of the most significant changes in DSM-V has been the revision of the diagnosis of autism. In previous versions, three different (sub)forms were recognized: autism (sometimes referred to as "true autism" or "low functioning autism"), Asperger syndrome (also referred to as "high functioning autism") and pervasive developmental disorders not otherwise specified (PDD-NOS), together comprising the group of the "pervasive developmental disorders". However, in the latest version (DSM-V), all subgroups are combined into Autism Spectrum Disorder (ASD), although the presence of intellectual impairment (a hallmark of the original "true autism") is specifically denoted. Nonetheless, within the scientific community, the grouping of all the different forms of ASD into one broad

BOX 8.2

THE DIAGNOSIS OF ASD (DSM-V)

1. A persistent deficit in social communication and social interaction across multiple contexts, as manifested by the following, currently or by history:
   - **a.** Deficits in social-emotional reciprocity, ranging, for example, from abnormal social approach and failure of normal back-and-forth conversation; to reduced sharing of interests, emotions, or affect; to failure to initiate or respond to social interactions.
   - **b.** Deficits in nonverbal communicative behaviours used for social interaction, ranging, for example, from poorly integrated verbal and nonverbal communication; to abnormalities in eye contact and body language or deficits in understanding and use of gestures: to a total lack of facial expressions and nonverbal communication.
   - **c.** Deficits in developing, maintaining, and understanding relationships, ranging, for example, from difficulties adjusting behaviour to suit various social contexts; to difficulties in sharing imaginative play or in making friends; to absence of interest in peers.
2. Restricted, repetitive patterns of behaviour, interests, or activities, as manifested by at least two of the following, currently or by history:
   - **a.** Stereotyped or repetitive motor movements, use of objects, or speech (eg, simple motor stereotypies, lining up toys or flipping objects, echolalia, idiosyncratic phrases).
   - **b.** Insistence on sameness, inflexible adherence to routines, or ritualized patterns of verbal or nonverbal behaviour (eg, extreme distress at small changes, difficulties with transitions, rigid thinking patterns, greeting rituals, need to take same route or eat same food every day).
   - **c.** Highly restricted, fixated interests that are abnormal in intensity or focus (eg, strong attachment to or preoccupation with unusual objects, excessively circumscribed or perseverative interests).
   - **d.** Hyper- or hypo-reactivity to sensory input or unusual interest in sensory aspects of the environment (eg, apparent indifference to pain/temperature, adverse response to specific sounds or textures, excessive smelling or touching of objects, visual fascination with lights or movement).
3. Symptoms must be present in the early developmental period (but may not become fully manifest until social demands exceed limited capacities, or may be masked by learned strategies in later life).
4. Symptoms cause clinically significant impairment in social, occupational, or other important areas of current functioning.

syndrome is generally viewed critically, and there is substantial evidence that fundamental differences exist, especially between Asperger's syndrome and true autism.

Although the core deficits of ASD encompass aspects of abnormal social behaviour, communication and repetitive behaviour, it is important to note that most patients with ASD also suffer from a large number of other symptoms (Levy et al., 2009). These symptoms fall into several different categories as illustrated in Fig. 8.11, and include behavioural (such as aggressive and self-injurious behaviour), psychiatric (anxiety and depression), gastro-intestinal (gastro-oesophageal reflux, food selectivity), neurological (epilepsy, sleep disruption), and developmental (language deficits, motor delay) symptoms. Many of these symptoms are common in patients, with for instance intellectual disability, anxiety and sleep disruption seen in up to 75% of the patients, while about half of all patients with ASD suffer from seizures, constipation or hypotonia.

Similar to ADHD, ASD often persists into adulthood with most people retaining the childhood diagnosis. Indeed only about 10% of patients with childhood ASD reach adjustment and functioning level similar to the normal population in adulthood (Taylor and Seltzer, 2010).

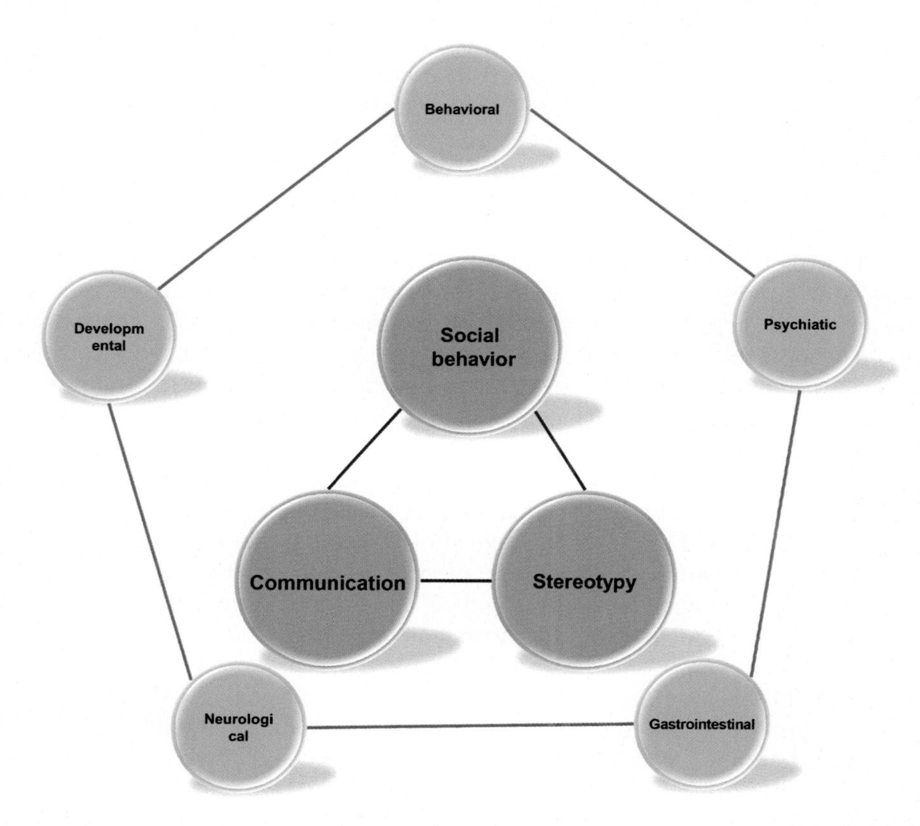

FIGURE 8.11 **The symptoms of ASD.** ASD is characterized by symptoms in three core domains: social behaviour, social communication and stereotypy. However, in addition, a plethora of other symptoms can be (and often are) present in patients, encompassing behavioural, psychiatric, gastro-intestinal, neurological and/or developmental symptoms.

Nonetheless, symptoms seem to change as children develop. For instance, both expressive and receptive language tend to improve with age, as does eye contact and reduced responsiveness. As a result, adults with ASD seem to suffer more from poor intonation and limited topics of interest rather than deficits in actually understanding what is spoken. In contrast to the improvements in the social domain, the severity of repetitive behaviors and restricted interests seem to be more stable over the course of the illness. Comorbidity also seems to change with age, with adults with ASD having a high comorbidity (up to 1/3 of all patients) with schizophrenia spectrum disorders (Vannucchi et al., 2014) as well as with ADHD. In a study investigating childhood predictor for adult outcome (Billstedt et al., 2007), it was found that childhood IQ and speech before year 5 were the best predictors for all three core aspects of ASD. In addition, epilepsy before the age of 5 years and female gender were negatively correlated with quality of social interaction and reciprocal communication.

The Epidemiology of ASD

One of the most interesting aspects of ASD is its apparent increase in prevalence. Studies done in several different countries consistently show that ASD is more commonly diagnosed now than in the past. In the USA the center for disease control and prevention has studied the prevalence of ASD over the years and confirmed this clear increase (see Fig. 8.12; http://www.cdc.gov/mmwr/pdf/ss/ss6302.pdf), showing a more than 4-fold increase between 2000 and 2010. Similarly high rates are found in other countries, with a total population study in South Korea reporting a prevalence rate of 2.6%, suggesting that a large number of patients may go unnoticed. Indeed, two thirds of all ASD cases were from mainstream schools, undiagnosed and untreated (Kim et al., 2011). Within the scientific, but also in the public health sector there is an intense discussion as to the reasons behind this increase and whether it represents an actual increase in the number of patients with ASD, an increased awareness of parents/teachers/clinicians or a change in the diagnostic criteria (or a combination of these factors). At present none of these factors can be ruled out, although the aforementioned study in South Korea suggests that even in recent years large numbers of patients are still undiagnosed.

ASD typically develops at an early age, with the majority of cases being detected before the age of 3, although deficits in social responsiveness, communication and play can be detected before the age of 1 and studies have shown that parents raise concerns about children around 14 to 19 months. Nonetheless, more than one third of all cases of ASD were not detected at 2 years of age (Tebruegge et al., 2004). As a result of this, the requirement in earlier versions of the DSM that the diagnosis has to be made before the age of 3 has been removed in DSM-V.

Similar to ADHD, ASD is more prominent in males than in females, with traditional figures suggesting it occurs 4 times more often in males (Gillberg et al., 2006) with the sex ratio being larger for high functioning ASD. However, in contrast to ADHD, there is some evidence that the difference in ASD is more quantitative than qualitative. In other words, it may be more severe in males than in females, and thus may be underdiagnosed in females. For instance the earlier mentioned prevalence study in South Korea (Kim et al., 2011) showed that within the high probability group (ie, those with a more severe phenotype) the male-to-female ratio was 5 : 1, while in the general population (including many undiagnosed children) the ratio

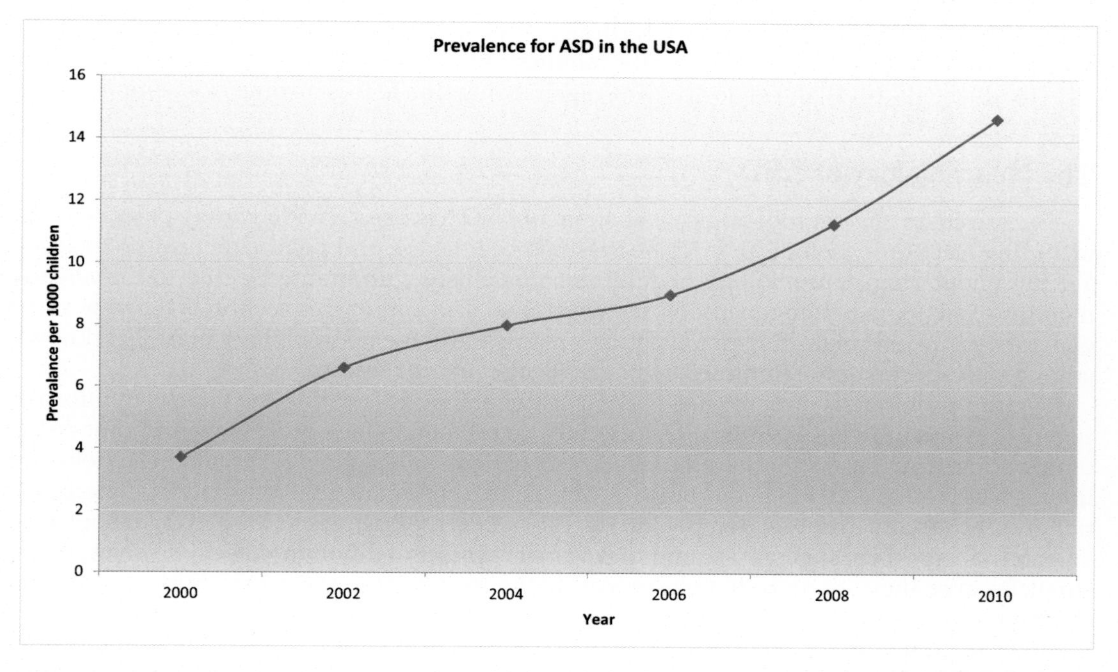

FIGURE 8.12 **The prevalence of ASD.** Data from several different countries have clearly indicated that the prevalence of ASD is rising. These data from the Centre for Disease Control in the USA confirm that in the last decade, the prevalence has quadrupled.

was only 2.5 : 1. In spite of this, there seems to be general agreement that ASD is more prevalent in males than females.

Given the early age of onset, and the relative persistence into adulthood, especially for those individuals with a low IQ, it is clear that ASD puts a significant burden on families as well as on society as a whole. Families of individuals with ASD usually work (and earn) less than families with healthy children. Indeed, a study in Sweden concluded that parents spend up to 1000 hours annually to care for their child with ASD. As a result, this often leads to emotional and financial problems for the families, reduces coherence and induces significant stress. In a recent study, 29 main caregivers of children with ASD were interviewed and virtually all the parents felt the burden was too much for them, as was the guilt of failing as a caregiver. As one caregiver put it "*If this were a partner relationship I would have quit ages ago*" (van Tongerloo et al., 2015).

In addition to this significant emotional burden, ASD places a heavy financial toll on families and society. Supporting an individual with ASD with an intellectual disability was estimated to costs US$2.4 million in the USA and US$2.2 in the UK, while the costs were somewhat lower for individuals with ASD without an intellectual disability (US$1.4 resp. US$0.92). The major costs came from special education services and parental productivity loss for children with ASD and residential care, supported living and individual loss of productivity for adults with ASD. This results in an aggregated societal costs of US$175 billion in the USA and US$43 billion in the UK (Buescher et al., 2014). A recent paper from Australia confirmed the high economic burden for families, with estimates of about AU$35,000 annually (Horlin et al., 2014), the majority (90%) again being due to loss income from employment.

The Neurobiology of ASD

The search for the neurobiological substrate of ASD has used a wide variety of techniques, including electrophysiological, magnetic resonance imaging and neurochemical procedures. Yet the underlying neurobiology of ASD remains elusive, undoubtedly due to the already mentioned extreme heterogeneity of the syndrome. The aim of this section is to draw some general conclusions regarding our current thinking about the neurobiology of ASD, acknowledging that several of the findings are not replicated in (all) studies.

Electrophysiological research has been instrumental in delineating the deficits in information processing in patients with ASD, especially in relation to face-processing and processing of novel information. As discussed earlier, ASD patients often show restricted behaviors and interests as well as actively avoid novelty. This is, to a certain degree reflected in altered EEG patterns, especially event related potentials (ERP). ERP are changes in the EEG in response to an external stimulus, such as a sound or a visual stimulus. Although these EEG changes are small, because they are reproducible, repeating the stimulus for a number of timed trials and overlaying the individual EEGs filters out the noise leading to a characteristic ERP.

One component that has been extensively studied in ASD is the so-called mismatch negativity (MMN). This ERP is the brain's response to a (rare) change in a series of repetitive auditory stimuli and is thought to be an unconscious component allowing the automatic detection of novelty. Several studies have found a decrease in MMN to vowels and a shorter latency to pitch change in patients with ASD, suggesting a deficit in (automatic) processing of novel information. This appears to be accompanied by a disturbance in attention to novel stimuli. This is typically measured using the so-called P_{300} (a positive EEG wave occurring about 300 ms after stimulus onset). P_{300} responses in ASD children are typically smaller in amplitude and longer in latency compared to normally developing children.

ERPs can also be detected after visual stimuli, and not surprisingly, much research in ASD has focused on processing faces, given the importance of face processing for social communication and interaction. Face processing is typically accompanied by several components, most prominently P_{100}, N_{170} and P_{400}, with the latter being associated with the cognitive processing of faces. Typically developing individuals show a strong N_{170} specific for faces (as opposed to objects) and a N_{290} specific for familiar faces (Rossion, 2014). Especially the N_{170} has received much attention and patients with ASD generally have longer N_{170} latencies to faces (but normal responses to objects). A peculiar effect of the N_{170} is a reduction in amplitude when the face is inverted, a phenomenon which seems to be absent in patients with ASD. Moreover several studies done in (genetically) high-risk children showed that even prior to the development of ASD, face processing ERPs were altered (Belger et al., 2011).

Electrophysiological techniques enable an almost real-time analysis of neural processes, but give only limited spatial information. Thus, in order to identify dysfunctional brain regions in ASD, imaging and post-mortem studies are necessary. These studies have revealed several key brain regions that show abnormal activity. As illustrated in Fig. 8.13, these brain regions can be (roughly) divided into two important circuits, one more related

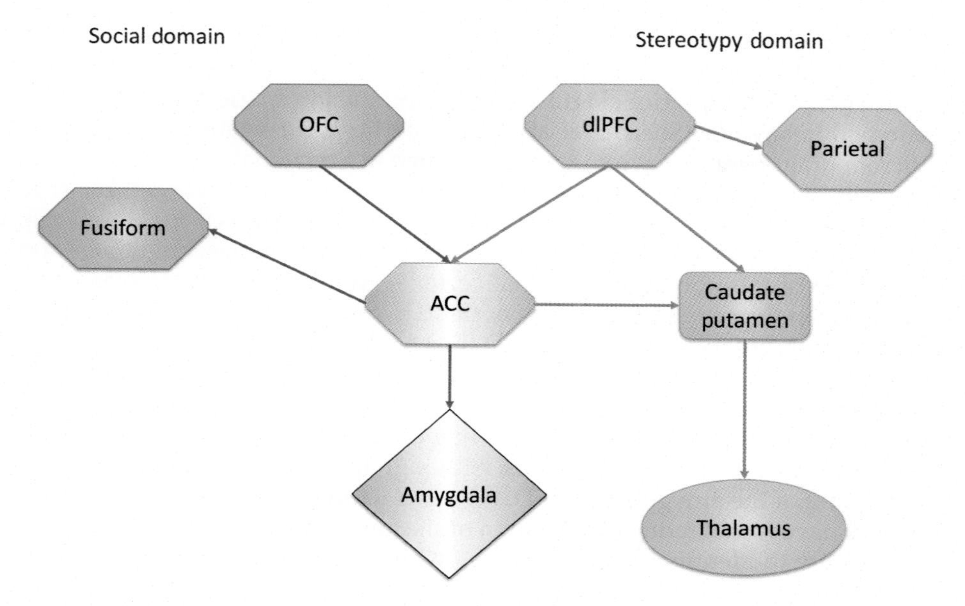

FIGURE 8.13 **The neurobiology of ASD.** Based on a large number of studies, there is ample evidence for the involvement of both cortical and subcortical structures in the neuropathology of ASD. Although these structures and networks can never be assigned a single function, there is evidence that the orbitofrontal cortex (OFC) and fusiform cortex play a more important role in the deficits seen in the social domain. The dorsolateral prefrontal cortex (dlPFC), parietal cortex, caudate putamen and thalamus are thought to be more involved in the stereotyped symptoms (including cognitive rigidity). Structures such as the amygdala and the anterior cingulate cortex (ACC) appear crucial as intermediate structures, linking both networks together.

to the social (interaction and communication) domain and one more related to the inflexibility and stereotypy domain. In addition, several regions, such as the amygdala and the anterior cingulate cortex can be seen as forming an interface between these two circuits. Interestingly, there is significant overlap between these brain regions and the ones putatively involved in ADHD (Fig. 8.5) which may underlie the large comorbidity between these two childhood disorders.

Within the social domain, studies have consistently shown that there is a significant reduction in the activity of the fusiform cortex, an area particularly involved in face processing. Activity in this area is related to gaze fixation, and a reduced activity in this area is therefore in line with a reduced face fixation seen in patients with ASD. In addition, both reduced activation and morphological changes have been reported, which may contribute to deficits in social cognition.

The stereotyped behaviour seen in patients with ASD has traditionally been linked to the circuitry involved in (motor) flexibility and executive functioning, involving the dorsolateral prefrontal and parietal cortex and the caudate-putamen. Studies in patients with frontal lobe dysfunction, schizophrenia and Parkinson's disease confirm the important role these structure play in motor as well as cognitive flexibility (Cools et al., 1984; Goldberg et al., 1990). Studies in patients with ASD confirm both a deficit in executive functioning (Van Eylen et al., 2015) as well as disturbances in the structures delineated in Fig. 8.13, especially reduced activity in the dorsolateral prefrontal, parietal and anterior cingulate cortex as well as the caudate-putamen and thalamus. Interestingly, one study found a significant negative correlation between activation of the anterior cingulate cortex and the intraparietal sulcus and restricted and stereotyped behaviour (Shafritz et al., 2008). Studies investigating the relationship between different brain regions have suggested a deficit in functional connectivity (see Box 9.2 for more details on functional and structural connectivity). For instance, in one study, both healthy volunteers and patients with ASD were tested in the "Tower of Hanoi" test, which assesses executive functioning. As expected patients performed worse, but more importantly, when the activity of the frontal and parietal cortex were compared there was a significant correlation between the activation patterns of these two regions in healthy volunteers, while this was virtually absent in patients with ASD (Just et al., 2007), a phenomenon referred to as underconnectivity which has been replicated in other experiments as well (Just et al., 2012).

Finally, several studies have investigated the role of the amygdala in ASD. The amygdala plays a key role in emotional processing, including identifying social emotions, one of the key areas where patients with ASD have deficits. Although there seems general agreement of aberrant functioning of the amygdala in ASD, the exact nature is less clear (Belger et al., 2011). Several studies have found reduced activation of the amygdala in patients (thought to reflect the reduced ability to identify social emotions), while some others have actually found enhanced activation (interpreted as heightened emotional response elicited by gaze fixation or social anxiety). In line with this, alterations in the structure of the amygdala, such as increased volume and post-mortem cytoarchitectural deficits have been reported (Zalla and Sperduti, 2013).

Studies investigating changes in neurotransmitter activity have focused predominantly on serotonin and oxytocin, though abnormalities in dopamine and glutamate have also been reported. One of the most reproduced neurochemical findings in ASD is an increase in extracellular serotonin concentrations in the plasma in patients with ASD (Lam et al., 2006). Given serotonin's role in development (Chapter 5), the fact that there is mounting evidence that prenatal exposure to selective serotonin reuptake inhibitors seems to increase the risk of ASD in the offspring (Gentile, 2015; Man et al., 2015) and the pivotal role serotonin plays in flexibility and social behaviour, makes this neurotransmitter a prime candidate for ASD. Unfortunately studies investigating alterations in serotonin transmission in the brain have been less consistent (Chugani, 2012; Zurcher et al., 2015). The majority of studies (especially those with strict exclusion criteria for serotonergic drugs) have found decreases in serotonin transporters and 5-HT2A receptors in various brain regions including several of those mentioned in Fig. 8.13 (anterior cingulate, dorsolateral prefrontal and parietal cortex). These data are consistent with an increase in extracellular 5-HT levels, and it has been shown that compared to typically developing individuals, children with ASD do not show a decrease in serotonin synthesis between the age of 5 and 14. Overall, there seems to be relatively convincing evidence for a role of serotonin in ASD. As we will discuss later, this is further supported by genetic studies.

Oxytocin is a hormone as well as a neuropeptide. In addition to its well-known role in the final stages of childbirth (stimulating uterine contractions) and stimulation of milk production, it has become increasingly clear that oxytocin plays an important role in social behaviour. Originally this was thought to be primarily in relation to maternal behaviour, but studies, especially in voles, have since underlined a more general role for oxytocin in social bonding (Insel, 2010) to the extent that, in the popular media oxytocin has now been referred to as the "love" neurotransmitter. Although few studies have evaluated changes in the oxytocin system, oxytocin (nasal) spray or infusion has been found to improve some of the core symptoms of ASD (Althaus et al., 2015). Moreover, it increases amygdala activity during face processing (Domes et al., 2013). Altered (both increased and decreased) peripheral (plasma) levels of oxytocin have been reported in patients with ASD, but a lack of central markers such as PET ligands have so far prohibited the evaluation of central oxytocin (receptor) levels in patients.

Finally, as mentioned earlier, both dopamine and glutamate have been implicated in ASD, but evidence for or against the involvement of these neurotransmitters is relatively scant (Zurcher et al., 2015). The extensive comorbidity between ASD and ADHD on the one hand, and ASD and schizophrenia on the other, have led people to investigate the role of dopamine as this neurotransmitter is crucially involved on both of these disorders (as discussed earlier and in Chapter 9). In agreement with this, there is genetic evidence linking the dopaminergic system to ASD (as discussed later; Nguyen et al., 2014), but neuroimaging studies are less convincing (Zurcher et al., 2015).

Glutamate has been implicated in the neurobiology of fragile X syndrome (as discussed later). Since up to 60% of fragile X patients show ASD like symptoms, it has been suggested that alterations in the glutamatergic system may also be observed in other ASD cases. Moreover, altered glutamate neurotransmission has been linked to epilepsy, which is observed in up to 50% of patients with ASD and ADHD (as discussed earlier). A recent review identified 16 studies using *in vivo* magnetic resonance spectroscopy (MRS) techniques to investigate the glutamatergic system in patients with ASD, most of which focused on the prefrontal cortex (Naaijen et al., 2015). Although some studies found increases in the glutamate content in the anterior cingulate cortex in children, the majority of studies seem to be unable to detect this. There was some evidence that in adulthood the reverse situation occurred. However, given the fact that most studies were done in relatively small groups of patients and that total glutamate was measured (the majority of which would be of glial origin) the functional significance of these findings has yet to be determined.

The Pharmacological Treatment of ASD

In spite of the tremendous emotional and economic burden, current treatment and especially pharmacological treatment for ASD is very limited. In fact, there are no drugs currently available that can successfully treat all the core symptoms of ASD. Of course, there are possibilities to treat many of the additional symptoms, such as atomoxetine for the treatment of ADHD-like symptoms, antidepressants for the treatment of depression and/or anti-anxiety and anticonvulsant drugs for the treatment of seizures.

In addition, the Food and Drug Administration in the USA has approved two drugs (aripiprazole and risperidone) specifically for the treatment of irritability in patients with ASD. These drugs are so-called second generation antipsychotics (see Chapter 9) and both have been well studied and found to be effective in patients with ASD. Risperidone was approved in 2006 but only for children above 5 years of age. Several double-blind placebo-controlled trials were performed and overall, risperidone was effective in reducing irritability, aggression and self-injurious behaviour at doses of 0.5–3.5 mg/day (which is much lower than typically used for the treatment of schizophrenia). In a long term (32 week) study, risperidone also reduced stereotypy and even improved social withdrawal and inappropriate speech

(Troost et al., 2005). A subsequent meta-analysis of open and double-blind trials concluded that risperidone was indeed effective with a reduction in problematic behaviour of about 1 standard deviation (Sharma and Shaw, 2012), with the most significant side effect being a rapid increase in body weight.

Similar to risperidone, there is ample clinical evidence that aripiprazole (approved by the FDA in 2009 for children between 6 and 17 years of age) is effective in the treatment of irritability and aggression in children with ASD. The effects of aripiprazole were evaluated in a 52 weeks double-blind placebo-controlled study in over 300 children. All patients started on a dose of 2 mg/day which was slowly increased to maximally 15 mg/day (Marcus et al., 2011). Aripiprazole was superior to placebo in reducing irritability and improving quality of life. The major side effects were extrapyramidal side effects, sedation, somnolence and weight gain. The effects of aripiprazole were confirmed in a meta-analysis (Ching and Pringsheim, 2012) which showed a mean improvement of 6.2 points in the ABC irritability subscale, 7.9 points on the ABC hyperactivity subscale and 2.7 points on the stereotypy subscale. Main side effects were weight gain, sedation and tremor.

Although several other treatments (including additional antipsychotics such as haloperidol) have been used off-label, so far no treatment improves the core symptoms of ASD, especially in relation to social behaviour and communication. Some recent studies have focused on the use of oxytocin in the treatment of ASD, but so far, these studies are hampered by the relatively limited brain penetrance of oxytocin and there is a need for an (ideally nonpeptidergic) oxytocin agonist that can cross the blood brain barrier and therefore reaches adequate brain exposure.

The Aetiology of ASD

Given the young age of onset, the focus of the aetiology of ASD has long been on genetic factors. Indeed, twin studies have suggested that ASD might have the highest heritability compared to other psychiatric disorders. Fig. 4.6 shows that the concordance rate in monozygotic twins may be as high as 90%, although more recent twin studies have suggested that the concordance rate is considerably lower. In a review of twin studies, it was suggested that the concordance rate depends strongly on the diagnosis with more narrow definitions leading to a higher concordance rate (Ronald and Hoekstra, 2011). Current estimates have suggested that the heritability of ASD is about 50% (Gauglerl et al., 2014; Sandin et al., 2014). The difference between the earlier and the later studies is mostly due to the fact that in more modern models nonadditive and *de novo* mutations are taken into account (see box 4.1. for a description of how to measure heritability). As illustrated in a recent large Swedish study (PAGES: Population-based Autism Genetics and Environment Study), incorporating these components into the model, suggests that the additive genetic effects (ie, h^2) account for about 52% of the total heritability, consisting of 49% common variations (defined as variants with a minor allele frequency of $> 1\%$) and 3% rare variations. In addition, 4% appear to be due to nonadditive and 3% to *de novo* variations (Gauglerl et al., 2014), leaving 41% for environmental effects.

Genetic Factors Contributing to ASD

Although there is ample evidence that ASD is polygenic, like all other psychiatric disorders there are some monogenetic disorders that show strong ASD-like features, most notably fragile X syndrome and tuberous sclerosis. Fragile X syndrome is caused by an unstable trinucleotide (CGG) repeat in the *FMR1* gene located on chromosome Xq27.3. If the repeat exceeds 200, it causes abnormal DNA methylation leading to transcriptional silencing and subsequently reduced FMRP protein expression in the brain. Fragile X leads to ASD in about 60% of the cases (especially true autism and PDD-NOS) (Hagerman et al., 2010). The autistic features of fragile X syndrome encompass deficits in peer interactions, while socio-emotional reciprocity is much less affected. Tuberous sclerosis is an autosomal dominant disease with high penetrance, resulting from an inactivating mutation in either *TSC1* (located on 9q34) or *TSC2* (located on 16p13.3). As with fragile X, autism is much more common in tuberous sclerosis than in the normal population (about 30%). However, although autism is much more common in these two genetic disorders, there is little evidence that alteration in these genes are related to other cases of autism.

Many candidate gene studies have focused on genes involved in neurodevelopment, especially those involved in synaptic plasticity, and genes related to neurotransmission, especially serotonin, oxytocin and dopamine (Table 8.5).

TABLE 8.5 Major Single Nucleotide Variants and Copy Number Variations Implicated in ASD

Genes/regions	Remarks
SINGLE NUCLEOTIDE VARIANTS	
DRD3	Dopamine D_3 receptor: the "A" allele of rs167771 has been associated to ASD, especially to repetitive behaviour and stereotypy.
EN2	Engrailed homeobox 2: protein involved in early development. The "C" allele of rs1861973 and "G" allele of rs1861972 have been associated with ASD
HOXA1	Homeobox A1: protein involved in early development. A frameshift mutation (175-176insG) has been reported in one family
MTHFR	Methylenetetrahydrofolate reductase: Protein involved in DNA synthesis. The "T" allele of rs1801133 has been associated with ASD
NRXN	Neurexins: Group of 3 proteins involved in synapse formation. Rare mutations in several members (esp. *NRXN1*) have been found in ASD.
OXTR	Oxytocin receptor: especially the "A" allele or rs7632287 and the "T" allele of rs2268491 have been associated with ASD
SHANK	SH3 and multiple ankyrin repeat domains: Group of 3 scaffolding proteins. Rare mutations in several members (esp. *SHANK3*) have been found in ASD.
SLC6A4	Serotonin transporter: Especially the 9/10 repeat of the VNTR in intron 2 has been associated with ASD
SLC25A12	Mitochondrial aspartate/glutamate carrier: Transporter involved in ATP synthesis
COPY NUMBER VARIATIONS	
1q21	Both deletion and duplication have been associated with ASD
2q15–2p16.1	Deletion has been associated with ASD
15q13	Both deletion and duplication have been associated with ASD
16p11.2	Both deletion and duplication have been associated with ASD

See text for further explanations.

Reelin is an extracellular matrix glycoprotein that plays an important role in neurodevelopment, especially in guiding the migration of specific subsets of neural cells, particularly in the cerebral cortex and the cerebellum. As lower reelin levels have been found in the human cortex and cerebellum in patients with ASD, the reelin gene (*RELN*) is an obvious candidate for ASD. Indeed several studies have found alterations in *RELN* associated with ASD, including a polymorphic GGC triplet repeat upstream of the initiator codon, as well as a C → G variant in exon 22 (rs362691) and T → C variant intron 59 (rs736707). In a recent meta-analysis on reelin's involvement in ASD, only the rs362691 variant was found to be significant (Wang et al., 2014), especially in ASD, which was confirmed in another more general meta-analysis on common genetic variants (Warrier et al., 2015).

Another group of proteins involved in neurodevelopment are the neuroligins, consisting of five different genes (*NLG1*, *NLG2*, *NLG3*, *NLG4X* and *NLG4Y*). Neuroligins are cell-adhesion molecules that play an important role in synapse formation. Interestingly, two of the neuroligins are located on the sex chromosome (*NLG4X* on the X- and *NLG4Y* on the Y-chromosome) which has increased interests in this gene, given the clear sex ratio of ASD. Several different mutations, some of which are *de novo* mutations, have been identified in cases of ASD, although the frequency of mutations is low (Persico and Napolioni, 2013). Related to the neuroligins, are three members of the SHANK family (SH3 and multiple ankyrin repeat domains, *SHANK1*, *SHANK2*, and *SHANK3*). They encode postsynaptic scaffolding proteins to which neuroligins bind. Especially *SHANK3* has received considerable interest since it is highly expressed in the cortex and SHANK3-deficient mice show ASD-like features (as discussed later). However, as with the neuroligins, although mutations in especially *SHANK2* and *SHANK3* have been linked to ASD, they are rare (incidences between 0.1 and 0.8%; Persico and Napolioni, 2013). Two other genes that have an established role in (especially early) neurodevelopment are *HOXA1* (homeobox A1) and *EN2* (engrailed homeobox 2) both of which have been implicated in ASD. In a recent meta-analysis two variants of *EN2*(rs1861973 T→ C and rs1861972 A → G) were associated with ASD (Warrier et al., 2015).

Given serotonin's prominent role in the neurobiology of ASD, candidate gene analyses have investigated several different components of the serotonergic system, most prominently the serotonin transporter, but also tryptophan hydroxylase (the enzyme involved in the synthesis of serotonin) and several of the serotonin receptors. Studies focusing on the serotonin transporter have focused on both the insertion/deletion of the promoter region (5-HTTLPR) and the variable number of tandem repeats in the second intron (STin2). Overall, the results of the 5-HTTLPR are inconclusive, with some studies reporting an association between ASD and the s-allele while others find an association with the l-allele and some finding no difference (Huang and Santangelo, 2008); although a recent paper found an association of the s-allele with Asperger syndrome (Nyffeler et al., 2014). This lack of consistency may well be due to the failure to consider environmental factors because the phenotypical expression of the 5-HTTLPR is appreciably modified by environmental factors. Results of investigations of the STin2 variant of the serotonin transporter have been more consistent, as evidenced by an association of the 9/10 repeats with ASD as confirmed in a meta-analysis (Warrier et al., 2015) with an effect size of 1.2. Some studies have also investigated the 5-HT_{2A} receptor gene (*HTR2A*) in relation to ASD, in particular the A → C variant of rs6311, but this association did not hold up in the aforementioned meta-analysis (Warrier et al., 2015).

The second neurotransmitter that has been implicated in the neurobiology of ASD is oxytocin especially given its role in social cognition. At least 11 independent studies have looked at variants within the oxytocin receptor gene (*OXTR*) studying in total 16 different variants. In a recent meta-analysis significant associations were found for 4 variants, most significantly the "A" allele of rs7632287 and the "T" allele of rs2268491 (LoParo and Waldman, 2015), with this latter variant also being identified in another meta-analysis (Warrier et al., 2015).

Genes associated with the dopaminergic system have also been investigated in patients with ASD (Nguyen et al., 2014) although with the exception of the dopamine D_3 receptor gene (*DRD3*) the associations have not been convincing. There have been several positive association studies between ASD and the rs167771 variant of the *DRD3 gene* with the "A" allele being the risk allele, with sample populations in the Netherlands, Spain and the UK (Staal, 2015). In line with this, in the already mentioned meta-analysis (Warrier et al., 2015), this variant showed the largest effect size (1.8). It has been suggested that this variant is especially related to stereotyped, repetitive behaviour in ASD (Staal et al., 2012).

In addition to candidate gene studies, the aetiology of ASD has also been investigated with genome wide association studies (GWAS) as well as whole exome sequencing (WES) studies. Given that the odds ratio of common variants is likely to be around 1.2 (Devlin et al., 2011), such studies would require large numbers of individuals. So far most GWAS appear to have been underpowered and so unable to reliably identify candidate loci (De Rubeis and Buxbaum, 2015). For instance, a recent study from Japan included 500 ASD and 624 volunteers and identified several loci, but none passed the usual threshold for GWAS significance ($p < 5.10^{-8}$) with the highest significance in the order of 10^{-6} (Liu et al., 2015). In 2012, four studies used a novel approach by performing large scale exome sequencing on about 1000 trios. These studies identified nine genes with a high linkage to ASD (De Rubeis and Buxbaum, 2015), although none of the candidate genes mentioned earlier were included.

Finally, several studies have investigated larger variations in the genome, such as copy number variations (CNVs, see Chapter 2) with several earlier studies indicating that CNVs are more common in ASD (6%–10%) than in the control population (1%–3%). As several *de novo* CNVs were found, it was suggested that ASD might be linked to genomic instability. However, later studies could not confirm this increased CNV frequency, although several CNVs (Table 8.5) have been found in patients with ASD.

Moderation of the Effects of Genes by Environmental Factors

In contrast to the extensive research done on identifying the genes involved in ASD, relatively few studies have investigated the moderating effects of environmental changes (Kim and Leventhal, 2015). This may, in part, be due to the (mis)conception that ASD is largely genetic in origin (see the discussion earlier). On the other hand, it has long been known that environmental factors such as prenatal infections, exposure to valproate or stress (and more recently to selective serotonin reuptake inhibitors) can appreciably increase the risk for ASD. Thus, the lack of gene–environmental interaction studies is somewhat surprising and represents a clear gap in our current understanding of ASD. In one study the influence of prenatal vitamin supplementation was investigated in 429 children with ASD compared to 279 controls (Schmidt et al., 2011). The authors found the risk of ASD increased when the mothers were carriers of the "T" allele of *MTHFR* rs1801133 or the "T" allele of rs234715 and did not take vitamins. A similar interaction was found with the child's genotype (the "A" allele of *MAOA* rs4680). However, these interactions were not significant after correction for multiple comparisons.

In another study, it was found that individuals homozygous for the "G" allele of rs1042714 of the *ADRB2* gene (which codes for a glutamate at position 27 of the β_2 adrenoceptor) have an increased risk for ASD, especially those individuals who were also exposed to prenatal stress (Cheslack-Postava et al., 2007).

Although not directly focused on patients with ASD, Nijmeijer and coworkers studied the role of prenatal smoking and low birth weight (considered to be predominantly determined by environmental factors) on ASD-related symptoms in children with ADHD (Nijmeijer et al., 2010). Using two different cohorts, the authors found interactions between the 5-HTTLPR and maternal smoking on social symptoms. Thus children with the s-allele exposed to maternal smoking showed the most symptoms. In addition, the 5-HTTLPR moderated the effects of low birth weight on several other subscales of the Children's Social Behavior Questionaire, again with the carriers of the s-allele being most influenced by low birth weight. Intriguingly, in none of the comparisons was there a significant main effect of 5-HTTLPR genotype, emphasizing the importance of simultaneously studying genetic and environmental factors. Finally, the authors found an interaction between prenatal smoking and the COMT gene, more specifically carriers of the "A" allele of rs4680 who were prenatally exposed to smoking showing more stereotyped behaviour.

Recently an interaction was also described between the presence of CNVs and maternal infection (Mazina et al., 2015), such that individuals with CNVs that were exposed to maternal infection had increased social communicative skills and repetitive/restricted behaviour. Interestingly, no such interaction was found in relation to cognition or adaptive functioning.

Animal Models for ASD

There are probably more genetic animal models for ASD than for any other disorder discussed in this book. Indeed, a recent review of the SFARI (Simons Foundation Autism Research Initiative) database deduced that more than 70 different genetic mouse models have been developed for ASD (Banerjee-Basu and Packer, 2010).

Several of these are based on genes that have been identified in candidate gene studies in patients, whereas others have been identified more serendipitously. Most of these models have focused on measuring the social aspects of ASD, especially as social behaviour can be fairly easily assessed in animals such as rats and mice, although species differences in social behaviour are undoubtedly present. Nonetheless, it is important to realize that social behaviour (and communication) is highly complex and therefore many different paradigms have been developed. In the following section we will discuss some that are used most frequently, as well as some other, potentially relevant techniques.

Measuring ASD-Like Features in Animals

One of the most obvious models is the social interaction paradigm in which multiple (typically 2) animals are placed together in an open field and social behaviour is assessed. This can be done either early in life or in adulthood. Although social behaviour is seen in both periods, play behaviour is typically confined to the period before adolescence (between postnatal days 20 to 35). Play behaviour has a characteristic, almost ritualistic structure, with each bout typically starting with one animal following (chasing) the other, followed by rough-and-tumble play culminating in pinning behaviour, with one animal lying on its back and the other holding down the first animal (Vanderschuren et al., 1995; Vanderschuren et al., 1997). Social behaviour also has multiple components, and includes both active and passive elements. A special form of social interaction is aggressive behaviour, which is often seen in ASD patients and should thus be considered as well. Whereas social interaction is technically an easy experiment to perform, there are several important considerations regarding the interpretability of the results. Most importantly, as the behaviour is interactive, the results are the combination of both animals. In other words, if one animal shows abnormal behaviour this will have consequences for the other animal as well. For that reason, social interaction is usually performed in pairs with the same treatment or genotype and gender. Moreover, it is usual not to use littermates. In addition, the exact conditions are important too, such as lighting intensity, size of the open field, whether the experiments are performed in the light or dark cycle, etc. Finally, it is customary to socially isolate rats for several hours before measuring social behaviour. This ensured that (normal) animals show adequate levels of interaction.

In order to overcome the issue that both animals contribute to the interaction, the social approach-avoidance paradigm has been developed (Silverman et al., 2010). This test is typically performed in a three-chamber apparatus, although other shapes have also been used. After a habituation period, a wire cylinder is placed in each of the side chambers (Fig. 8.14). Under one of these cylinders, a conspecific animal is placed while the other cylinder remains empty. The experimental animal is then allowed to explore the cage. The relative time spent in the chamber with the other animals (or sometimes in a more narrowly defined zone around the cylinder) is used as an indication of the sociability of the experiment animal. Often, this is followed by a third phase in which a novel animal is placed under the originally empty cylinder. Most normal animals will, in the second phase, spent more time in the area with the animal than the empty cylinder, and in the third phase, in the area with the novel rather than the familiar animal. In addition to measuring the time spent in each chamber, the time sniffing the wire cylinders is often measured as well, which may represent a more accurate measure of sociability.

A paradigm which has not been used very often, but may have special relevance for ASD is the empathy paradigm (Ben-Ami Bartal et al., 2011; Sato et al., 2015). Although different

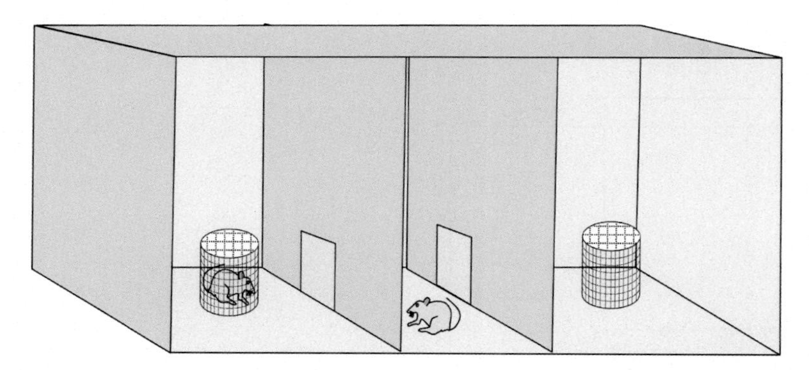

FIGURE 8.14 **The social approach avoidance paradigm.** One of the most often used paradigms to assess sociability is the social approach avoidance paradigm. In this test rodents are typically placed in a three chamber box in which the two external chambers contain a wire meshed cylinder. After a brief habituation phase, an animal is placed underneath one cylinder while the other one remains empty. The experimental animal is placed in the central chamber and the time spent in each compartment (and often the time sniffing each cylinder) is recorded. In a third phase, usually following immediately after the second phase, a novel animal is placed in the empty wire cylinder. Thus this test enables measurement of social approach (in phase 2) and preference for social novelty (in phase 3).

paradigms exist, the basic premise is that an animal willingly helps another (distressed) animal, despite this not leading to an obvious reward. This requires higher order social cognition of an animal, something which is significantly reduced in ASD and thus would add an important dimension to the face validity of any animal model. Another type of empathy test measures the response in animals which are made to observe distress in conspecifics (Langford et al., 2006). Interestingly, it was found that the empathy measure differs among different mice strains with BALB/cJ mice failing to exhibit increased freezing response after being pre-exposed to observing distress in other mice (Chen et al., 2009). BALB/cJ mice have likewise been shown to exhibit less sociability compared to the standard 6J mice in other tests as well (Brodkin, 2007; as discussed later).

In the area of social communication, researchers have predominantly focused on the auditory domain, although research in the olfactory domain is increasing as well. Most rodent communication occurs in the ultrasonic range, above normal human hearing and thus requires specific equipment and software to detect and analyze calls. Although differences exist between rats and mice, significant progress has been made in interpreting these so-called ultrasonic vocalizations (Burgdorf et al., 2011; Wohr and Schwarting, 2013). Broadly speaking, rats can vocalize in three different ranges (Fig. 8.15). Calls within the 20–25 kHz range are typically made in stressful, anxious environments, while calls in the 50–80 kHz

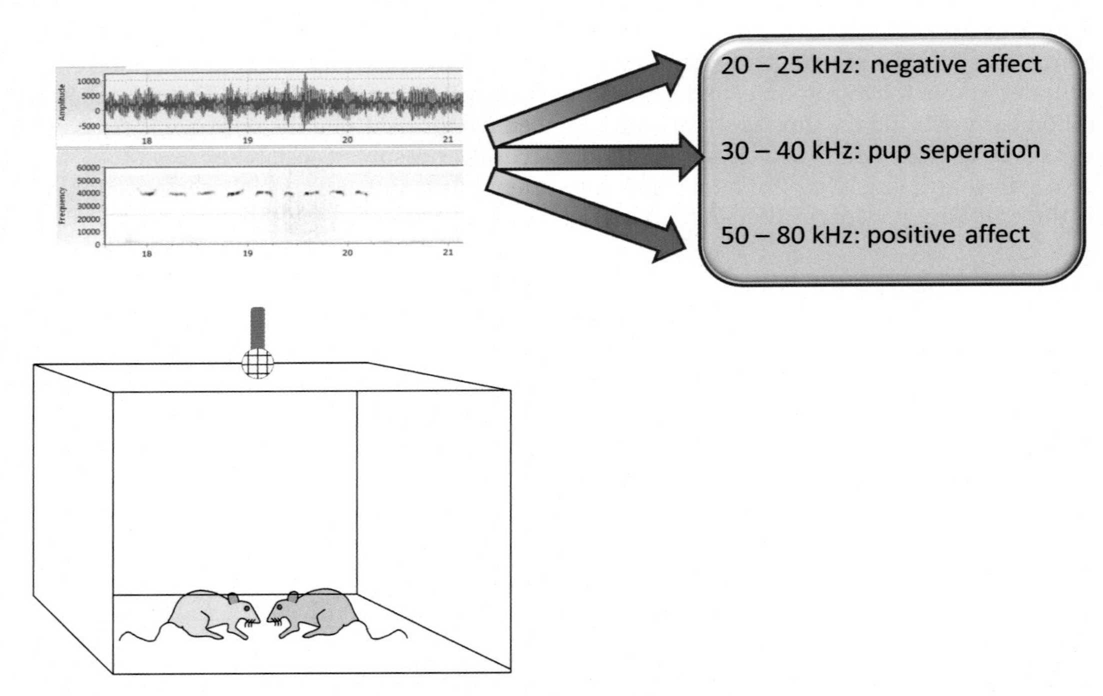

FIGURE 8.15 **Ultrasonic vocalizations (USVs).** Rodents communicate with each other predominantly in the ultrasonic range (above our human hearing, typically >20 kHz). Such USVs can be subdivided into three broad ranges: 20–25 kHz, generally associated with negative affect; > 50 kHz, generally associated with positive affect, and 30–40 kHz, typically produced by young pups separated from their mothers. In addition, within each band different types of calls can be distinguished, though the exact nature of each of these calls has yet to be determined.

are predominantly related to positive affect and can be detected during social behaviour, including sexual and play behaviour. Finally, young pups when they are separated from their mothers produce predominantly calls within the 30–40 kHz range. In addition to these broad categories, it is customary to subdivide calls within, for instance, the 50 kHz range into separate categories, based on the frequency modulation of individual calls, with flat calls encoding less affective and trills encoding more affective calls (Fig. 8.15).

Rodents, in contrast to humans, have an excellent sense of smell. Indeed the relative size of the olfactory bulbs is much larger in rats and mice compared to humans. As a result, rodents rely strongly on scent as a measure of communication. In relation to animal models for ASD this is most often assessed using the scent marking paradigm (Fig. 8.16). In this paradigm, after a habituation period, rodents (typically males) are exposed to one or two scents, a nonsocial scent (such as lemon) and a social scent. This latter one is typically urine from females

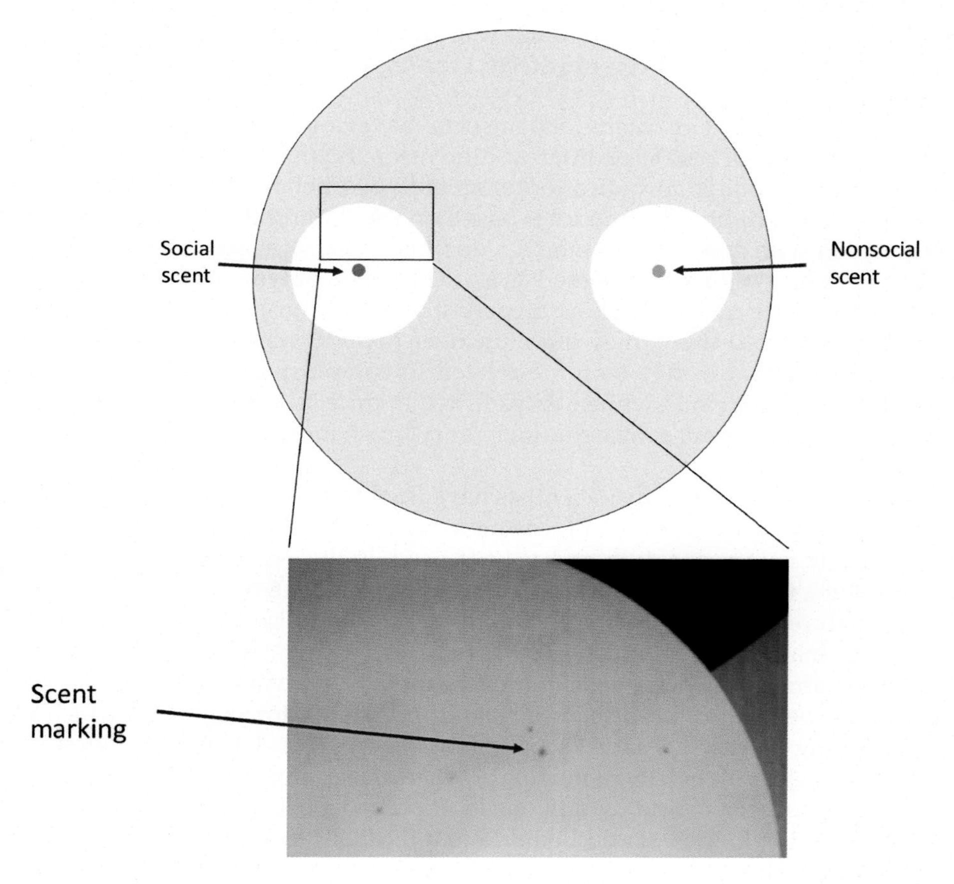

FIGURE 8.16 **Scent marking in rodents.** In addition to vocalizing, rodents also communicate through scents. In a typical scent marking experiment, rodents are placed in an open field in which one or two scents are available: a social (typically female urine) and a nonsocial. Animals are allowed to explore the environment and normal animals will produce specific scent markings especially around the social scent. The small markings can subsequently be visualized using staining with ninhydrin and counted.

in estrus. Normal animals will typically produce much more urine scent markings around the social scent (Wohr et al., 2011).

A special case of olfactory communication is the food preference transfer (Galef and Whiskin, 2003). This paradigm is based, in part, on the lack of a vomiting reflex in rodents. As a result, they are relatively careful in eating new food, observing other's behaviour first, before trying. To capture this behaviour in an experimental paradigm, one animal is food-restricted and, for several consecutive days exposed to normal animal feed that is supplemented with for instance chocolate or cinnamon. As the so-called demonstrator animal has no choice, it will consume this food. After a few days, at the end of the feeding session, the food is removed and the animal is housed for a short period with a so-called observer animal. Immediately after this, the observer animal will be given a choice between two flavored foods, one identical to the one the demonstrator animal has consumed and one with a novel flavor. Normal animals will consume significantly more from the food the demonstrator animal consumed. As the observer animal was never in contact with this food, the preference for this food must have been transferred through olfactory cues (ie, the demonstrator animal's smell).

In contrast to the study of alterations in social behaviour, much less attention has been paid to the stereotyped, restricted interest dimension. Some researchers have argued that excessive self-grooming is an indication of stereotyped behaviour. Others have used the marble-burying paradigm. This model is based on the finding that mice, when confronted with a box filled with sawdust, on which a number of marbles are placed, rapidly start to bury these marbles. Originally designed as a screen for anxiolytic drugs, it is now generally accepted that mice do not really show anxiety-like behaviour towards the marbles. In fact after they have covered them, they often uncover them to start burying them again. This has led to the idea that it may be more related to compulsive or stereotyped behaviour. Unfortunately this behaviour, while easily elicited in mice, is rare in rats, making the model less general, possibly because mice routinely bury much more in their natural environment than rats.

In addition to these paradigms, scientists have tried to study stereotyped motor patterns, and have developed procedures that focus more on rigidity in cognition. With respect to the former, especially hole-board dipping has been used. In the experiment an animal is placed in a box in which the base has a number of holes (typically 16 or more). As rodents are inquisitive by nature, they will not only explore the box, but also the holes. Normal animals will investigate many different holes, making only few re-visits, while animals with an ASD-like phenotype are expected to make more repeat visits. Patients with ASD not only show restricted motor behaviour but also an increased preference for sameness, which results in deficits in learning new things and executive functioning. This deficit is also increasingly assessed in ASD models, especially in models of reversal learning, where, for instance, animals are first required to learn to find food in one location in a maze and, after they have successfully learned this, the food is placed in a different location, requiring the animal to inhibit a previously learned strategy and re-learn a new one.

Given that ASD is accompanied by a large number of additional symptoms, as discussed earlier and illustrated in Fig. 8.11, many of the animal models have also been tested for the presence or absence of symptoms such as increased anxiety, decreased threshold for seizures etc. (Pasciuto et al., 2015).

Genetic Models of ASD

As mentioned earlier, a large number of genetic animal models have been proposed for ASD (in addition to a number of environmental models, such as the maternal immune challenge and the prenatal valproate model). Most of these models have been based on the principle of reverse genetics (Chapter 3) meaning that a single gene (or a group of genes) has been specific altered. In a recent paper, neurochemical similarities and differences were assessed in 26 different genetic mouse models for ASD (Ellegood et al., 2015b), most of which are based on a single gene alterations (such as the *SHANK3, SLC6A4, EN2 and NRXN1α* knock-outs) or on CNVs (such as 15q11-13 and 16p11.2). As discussed by the authors, virtually all of these individual genetic factors have odds ratio of higher than 1.2 in ASD association studies. This implies that, from a construct validity point of view, these reverse models have limited validity, in spite of the fact that they often have good face validity (Ellegood and Crawley, 2015a). Thus deficits in the social approach avoidance test have been described for, among others, the *SHANK3, PTEN, EN2,* and the *OXTR* knockout mice. These same transgenic mice also show deficits in social interaction and the *SHANK3* knock-out mice, at least, also show increased stereotyped behaviour.

Intriguingly, in the aforementioned comparative neuroanatomical studies of 26 models, substantial differences were seen between these models, in spite of their similar face validity (Ellegood et al., 2015b). While some models had significantly smaller brains, such as the *SHANK3* and the *SLC6A4* knock out mice, others had significantly larger brains, such as the *EN2* and *NRXN1α* knock out mice (all compared to their own controls). Similar widespread differences were seen when the sizes of individual brain regions were compared. Overall, this analysis emphasized the heterogeneity of ASD which is apparent also in the different animal models. Moreover, the data suggest that dysfunction, rather than hyper- or hypofunctioning of brain regions may be a key feature of ASD. This has also been suggested on the basis of other results. For instance ASD has been linked to both the s- and the l-allele of the 5-HTTLPR. Nonetheless, an analysis of the structure of 62 different brain regions in the 26 different models led to the identification of three separate circuitries, roughly suggested to be involved in [1] repetitive behaviour, executive functioning and communication (involving among others the cerebral cortex and striatum), [2] social perception and autonomic regulation (including the amygdala and hippocampus) and [3] a cerebellar cluster (including the cerebellar, peduncle, and the cerebellar cortex).

While evaluating a series of inbred mouse strains, several strains were identified that showed ASD-like deficits. The BTBR (or in full BTBR T + tf/J) mouse was originally derived in 1956 from a tufted mutation T in 129S mice but not investigated in any detail until the beginning of this century when it was found that these mice show significant deficits in social behaviour, including reduced social play behaviour, social interaction and social approach avoidance behaviour. In addition, a reduction of food preference transfer was found, which was accompanied by a reduction in facial sniffing of the demonstrator mice (McFarlane et al., 2008). In addition, male BTBR mice show reduced scent marking and reduced ultrasonic vocalizations around female urine compared to C57BL/6J mice (Wohr et al., 2011). Finally, an increase in self-grooming, as an indication of repetitive behaviour, was found in BTBR mice (McFarlane et al., 2008). Interestingly, intranasal oxytocin did not improve any of the ASD-related deficits, with the exception of female sniffing in the social approach avoidance test (Bales et al., 2014). Although the efficacy of intranasal oxytocin for the treatment of ASD still has to be confirmed, so far these data shed some doubt with the respect to the predictive validity of the BTBR model for ASD. From a structural anatomical point of view, the BTBR appears to be one of the "small-brained" animal models. Ellegood and coworkers actually investigated two different BTBR mice (on a FVB and a C57BL/6J background) but both models were among the 5 genetic models with the smallest total brain volume. Moreover, with respect to the striatum and corpus callosum both mice had the smallest volumes of all 26 mice models.

The second forward genetics mouse model is the BALB/cByj mouse. In fact, there are several different substrains, some showing increased aggression, some a reduced threshold for (audiogenic) seizure, and some showing considerable abnormalities in sociability. The most commonly studied model is the BALB/cByJ mouse. In a study comparing 6 different strains of mice, the BALB/cByJ mouse showed the greatest amount of social avoidance in the social approach avoidance task (Brodkin, 2007). However, this study was performed with prepubescent female mice, while as we have seen earlier, ASD is much more prevalent in males. However, in another study comparing males of 10 different strains of mice, BALB/cByJ mice were again shown to have a deficit in the social approach avoidance paradigm: that is, in contrast to the C57BL/6J mice, these mice spent about equal time in the compartment with the mouse and the empty cylinder (as discussed earlier and Fig. 8.14) in phase 2 (Moy et al., 2007). Moreover, of all 10 strains, the BALB/cByJ mouse spent the least amount of time in the compartment with the new mouse. However, in phase 3, the BALB/cByJ mice spent more time in the compartment with the novel mouse, compared to the mouse from phase 2. Interestingly, in contrast to all the other strains, the BALB/cByJ mouse spent most of their time in the middle chamber, especially in phase 2, suggesting low mobility and/or increased anxiety. Indeed, as the authors and others have shown, the BALB/cByJ mouse show reduced locomotor activity in a new environment and a reduced time spent on the open arms in the elevated plus maze. Thus, the enhanced time spent in the central chamber, and subsequently low time spent in the novel chamber, may be an artefact. In line with this, when the time sniffing the wire cylinders was compared, the BALB/cByJ mice spent more time sniffing the wire cylinder with the novel mouse. On the other hand, it is important to realize that anxiety is a frequent comorbid symptom in ASD, and thus this may actually increase the face validity of the BALB/cByJ mouse model. The mice were also tested in a T-maze but the BALB/cByJ mic did not show any deficit in either acquisition or reversal learning. Thus, additional research is required to further evaluate the face validity of the BALB/cByJ mice for ASD.

The third forward genetic model with ASD-like characteristics was also identified in a strain comparison approach (Moy et al., 2008). Like the BTBR and the BALB/cByJ mouse, the C58/J strain showed no difference in time spent in the chamber with the mice in phase 2, and decreased time spent in the chamber with the novel mouse in phase 3. However, in contrast to the two other strains, the C58/J does not seem to show significant anxiety, as measured in the elevated plus maze and the open field. Likewise, although total locomotor activity is somewhat lower for the C58/J mice, it is unlikely to impact performance in the social tasks. Interestingly, in the Morris water maze, the C58/J mice showed poor performance, predominantly due to poor swimming performance. In addition to this, these mice showed retarded acquisition of the T-maze learning and increased stereotyped behaviour, in the form of increased hyperactivity, repeated jumping and back-flipping behaviour. In a subsequent study the C58/J mice were found to show increased motor stereotypy, but not excessive grooming.

In addition, the C58/J strain showed significant deficits in marble burying, and reduced hole-board exploration (Moy et al., 2014). On the other hand, these mice did not show cognitive rigidity.

Summarizing these strain comparison studies, it becomes clear that quite a number of strains show deficits in social approach avoidance. In fact, a comparison of seventeen strains in the social approach avoidance showed that only 41% (7 strains) actually showed a significant preference for the compartment with the mouse in phase 2, while 59% (10 strains) showed a preference for the novel mouse in phase 3. Thus, it is important to not restrict the analysis of social functioning to the social approach avoidance paradigm but to also incorporate other aspects of social functioning such as social interaction and communication. Moreover it is important to consider (potential) species-specific differences. Thus it is generally accepted that rats show more affiliative while mice show more aggressive social behaviour.

Especially given that the BTBR and the C58/J mice show substantial face validity for ASD, it becomes interesting to identify the genes underlying these ASD-like characteristics. Although this process is only in its preliminary stages, several interesting genes have already been identified. For instance both the C58/J and the BALB/cByJ mice have a mutation in the *SLC6A4* gene, coding for the serotonin transporter (Moy et al., 2014), and the C58/J also shows a mutation in *TPH2*, coding for tryptophan hydroxylase-2, the enzyme involved in serotonin synthesis, thus clearly pointing to a role of the serotonergic system in the phenotype of ASD, although it must be recognized that other genes are also altered.

Moderation of the Effects of Genes by Environmental Factors

Similar to the limited number of studies investigating gene–environment interactions in patients with ASD, few animal studies have combined both factors, in spite of the fact that in addition to the large number of genetic animal models (as discussed earlier) several environmental factors such as prenatal exposure to stress or valproate or maternal immune activation are known to induce ASD-like symptoms.

One study investigated the interaction between prenatal stress (using a chronic variable stress regime from gestational day 6 until birth) and the serotonin transporter gene (Jones et al., 2010) on anxiety and social behaviour (as assessed with the social approach avoidance paradigm). Although this study did show some interactive effects, only the genotype of the mothers was used as an independent variable. As heterozygous females were mated with wildtype males, 50% of the offspring would be WT and 50% would be heterozygous, and so the relatively small effects seen could be due to the fact that both genotypes of the offspring were combined.

In another study, heterozygous *TSC2* knock-out mice (an established model for tuberous sclerosis, as discussed earlier) and wildtype mice were exposed to maternal immune activation, using the viral mimetic polyI:C (Chapter 4) at embryonic day 12.5 (Ehninger et al., 2012). The *TSC2* knock-out mice were more sensitive to effects of polyI:C on pup survival (with more pups dying in the *TSC2* knock-out mice exposed to polyI:C). Moreover, in the social approach avoidance paradigm, all groups, except the *TSC2* knock-out mice exposed to polyI:C, showed a preference for the chamber with the mouse, suggestive of an interaction between the gene and environment.

Finally, Laviola and coworkers investigated the effects of different environmental challenges on the behaviour of reeler mice (heterozygous knock-out of the *RELN* gene). Interestingly they found that while reeler pups show reduced USVs when separated from their mothers, this effect disappeared in offspring of mothers treated with the pesticide chlorpyrifos (Laviola et al., 2006). On the other hand the normal righting reflexes was significantly prolonged and the amphetamine-induced stereotypy significantly reduced by chlorpyrifos, an effect specific for the reeler mice. Somewhat surprisingly, repeated maternal separation showed a similar normalizing effect on pup USVs as prenatal exposure to chlorpyrifos, while at the same time leading to significant reductions in BDNF levels in the striatum and prefrontal cortex in reeler mice (Ognibene et al., 2008).

Summarizing, although a large number of studies both in patients and in animals have investigated the genetic factors underlying ASD, relatively few have investigated the interaction between genetic and environmental factors. The main reason for this is presumably the fact that for a long time, the heritability of ASD was estimated at about 90%. However, as discussed earlier, more recent studies suggest that the heritability is likely to be closer to 50%, thus making the study of gene–environment even more pressing than it already was.

References

Alemany, S., Ribases, M., Vilor-Tejedor, N., et al., 2015. New suggestive genetic loci and biological pathways for attention function in adult attention-deficit/hyperactivity disorder. Am. J. Med. Genet. B.

Alexander, D.C., 2005. Multiple-fiber reconstruction algorithms for diffusion MRI. Ann. NY Acad. Sci. 1064, 113–133.

Alexander, G.E., DeLong, M.R., Strick, P.L., 1986. Parallel organization of functionally segregated circuits linking basal ganglia and cortex. Ann. Rev. Neurosci. 9, 357–381.

Althaus, M., Groen, Y., Wijers, A.A., et al., 2015. Oxytocin enhances orienting to social information in a selective group of high-functioning male adults with autism spectrum disorder. Neuropsychologia 79, 53–69.

Altink, M.E., Arias-Vasquez, A., Franke, B., et al., 2008. The dopamine receptor D4 7-repeat allele and prenatal smoking in ADHD-affected children and their unaffected siblings: no gene-environment interaction. J. Child Psychol. Psychiatr. Allied Discip. 49, 1053–1060.

Arnsten, A.F., Rubia, K., 2012. Neurobiological circuits regulating attention, cognitive control, motivation, and emotion: disruptions in neurodevelopmental psychiatric disorders. J. Am. Acad. Child Adolesc. Psychiatr. 51, 356–367.

Bales, K.L., Solomon, M., Jacob, S., et al., 2014. Long-term exposure to intranasal oxytocin in a mouse autism model. Trans. Psychiatr. 4.

Banerjee-Basu, S., Packer, A., 2010. SFARI Gene: an evolving database for the autism research community. Dis Model Mech B. 3, 133–135.

Barkley, R.A., Cox, D., 2007. A review of driving risks and impairments associated with attention-deficit/hyperactivity disorder and the effects of stimulant medication on driving performance. J. Safety Res. 38, 113–128.

Barkley, R.A., Fischer, M., 2010. The unique contribution of emotional impulsiveness to impairment in major life activities in hyperactive children as adults. J. Am. Acad. Child Adolesc. Psychiatr. 49, 503–513.

Becker, K., El-Faddagh, M., Schmidt, M.H., et al., 2008. Interaction of dopamine transporter genotype with prenatal smoke exposure on ADHD symptoms. J. Pediatr. 152, 263–269.

Belger, A., Carpenter, K.L., Yucel, G.H., et al., 2011. The neural circuitry of autism. Neurotox. Res. 20, 201–214.

Ben-Ami Bartal, I., Decety, J., Mason, P., 2011. Empathy and pro-social behavior in rats. Science 334, 1427–1430.

Berridge, C.W., Arnsten, A.F.T., 2015. Catecholamine mechanisms in the prefrontal cortex: proven strategies for enhancing higher cognitive function. Curr. Op. Behav. Sci. 4, 33–40.

Billstedt, E., Gillberg, I.C., Gillberg, C., 2007. Autism in adults: symptom patterns and early childhood predictors. Use of the DISCO in a community sample followed from childhood. J. Child Psychol. Psychiatr. Allied Discip. 48, 1102–1110.

Brikell, I., Kuja-Halkola, R., Larsson, H., 2015. Heritability of attention-deficit hyperactivity disorder in adults. Am. J. Med. Genet. B.

Brodkin, E.S., 2007. BALB/c mice: low sociability and other phenotypes that may be relevant to autism. Behav. Brain Res. 176, 53–65.

Brookes, K.J., Mill, J., Guindalini, C., et al., 2006. A common haplotype of the dopamine transporter gene associated with attention-deficit/hyperactivity disorder and interacting with maternal use of alcohol during pregnancy. Archiv. General Psychiatr. 63, 74–81.

Bruno, K.J., Freet, C.S., Twining, R.C., et al., 2007. Abnormal latent inhibition and impulsivity in coloboma mice, a model of ADHD. Neurobiol. Dis. 25, 206–216.

Buescher, A.V.S., Cidav, Z., Knapp, M., et al., 2014. Costs of autism spectrum disorders in the united kingdom and the united states. Jama Pediatr. 168, 721–728.

Burgdorf, J., Panksepp, J., Moskal, J.R., 2011. Frequency-modulated 50 kHz ultrasonic vocalizations: a tool for uncovering the molecular substrates of positive affect. Neurosci. Biobehav. Rev. 35, 1831–1836.

Burton, P.R., Hansell, A.L., Fortier, I., et al., 2009. Size matters: just how big is BIG?: Quantifying realistic sample size requirements for human genome epidemiology. Int. J. Epidemiol. 38, 263–273.

Byrne, J.M., Bawden, H.N., DeWolfe, N.A., et al., 1998. Clinical assessment of psychopharmacological treatment of preschoolers with ADHD. J. Clin. Exp. Neuropsychol. 20, 613–627.

Carrey, N., Bernier, D., Emms, M., et al., 2012. Smaller volumes of caudate nuclei in prepubertal children with ADHD: impact of age. J. Psychiatr. Res. 46, 1066–1072.

Castellanos, F.X., Lee, P.P., Sharp, W., et al., 2002. Developmental trajectories of brain volume abnormalities in children and adolescents with attention-deficit/hyperactivity disorder. Jama J. Am. Med. Assoc. 288, 1740–1748.

Castellanos, F.X., Margulies, D.S., Kelly, C., et al., 2008. Cingulate-precuneus interactions: a new locus of dysfunction in adult attention-deficit/hyperactivity disorder. Biol. Psychiatr. 63, 332–337.

Chen, Q., Panksepp, J.B., Lahvis, G.P., 2009. Empathy is moderated by genetic background in mice. PloS one 4, e4387.

Cheslack-Postava, K., Fallin, M.D., Avramopoulos, D., et al., 2007. beta2-Adrenergic receptor gene variants and risk for autism in the AGRE cohort. Mol. Psychiatr. 12, 283–291.

Ching, H., Pringsheim, T., 2012. Aripiprazole for autism spectrum disorders (ASD). The Cochrane database of systematic reviews 5, CD009043.

Chugani, D.C., 2012. Neuroimaging and neurochemistry of autism. Pediatr. Clin. North America 59, 63-73, x.

Clemow, D.B., Bushe, C.J., 2015. Atomoxetine in patients with ADHD: A clinical and pharmacological review of the onset, trajectory, duration of response and implications for patients. J. Psychopharmacol. 29, 1221–1230.

Coccini, T., Crevani, A., Rossi, G., et al., 2009. Reduced platelet monoamine oxidase type B activity and lymphocyte muscarinic receptor binding in unmedicated children with attention deficit hyperactivity disorder. Biomarkers : biochemical indicators of exposure, response, and susceptibility to chemicals 14, 513–522.

Cook, Jr., E.H., Stein, M.A., Krasowski, M.D., et al., 1995. Association of attention-deficit disorder and the dopamine transporter gene. Am. J. Hum. Genet. 56, 993–998.

Cools, A.R., van-den, B.J.H., Horstink, M.W., et al., 1984. Cognitive and motor shifting aptitude disorder in Parkinson's disease. J. Neurol.Neurosurg.Psychiatry 47, 443–453.

Cortese, S., 2012. The neurobiology and genetics of Attention-Deficit/Hyperactivity Disorder (ADHD): what every clinician should know. EJPN Official J. Eur. Paediatr. Neurol. Soc. 16, 422–433.

Cortese, S., Kelly, C., Chabernaud, C., et al., 2012. Toward systems neuroscience of ADHD: a meta-analysis of 55 fMRI studies. Am. J. Psychiatr. 169, 1038–1055.

Cubillo, A., Halari, R., Smith, A., et al., 2012. A review of fronto-striatal and fronto-cortical brain abnormalities in children and adults with Attention Deficit Hyperactivity Disorder (ADHD) and new evidence for dysfunction in adults with ADHD during motivation and attention. Cortex 48, 194–215.

De Rubeis, S., Buxbaum, J.D., 2015. Recent advances in the genetics of autism spectrum disorder. Curr. Neurol. Neurosci. Rep., 15.

Devlin, B., Melhem, N., Roeder, K., 2011. Do common variants play a role in risk for autism? Evidence and theoretical musings. Brain Res. 1380, 78–84.

Domes, G., Heinrichs, M., Kumbier, E., et al., 2013. Effects of intranasal oxytocin on the neural basis of face processing in autism spectrum disorder. Biol. Psychiatry 74, 164–171.

Dommett, E.J., 2014. Using the five-choice serial reaction time task to examine the effects of atomoxetine and methylphenidate in the male spontaneously hypertensive rat. Pharmacol. Biochem. Behav. 124, 196–203.

Ebstein, R.P., Novick, O., Umansky, R., et al., 1996. Dopamine D4 receptor (D4DR) exon III polymorphism associated with the human personality trait of novelty seeking. Nat. Genet. 12, 78–80.

Ehninger, D., Sano, Y., de Vries, P.J., et al., 2012. Gestational immune activation and Tsc2 haploinsufficiency cooperate to disrupt fetal survival and may perturb social behavior in adult mice. Mol. Psychiatr. 17, 62–70.

Elia, J., Glessner, J.T., Wang, K., et al., 2012. Genome-wide copy number variation study associates metabotropic glutamate receptor gene networks with attention deficit hyperactivity disorder. Nat. Genet. 44, 78–84.

Ellegood, J., Crawley, J.N., 2015a. Behavioral and neuroanatomical phenotypes in mouse models of autism. Neurotherapeutics 12, 521–533.

Ellegood, J., Anagnostou, E., Babineau, B.A., et al., 2015b. Clustering autism: using neuroanatomical differences in 26 mouse models to gain insight into the heterogeneity. Mol. Psychiatr. 20, 118–125.

Elmore, A.L., Nigg, J.T., Friderici, K.H., et al., 2015. Does 5HTTLPR Genotype Moderate the Association of Family Environment With Child Attention-Deficit Hyperactivity Disorder Symptomatology? J. Clin. Child Adolescent Psychol. Official J. Soc. Clin. Child Adolescent Psychol. Am. Psychol. Assoc. Div. 53, 1–13.

Faedda, G.L., Serra, G., Marangoni, C., et al., 2014. Clinical risk factors for bipolar disorders: a systematic review of prospective studies. J. Affect Disord. 168, 314–321.

Faraone, S.V., Upadhyaya, H.P., 2007. The effect of stimulant treatment for ADHD on later substance abuse and the potential for medication misuse, abuse, and diversion. J. Clin. Psychiatr. 68, e28.

Faraone, S.V., Biederman, J., Roe, C., 2002. Comparative efficacy of Adderall and methylphenidate in attention-deficit/hyperactivity disorder: a meta-analysis. J. Clin. Psychopharmacol. 22, 468–473.

Faraone, S.V., Biederman, J., Mick, E., 2006. The age-dependent decline of attention deficit hyperactivity disorder: a meta-analysis of follow-up studies. Psychol. Med. 36, 159–165.

Forssberg, H., Fernell, E., Waters, S., et al., 2006. Altered pattern of brain dopamine synthesis in male adolescents with attention deficit hyperactivity disorder. BBF 2, 40.

Franke, B., Neale, B.M., Faraone, S.V., 2009. Genome-wide association studies in ADHD. Hum. Genet. 126, 13–50.

Fusar-Poli, P., Rubia, K., Rossi, G., et al., 2012. Striatal dopamine transporter alterations in ADHD: pathophysiology or adaptation to psychostimulants? A meta-analysis. Am. J. Psychiatr. 169, 264–272.

Galef, B.G., Whiskin, E.E., 2003. Socially transmitted food preferences can be used to study long-term memory in rats. Learning Behav. 31, 160–164.

Gao, Q., Qian, Y., He, X.X., et al., 2015. Childhood predictors of persistent ADHD in early adulthood: results from the first follow-up study in China. Psychiatr. Res. 230, 905–912.

Gauglerl, T., Klei, L., Sanders, S.J., et al., 2014. Most genetic risk for autism resides with common variation. Nature Genetic. 46, 881–885.

Gentile, S., 2015. Prenatal antidepressant exposure and the risk of autism spectrum disorders in children. Are we looking at the fall of Gods? J. Affect. Disorder. 182, 132–137.

Gillberg, C., Cederlund, M., Lamberg, K., et al., 2006. Brief report: "the autism epidemic". The registered prevalence of autism in a Swedish urban area. J. Autism Dev. Disord. 36, 429–435.

Gizer, I.R., Ficks, C., Waldman ID, 2009. Candidate gene studies of ADHD: a meta-analytic review. Hum. Genet. 126, 51–90.

Goldberg, T.E., Berman, K.F., Mohr, E., et al., 1990. Regional cerebral blood flow and cognitive function in Huntington's disease and schizophrenia. A comparison of patients matched for performance on a prefrontal-type task. Arch. Neurol. 47, 418–422.

Gunn, R.K., Keenan, M.E., Brown, R.E., 2011. Analysis of sensory, motor and cognitive functions of the coloboma (C3Sn.Cg-Cm/J) mutant mouse. Genes Brain Behav. 10, 579–588.

Gustavsson, A., Svensson, M., Jacobi, F., et al., 2011. Cost of disorders of the brain in Europe 2010. Eur. Neuropsychopharmacol. J. Eur. Coll. Neuropsychopharmacol. 21, 718–779.

Hagerman, R., Hoem, G., Hagerman, P., 2010. Fragile X and autism: intertwined at the molecular level leading to targeted treatments. Mol. Autism 1, 12.

Hannestad, J., Gallezot, J.D., Planeta-Wilson, B., et al., 2010. Clinically relevant doses of methylphenidate significantly occupy norepinephrine transporters in humans in vivo. Biol. Psychiatr. 68, 854–860.

Hanwella, R., Senanayake, M., de Silva, V., 2011. Comparative efficacy and acceptability of methylphenidate and atomoxetine in treatment of attention deficit hyperactivity disorder in children and adolescents: a meta-analysis. BMC Psychiatr. 11, 176.

Helms, C.M., Gubner, N.R., Wilhelm, C.J., et al., 2008. D4 receptor deficiency in mice has limited effects on impulsivity and novelty seeking. Pharmacol. Biochem. Behav. 90, 387–393.

Hesse, S., Ballaschke, O., Barthel, H., et al., 2009. Dopamine transporter imaging in adult patients with attention-deficit/hyperactivity disorder. Psychiatry Res 171, 120–128.

Hirota, T., Schwartz, S., Correll, C.U., 2014. Alpha-2 agonists for attention-deficit/hyperactivity disorder in youth: a systematic review and meta-analysis of monotherapy and add-on trials to stimulant therapy. J. Am. Acad. Child Adolesc. Psychiatr. 53, 153–173.

Horlin, C., Falkmer, M., Parsons, R., et al., 2014. The cost of autism spectrum disorders. Plos One 9.

Huang, C.H., Santangelo, S.L., 2008. Autism and serotonin transporter gene polymorphisms: a systematic review and meta-analysis. Am. J. Med. Genetic. Part B 147B, 903–913.

Insel, T.R., 2010. The challenge of translation in social neuroscience: a review of oxytocin, vasopressin, and affiliative behavior. Neuron 65, 768–779.

Jacobsen, L.K., Staley, J.K., Zoghbi, S.S., et al., 2000. Prediction of dopamine transporter binding availability by genotype: a preliminary report. Am. J. Psychiatr. 157, 1700–1703.

Jones, K.L., Smith, R.M., Edwards, K.S., et al., 2010. Combined effect of maternal serotonin transporter genotype and prenatal stress in modulating offspring social interaction in mice. Int. J. Dev. Neurosci. 28, 529–536.

Just, M.A., Cherkassky, V.L., Keller, T.A., et al., 2007. Functional and anatomical cortical underconnectivity in autism: evidence from an FMRI study of an executive function task and corpus callosum morphometry. Cereb. Cortex 17, 951–961.

Just, M.A., Keller, T.A., Malave, V.L., et al., 2012. Autism as a neural systems disorder: a theory of frontal-posterior underconnectivity. Neurosci. Biobehav. Rev. 36, 1292–1313.

Kahn, R.S., Khoury, J., Nichols, W.C., et al., 2003. Role of dopamine transporter genotype and maternal prenatal smoking in childhood hyperactive-impulsive, inattentive, and oppositional behaviors. J. Pediatr. 143, 104–110.

Kessler, R.C., Adler, L.A., Barkley, R., et al., 2005. Patterns and predictors of attention-deficit/hyperactivity disorder persistence into adulthood: results from the national comorbidity survey replication. Biol. Psychiatr. 57, 1442–1451.

Kessler, R.C., Adler, L., Barkley, R., et al., 2006. The prevalence and correlates of adult ADHD in the United States: results from the National Comorbidity Survey Replication. Am. J. Psychiatr. 163, 716–723.

Kieling, C., Hutz, M.H., Genro, J.P., et al., 2013. Gene-environment interaction in externalizing problems among adolescents: evidence from the Pelotas 1993 Birth Cohort Study. J. Child Psychol. Psychiatr. Allied Discip. 54, 298–304.

Kim, B.N., Kim, J.W., Kang, H., et al., 2010. Regional differences in cerebral perfusion associated with the alpha-2A-adrenergic receptor genotypes in attention deficit hyperactivity disorder. J. Psychiatr. Neurosci. 35, 330–336.

Kim, Y.S., Leventhal, B.L., 2015. Genetic epidemiology and insights into interactive genetic and environmental effects in autism spectrum disorders. Biol. Psychiatr. 77, 66–74.

Kim, Y.S., Leventhal, B.L., Koh, Y.J., et al., 2011. Prevalence of autism spectrum disorders in a total population sample. Am. J. Psychiatr. 168, 904–912.

Konrad, K., Eickhoff, S.B., 2010. Is the ADHD brain wired differently? A review on structural and functional connectivity in attention deficit hyperactivity disorder. Hum. Brain Mapping 31, 904–916.

Lam, K.S., Aman, M.G., Arnold, L.E., 2006. Neurochemical correlates of autistic disorder: a review of the literature. Res. Dev. Disab. 27, 254–289.

Langford, D.J., Crager, S.E., Shehzad, Z., et al., 2006. Social modulation of pain as evidence for empathy in mice. Science 312, 1967–1970.

Lasky-Su, J., Faraone, S.V., Lange, C., et al., 2007. A study of how socioeconomic status moderates the relationship between SNPs encompassing BDNF and ADHD symptom counts in ADHD families. Behav. Genet. 37, 487–497.

Lasky-Su, J., Lange, C., Biederman, J., et al., 2008. Family-based association analysis of a statistically derived quantitative traits for ADHD reveal an association in DRD4 with inattentive symptoms in ADHD individuals. Am. J. Med. Genet. B 147B, 100–106.

Laucht, M., Skowronek, M.H., Becker, K., et al., 2007. Interacting effects of the dopamine transporter gene and psychosocial adversity on attention-deficit/hyperactivity disorder symptoms among 15-year-olds from a high-risk community sample. Arch. Gen. Psychiatr. 64, 585–590.

Laviola, G., Adriani, W., Gaudino, C., et al., 2006. Paradoxical effects of prenatal acetylcholinesterase blockade on neuro-behavioral development and drug-induced stereotypies in reeler mutant mice. Psychopharmacology (Berlin) 187, 331–344.

Leo, D., Gainetdinov, R.R., 2013. Transgenic mouse models for ADHD. Cell Tissue Res. 354, 259–271.

Lesch, K.P., Bengel, D., Heils, A., et al., 1996. Association of anxiety-related traits with a polymorphism in the serotonin transporter gene regulatory region. Science 274, 1527–1531.

Levy, S.E., Mandell, D.S., Schultz, R.T., 2009. Autism. Lancet 374, 1627–1638.

Li, Z., Chang, S.H., Zhang, L.Y., et al., 2014. Molecular genetic studies of ADHD and its candidate genes: a review. Psychiatry Res. 219, 10–24.

Liu, X., Shimada, T., Otowa, T., et al., 2015. Genome-wide association study of autism spectrum disorder in the east asian populations. Autism Res.

LoParo, D., Waldman, I.D., 2015. The oxytocin receptor gene (OXTR) is associated with autism spectrum disorder: a meta-analysis. Mol. Psychiatr. 20, 640–646.

Ludolph, A.G., Kassubek, J., Schmeck, K., et al., 2008. Dopaminergic dysfunction in attention deficit hyperactivity disorder (ADHD), differences between pharmacologically treated and never treated young adults: a 3,4-dihdroxy-6-[18F]fluorophenyl-l-alanine PET study. Neuroimage 41, 718–727.

Man, K.K.C., Tong, H.H.Y., Wong, L.Y.L., et al., 2015. Exposure to selective serotonin reuptake inhibitors during pregnancy and risk of autism spectrum disorder in children: a systematic review and meta-analysis of observational studies. Neurosci. Biobehav. Rev. 49, 82–89.

Marcus, R.N., Owen, R., Manos, G., et al., 2011. Safety and tolerability of aripiprazole for irritability in pediatric patients with autistic disorder: a 52-week, open-label, multicenter study. J. Clin. Psychiatr. 72, 1270–1276.

Mazina, V., Gerdts, J., Trinh, S., et al., 2015. Epigenetics of autism-related impairment: copy number variation and maternal infection. J. Developmental Behav. Pediatr. 36, 61–67.

McFarlane, H.G., Kusek, G.K., Yang, M., et al., 2008. Autism-like behavioral phenotypes in BTBR T + tf/J mice. Genes Brain Behav. 7, 152–163.

Moffitt, T.E., Houts, R., Asherson, P., et al., 2015. Is adult ADHD a childhood-onset neurodevelopmental disorder? Evidence from a four-decade longitudinal cohort study. Am. J. Psychiatr. 172, 967–977.

Moron, J.A., Brockington, A., Wise, R.A., et al., 2002. Dopamine uptake through the norepinephrine transporter in brain regions with low levels of the dopamine transporter: evidence from knock-out mouse lines. J. Neurosci. 22, 389–395.

Moy, S.S., Nadler, J.J., Young, N.B., et al., 2008. Social approach and repetitive behavior in eleven inbred mouse strains. Behav. Brain Res. 191, 118–129.

Moy, S.S., Riddick, N.V., Nikolova, V.D., et al., 2014. Repetitive behavior profile and supersensitivity to amphetamine in the C58/J mouse model of autism. Behav. Brain Res. 259, 200–214.

Moy, S.S., Nadler, J.J., Young, N.B., et al., 2007. Mouse behavioral tasks relevant to autism: phenotypes of 10 inbred strains. Behav. Brain Res. 176, 4–20.

Muller, D.J., Mandelli, L., Serretti, A., et al., 2008. Serotonin transporter gene and adverse life events in adult ADHD. Am. J. Med. Genet. B 147B, 1461–1469.

Naaijen, J., Lythgoe, D.J., Amiri, H., et al., 2015. Fronto-striatal glutamatergic compounds in compulsive and impulsive syndromes: a review of magnetic resonance spectroscopy studies. Neurosci. Biobehav. Rev. 52, 74–88.

Nakao, T., Radua, J., Rubia, K., et al., 2011. Gray matter volume abnormalities in ADHD: voxel-based meta-analysis exploring the effects of age and stimulant medication. Am. J. Psychiatr. 168, 1154–1163.

Neale, B.M., Purcell, S., 2008. The positives, protocols, and perils of genome-wide association. Am. J. Med. Genet. B 147B, 1288–1294.

Nguyen, M., Roth, A., Kyzar, E.J., et al., 2014. Decoding the contribution of dopaminergic genes and pathways to autism spectrum disorder (ASD). Neurochem. Int. 66, 15–26.

Niarchou, M., Martin, J., Thapar, A., et al., 2015. The clinical presentation of attention deficit-hyperactivity disorder (ADHD) in children with 22q11.2 deletion syndrome. Am. J. Med. Genet. B 168, 730–738.

Nigg, J., Nikolas, M., Burt, S.A., 2010. Measured gene-by-environment interaction in relation to attention-deficit/hyperactivity disorder. J. Am. Acad. Child Adolesc. Psychiatr. 49, 863–873.

Nijmeijer, J.S., Hartman, C.A., Rommelse, N.N., et al., 2010. Perinatal risk factors interacting with catechol O-methyltransferase and the serotonin transporter gene predict ASD symptoms in children with ADHD. J. Child Psychol. Psychiatr. Allied Discip. 51, 1242–1250.

Nyffeler, J., Walitza, S., Bobrowski, E., et al., 2014. Association study in siblings and case-controls of serotonin- and oxytocin-related genes with high functioning autism. J. Mol. Psychiatr. 2, 1.

O'Brien, T.C., Mustanski, B.S., Skol, A., et al., 2013. Do dopamine gene variants and prenatal smoking interactively predict youth externalizing behavior? Neurotoxicol. Teratol. 40, 67–73.

Ognibene, E., Adriani, W., Caprioli, A., et al., 2008. The effect of early maternal separation on brain derived neurotrophic factor and monoamine levels in adult heterozygous reeler mice. Prog. Neuropsychopharmacol. Biol. Psychiatr. 32, 1269–1276.

Okuyama, Y., Ishiguro, H., Toru, M., et al., 1999. A genetic polymorphism in the promoter region of DRD4 associated with expression and schizophrenia. Biochem. Biophys. Res. Comm. 258, 292–295.

Pamplona, F.A., Pandolfo, P., Savoldi, R., et al., 2009. Environmental enrichment improves cognitive deficits in spontaneously hypertensive rats (shr): relevance for attention deficit/hyperactivity disorder (ADHD). Prog. Neuropsychopharmacol. Biol. Psychiatr. 33, 1153–1160.

Papaleo, F., Erickson, L., Liu, G., et al., 2012. Effects of sex and COMT genotype on environmentally modulated cognitive control in mice. Proc. Natl. Acad. Sci. USA 109, 20160–20165.

Pasciuto, E., Borrie, S.C., Kanellopoulos, A.K., et al., 2015. Autism spectrum disorders: translating human deficits into mouse behavior. Neurobiol. Learning Memory 124, 71–87.

Paus, T., Keshavan, M., Giedd, J.N., 2008. Why do many psychiatric disorders emerge during adolescence? Nature Rev. Neurosci. 9, 947–957.

Persico, A.M., Napolioni, V., 2013. Autism genetics. Behav. Brain Res. 251, 95–112.

Pillidge, K., Porter, A.J., Vasili, T., et al., 2014a. Atomoxetine reduces hyperactive/impulsive behaviours in neurokinin-1 receptor 'knockout' mice. Pharmacol. Biochem. Behav. 127, 56–61.

Pillidge, K., Porter, A.J., Dudley, J.A., et al., 2014b. The behavioural response of mice lacking NK(1) receptors to guanfacine resembles its clinical profile in treatment of ADHD. Br. J. Pharmacol. 171, 4785–4796.

Pingault, J.B., Cote, S.M., Vitaro, F., et al., 2014. The developmental course of childhood inattention symptoms uniquely predicts educational attainment: a 16-year longitudinal study. Psychiatr. Res. 219, 707–709.

Pingault, J.B., Tremblay, R.E., Vitaro, F., et al., 2011. Childhood trajectories of inattention and hyperactivity and prediction of educational attainment in early adulthood: a 16-year longitudinal population-based study. Am. J. Psychiatr. 168, 1164–1170.

Pingault, J.B., Cote, S.M., Galera, C., et al., 2013. Childhood trajectories of inattention, hyperactivity and oppositional behaviors and prediction of substance abuse/dependence: a 15-year longitudinal population-based study. Mol. Psychiatr. 18, 806–812.

Polanczyk, G., de Lima, M.S., Horta, B.L., et al., 2007. The worldwide prevalence of ADHD: a systematic review and metaregression analysis. Am. J. Psychiatr. 164, 942–948.

Polanczyk, G.V., Willcutt, E.G., Salum, G.A., et al., 2014. ADHD prevalence estimates across three decades: an updated systematic review and meta-regression analysis. Int. J. Epidemiol. 43, 434–442.

Porter, A.J., Pillidge, K., Stanford, S.C., et al., 2016. Differences in the performance of NK1R-/- ('knockout') and wild-type mice in the 5choice continuous performance test. Behav. Brain Res. 298, 268–277.

Porter, A.J., Pillidge, K., Tsai, Y.C., et al., 2015. A lack of functional NK1 receptors explains most, but not all, abnormal behaviours of NK1R-/- mice(1). Genes Brain Behav. 14, 189–199.

Punja, S., Zorzela, L., Hartling, L., et al., 2013. Long-acting versus short-acting methylphenidate for paediatric ADHD: a systematic review and meta-analysis of comparative efficacy. BMJ Open 3.

Ramos-Quiroga, J.A., Sanchez-Mora, C., Casas, M., et al., 2014. Genome-wide copy number variation analysis in adult attention-deficit and hyperactivity disorder. J. Psychiatr. Res. 49, 60–67.

Remschmidt, H.E., Schulz, E., Martin, M., et al., 1994. Childhood-onset schizophrenia: history of the concept and recent studies. Schizophr. Bull. 20, 727–745.

Retz, W., Freitag, C.M., Retz-Junginger, P., et al., 2008. A functional serotonin transporter promoter gene polymorphism increases ADHD symptoms in delinquents: interaction with adverse childhood environment. Psychiatr. Res. 158, 123–131.

Risch, N., Merikangas, K., 1996. The future of genetic studies of complex human diseases. Science 273, 1516–1517.

Ronald, A., Hoekstra, R.A., 2011. Autism spectrum disorders and autistic traits: a decade of new twin studies. Am. J. Med. Genet. B 156B, 255–274.

Rossion, B., 2014. Understanding face perception by means of human electrophysiology. Trends Cogn. Sci. 18, 310–318.

Rothman, R.B., Baumann, M.H., 2003. Monoamine transporters and psychostimulant drugs. Eur. J. Pharmacol. 479, 23–40.

Rubia, K., Cubillo, A., Woolley, J., et al., 2011. Disorder-specific dysfunctions in patients with attention-deficit/hyperactivity disorder compared to patients with obsessive-compulsive disorder during interference inhibition and attention allocation. Hum. Brain Mapping 32, 601–611.

Ruggiero, S., Clavenna, A., Reale, L., et al., 2014. Guanfacine for attention deficit and hyperactivity disorder in pediatrics: a systematic review and meta-analysis. Eur. Neuropsychopharmacol. J. Eur. Coll. Neuropsychopharmacol. 24, 1578–1590.

Sagvolden, T., 2000. Behavioral validation of the spontaneous hypertensive rat (SHR) as an animal model of attention-deficit/hyperactivity disorder (AD/HD). Neurosci. Biobehav. Rev. 24, 31–39.

Sagvolden, T., Metzger, M.A., Schiorbeck, H.K., et al., 1992. The spontaneously hypertensive rats (SHR) as an animal model of childhood hyperactivity (ADHD): Changed reactivity to reinforcers and to psychomotor stimulation. Behav. Neural Biol. 58, 103–112.

Sallee, F.R., 2010. The role of alpha2-adrenergic agonists in attention-deficit/hyperactivity disorder. Postgrad. Med. 122, 78–87.

Sandin, S., Lichtenstein, P., Kuja-Halkola, R., et al., 2014. The familial risk of autism. Jama J. Am. Med. Assoc. 311, 1770–1777.

Sato, N., Tan, L., Tate, K., et al., 2015. Rats demonstrate helping behavior toward a soaked conspecific. Anim. Cogn. 18, 1039–1047.

Scassellati, C., Bonvicini, C., Faraone, S.V., et al., 2012. Biomarkers and attention-deficit/hyperactivity disorder: a systematic review and meta-analyses. J. Am. Acad. Child Adolesc. Psychiatr. 51, 1003-1019 e1020.

Schilbach, L., Eickhoff, S.B., Rotarska-Jagiela, A., et al., 2008. Minds at rest? Social cognition as the default mode of cognizing and its putative relationship to the "default system" of the brain. Conscious Cogn. 17, 457–467.

Schmidt, R.J., Hansen, R.L., Hartiala, J., et al., 2011. Prenatal vitamins, one-carbon metabolism gene variants, and risk for autism. Epidemiology 22, 476–485.

Seeger, G., Schloss, P., Schmidt, M.H., et al., 2004. Gene-environment interaction in hyperkinetic conduct disorder (HD + CD) as indicated by season of birth variations in dopamine receptor (DRD4) gene polymorphism. Neurosci. Lett. 366, 282–286.

Shafritz, K.M., Dichter, G.S., Baranek, G.T., et al., 2008. The neural circuitry mediating shifts in behavioral response and cognitive set in autism. Biol. Psychiatr. 63, 974–980.

Sharma, A., Shaw, S.R., 2012. Efficacy of risperidone in managing maladaptive behaviors for children with autistic spectrum disorder: a meta-analysis. J. Pediatr. Health Care Official Pub. Nat. Assoc. Pediatr. Nurse Assoc. Practitioners 26, 291–299.

Sharp, S.I., McQuillin, A., Marks, M., et al., 2014. Genetic association of the tachykinin receptor 1 TACR1 gene in bipolar disorder, attention deficit hyperactivity disorder, and the alcohol dependence syndrome. Am. J. Med. Genet. B 165B, 373–380.

Shekim, W.O., Bylund, D.B., Alexson, J., et al., 1986. Platelet MAO and measures of attention and impulsivity in boys with attention deficit disorder and hyperactivity. Psychiatry Res. 18, 179–188.

Sherrington, R., Mankoo, B., Attwood, J., et al., 1993. Cloning of the human dopamine D5 receptor gene and identification of a highly polymorphic microsatellite for the DRD5 locus that shows tight linkage to the chromosome 4p reference marker RAF1P1. Genomics 18, 423–425.

Silverman, J.L., Yang, M., Lord, C., et al., 2010. Behavioural phenotyping assays for mouse models of autism. Nature Rev. Neurosci. 11, 490–502.

Simon, V., Czobor, P., Balint, S., et al., 2009. Prevalence and correlates of adult attention-deficit hyperactivity disorder: meta-analysis. Br. J. Psychiatr. 194, 204–211.

Singh, I., 2008. Beyond polemics: science and ethics of ADHD. Nature Rev. Neurosci. 9, 957–964.

Sollner, T., Whiteheart, S.W., Brunner, M., et al., 1993. SNAP receptors implicated in vesicle targeting and fusion. Nature 362, 318–324.

Spencer, R.C., Devilbiss, D.M., Berridge, C.W., 2015. The cognition-enhancing effects of psychostimulants involve direct action in the prefrontal cortex. Biol. Psychiatr. 77, 940–950.

Staal, W.G., 2015. Autism, DRD3 and repetitive and stereotyped behavior, an overview of the current knowledge. Eur. Neuropsychopharmacol. 25, 1421–1426.

Staal, W.G., de Krom, M., de Jonge, M.V., 2012. Brief report: the dopamine-3-receptor gene (DRD3) is associated with specific repetitive behavior in autism spectrum disorder (ASD). J. Autism Dev. Disord. 42, 885–888.

Stuhec, M., Munda, B., Svab, V., et al., 2015. Comparative efficacy and acceptability of atomoxetine, lisdexamfetamine, bupropion and methylphenidate in treatment of attention deficit hyperactivity disorder in children and adolescents: a meta-analysis with focus on bupropion. J. Affect Disord. 178, 149–159.

Taylor, J.L., Seltzer, M.M., 2010. Changes in the autism behavioral phenotype during the transition to adulthood. J. Autism Dev. Disord. 40, 1431–1446.

Tebruegge, M., Nandini, V., Ritchie, J., 2004. Does routine child health surveillance contribute to the early detection of children with pervasive developmental disorders? An epidemiological study in Kent, UK. BMC Pediatr. 4, 4.

Tian, L., Jiang, T., Wang, Y., et al., 2006. Altered resting-state functional connectivity patterns of anterior cingulate cortex in adolescents with attention deficit hyperactivity disorder. Neurosci. Lett. 400, 39–43.

Timimi, S., Taylor, E., 2004. ADHD is best understood as a cultural construct. Br. J. Psychiatr. 184, 8–9.

Todd, R.D., Neuman, R.J., 2007. Gene-environment interactions in the development of combined type ADHD: evidence for a synapse-based model. Am. J. Med. Genet. B 144B, 971–975.

Troost, P.W., Lahuis, B.E., Steenhuis, M.P., et al., 2005. Long-term effects of risperidone in children with autism spectrum disorders: a placebo discontinuation study. J. Am. Acad. Child Adolesc. Psychiatr. 44, 1137–1144.

Turner, M., Wilding, E., Cassidy, E., et al., 2013. Effects of atomoxetine on locomotor activity and impulsivity in the spontaneously hypertensive rat. Behav. Brain Res. 243, 28–37.

van der Meer, D., Hartman, C.A., Richards, J., et al., 2014. The serotonin transporter gene polymorphism 5-HTTLPR moderates the effects of stress on attention-deficit/hyperactivity disorder. J. Child Psychol. Psychiatr. 55, 1363–1371.

van Ewijk, H., Heslenfeld, D.J., Zwiers, M.P., et al., 2012. Diffusion tensor imaging in attention deficit/hyperactivity disorder: a systematic review and meta-analysis. Neurosci. Biobehav. Rev. 36, 1093–1106.

Van Eylen, L., Boets, B., Steyaert, J., et al., 2015. Executive functioning in autism spectrum disorders: influence of task and sample characteristics and relation to symptom severity. Eur. Child Adolescent Psychiatr. 24, 1399–1417.

van Tongerloo, M., van Wijngaarden, P.J.M., van der Gaag, R.J., et al., 2015. Raising a child with an autism spectrum disorder: 'if this were a partner relationship, i would have quit ages ago'. Family Practice 32, 88–93.

Vanderschuren, L.J.M.J., Niesink, R.J.M., Van Ree, J.M., 1997. The neurobiologgy of social play behavior in rats. Neurosci. Biobehav. Rev. 21, 309–326.

Vanderschuren, L.J.M.J., Spruijt, B.M., Hol, T., et al., 1995. Sequential analysis of social play behavior in juvenile rats: effects of morphine. Behav. Brain Res. 72, 89–95.

Vannucchi, G., Masi, G., Toni, C., et al., 2014. Clinical features, developmental course, and psychiatric comorbidity of adult autism spectrum disorders. CNS Spectr. 19, 157–164.

Vanyukov, M.M., Maher, B.S., Devlin, B., et al., 2007. The MAOA promoter polymorphism, disruptive behavior disorders, and early onset substance use disorder: gene-environment interaction. Psychiatr. Genetics 17, 323–332.

Vloet, T.D., Gilsbach, S., Neufang, S., et al., 2010. Neural mechanisms of interference control and time discrimination in attention-deficit/hyperactivity disorder. J. Am. Acad. Child Adolesc. Psychiatr. 49, 356–367.

Volkow, N.D., Wang, G.J., Newcorn, J., et al., 2007. Depressed dopamine activity in caudate and preliminary evidence of limbic involvement in adults with attention-deficit/hyperactivity disorder. Arch. Gen. Psychiatr. 64, 932–940.

Volkow, N.D., Wang, G.J., Kollins, S.H., et al., 2009. Evaluating dopamine reward pathway in ADHD: clinical implications. JAMA 302, 1084–1091.

Waldman, I.D., 2007. Gene-environment interactions reexamined: does mother's marital stability interact with the dopamine receptor D2 gene in the etiology of childhood attention-deficit/hyperactivity disorder? Dev. Psychopathol. 19, 1117–1128.

Waldman, I.D., Rowe, D.C., Abramowitz, A., et al., 1998. Association and linkage of the dopamine transporter gene and attention-deficit hyperactivity disorder in children: heterogeneity owing to diagnostic subtype and severity. Am. J. Hum. Genet. 63, 1767–1776.

Wang, Z.L., Hong, Y., Zou, L., et al., 2014. Reelin gene variants and risk of autism spectrum disorders: an integrated meta-analysis. Am. J. Med. Genetics Part B 165, 192–200.

Warrier, V., Chee, V., Smith, P., et al., 2015. A comprehensive meta-analysis of common genetic variants in autism spectrum conditions. Mol. Autism 6.

Weyandt, L., Swentosky, A., Gudmundsdottir, B.G., 2013. Neuroimaging and ADHD: fMRI, PET, DTI findings, and methodological limitations. Dev. Neuropsychol. 38, 211–225.

Willcutt, E.G., 2012. The prevalence of DSM-IV attention-deficit/hyperactivity disorder: a meta-analytic review. Neurotherapeutics 9, 490–499.

Willcutt, E.G., Nigg, J.T., Pennington, B.F., et al., 2012. Validity of DSM-IV attention deficit/hyperactivity disorder symptom dimensions and subtypes. J. Abnorm. Psychol. 121, 991–1010.

Wohr, M., Schwarting, R.K.W., 2013. Affective communication in rodents: ultrasonic vocalizations as a tool for research on emotion and motivation. Cell Tissue Res. 354, 81–97.
Wohr, M., Roullet, F.I., Crawley, J.N., 2011. Reduced scent marking and ultrasonic vocalizations in the BTBR T + tf/J mouse model of autism. Genes Brain Behav. 10, 35–43.
Wolf, R.C., Plichta, M.M., Sambataro, F., et al., 2009. Regional brain activation changes and abnormal functional connectivity of the ventrolateral prefrontal cortex during working memory processing in adults with attention-deficit/hyperactivity disorder. Hum. Brain Mapping 30, 2252–2266.
Yan, T.C., Hunt, S.P., Stanford, S.C., 2009. Behavioural and neurochemical abnormalities in mice lacking functional tachykinin-1 (NK1) receptors: a model of attention deficit hyperactivity disorder. Neuropharmacology 57, 627–635.
Yan, T.C., McQuillin, A., Thapar, A., et al., 2010. NK1 (TACR1) receptor gene 'knockout' mouse phenotype predicts genetic association with ADHD. J. Psychopharmacol. 24, 27–38.
Zalla, T., Sperduti, M., 2013. The amygdala and the relevance detection theory of autism: an evolutionary perspective. Frontiers Hum. Neurosci. 7.
Zimmermann, A.M., Jene, T., Wolf, M., et al., 2015. Attention-deficit/hyperactivity disorder-like phenotype in a mouse model with impaired actin dynamics. Biol. Psychiatr. 78, 95–106.
Zurcher, N.R., Bhanot, A., McDougle, C.J., et al., 2015. A systematic review of molecular imaging (PET and SPECT) in autism spectrum disorder: current state and future research opportunities. Neurosci. Biobehav. Rev. 52, 56–73.

CHAPTER

9

Schizophrenia

INTRODUCTION

In the 1896 edition of his textbook on psychiatry, Emil Kraepelin distinguished between manic depressive psychoses and dementia praecox. Although in an earlier version of his book he equated dementia praecox with hebephrenia, in this edition he combined hebephrenia, dementia paranoias and catatonia in his new diagnosis. He considered a progressively deteriorating disease process with a poor outcome to be the most important aspect of dementia praecox. The term schizophrenia was introduced in 1908 by the Swiss psychiatrist Eugen Bleuler. He was dissatisfied with the poor outcome criterion of dementia praecox and therefore emphasized the split (Greek "schizein") of the mind (Greek "phren"). Unfortunately, this is often interpreted (especially in the popular media) as split personality, but Bleuler considered schizophrenia as a separation of personality, perception, memory and thinking. Importantly, in his textbook, Bleuler spoke of the "group of the schizophrenias" thus emphasizing the heterogeneity of the disorder. Moreover, he also suggested that hallucinations and delusions were secondary symptoms resulting from the primary deficit in associations. This idea of primary and secondary symptoms has been very influential beyond the diagnosis of schizophrenia.

DIAGNOSIS AND SYMPTOMS

Since Bleuler's original description, many different diagnostic schemes have been proposed for schizophrenia. The currently most used scheme is the Diagnostic and Statistical Manual published by the American Psychiatric Association, which the latest version (DSM-V) published in 2013 (APA, 2013). In this diagnostic scheme, a diagnosis of schizophrenia depends on the presence of at least 2 symptoms out of a total of 5, including both positive and negative symptoms (Box 9.1), with at least one of the symptoms being hallucinations, delusions or disorganized speech. Positive symptoms are defined as features that are not normally present but are induced by the disorder, while negative symptoms are defined as features normally present but significantly reduced or absent as a result of the disorder. In addition to the basic symptoms, it is important to separate schizophrenia from other conditions such as schizoaffective and bipolar disorders, and to ensure the patient is not suffering from an acute psychotic episode. For that reason the symptoms

Gene-Environment Interactions in Psychiatry. http://dx.doi.org/10.1016/B978-0-12-801657-2.00009-4

BOX 9.1

THE DIAGNOSIS OF SCHIZOPHRENIA (DSM V)

1. Two or more of the following symptoms should be present for a significant portion of the time during a period of at least 1 month (with at least one of the symptoms being a, b or c):
   a. Delusions
   b. Hallucination
   c. Disorganized speech
   d. Grossly disorganized or catatonic behaviour
   e. Negative symptoms
2. The level of functioning in areas such as work, interpersonal relationships etc. should be significantly impaired since onset of the symptoms.
3. Continuous signs of the disturbance should be present for at least 6 months.
4. Other disorders (such as schizoaffective, bipolar, physical illness) are ruled out in spite of having persistent negative or recurrent social or interpersonal problems.

should be present for a considerable period of time, and, as is almost always the case in schizophrenia, the patient almost never returns to the normal level of functioning. Fig. 9.1 illustrates the typical course of schizophrenia, with symptoms often coming in specific episodes and (usually after treatment) subside, but the patient's overall level of functioning rarely reaching the level prior to the psychotic episode. As a result, over time the patient's functioning becomes less and less.

In addition to DSM, many other diagnostic schemes have been developed including a number of scales more focused on measuring the severity of the symptoms rather than simply the presence or absence of individual symptoms. Currently, the most often used of these rating

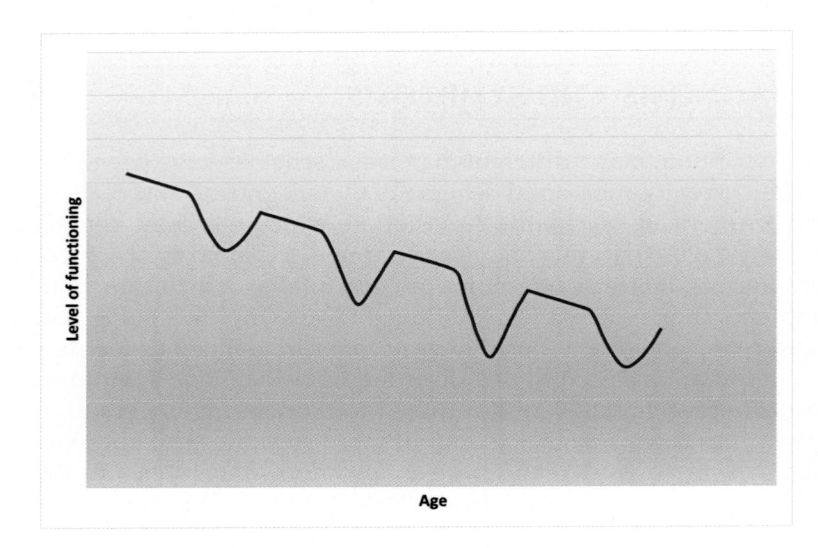

FIGURE 9.1 **Typical course and outcome of schizophrenia.** Schizophrenia is typically characterized by psychotic episodes during which the level of functioning significantly decreases. These episodes can vary in length but often a certain level of remission occurs (mostly after antipsychotic treatment). However, in most cases after remission the level of functioning is lower than before, inevitably leading to a more severe deficit after each episode.

scales is the Positive And Negative Syndrome Scale (PANSS). The symptoms assessed in the scale originally consisted of 30 items: seven covering the positive symptoms, seven the negative symptoms and sixteen items on general psychopathology (Kay et al., 1987). Each item was again rated on a scale of 1 (absent) to 7 (severe), thus leading to a score between 30 and 210. Overall the assessment takes about 40 min. Subsequent analyses of these 30 items with large cohort of patients have suggested that the symptoms can be systematically grouped into five independent clusters. This was recently confirmed in an evaluation of patients from the USA, Brazil and China, confirming five independent clusters while simultaneously reducing the number of symptoms to 20: These five factors are labelled: [a] Positive; [b] Negative; [c] Disorganized; [d] Excited; and [e] Depressed (Stefanovics et al., 2014). This five cluster model and some representative symptoms are shown graphically in Fig. 9.2.

In addition to the symptoms assessed by the DSM-V and the PANSS, schizophrenic patients also suffer from cognitive deficits. As illustrated in Fig. 9.2. Much research has been devoted

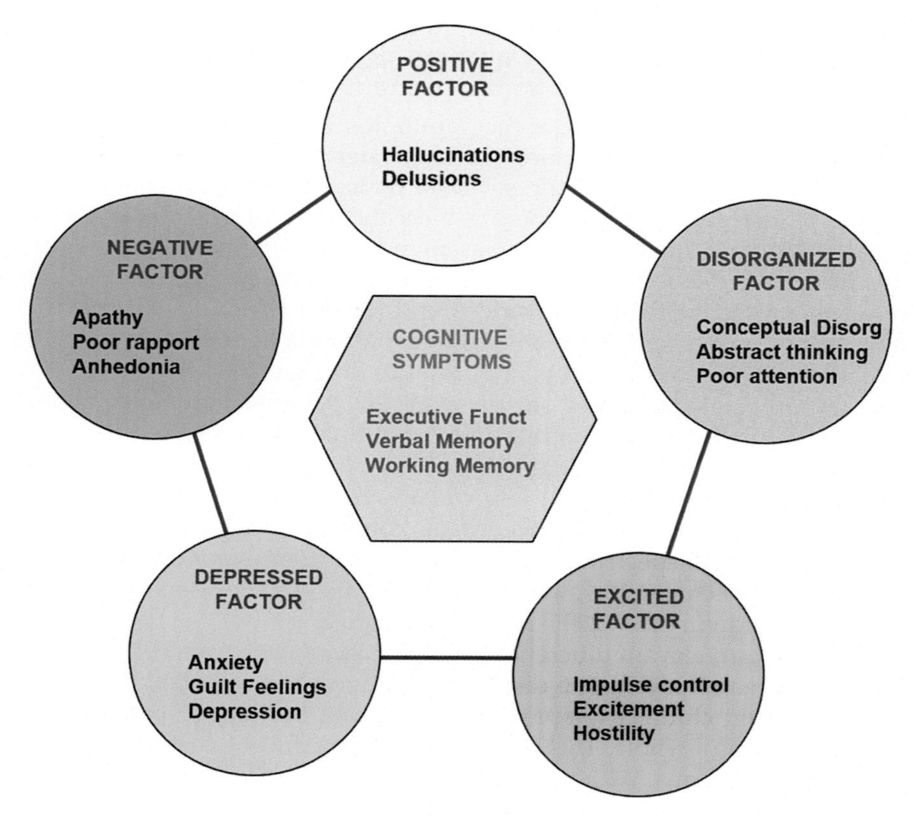

FIGURE 9.2 **The symptoms of schizophrenia.** The PANSS (Positive and negative syndrome scale) is a scale developed for measuring the severity of the symptoms of schizophrenia. It was originally composed of 30 items, measured on a scale of 1 to 7. More recently, using statistical analyses five different clusters of symptoms could be ascertained, and the total number of items was reduced to 20. In addition, although not measured in the PANSS, patients with schizophrenia also suffer from cognitive deficits.

in assessing these cognitive deficits, and it seems safe to say that patients with schizophrenia have a global cognitive deficit, with reduced functioning in virtually every cognitive domain (Marder and Fenton, 2004; Marder et al., 2004; Nuechterlein et al., 2004). However, superimposed on this general cognitive deficit is a more specific disturbance in especially executive functioning, attention and verbal and working memory (Heinrichs and Zakzanis, 1998; Mesholam-Gately et al., 2009). An important research question is whether these cognitive deficits are stable, or (as suggested in Fig. 9.1) slowly deteriorate over time. There is no doubt that cognitive deficits are present very early in the course of the illness and are seen in drug naïve patients (Fatouros-Bergman et al., 2014). Moreover, in a recent longitudinal study the cognitive deficits were stable over a 10-year period (Dickerson et al., 2014). However, it is important to note that patients in this study had a mean duration of illness of 22 years. Thus it is possible that a cognitive decline occurs within the first years after disease onset. Interestingly, cognitive deficits are also found in non-affected siblings of patients with schizophrenia, although to a lesser degree (Cannon et al., 1994).

THE EPIDEMIOLOGY OF SCHIZOPHRENIA

Schizophrenia occurs in about 1% of the population with the risk being more or less the same for males and females. Nonetheless, some gender differences do occur. As illustrated in Fig. 9.3, the age of onset is slightly earlier in males (18–25 year) compared to females (20–30 years). Moreover, in females there is a second peak of onset after the menopause (Hafner et al., 1993), which has been taken as an indication that hormones such as estrogen may play a protective role in females with schizophrenia. As estrogen levels drop after the menopause this protective factor is removed and (latent) cases of schizophrenia emerge. In line with this hypothesis, the outcome of schizophrenia appears somewhat less severe in female patients (Seeman, 1986). It is important to distinguish between age of onset, and "age of first presentation/diagnosis" (Scully et al., 2002). Given that most patients often do not visit a psychiatrist for a considerable amount of time (often referred to as the duration of untreated psychosis) the actual age of onset may precede the age of first diagnosis by several years. In the study by Hafner and coworkers, the authors therefore used a retrospective interview to determine the "true" age of onset. Overall, although the effect is not as significant as in autism spectrum disorders (Chapter 8) or major depression (Chapter 7) some differences between genders, with males more likely to develop (severe) schizophrenia than females (Baldwin et al., 2005) has been reported.

In spite of subtle but relevant differences in prevalence rates across countries and regions, schizophrenia is a major psychiatric disorder all around the world, and given its relative early age of onset and chronicity represent a significant burden for patients and families, as well as for society as a whole. This is evident both in terms of years lived with disability as well as in economic terms. With respect to the former, the global burden of disease study calculated the disability adjusted life years (DALY) in 2010 to be 15 million, an increase of 43% in 20 years (Murray et al., 2012). Although this is significantly less than disorders such as substance abuse and major depression, it leads to significant costs for society. In a recent study in Europe, the total costs of psychotic disorders was estimated to be over 93 billion Euro (Gustavsson et al., 2011), making it the third most expensive brain disorder after dementia

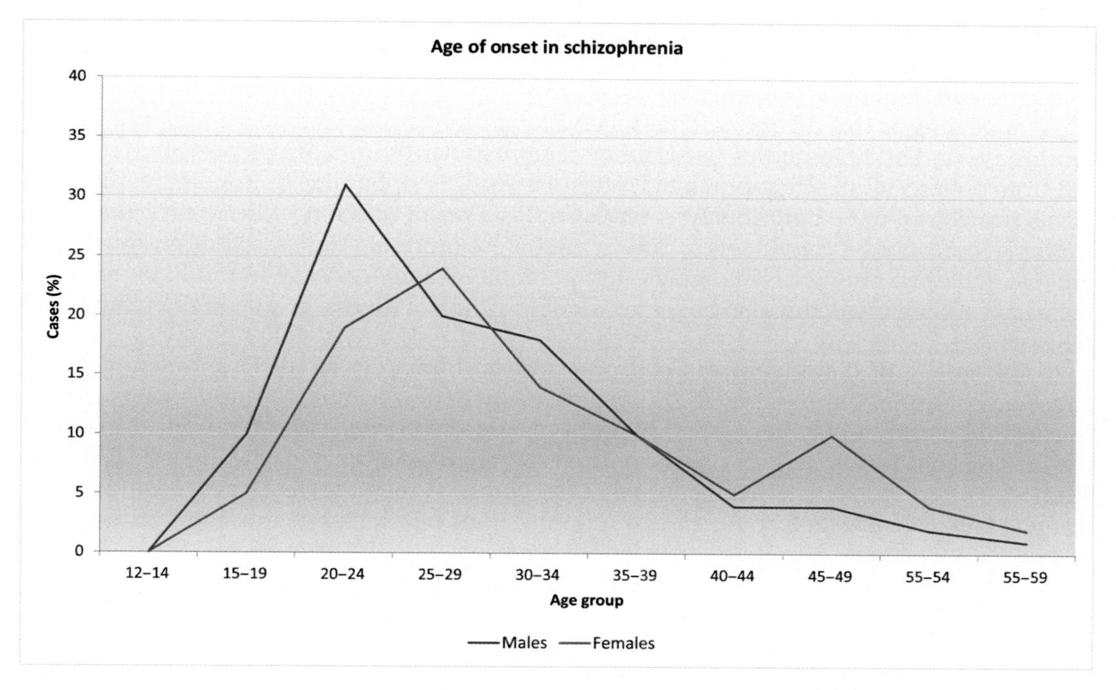

FIGURE 9.3 **The age of onset of schizophrenia.** Hafner and his coworkers (1993) studied the age of onset of schizophrenia in both male and female patients. The results show that the onset is slightly earlier in males, and has a slightly sharper peak (18–25 years in males vs 20–30 years in females). In addition, there was a second (albeit smaller) peak in females short after the menopause. Overall, there seems to be little difference in incidence between males and females.

(105 billion) and mood disorders (113 billion). These high costs are especially due to the so-called indirect costs (such as lost production due to absence of work or early retirement). In fact, for psychotic disorders the indirect costs alone were estimated at almost 65 billion euro.

THE NEUROBIOLOGY OF SCHIZOPHRENIA

From the description of the symptoms it is obvious that patients with schizophrenia suffer from symptoms belonging to several different classes or clusters, including positive, negative and cognitive symptoms. The factor analysis of the PANSS (Fig. 9.2) further supported this subdivision. It is therefore logical to assume that the underlying neurobiological substrate is also different for these different classes.

With respect to the positive symptoms, it has long been thought that dopamine was a key neurotransmitter. This hypothesis was based on two principle lines of research: First of all as early as 1958, it was well known that amphetamine (a dopamine releasing drug) could induce psychotic symptoms closely resembling the positive symptoms of schizophrenia (Connell, 1958). Secondly, antipsychotics, which have a therapeutic effect especially on the positive symptoms (as discussed later) have one common denominator: they block the dopamine D_2 receptor. These two sets of data together suggest that positive symptoms result from an increase in dopamine transmission.

As appealing as this theory is, it has taken a long time to substantiate this, mainly for technical reasons. The first studies used either cerebrospinal fluid (CSF) or post-mortem material from patients with schizophrenia. Unfortunately, CSF dopamine or dopamine metabolite levels turned out to be notoriously variable and as a result differences between patients and controls could not be convincingly determined. Post-mortem studies, although much more reliable, have the strong disadvantage that potential differences may not only be related to the cause of the illness, but also be the result of the chronic illness, or due to the treatment or other causes.

It was not until the advent of positron emission tomography (PET) and single photon emission computer tomography (SPECT) that researchers were able to study changes in neurotransmitter activity (such as dopamine) in patients with schizophrenia from the onset of the diseases (and even in high risk individuals before the outbreak of the illness). The initial studies focused on changes in dopamine D2 receptors as they are the target of antipsychotic drugs. However, although some groups found a significant increase, most studies especially in drug naïve patients found no consistent difference.

The real breakthrough came with studies looking at the presynaptic aspect of the dopamine transmission. These studies are based on the rationale that the endogenous dopamine presynaptically released will compete with a PET or SPECT ligand that selectively binds to the dopamine D_2 receptor (Fig. 9.4). Thus, under normal circumstances the radioactivity seen in the SPECT/PET scan is a measure of only that part of the D_2 receptors occupied by the ligand. When individuals are treated with the drug α-methyl-para-tyrosine (αMpT), dopamine synthesis is inhibited as the drug selectively blocks tyrosine hydroxylase, the rate-limiting enzyme in the dopamine synthesis. Thus, the amount of dopamine able to compete with the radioactive ligand decreases, and the same concentration of the ligand will now be able to bind more D_2 receptors, and thus radioactivity increases. The difference between before and after αMpT is an indication of the amount of baseline dopamine release (see the upper part of Fig. 9.4). Using this technique it was shown that the increase in radioactivity was larger in patients with schizophrenia than in control subjects, indicating higher baseline dopamine release (Hietala et al., 1995). Subsequently, it was shown that the amphetamine-induced release of dopamine was also higher in patients with schizophrenia (Laruelle et al., 1996), using a similar rationale, that is, the amphetamine-induced dopamine release competes with the radioactive ligand (see the lower part of Fig. 9.4). Subsequent studies have found that baseline and amphetamine-induced dopamine release are correlated (Abi-Dargham et al., 2009). Interestingly, the enhanced amphetamine-induced release in patients was also seen in drug naïve patients and was strongly correlated with the positive (but not the negative) symptoms). In line with this, the release returned to normal during remission (Laruelle et al., 1999). These data clearly show that dopamine release is altered in patients with schizophrenia, and there is even some evidence that it may already be enhanced in individuals at high risk. In a longitudinal study it was shown that such high-risk individuals have high baseline dopamine release, which increases further when florid psychotic symptoms occur (Howes et al., 2011).

Altogether, there is now overwhelming support for an increased presynaptic dopaminergic capacity in patients with schizophrenia (Howes et al., 2012), the question that remains

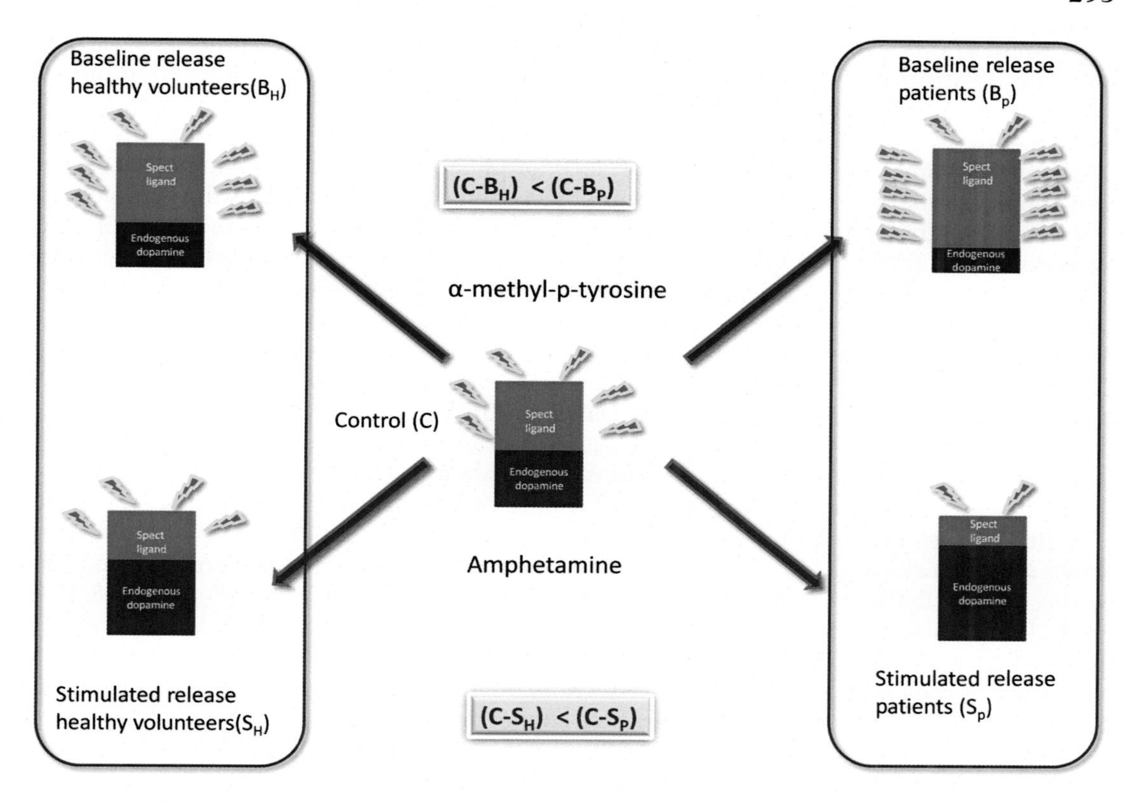

FIGURE 9.4 **Testing the dopamine hypothesis of schizophrenia.** Using brain imaging techniques it has now been convincingly shown that schizophrenia is accompanied by an increased baseline (top part) and amphetamine-induced (lower part) dopamine release. The methodology is based on the principle of competition between endogenous dopamine (blue) and a radioactive ligand (red). Under control conditions (center), part of the dopamine receptors will be occupied by the radioactive ligand and part of the receptors by dopamine. When treated with α-methyl-p-tyrosine the dopamine synthesis is blocked leading to a significantly reduced release of dopamine. As a result, the radioactive ligand has less competition and can bind more receptors. The difference between before and after α-methyl-p-tyrosine is a measure of the normal (baseline) release of dopamine. Conversely, after treatment with amphetamine, dopamine is released, meaning there is a stronger competition and less radioactive ligand can bind to the receptor. The difference before and after amphetamine is a measure of the stimulated release of dopamine. Studies have shown that both the baseline and stimulated release of dopamine is larger in patients with schizophrenia.

is whether the increase is seen throughout the brain or whether it is limited to a specific subset of dopaminergic cells (and terminals). With an increase in spatial resolution of SPECT analysis in recent years, it has been possible to address this question and it now appears that the increased dopaminergic activity is indeed restricted to a specific set of neurons in the anterodorsal part of the caudate nucleus (part of the striatal complex). Somewhat surprisingly, there was no increase in the ventral striatum (Kegeles et al., 2010), a region traditionally linked to schizophrenia. However, dopamine release in this area seems to be much more related to euphoria than to the positive symptoms of schizophrenia (Drevets et al., 2001).

In spite of this strong evidence linking dopamine to (the positive symptoms of) schizophrenia, the disorder is much too complex to be due to only one neurotransmitter. Indeed, many other neurotransmitters and neuropeptides have been implicated in the disorder, most prominently the glutamatergic system. This is primarily based on findings that drugs that block the glutamatergic NMDA receptor such as phencyclidine and ketamine can induce schizophrenia-like symptoms in healthy volunteers and can exacerbate existing symptoms in patients with schizophrenia (Luby et al., 1959; Malhotra et al., 1997). These and other findings have suggested that schizophrenia might (in part) be related to a reduced glutamatergic functioning (Jentsch and Roth, 1999; Marsman et al., 2013). Unfortunately, studying glutamatergic functioning in humans has proven elusive, and the lack of a specific radioactive ligand that binds to glutamatergic receptors has hampered research into this area. Moreover, pharmacological studies using drugs that enhance glutamate neurotransmission have so far not led to any significant breakthrough (Porter and Dawson, 2015).

The (potential) alteration in glutamatergic neurotransmission may, however, be related to changes in brain connectivity (Box 9.2). In recent years it has become increasingly clear that patients with schizophrenia show abnormal communication between different brain regions. The first studies used regional cerebral blood flow to investigate the activity of brain structures during tasks and found clear disturbances in the correlational patterns between the cortex and the thalamus and between frontal and temporal regions (Volkow et al., 1988; Frith et al., 1995), leading to the disconnection hypothesis (Friston and Frith, 1995), later more aptly referred to as the dysconnectivity model (Stephan et al., 2006). The renaming was deemed necessary as the Latin "dis" refers to a reduction, while the Greek "dys" refers to"bad", thus allowing for both reductions and increases in connectivity.

Studies in patients with schizophrenia have provided ample support for the dysconnectivity hypothesis, both in terms of structural and functional connectivity. Whereas earlier studies had already identified consistent reductions in white matter, more recent studies confirmed reduced levels of overall structural connectivity in schizophrenic patients (Zalesky et al., 2011), especially between frontal, temporal and parietal cortical areas. Overall, the analysis of structural networks leads to the hypothesis that the brains of patients with schizophrenia are more segregated with an overall decrease in global communication (van den Heuvel and Fornito, 2014). Studies investigating functional connectivity also overwhelmingly support altered connectivity, emphasizing especially deficient prefrontal connectivity, including prefrontal-subcortical, prefrontal-temporal and prefrontal-parietal functional connectivity.

An interesting discrepancy between structural and functional connectivity studies is that while structural analyses almost without exception show reduced connectivity, functional studies have reported both increased and decreased connectivity. In addition, whereas structural connectivity studies emphasized reduced integration and enhanced segregation, many functional studies seem to point to reduced segregation. There are several possible explanations for these seemingly contradictory findings, including the fact that (as mentioned above) structural and functional connectivity are not necessarily correlated. For instance, many long connection fibers contain the excitatory neurotransmitter glutamate or the inhibitory neurotransmitter GABA. One can imagine that a reduction in the structural connectivity of either can lead to different effects in the functional connectivity. Moreover, there may be a fundamental difference between resting state connectivity and connectivity during cognitive

BOX 9.2

BRAIN CONNECTIVITY

An important new concept in our understanding of the functioning of the brain is the concept of connectivity. Brain connectivity is directly related to the cooperative activity between different brain regions. Thus in contrast to the more traditional view that one specific brain region is dysfunctional (for instance the degeneration of dopaminergic cell bodies in Parkinson's Disease), the assumption is that the normal communication between brain regions is disturbed.

Currently, connectivity is often subdivided into structural and functional connectivity (van den Heuvel and Fornito, 2014). Structural connectivity is usually measured using diffuse weighted imaging, which measures how water molecules are constraint by white matter fibre tracts (Basser et al., 2000). Several different metrics have been developed to assess structural connectivity and the reader is referred to a recent review for more details (van den Heuvel and Fornito, 2014). Although structural connectivity can be assessed fairly quickly (in about 20 min), its low spatial resolution (mm) and the relatively indirect nature of the connectivity estimates are obvious limitations of the technique.

Functional connectivity refers to the statistical relationship between specific physiological signals in time and are generally assessed using techniques such as functional magnetic resonance imaging (fMRI), electroencephalography (EEG) or magnetic electroencephalography (MEG). The rationale behind this technique is based on the construct that synchronization between two brain structures (either at rest or during the performance of a specific task) reflects communication between these regions. It is important to realize that, although functional connectivity is somehow constraint by anatomical connectivity, the two are not necessarily correlated with each other. In fact in many cases, strong functional connectivity can occur in the absence of a direct anatomical connection.

tasks. Indeed studies have shown that frontal-parietal networks are especially relevant from cognitive functioning, while the frontal-accumbal network is more related to reward assessment (Jiang et al., 2013; Fischer et al., 2014). One of the major challenges will therefore be to investigate the relationship between disturbances in connectivity (functional and structural) and specific symptoms in patients with schizophrenia. In a recent study this was attempted, at least at the level of volumetric brain region analysis (Zhang et al., 2015). In this study the researchers subdivided patients in clusters with predominantly positive, predominantly negative or predominantly disorganized symptoms (using a three factor model of the PANSS, rather than the 5 factor model discussed in Fig. 9.2). Subsequent grey matter volume was measured and subgroups compared. The analysis showed the largest differences between the groups for the orbitofrontal cortex, being smallest in patients with predominantly positive symptoms, intermediate in patients with predominantly negative symptoms, and largest in the group with predominantly disorganized symptoms. In addition, the cerebellum was significantly smaller in the group with prominent negative vs positive symptoms.

THE PHARMACOLOGICAL TREATMENT OF SCHIZOPHRENIA

Until the early 1950s the treatment of schizophrenia consisted predominantly of sedating the patients (either chemically or through physical restraint) or using shock therapy such as cardiozol or insulin. The rationale behind this type of therapy (introduced by the Austrian-American psychiatrist Manfred Sakel in 1927) was based on the (erroneous) theory that epilepsy and schizophrenia rarely co-occur.

From 1953 onwards, everything would change for patients with schizophrenia. While experimenting with methods to improve the anaesthesia, the French surgeon Henri Laborit had already experimented with promethazine (a drug belonging to the class of the phenothiazines), and turned to the pharmaceutical company Rhone-Poulenc for a more potent sedative. Shortly before this, Paul Carpentier had synthesized chlorpromazine and Laborit experimented with it. His results were published in two papers (Laborit et al., 1952 ; Laborit and Huguenard, 1951) and especially his description of "disinterest" that he noted in his patients attracted the attention of the psychiatrist Jean Delay and Pierre Deniker, who subsequently tried it in patients with schizophrenia. The results were astounding as patients were not just sedated, but actually showed a reduction in hallucinations and delusions (Deniker, 1983). Chlorpromazine rapidly became the drug of choice for the treatment and even 30 years later it was still the second most prescribed antipsychotic (Ban et al., 1984). The introduction of chlorpromazine was rapidly followed by a large number of new antipsychotics, belonging to several different chemical classes such as the thiothixenes (tricyclic compounds, like the phenothiazines but with the central ring nitrogen replaced by carbon) such as chlorprothixene and flupenthixol; butyrophenones such as haloperidol and diphenylbutylamines (structurally related to the butyrophenones) such as pimozide and fluspirilene. Although large structural differences between these drug are apparent, they all showed significant improvement of the psychotic symptoms in patients with schizophrenia. However, it was also soon realized that these drugs had significant side effects, reminiscent of the symptoms of Parkinson's disease, including rigidity or bradykinesia (collectively referred to as extrapyramidal side effects). For this particular reason, these drugs were referred to as "neuroleptic" drugs (from the Greek "that takes hold of the nerves").

A second break-through in the treatment of patients with schizophrenia came with the re-introduction of clozapine in 1989. Originally synthesized and tested in the 1960s this drug showed some unique properties. A major comparison with chlorpromazine showed that it was effective in otherwise therapy resistant patients (Kane et al., 1988). This particular characteristic, in addition to its significantly lower potency to induce extrapyramidal side effects led to the reintroduction of clozapine in most countries in the world, albeit under strict conditions where white blood cell counts need to be checked regularly.

The findings that clozapine induce virtually no extrapyramidal side effects echoed some earlier studies showing that some antipsychotics such as thioridazine, induces less extrapyramidal side effects than others (Cole and Clyde, 1961; Laskey et al., 1962). This defied the idea voiced by Delay and Deniker and later also by Haasse that extrapyramidal side effects were an integral part of antipsychotic drugs. As a result, such drugs were called "atypical antipsychotics" as opposed to the typical or classical antipsychotics. However, as we will see later, the differences between classical and atypical antipsychotics is less obvious as previously thought and many researchers and clinicians now have abandoned this terminology.

Nowadays, antipsychotic are most commonly subdivided into first- and second-generation, with 1989 (ie, the re-introduction of clozapine) as the dividing point. All drugs introduced before this data are referred to as first-generation, while all those after that are referred to as second-generation. To this latter group belong drugs such as risperidone, olanzapine, quetiapine and of course clozapine. In addition to the difference in time, many of these second generation antipsychotics are also characterized by a rich pharmacology, that is, they affect multiple neurotransmitter receptors (as discussed later).

With such a large number of different antipsychotic drugs and their subdivision in classical/atypical or first/second generation, an important question is whether there are fundamental differences in the therapeutic (beneficial or side) effects between these drugs. This is an issue that has been investigated in many different trials, both small scale and large multicenter studies, including the CATIE (Clinical Antipsychotics Trial Intervention Effectiveness), CUtLASS (Cost Utility of the Latest Antipsychotics drugs in Schizophrenia Study), SOHO (Schizophrenia Outpatients Health Outcomes), CAFÉ (Comparison of Atypicals in First Episode) and EUFEST (European First Episode Schizophrenia Trial) studies. Although there are clear differences between these studies, in terms of drugs used, patients groups and outcome parameters (Ellenbroek, 2012), the overall conclusions are that while there are subtle differences between different drugs, the similarities outweigh the differences. The differences seem predominantly related to the side effects where first generation antipsychotics were more liable to induce extrapyramidal side effects, while second generation drugs were more likely to cause weight gain and metabolic problems. In an extensive meta-analysis of over 43,000 patients Leucht and colleagues analysed all studies in which multiple antipsychotics were compared in head to head comparisons (Leucht et al., 2013). As is evident from Fig. 9.5, the therapeutic differences (as measured by symptoms improvement) did not differ that much: the standardized mean differences range from 0.33 for iloperidone to 0.88 for clozapine, while the extrapyramidal side effects differed significantly more (from 0.30 for clozapine to 4.76 for haloperidol). Moreover, the analysis confirmed that clozapine was the most effective in symptom reduction and least liable to induce extrapyramidal side effects. However, it does induce other side effects such as weight gain and sedation (although when patients are very agitated this may also be an advantage) and most importantly agranulocytosis, a significant reduction in white blood cell count which can be fatal.

A final aspect of the therapeutic effects of antipsychotics that needs to be addressed is which of the different symptoms of schizophrenia are influenced by these drugs. As discussed above (Fig. 9.2) the symptoms of schizophrenia can be subdivided in several different clusters. Most of the studies so far have focused only on the positive, negative, and cognitive symptoms. Although there again appear to be some subtle differences, overall the consensus is that whereas antipsychotics significantly reduce the positive symptoms, their effect on the negative and cognitive symptoms are much less. Again clozapine seems to be the outlier, being more effective against the negative symptoms, and in addition, only clozapine appeared effective in otherwise therapy resistant patients (McEvoy et al., 2006).

Given the overall similarity in therapeutic efficacy, it seems logical to assume that all antipsychotic drugs also have a similar mode of action. Moreover, given the dominant effect on the positive symptoms, which have been linked to increases in dopaminergic neurotransmission (as discussed earlier), a dopamine blocking action of antipsychotic drugs has long been suspected. The first indications that this might be the case go back to the early

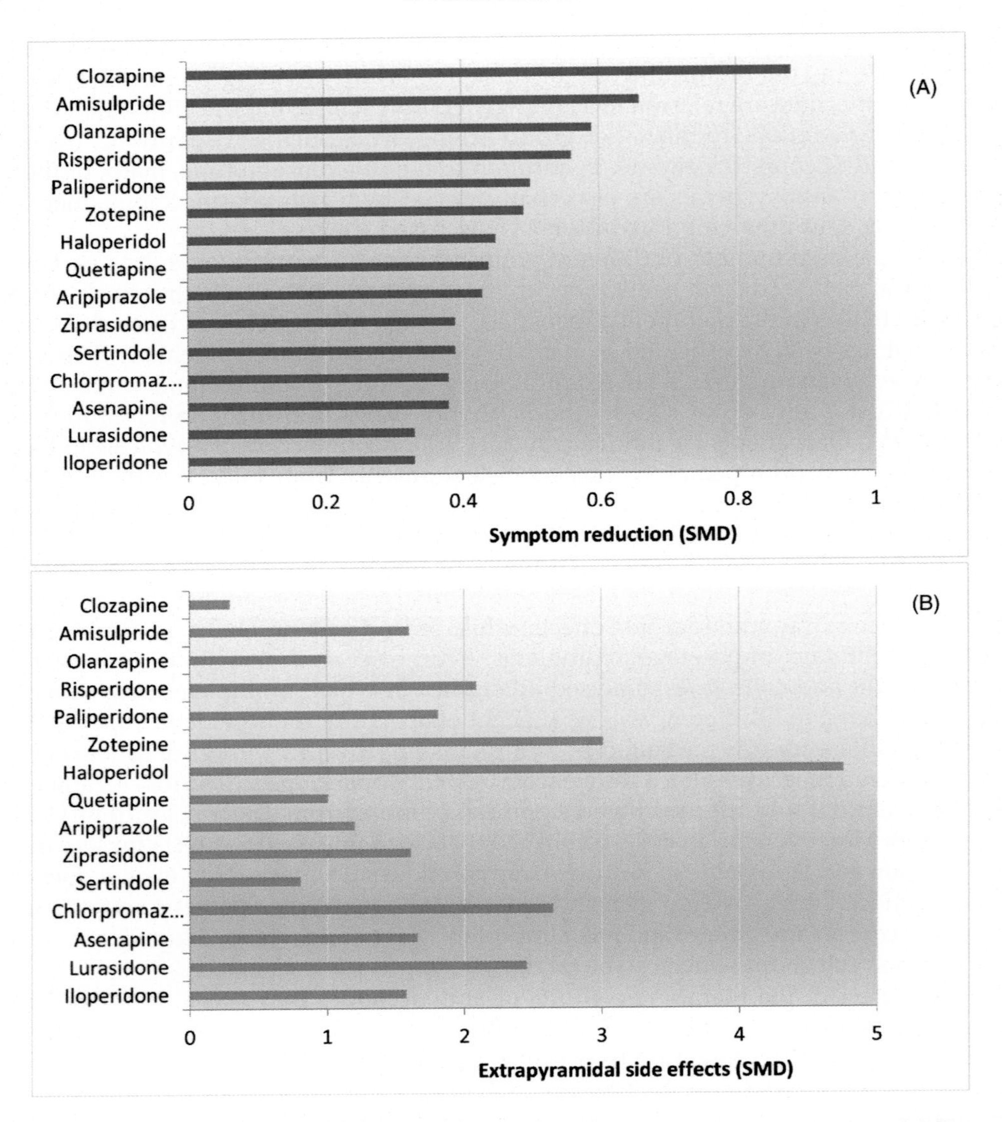

FIGURE 9.5 **The differences between antipsychotic drugs.** (A) In spite of the large number of different antipsychotic drugs, the differences between them are relatively small, especially in relation to symptoms reduction. (B) Larger differences exist with respect to the induction of extrapyramidal side effects (notice the difference in the X-axis). *Data from Leucht et al. (2013).*

and mid1960s when Carlsson and Lindqvist found that antipsychotics enhanced the production of catechoaminergic (dopaminergic and noradrenergic) metabolites (Carlsson and Lindqvist, 1963). Three years later it was van Rossum who for the first time suggested that all antipsychotics block dopamine receptors, as they all reversed L-DOPA induced behaviours (van Rossum, 1966). A decade later two groups independently showed that there was a significant correlation between the binding of antipsychotics to dopamine receptors and their clinical daily dose (Creese et al., 1976; Seeman et al., 1976). Another decade later, using PET and SPECT imaging, it was shown that antipsychotics also bind to dopamine receptors in the brains of patients with schizophrenia (Farde et al., 1988; Farde et al., 1989; Nord and Farde, 2011; Yilmaz et al., 2012). After the initial studies by Creese and coworkers, it has become clear that there were multiple dopamine receptors, and studies in humans showed that antipsychotics predominantly bind to the D_2 receptor (Farde et al., 1989). When comparing the dopamine receptor occupancy of different antipsychotics it became clear that for the vast majority of drugs, occupancy of 60%–75% was necessary to obtain a robust therapeutic effect. Unfortunately, increasing occupancy above 80% in most cases led to the induction of extrapyramidal side effects. Thus the therapeutic window for most antipsychotics is very limited. This suggests that the main difference between first and second generation of antipsychotics is due to the steepness of the dose response curve (Fig. 9.6). Again, a notable exception to this general rule is clozapine, which appears to be therapeutically active at occupancy

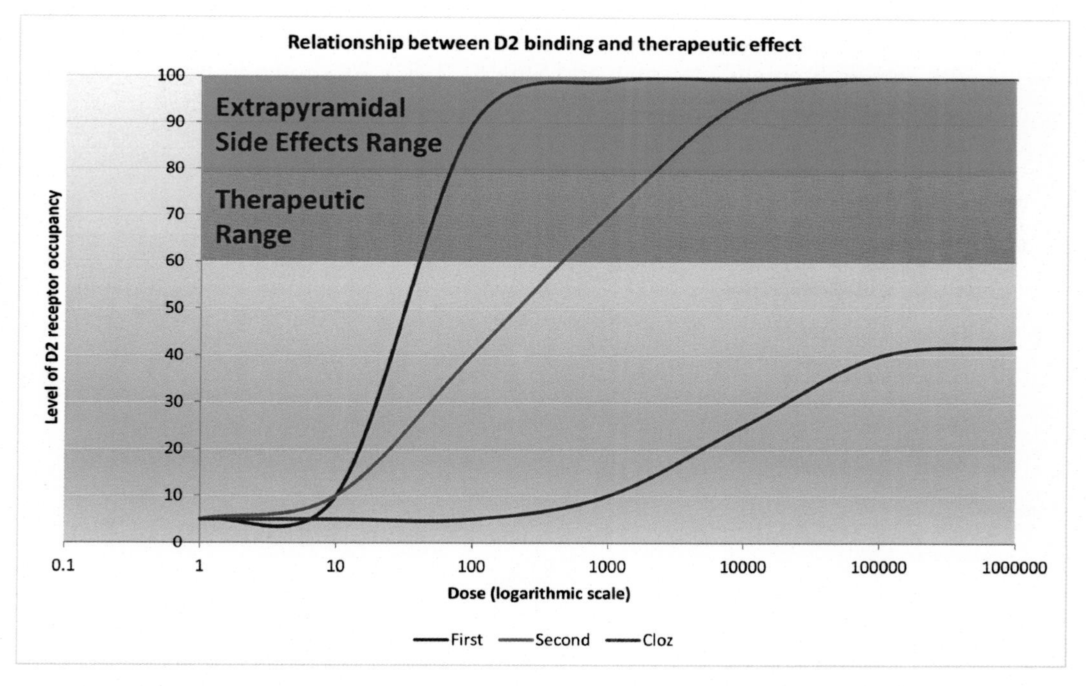

FIGURE 9.6 **The relationship between dopamine D_2 receptor blockade and clinical effects.** Brain imaging studies have clearly shown that blockade of the dopamine D_2 receptor is responsible for both the therapeutic effects on the positive symptoms and the extrapyramidal side effects. However, the therapeutic effects require a lower level of blockade. As second generation antipsychotic are somewhat less likely to induce extrapyramidal side effects (Fig. 9.5) their dose response curve is considered to be less steep than for first generation antipsychotics. Imaging studies with clozapine have shown that it generally only blocks about 35%–40% of the dopamine receptors, and hence is virtually devoid of extrapyramidal side effect. What additional receptors are involved in the therapeutic effects of clozapine is largely unknown.

levels much lower than the other antipsychotics (Fig. 9.6), suggesting that indeed clozapine has a unique mode of action.

So far the discussion of the effects of antipsychotics on the brain have been limited to dopamine and in particular the dopamine D_2 receptor. Most antipsychotics, however, affect many other receptors. Most of the first generation antipsychotics also influence serotonergic (especially the 5-HT_{2A}) and noradrenergic (especially the α_1) receptor. Second generation antipsychotics in addition affect a multitude of other receptors, including histaminic (H_1) and muscarinic (especially m_1 and m_2) receptors. Although some of these additional receptors may contribute to the therapeutic effect, most of these have been linked more to the side effects. Thus blockade of α_1 receptors is known to cause orthostatic hypotension, blockade of H_1 receptors is known to contribute to sedation and weight gain, and blockade of m_1 receptors can cause dry mouth, blurred vision and constipation.

THE AETIOLOGY OF SCHIZOPHRENIA

Family studies have clearly shown that schizophrenia has a strong genetic component. In a famous study, the risk of schizophrenia was estimated in 1st, 2nd and 3rd degree relatives of patients with schizophrenia (Gottesman and Shields, 1982). As shown in Fig. 9.7, the risk of

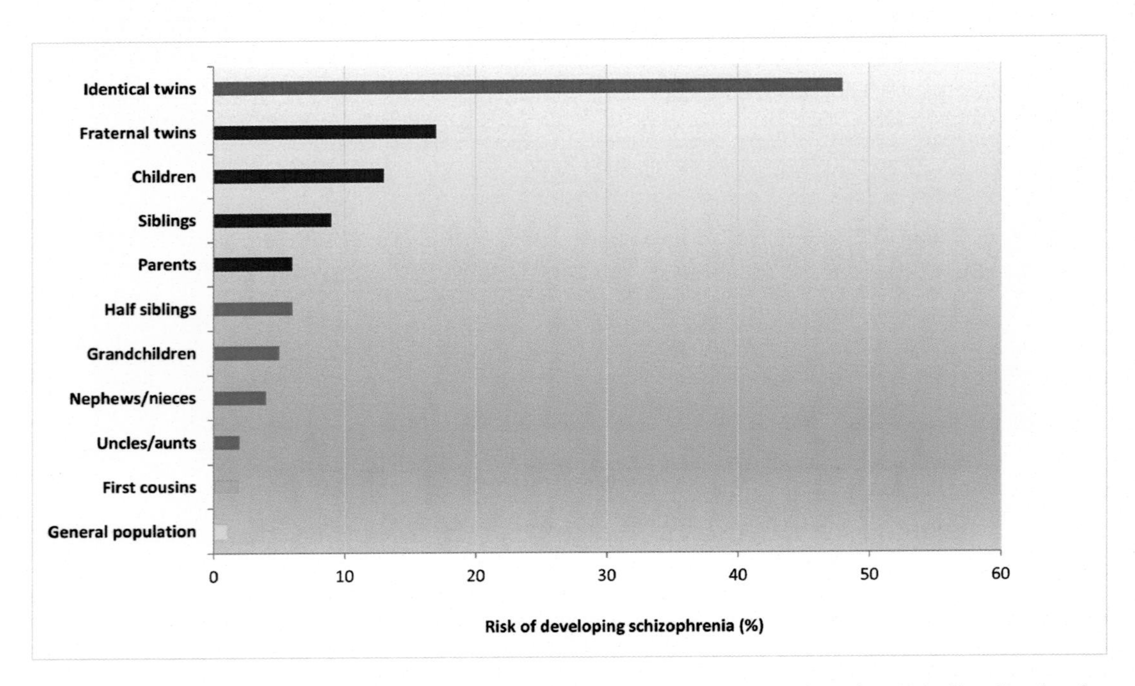

FIGURE 9.7 **The risk of developing schizophrenia.** The risk of developing schizophrenia is directly related to the percentage of genes shared with an individual suffering from the disease. Thus while the risk in the general population (yellow) is about 1%, it increased in third (orange), second (green) and first (blue) degree relatives. Finally, the largest risk occurs in monozygotic twins (red). *Data from Gottesman and Shields (1982).*

developing schizophrenia increases with an increase in shared genetic background, with by far the largest risk in monozygotic twins. However, at the same time, the concordance rate for monozygotic twins is only 50%, implying that schizophrenia is not "determined" by genes, but that certain genes enhance the susceptibility and that non-genetic factors are also important. An interesting and often underestimated aspect of Fig. 9.7 is the difference between fraternal twins (17%) and siblings (9%). The much higher rate in fraternal twins suggests that at least some of the environmental factors impinge early in life when the shared environment is largest. Adoption studies have further underlined the role of genetics (with the caveat discussed in Chapter 4 regarding the shared prenatal environment) as the risk of developing schizophrenia in adopted away children is related to the presence of illness in the biological and not the adopted parents (Kety 1968). Twin and adoption studies have also been used to calculate the heritability of schizophrenia (Box 4.1 for a discussion about genetic heritability) leading to estimates of h^2 of about 0.80 (Sullivan et al., 2003). Recently, however, this estimate has been challenged (Fosse et al., 2015), on the basis that classical twin studies assume that the shared environment is the same for monozygotic and dizygotic twins. Their analysis of the literature, however, suggested that for factors such as bullying, sexual abuse, physical maltreatment and emotional neglect or abuse, monozygotic twins showed significant higher correlations than dizygotic twins, thus violating the assumption that both sets of twins share the same environment.

Genetic Factors Contributing to Schizophrenia

Irrespective of the discussion about the exact value of the heritability, there is little doubt that schizophrenia has a significant genetic component. However, the exact nature of this genetic component is so far virtually unknown. One thing that is clear though is that there is no "schizophrenia" gene, that is, a single gene that can explain a significant proportion (ie, 50% or more) of the variance of schizophrenia (Giusti-Rodriguez and Sullivan, 2013). Although originally several mathematical models based on recessive genes with limited penetrance had been developed to explain the mode of transmission of schizophrenia [see (Kendler, 2015) for a historical analysis], it soon became clear that none of these models could adequately describe the genetic complexities of schizophrenia, leading to the polygenic theory of schizophrenia (Gottesman and Shields, 1967). This then leads to the question of how many genes are involved in the aetiology of schizophrenia. A recent mathematical analysis estimated the number of genes at over 100 (Paek and Kang, 2012). However, the model is based on a number of assumptions that are likely incorrect, such as all the genes are transmitted independently and environmental factors do not play a role. Indeed as we will as discussed later, in a recent study it was found that sets of between 4 and 19 were enough to account for ≥ 90% risk for specific sets of schizophrenic patients and a single set of 9 single nucleotide polymorphisms was responsible for 100% risk in a group of 9 patients (Arnedo et al., 2015).

Identifying the genes related to schizophrenia has proven to be a very difficult task. Indeed, probably no other disorder has been subjected to so many genetic studies, both with respect to targeted genetic analysis as well as to genome wide association studies. The results have led to a plethora of genes that have been (repeatedly) associated with schizophrenia. Most of these genes are relatively rare and/or confer only a small genetic risk (Giusti-Rodriguez and Sullivan, 2013), explaining the need for very large cohorts. The largest cohort currently being investigated is undoubtedly performed by the Psychiatric Genomic Consortium, with over 375 investigators in 22 countries. Together they analyze approximately 125,000 patients and controls worldwide, focusing on several different psychiatric disorders, including schizophrenia. The consortium has published several papers to date (Ripke et al., 2011; Ripke et al., 2013; Ripke et al., 2014), with the last one involving 36,989 patients and 113,075 controls. Overall this analysis identified 108 schizophrenia-associated genetic loci (83 of which had not been reported previously) which exceeded the genome-wide significance (set at $p \leq 5 \times 10^{-8}$). These loci were distributed throughout the genome with only chromosome 13 as an exception and were in line with the polygenic hypothesis of schizophrenia. Of the 108 loci, 75% included protein-coding genes and a further 8% were within 20 kilobases of a gene. However, only 10 loci could reliably be associated with a non-synonymous exonic polymorphism, suggesting that most of these variants exert their effect though altering gene expression rather than protein structure. Among the most interesting genes identified (Table 9.1) were the dopamine D_2 receptor (*DRD2*), several genes related to glutamate neurotransmission as well as several

TABLE 9.1 Major Single Nucleotide Variants and Copy Number Variations Implicated in Schizophrenia

Genes/regions	Remarks
SINGLE NUCLEOTIDE VARIANTS	
ANKK1	The "G" allele of rs1801028 is associated with enhanced risk
ANK3	Ankyrin 3 or ankyrin G: the "T" allele of rs10761482 is associated with enhanced risk
CACNA1C	Voltage gated Calcium channel ($Ca_v1.2$): the "A" allele of rs1006737 is associated with enhanced risk
CACNB2	β-subunit of Voltage gated Calcium channels: the "T" allele of rs2799573 is associated with enhanced risk
DRD2	Dopamine D2 receptor: the "C" allele of rs2514218 is associated with enhanced risk, reduced sensitivity to the therapeutic effects of antipsychotic drugs and enhanced side effects sensitivity
MHC	Major Histocompatibility Complex: the "G" allele of rs6904071 is associated with enhanced risk, disturbed episodic memory and reduced hippocampal volume
MIR137	microRNA 137: the "T" allele of rs1625579 is associated with enhanced risk
ZNF804A	Zinc Finger Protein 804A: the "A" allele of rs1344706 is associated with enhanced risk
COPY NUMBER VARIATIONS	
1q21.1	Deletion/duplication of 34 genes
3q29	Deletion of 19 genes
15q13.3	Deletion /duplication of 12 genes
16p11.2	Deletion /duplication of 29 genes
17q12	Deletion of 18 genes
22q11.2	Deletion /duplication of 53 genes
NRXN1	Deletion of one gene (Neurexin 1)
VIPR2	Duplication of one gene (VIPR2)

See text for further explanations.

genes related calcium channel signaling. A recent meta-analysis of single nucleotide variants in the D_2 receptor gene (as well as the TaqA1 variant originally associated with the D_2 but which is actually located in the nearby *ANKK1* (ankyrin repeat and kinase domain containing 1 gene) confirmed at least one variant (rs1801028,a C - G transversion in codon 311, resulting in a cysteine for serine substitution) which was significantly associated with schizophrenia (Yao et al., 2015). Especially the identification of the dopamine and glutamate neurotransmission related genes has generated quite substantial enthusiasm within the scientific world. However, it is sobering to realize, as the authors point out that there are an estimated 8,300 independent, mostly common, single nucleotide variants that contribute to the risk for schizophrenia. This further emphasizes the heterogeneity of the disorder and suggests that many different pathways may lead to schizophrenia. In this respect, a study using the Danish national registers, recently showed that the risk of schizophrenia (and schizophrenia spectrum disorders) was increased in children that had been diagnosed with a child or adolescent psychiatric illness (Maibing et al., 2015), irrespective of which diagnosis the child or adolescent received (including autism, spectrum disorder, attention deficit hyperactivity disorder, eating disorders, affective disorders and obsessive compulsive disorders).

In addition to the single nucleotide variations identified in genome wide association studies, copy number variations have also been identified in patients with schizophrenia. As discussed in more detail in Chapter 2, copy number variations are duplications or deletions of larger parts of the genome. Although much rarer than single nucleotide variants, they confer a significantly greater risk for schizophrenia (between 5 and 20%) than single nucleotide variants (Giusti-Rodriguez and Sullivan, 2013). One of the best known copy number variations is a deletion on 22q11.2, involving 53 genes. Patient with a 22q11.2 deletion suffer from what is known as velocardiofacial syndrome, involving cardiac and facial abnormalities, thymic aplasia, palate cleft and hypocalcemia. In addition, these patients often develop psychotic symptoms and have a 20% risk of developing schizophrenia (Debbane et al., 2006).

Finally, several non-coding genes have been associated with schizophrenia, including several long noncoding RNAs as well as micro RNAs. Especially the miRNA 137 (encoded by *miR137*) has emerged as a strong candidate in genome wide association studies (Ripke et al., 2013). As detailed on Chapter 5, miRNA molecules act by inhibiting the expressions of other mRNAs and miRNA 137 has a long list of known targets, including several of the genes also implicated in schizophrenia (such as *CACNA1C*). Moreover, miRNA is involved in neurodevelopment, including neural stem cell proliferation and differentiation as well as dendritic arborisation (Szulwach et al., 2010).

In addition to providing information on the genes related to schizophrenia, Genome Wide Association Studies have also been useful in excluding genes previously (strongly) associated with schizophrenia. For instance reelin is a large extracellular matrix glycoprotein that is involved in neuronal migration. Given the disturbance in neurodevelopment in patients with schizophrenia, coupled with some evidence that patients have reduced levels of reelin, has led investigators to search for single nucleotide variants in the *RELN* gene. Initially a common variant was associated with schizophrenia in women (Shifman et al., 2008) and with working memory (Wedenoja et al., 2010). However, most genome wide association studies have failed to replicate an association between the *RELN* gene and schizophrenia. Another gene that has been implicated in schizophrenia is *DISC1* (disrupted in schizophrenia 1). Like reelin, DISC1 is assumed to be involved in neurodevelopment, including neurite outgrowth, neuronal migration and synaptogenesis. Moreover, it is involved in glutamatergic neurotransmission, thus linking two major theories (the neurodevelopmental and the glutamate theory) in schizophrenia research (Hodgkinson et al., 2004). However, genome wide analyses again failed to substantiate this claim (Sullivan, 2013).

Thus the search for genes associated with schizophrenia continues and as mentioned earlier, given the small effect sizes (typically around 1 - 2%) requires very large numbers of patients (Sun et al., 2008). An alternative to using such large cohorts of patients, is to search for more homogenous populations, or more specific symptoms clusters. As an example, a recent study investigated the role of single nucleotide variants in *SLC6A4* gene coding for the serotonin transporter in patients with schizophrenia using the five cluster analysis of the PANSS (Fig. 9.2). The authors found a single variant (rs140700, a G–A transition) that was highly associated with schizophrenia, but only with the negative, depressive and anxiety clusters (Li et al., 2013). Likewise a single variant of the *COMT* gene (rs4680, a G–A transition, leading to a val–met substitution) has been specifically linked to alterations in executive function (Solis-Ortiz et al., 2010). Arnedo and coworkers recently attempted a new approach in which they combined genomics and phenomics (Arnedo et al., 2015). Rather than using the traditional genome wide association approach categorizing subjects simply as having schizophrenia or not, the authors analyzed the symptoms of their patients in great detail (i.e., in their largest cohort they analyzed 93 different symptoms). They first identified single nucleotide variant sets (i.e., genetic variants that cluster together within individuals regardless of the diagnosis) leading to a total of 723 non-identical sets, many of which consisted of a large number of genes. Interestingly, 42 sets were found that accounted for ≥70% risk for schizophrenia. These individual sets consisted of between 3 and 32 genes and were found in groups of 9 and 877 patients. Several of these sets predicted more than 90% of the risk for schizophrenia. For instance in a group of 9 patients, a set of 9 genes conferred a 100% risk for schizophrenia. In a subsequent analysis, these sets were studied for overlap between patients, with the idea that if completely independent gene set networks were found, this would be an indication of heterogeneity in aetiology. Indeed, these 42 sets could be combined into 17 distinct non-overlapping networks, thus providing strong evidence for aetiological heterogeneity. Likewise, using cluster analysis of the symptoms, the authors came to the conclusion that genotypic and phenotypic relationships could be grouped into 8 classes. A particular strong point of the study was that the results were tested in two different cohorts and the results were very similar.

Moderation of the Effects of Genes by Environmental Factors

The earlier discussion clearly indicates that no single gene can account for the heritability of schizophrenia, even if we accept a more conservative estimate like the 80% found in twin studies. One of the main shortcoming of genome wide association studies is the fact that they ignore the role of environmental influences. Especially if the influence of the gene is only apparent in interaction with an environmental factor, it may not be detected in genome wide association studies. Fortunately, there is increased research aimed at identifying such gene * environment interactions (Table 9.2). Among the environmental factors, cannabis use is probably studied the most. One of the first studies in this respect was performed by Caspi and coworkers (Caspi et al., 2005). They studied the well-known COMTval158met variant and found that val carriers were significantly more sensitive to the effects of adolescent cannabis use on inducing schizophreniform disorders and hallucinations (but not delusions). These data are similar to those obtained by others, showing enhanced cannabis induced psychotic experiences in val carriers (Henquet et al., 2006). However, others failed to confirm these interactions (Zammit et al., 2007; Kantrowitz et al., 2009; Zammit et al., 2011). Another gene that has been associated with an enhanced sensitivity to cannabis is the *AKT1* gene. Indeed several studies have found that carriers of the "C" allele in rs2494732, when exposed to cannabis, are more sensitive in terms of risk of for schizophrenia (van Winkel et al., 2011; Di Forti et al., 2012). This latter study illustrates the relevance of studying gene–environment interactions, as the authors found that the "C" allele itself was not more prevalent in patients with schizophrenia. In other words, if only the genetic component was investigated, the conclusion would be that the *AKT1* gene is not associated with schizophrenia. In an interesting study, a gene * gene * environment interaction was found between the *AKT1* gene (a different variant rs1130233), the dopamine transporter *SLC6A3* gene (a variable number of tandem repeats variant in the 3′ UTR) and cannabis use. After exposure of Δ^9-tetrahydrocannabinol, the risk of psychotic experiences was highest in individuals that carried both the "G" allele

TABLE 9.2 Examples of Gene * Environment Interactions in Schizophrenia

Genes	Environments	Effects	References
AKT1	Cannabis	rs2494732 C carriers had increased odds for schizophrenia with cannabis	Di Forti et al. (2012)
AKT1	OC	Individuals with OC were more likely to have the minor allele at the different variants	Nicodemus et al. (2008)
BDNF	OC	Individuals with OC were more likely to have the minor allele at rs2049046	Nicodemus et al. (2008)
BDNF	Childhood abuse	Met carriers exposed to childhood abuse had more psychotic like behaviors compared to val/val	Alemany et al. (2011)
BDNF	Childhood abuse	Met carriers exposed to childhood abuse had more cognitive deficits and smaller hippocampal volume compared to val/val	Aas et al. (2013)
COMT	Adolescent Cannabis	Cannabis use increased schizophreniform symptoms in val/val and val/met carriers	Caspi et al. (2005)
COMT	Cannabis	Cannabis induced psychotic symptoms strongest in val/val carriers	Henquet et al. (2006)
COMT	Stress	Stress increased more severe psychotic reactivity in met/met carriers	Collip et al. (2011)
DRD2	Cannabis	Carrier of the T allele at rs1076560 that use cannabis had higher risk of psychosis	Colizzi et al. (2015)
FOXP2	Childhood emotional abuse	Homozygous CC carrier at rs1456031 exposed to emotional abuse were more likely to experience auditory hallucinations	McCarthy-Jones et al. (2014)
HLA-DR1	Season of birth	Increased winter birth in patients with HLA-DR1	Narita et al. (2000)

See text for further explanations.
OC: Obstetric complications.

of the rs1130233 of the *AKT1* gene and the 9 repeat of the *SLC6A3* variable number of tandem repeat (Bhattacharyya et al., 2012).

Another often studied environmental factor is the season of birth effect. As discussed earlier (Table 4.2), patients with schizophrenia are more likely to be born in the winter and spring than in other seasons. Studies that looked at interactions with genetic factors found positive interactions with the 7-repeat allele of the *DRD4* gene (Chotai et al., 2003) and the *HLA-DR1* genotype (Narita et al., 2000). No influence was seen on the season of birth effect for other HLA variants [*HLA-A24* and *A26*, (Tochigi et al., 2002)] or for a variant (rs1801133) in the *MTHFR* gene (Muntjewerff et al., 2011).

The *COMT* val158met variant has not only been investigated in relation to cannabis use, but also in relation to stressors. Non-schizophrenic val carriers exposed to stressors showed increased levels of psychotic symptoms (Stefanis et al., 2007) and reported more feelings of paranoia (Simons et al., 2009). Interestingly, opposite patterns have also been reported for patients with schizophrenia, such as a greater increase in psychotic episodes in met carriers exposed to daily stressors (van Winkel et al., 2008). One possible explanation might be a gene * gene interaction, as some studies have shown that the moderating effect of stressors on the *COMT* met allele is dependent on the rs1801133 variation in the *MTHFR* gene (with only the "T" carriers showing the *COMT* * stress interaction. Additionally the influence of the COMT protein may differ between patients and healthy volunteers.

Another gene that has been studied in gene * environment interactions in relation to schizophrenia is the *BDNF* gene. This gene has a common variant (rs6265 G/A), a val/met substitution leading to decrease expression in the val/val genotype. Given that BDNF levels are negatively influenced by stress, the relation between the *BDNF*val66met genotype and stress has been the focus of several papers both in the field of schizophrenia as well as in other psychiatric disorders. Thus carriers of the met allele reported more psychotic episodes after childhood abuse than homozygous val/val carriers (Alemany et al., 2011) along with poorer cognitive performance and smaller hippocampal volume (Aas et al., 2013). When healthy volunteers were exposed to event stress, val/met carriers also showed more feelings of paranoia than met/met carriers (Simons et al., 2009). Interestingly, when exposed to social stressors, no such interaction was found.

In summary, several gene * environment interactions have been reported. Interestingly, and encouragingly, many of these interactions are with genes that have not reliably shown up in genome wide association studies (such as the *BDNF* and *COMT* gene), suggesting that a combined gene * environment analysis of schizophrenia (so called Genome Environment Wide Interaction Studies) has a much larger potential of identifying relevant aetiological factors compared to (now traditional) genome wide association studies.

ANIMAL MODELS FOR SCHIZOPHRENIA

Measuring Schizophrenia-Like Features in Animals

Due to the heterogeneity of the aetiology, the pathology and the symptoms schizophrenia, it has proven extremely difficult to develop animal models for schizophrenia. Moreover, many of the symptoms (Box 9.1) can only be determined by interviewing patients and are thus impossible to mimic in animals (hallucinations, delusions, disorganized speech etc.). Thus, especially with respect to the positive symptoms, our possibilities of mimicking schizophrenia are limited. One possible alternative has therefore been to look at biomarkers or endophenotypes. We will discuss the concept of endophenotypes more fully in Chapter 10. For the moment it is important to note that endophenotypes refer to signs or symptoms that can be objectively quantified and that are related to the disease in question. With respect to schizophrenia, a number of clinical (and preclinical) endophenotypes have been developed especially in the cognitive domain (Braff, 2015). One endophenotype or biomarker for the positive symptoms is an enhanced dopamine release. As mentioned in a previous section, patients with schizophrenia suffer from an increased basal and amphetamine-induced release, both of which are specifically related to the positive symptoms. Enhanced dopamine release can be measured using microdialysis, in which a probe is implanted directly into the brain allowing the measurement of extracellular levels of neurotransmitters such as dopamine. An alternative approach has been to study the amphetamine-induced increase in locomotion. Although this is indeed directly related to dopamine release, it should be noted that the enhanced dopamine release in patients is limited to the anterodorsal part of the caudate, while the amphetamine-induced hyperactivity is more related to the enhanced release from the nucleus accumbens. The behavioural consequences of enhanced dopamine release in the (rodent equivalent) of the anterodorsal caudate are as yet unknown. Thus it is at present unclear whether an enhanced amphetamine induced hyperactivity is an appropriate endophenotype/biomarker for the positive symptoms of schizophrenia.

Probably one of the most often used endophenotypes in schizophrenia research is a disturbance in prepulse inhibition. Prepulse inhibition (Fig. 9.8) refers to the reduction of the startle response to an intense stimulus, when this stimulus is preceded by a weaker stimulus.

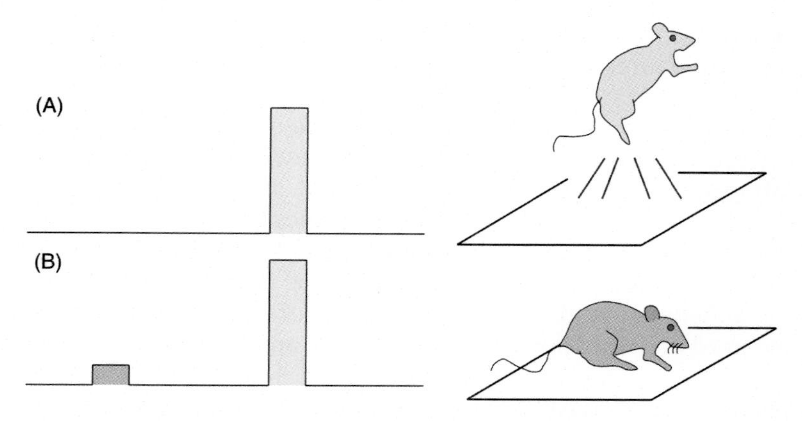

FIGURE 9.8 **The prepulse inhibition paradigm.** Prepulse inhibition refers to the reduction in startle response when the startle stimulus is preceded by a weak stimulus. (A) The procedure therefore involves two different types of trial. In the first trial (startle trial) the startle stimulus is presented alone and the startle response in recorded. (B) In the second trial (prepulse trial) a weak prepulse (which usually does not induce a measurable startle response itself) is given immediately (typically 60–100 ms) prior to the startle stimulus. The ratio of the response in (B) over (A) is an indication of the degree of prepulse inhibition.

Although prepulse inhibition can be observed in many sensory domains (and even using cross domain stimuli, ie, the prepulse is in a different domain than the startle stimulus) it is most often measured using acoustic stimuli. Ever since the seminal paper by David Braff and coworkers (Braff et al., 1978), prepulse inhibition has repeatedly been shown to be reduced in patients with schizophrenia (Braff et al., 2001). However, it should be pointed out that prepulse inhibition is also reduced in many other psychiatric and neurological disorders (Kohl et al., 2013). Nonetheless, the fact that prepulse inhibition can be assessed in humans and animals with virtually identical methods has made it increasingly popular as a tool in animal research. Related to prepulse inhibition is a phenomenon called P_{50} gating (Adler et al., 1982). Like prepulse inhibition, this paradigm involves two (usually acoustic) stimuli, where the response to the second stimulus is reduced by the presence of the first. However, unlike prepulse inhibition (where the first stimulus is weaker and does not induce a response), in P50 gating both stimuli are equally strong and the response is measured as an EEG evoked potential (as it is a positive wave occurring 50 ms after the stimulus it is referred to as P_{50}). As discussed in other papers, there are additional differences between prepulse inhibition and P_{50} gating (Ellenbroek et al., 1999; Braff et al., 2004).

In contrast to the positive symptoms, more success has been made in modelling the negative and cognitive symptoms of schizophrenia. With respect to the former, deficits in social behaviour and anhedonia are most often studied. Social behaviour is often studied in dyadic interaction (Sams Dodd, 1996) although it can also be studied in larger groups for instance in monkeys (Ellenbroek et al., 1996). Other social behavioural procedures are also used, including the social approach avoidance paradigm which is discussed in more detail in Chapter 8 (fig 8.14). With respect to anhedonia, most of the paradigms are related to alterations in reinforcement learning and motivation (Markou et al., 2013). One of the earliest paradigms, originally developed for depression is the consumption of a palatable solution, such as sucrose or saccharine. The idea behind this is that under normal circumstances, animals prefer drinking sucrose or saccharine and consider it pleasurable. Hence a reduction is indicative of a reduction in pleasure (anhedonia). An alternative, but related procedure is intracranial self-stimulation, which is also based on the same principle, namely that electrical stimulation of (specific areas of) the brain is inherently rewarding, and therefore a reduction is indicative of anhedonia. The advantage of the second paradigm is that it can also be used to measure the effort an animal is willing to invest to obtain the reward by using a progressive ratio approach (as discussed in Chapter 6). Similar to this are approaches where animals have to choose between two different outcomes: an easy, low reward or a more difficult but higher reward condition (Salamone et al., 1994). Deficits in motivation or reward should shift the balance between the two choices to the more towards the easy, low reward. Importantly, as anhedonia is also a crucial element of major depression, several of these models are also used in depression and some of them are therefore discussed in more detail in Chapter 7.

Schizophrenia, as discussed earlier in this chapter, is accompanied by severe cognitive deficits. Although patients with schizophrenia score poorly at virtually all cognitive tests, they show a more significant deficit in certain domains, especially attention, working memory and executive functioning. This has also been recognized in human and animal research and has led to several initiatives such as CNTRICS (Cognitive Neuroscience Treatment Research to Improve Cognition in Schizophrenia). This initiative aimed to develop a comprehensive battery of tests to assess cognitive functioning in patients with schizophrenia and to develop (or identify) test in animals that assess the same constructs. The first results regarding the animal paradigms were summarized in a special issue of Neuroscience and Biobehavioural Reviews (volume 37B, 2013), pp 2087-2193). Specifically, tasks were selected that have cross-species validity for domains such as attention, working memory, long-term memory and executive functioning (Table 9.3).

The consensus around the selection of the tasks should provide a strong impetus for animal research in the field of cognitive deficits in schizophrenia. So far, however, with the possible exception of prepulse inhibition, few of these tasks have yet been applied to models based on genetic factors or gene environment interactions.

Genetic Animal Models for Schizophrenia

As mentioned previously in this chapter, most of the single genetic variants only have a very small effect size and schizophrenia is therefore thought to have a multi-genetic origin, a hypothesis confirmed by the recent study of Arnedo and coworkers (Arnedo et al., 2015). From an animal modelling perspective, it therefore seems more logical to use a forward genetic approach (see Chapter 3 for details about the difference between forward and backward genetic approaches) rather than altering a single gene. Several examples of forward genetic approaches, especially selective breeding strategies have been reported in relation to schizophrenia. The APO-SUS/UNSUS rat lines were originally selected on the basis of their stereotyped gnawing response to the dopamine agonist apomorphine (Cools et al., 1990). Given that stereotyped behaviour and enhanced dopamine are intricately linked to schizophrenia, the model was subsequently assessed for schizophrenia-like features (Ellenbroek and Cools, 2002). The APO-SUS rats have deficits in prepulse inhibition and latent inhibition, as well as enhanced dopamine release and transmission in the dorsal and ventral striatum. Intriguingly, the APO-SUS rats also show similarities with schizophrenia in the endocrine and immune system and even showed changes in tumor sensitivity reminiscent of schizophrenia (Jones et al., 2011). Two other rat strains that have been linked to schizophrenia are the Brown Norway and the Brattleboro rat. Compared to Wistar Kyota rats, Brown Norway rats were found to have significantly reduced prepulse inhibition (Palmer et al., 2000) and latent inhibition (Conti et al., 2001). However, P_{50} gating was found to be normal in these animals, although they did show alterations in the theta and gamma band of the EEG (Tomimatsu et al., 2015). The Brattleboro rat was originally derived from a strain of Long Evans rats as an animal with severe polydipsia and polyuria. Subsequent breeding and analysis showed that this phenotype resulted from a homozygous frameshift mutation leading to a reduction in vasopressin release. Interest in the model in relation to schizophrenia came after the identification of reductions in prepulse inhibition, startle habituation and latent inhibition, as well as an enhanced levels of dopamine and D_2 receptors in the striatum (Feifel and Shilling, 2013). Interestingly, similar prepulse inhibition deficits were recently reported in animals heterozygous for the frameshift mutation. As these animals do not show signs of polydipsia and polyuria, this might represent a more valid model for schizophrenia. In addition to these more well-known strains, the spontaneous hypertensive rat strain (Almeida et al., 2014) has been proposed as a model for schizophrenia. However, this strain is better known as a model for ADHD and is discussed in Chapter 8.

TABLE 9.3 Examples of Cognitive Test Used in Animal Models for Schizophrenia

Domains	Animal tests	Descriptions	References
Attention	5-CSRTT	*5 Choice serial reaction time task*: Animals are requires to nose poke one of 5 holes after a brief visual cue.	Lustig et al. (2013)
	5C-CPT	*5 Choice continuous performance task*: adaptation of 5-CSRTT to include trials in which all holes are illuminated and responses should be inhibited	Lustig et al. (2013)
	dSAT	*Distractor condition sustained attention task*: one stimulus discrimination task with or without distractors.	Lustig et al. (2013)
Executive functioning	RL	*Reversal learning*: Different paradigms, usually animals learn to make one of two choices and after acquisition the learning rule is reversed	Gilmour et al. (2013)
	ID/ED shift	*Intradimensional/extradimensional shift task*: Different paradigms, involving two choice discrimination of compound stimuli that contain (at least) two dimensions. The task usually involves reversal learning steps, as well as intra- and extradimensional shifts	Gilmour et al. (2013)
	SST	*Stop signal task:* animals are trained to press two levers sequentially, but have to inhibit the response on the second lever if a signal is given	Gilmour et al. (2013)
Long term memory	PAL	*Paired associate learning*: animals typically have to learn to associate three objects with three locations. During testing two objects are presented with only one in the correct position	Bussey et al. (2013)
Perceptual processing	PPI	*Prepulse inhibition*: discussed in the text	Siegel et al. (2013)
	P50 gating	*P50 gating*: also discussed in the text	Siegel et al. (2013)
	MMN	*Mismatch negativity*: Change in EEG evoked potentials when a series of regular stimuli is interrupted with a different stimulus	Siegel et al. (2013)
Working memory	DNMTP	*Delayed non-matching to sample*: different paradigms. Basically animals are presented with two choices (such as levers in an operant chamber). Animal are required to make one forced choice and, after a variable delay, have to make the opposite choice for reward.	Dudchenko et al. (2013)
	RST	*Rodent span test*: rodents are sequentially presented with a increasing of cups filed with scented sand and a reward. In each trial the animal has to recognize the novel scent	Dudchenko et al. (2013)
	CTRM	*Continuous temporal relational memory task*: in this task animals have to remember a specific sequence of events and respond to the penultimate rather than the last event	Dudchenko et al. (2013)

Given the plethora of (potential) genes that have been linked to schizophrenia, it is not surprising that a large number of different animal (especially mouse) models based on the reverse genetic approach have been developed. For instance several different transgenic models have been developed for the *DISC1* gene and the *Neuregulin* gene. In addition, models targeting the *dysbindin* gene and the *reelin* gene have been created (Harrison et al., 2012). Virtually all of these models show deficits in the prepulse inhibition and most show an increase in spontaneous and/or drug-induced locomotor activity. In addition, deficits in social behaviour and cognitive performance are often found (Jones et al., 2011). An interesting animal model is the mouse variant of the 22q11.2 deletion. As discussed earlier, in humans this copy number variation leads to velocardiofacial syndrome and is one of the highest (albeit rare) risk factors for schizophrenia. The syntenic region of human 22q11.2 (see Chapter 2 for an explanation of synteny) is on mouse chromosome 16. Intriguingly, although the 22q11.2 (Karayiorgou et al., 2010) region harbors about 53 genes, the mouse region 16 contains almost exactly the same genes (with only two exceptions), thus allowing for a strong aetiological similarity between the mouse and human conditions. Several different approaches have been used to model 22q11.2 deletions in the mouse, including long deletions that encompass the whole region (1.5 Mb), short deletions (of about 150 Kb) or deletions of specific genes. Detailed phenotypical analysis of these mouse models show reduced prepulse inhibition, and deficits in cognition (Kvajo et al., 2012).

Moderation of the Effects of Genes by Environmental Factors

Both the forward and reverse genetic animal models have been studied in relation to environmental challenges. Table 9.4 lists several of the major gene environmental interactions. Studies using the APO-SUS rats have shown that the sensitivity for apomorphine was significantly reduced when the rats were reared by APO-UNSUS mothers. However, whether this also leads to changes in prepulse inhibition or latent inhibition is not known. Several other interaction studies have focused on the effects of specific environmental factors in different strains of rats. Interestingly, in studies using maternal deprivation and isolation rearing, it was found that the Lewis strain was significantly less sensitive, as measured by a deficit in prepulse inhibition (Varty and Geyer, 1998; Ellenbroek and Cools, 2000). Likewise, in another developmental model used from schizophrenia, the neonatal ventral hippocampal lesioned rat, the Lewis rat seems rather resistant (Lipska and Weinberger, 1995). Although the ventral hippocampal lesion technique cannot be considered an aetiological model for schizophrenia, it would be interesting to investigate why the Lewis rats are relatively insensitive to these varying environmental challenges.

The DISC1, neuregulin and Nurr1 genetic mouse models have most often been studied in relation to (early) environmental factors (Table 9.4). Thus at least two papers have investigated the interaction between a DISC1 mutation and early viral infections (modelled using PolyI:C). When neonatally (between postnatal days 2 and 6) exposed to polyI:C, DN-DISC1 mice show a decrease in spontaneous alternation in a Y maze as well as a deficit in novel object recognition (Ibi et al., 2010). No alterations were found in prepulse inhibition (either by the mutation, the immune challenge or the combination). Strangely enough, although the authors report some additional behavioural differences (such as changes in fear extinction, locomotor activity and social behaviour) between wildtype animals treated with vehicle

TABLE 9.4 Examples of Gene Environment Interactions in Animal Models for Schizophrenia

Genes/strains	EF	Results	References
FORWARD GENETICS			
APO-SUS/ UNSUS	CF	Reduced dopamine sensitivity	Ellenbroek et al. (2000)
Lewis/Wistar	MD	MD disrupts PPI in Wistar, not in Lewis	Ellenbroek and Cools (2000)
Lewis/Sprague Dawley (SD)	Isolation rearing	Isolation rearing disrupts PPI and startle habituation only in SD rats	Varty and Geyer (1998)
REVERSE GENETICS			
DISC1	Prenatal polyI:C	Decreased social behaviour and increased anxiety & immobility in forced swim test in combined model	Abazyan et al. (2010)
DISC1	Neonatal polyI:C	Combined model showed deficits in cognition and spontaneous alternation	Ibi et al. (2010)
DISC1	Chronic social defeat	Social defeat reduced hyperactivity, and PPI and LI deficits induced by the *DISC1* mutation. Enhanced anxiety like behaviour	Haque et al. (2012)
DISC1	Adolescent Δ^9-THC	Combined model showed reduced cued fear memory	Ballinger et al. (2015)
Neuregulin-1	Chronic social defeat	Combined model showed increased short term memory deficits and reduced aggression	Desbonnet et al. (2012)
Neuregulin-1	Acute Δ^9-THC	Combined model showed stronger locomotor suppression and anxiogenic effect	Boucher et al. (2007)
Neuregulin-1	Chronic CP 55940	Combined model showed enhanced anxiety	Boucher et al. (2011)
Neuregulin-1	Adolescent stress	Combined model showed deficits in PPI and stress coping	Chohan et al. (2014)
Nurr1	Prenatal polyI:C	Combined model showed deficits in LI and sustained visual attention	Vuillermot et al. (2012)
Nurr1	*Toxoplasma gondii* infection	Combined model showed increased novelty induced exploration.	Eells et al. (2015)
Nurr1	Isolation rearing	Combined model showed increased PPI deficit	Eells et al. (2006)
Reelin	Chronic corticosterone	Combined model showed enhanced spatial memory deficits but reduced PPI deficit	Schroeder et al. (2015)
Reelin	Prenatal hypoxia	No interaction was found	Howell and Pillai (2014)
SNAP-25	Prenatal stress	Combined model showed enhanced social and PPI deficits	Oliver and Davies (2009)

EF, environmental factor; CF, cross fostering; LI, latent inhibition; MD, maternal deprivation; PPI, prepulse inhibition; Δ9-THC, Delta-9-tetrahydrocannabinol.

and DISC1 mutant treated with polyI:C, they did not include the other two groups, thereby precluding an analysis of the interaction between the two factors. However, a gene - environment interaction was found on neurogenesis in the granule cell layer of the hippocampus with only the combined model (DISC1 mutation plus neonatal polyI:C) having more BrdU positive cells. 5-bromo-2-deoxyuridine is a synthetic analogue of thymidine and during DNA duplication (Chapter 2) is incorporated into new DNA strands. Using specific antibodies against BrdU, new cells can be detected. At the same time, the combined model led to a significant reduction in the number of parvalbumin positive interneurons in the medial prefrontal cortex, a finding reminiscent of results obtained in post-mortem tissue with schizophrenic patients. In another study, the human mutant Disc1 mouse model was combined with prenatal (gestational day 9) polyI:C challenge (Abazyan et al., 2010). In these animals, anxiety like behaviour was again strongest in the combined model as was a reduction in sociability (as measured using the three chamber social approach avoidance task, see Chapter 8 for a more detailed description of this test). An interesting pattern occurred when the stress-sensitivity of these animals was assessed. Whereas the acute corticosterone release was reduced in the combined model, the levels after 60 min of recovery were significantly higher compared to the other groups.

Another interesting model has focused on the interaction between the DISC1 mutation and adolescent cannabis use (Ballinger et al., 2015), in line with the moderating effects of cannabis on several other schizophrenia-related genes (see Table 9.2). In this study Disc1 mutant mice were exposed to Δ^9-tetrahydrocannabinol (Δ^9-THC) between postnatal days 28 and 48 (adolescence) and tested on postnatal day 68 (young adulthood). Overall the study provided little evidence for a gene–environment interaction. Thus no clear effect was seen on spontaneous alternation, novel object recognition nor on fear conditioning in general. The only gene * treatment interaction on behaviour was seen in the cued version of the fear conditioning paradigm, in which the combined model showed significantly less freezing compared to all other groups. Biochemically, several additional DISC1 - Δ^9-THC interaction effects were also observed. However, in most cases, the effects of Δ^9-THC was actually stronger in the controls compared to the DISC1 mutant mice. Overall, although several studies have looked at gene environment interactions in (several different) DISC1 mutant mice (Cash-Padgett and Jaaro-Peled, 2013) few schizophrenia relevant interactions were found (see also Table 9.3)

Prenatal immune challenge and Δ^9-THC treatment has also been combined with another schizophrenia related gene, the neuregulin mutant mice. Since homozygous neuregulin knock-out mice are not viable, all these experiments are performed using heterozygous mice. In a design similar to that described earlier for DISC1, heterozygous neuregulin and wildtype mice were prenatally injected with polyI:C on gestational day 9 and subjected to behavioural experiments during adolescence or in adulthood (O'Leary et al., 2014). Several different gene * environment interactions were found, but the conclusions are complicated by the fact that animals were cross-fostered after birth. Although from a theoretical point of view, this allows for the separation of the "true" effect of maternal immune response from the potential effect of maternal infections on maternal behaviour, translationally this may be less relevant. Several studies have also investigated the relation between adolescent stimulation of the cannabinoid receptors and the neuregulin mutation in mice (Karl, 2013). The results show that the neuregulin mice are generally more sensitive to cannabinoids (as evidenced for instance in an increased development of tolerance for the hypolocomotion and hypothermia effects), although the relevance for schizophrenia is less clear (Boucher et al., 2011). For instance, neuregulin mutant mice treated acutely with Δ^9-THC actually showed an improvement of prepulse inhibition compared to wildtype mice treated with Δ^9-THC (Boucher et al., 2007). Moreover, chronic Δ^9-THC led to an increased hypolocomotion upon withdrawal in neuregulin but not in control (WT) mice, while the WT mice seemed more sensitive to the acute anxiolytic effects of Δ^9-THC (Long et al., 2013).

In a recent study the interactive effects of adolescent stress (30 min/day restraint stress from postnatal day 36 to 49) and the neuregulin mutation was studied. Interestingly, a strong gene environment interaction was found on prepulse inhibition, with only the combined model showing the schizophrenia-like deficit. In addition, the combined model showed significant alterations in dendritic morphology (including an increase in dendritic spine density, combined with a reduction in dendritic length and complexity) in the medial prefrontal cortex (Chohan et al., 2014).

Nurr1 (NR4A2) is an orphan nuclear receptor that appears to be essential for the development and survival of dopaminergic cells (Castillo et al., 1998) and while homozygous Nurr1 deletions are lethal at birth, heterozygous Nurr1 null mice survive normally. Although somewhat counterintuitive (as schizophrenia is more linked to dopaminergic hyperactivity, as discussed earlier), rare mutations in Nurr1 have been associated with schizophrenia (Buervenich et al., 2000), especially with attention deficits (Ancin et al., 2013). In line with this clinical finding, heterozygous Nurr1 null mice show deficits reminiscent of schizophrenia such as a hypersensitivity to stress (Rojas et al., 2007), a deficit in prepulse inhibition and an increased sensitivity to NMDA antagonists such as MK801 (Vuillermot et al., 2011). However, in many other aspects of behaviour such as social interaction and cognition these animals appeared to be normal. This makes the animals an interesting model for gene–environment interactions. In this respect several studies have been reported looking at the effects of isolation rearing as well as pre- and postnatal exposure to infectious agents. When heterozygous Nurr1 null mutant mice were prenatally exposed to PolyI:C, several additive effects were seen in the combination such as an increase in hyperactivity, a deficit in prepulse inhibition and in spatial working memory (Vuillermot et al., 2012). However, importantly, these were additive effects and not interactions effects (ie, the effects of polyI:C were similar in both the WT and the Nurr1 mutant mice). Interestingly, interactive effects were found for latent inhibition and a sustained visual attention task, where only the mutants treated with polyI:C showed a significant deficit. Isolation rearing, on the other hand led to a significant deficit in prepulse inhibition that was only observed in Nurr1 mutant but not wild type control mice (Eells et al., 2006).

In conclusion, important advances have been made in the field of animal modelling of schizophrenia, focusing on gene * environment interactions. However, most of the studies have used a reverse genetic approach. From a mechanistic point of view this is a useful strategy and many models have focused on genes with a reasonably sound association with schizophrenia (see Tables 9.1, 9.2, and 9.4). However, given the overwhelming evidence pointing to schizophrenia having a polygenic origin, combining environmental factors with a forward genetic approach may ultimately prove more beneficial in trying to understand the aetiology of schizophrenia.

References

Aas, M., Haukvik, U.K., Djurovic, S., et al., 2013. BDNF val66met modulates the association between childhood trauma, cognitive and brain abnormalities in psychoses. Prog. Neuro-Psychopharmacol. Biol. Psychiatr. 46, 181–188.

Abazyan, B., Nomura, J., Kannan, G., et al., 2010. Prenatal interaction of mutant DISC1 and immune activation produces adult psychopathology. Biol. Psychiatr. 68, 1172–1181.

Abi-Dargham, A., van de Giessen, E., Slifstein, M., et al., 2009. Baseline and amphetamine-stimulated dopamine activity are related in drug-naive schizophrenic subjects. Biol. Psychiatr. 65, 1091–1093.

Adler, L.E., Pachtman, E., Franks, R.D., et al., 1982. Neurophysiological evidence for a defect in neuronal mechanisms involved in sensory gating in schizophrenia. Biol. Psychiatry 17, 649–654.

Alemany, S., Arias, B., Aguilera, M., et al., 2011. Childhood abuse, the BDNF-Val66Met polymorphism and adult psychotic-like experiences. Br. J. Psychiatr. 199, 38–42.

Almeida, V., Peres, F.F., Levin, R., et al., 2014. Effects of cannabinoid and vanilloid drugs on positive and negative-like symptoms on an animal model of schizophrenia: The SHR strain. Schizophrenia Res. 153, 150–159.

Ancin, I., Cabranes, J.A., Vazquez-Alvarez, B., et al., 2013. NR4A2: effects of an "orphan" receptor on sustained attention in a schizophrenic population. Schizophr Bull. 39, 555–563.

APA, 2013. Diagnostic and Statistical Manual of mental disorders, Fifth Edition. edn American Psychiatric Publishing, Arlington VA.

Arnedo, J., Svrakic, D.M., del Val, C., et al., 2015. Uncovering the hidden risk architecture of the schizophrenias: confirmation in three independent genome-wide association studies. Am. J. Psychiatr. 172, 139–153.

Baldwin, P., Browne, D., Scully, P.J., et al., 2005. Epidemiology of first-episode psychosis: Illustrating the challenges across diagnostic boundaries through the Cavan-Monaghan study at 8 years. Schizophrenia Bull. 31, 624–638.

Ballinger, M.D., Saito, A., Abazyan, B., et al., 2015. Adolescent cannabis exposure interacts with mutant DISC1 to produce impaired adult emotional memory. Neurobiol. Dis. 82, 176–184.

Ban, T.A., Guy, W., Wilson, W.H., 1984. Pharmacotherapy of chronic hospitalized schizophrenics: diagnosis and treatment. Psychiatr. J. Univ. Ott. 9, 191–195.

Basser, P.J., Pajevic, S., Pierpaoli, C., et al., 2000. In vivo fiber tractography using DT-MRI data. Magnetic resonance in medicine. Official J. Soc. Magnetic Resonance Medicine/Soc. Magnetic Resonance Medicine 44, 625–632.

Bhattacharyya, S., Atakan, Z., Martin-Santos, R., et al., 2012. Preliminary report of biological basis of sensitivity to the effects of cannabis on psychosis: AKT1 and DAT1 genotype modulates the effects of delta-9-tetrahydrocannabinol on midbrain and striatal function. Mol. Psychiatr. 17, 1152–1155.

Boucher, A.A., Arnold, J.C., Duffy, L., et al., 2007. Heterozygous neuregulin 1 mice are more sensitive to the behavioural effects of Delta(9)-tetrahydrocannabinol. Psychopharmacology 192, 325–336.

Boucher, A.A., Hunt, G.E., Micheau, J., et al., 2011. The schizophrenia susceptibility gene neuregulin 1 modulates tolerance to the effects of cannabinoids. Int. J. Neuropsychopharmacol. 14, 631–643.

Braff, D., Stone, C., Callaway, E., et al., 1978. Prestimulus effects of human startle reflex in normals and schizophrenics. Psychophysiology 15, 339–343.

Braff, D., Light, G.A., Cadenhead, K.S., et al., 2004. Divergence of PPI and P50 suppression deficits in schizophrenia patients. Neuropsychopharmacol. 29, S108–S109.

Braff, D.L., 2015. The importance of endophenotypes in schizophrenia research. Schizophrenia Res. 163, 1–8.

Braff, D.L., Geyer, M.A., Swerdlow, N.R., 2001. Human studies of prepulse inhibition of startle: normal subjects, patient groups, and pharmacological studies. Psychopharmacology 156, 234–258.

Buervenich, S., Carmine, A., Arvidsson, M., et al., 2000. NURR1 mutations in cases of schizophrenia and manic-depressive disorder. Am. J. Med. Genetics 96, 813.

Bussey, T.J., Barch, D.M., Baxter, M.G., 2013. Testing long-term memory in animal models of schizophrenia: suggestions from CNTRICS. Neurosci. Biobehav. Rev. 37, 2141–2148.

Cannon, T.D., Zorrilla, L.E., Shtasel, D., et al., 1994. Neuropsychological functioning in siblings discordant for schizophrenia and healthy-volunteers. Archives General Psychiatr. 51, 651–661.

Carlsson, A., Lindqvist, M., 1963. Effects of chlorpromazine or haloperidol on formation of 3-methoxytyramin and normetanephrine in mouse brain. Acta Pharmacol.Toxicol. 20, 140–144.

Cash-Padgett, T., Jaaro-Peled, H., 2013. DISC1 mouse models as a tool to decipher gene-environment interactions in psychiatric disorders. Frontiers Behav. Neurosci. 7.

Caspi, A., Moffitt, T.E., Cannon, M., et al., 2005. Moderation of the effect of adolescent-onset cannabis use on adult psychosis by a functional polymorphism in the catechol-O-methyltransferase gene: Longitudinal evidence of a gene X environment interaction. Biol. Psychiatr. 57, 1117–1127.

Castillo, S.O., Baffi, J.S., Palkovits, M., et al., 1998. Dopamine biosynthesis is selectively abolished in substantia nigra/ventral tegmental area but not in hypothalamic neurons in mice with targeted disruption of the Nurr1 gene. Mol. Cell. Neurosci. 11, 36–46.

Chohan, T.W., Boucher, A.A., Spencer, J.R., et al., 2014. Partial genetic deletion of neuregulin 1 modulates the effects of stress on sensorimotor gating, dendritic morphology, and HPA axis activity in adolescent mice. Schizophrenia Bull. 40, 1272–1284.

Chotai, J., Serretti, A., Lattuada, E., et al., 2003. Gene-environment interaction in psychiatric disorders as indicated by season of birth variations in tryptophan hydroxylase (TPH), serotonin transporter (5-HTTLPR) and dopamine receptor (DRD4) gene polymorphisms. Psychiatry Res. 119, 99–111.

Cole, J., Clyde, D., 1961. Extrapyramidal side effects and clinical response to the phenothazines. Rev. Can. Biol 20, 565–574.

Colizzi, M., Iyegbe, C., Powell, J., et al., 2015. Interaction between functional genetic variation of DRD2 and cannabis use on risk of psychosis. Schizophr. Bull.

Collip, D., van Winkel, R., Peerbooms, O., et al., 2011. COMT Val158Met-stress interaction in psychosis: role of background psychosis risk. CNS Neurosci. Ther. 17, 612–619.

Connell, P.H., 1958. Amphetamine psychosis, edn Oxford University Press, London.

Conti, L.H., Palmer, A.A., Vanella, J.J., et al., 2001. Latent inhibition and conditioning in rat strains which show differential prepulse inhibition. Behav. Genetics 31, 325–333.

Cools, A.R., Brachten, R., Heeren, D., et al., 1990. Search after neurobiological profile of individual-specific features of Wistar rats. Brain Res. Bull. 24, 49–69.

Creese, I., Burt, D., Snyder, S.H., 1976. Dopamine receptor binding predicts clinical and pharmacological potenties of antischizophrenic drugs. Science 192, 481–483.

Debbane, M., Glaser, B., David, M.K., et al., 2006. Psychotic symptoms in children and adolescents with 22q11.2 deletion syndrome: Neuropsychological and behavioral implications. Schizophrenia Res. 84, 187–193.

Deniker, P., 1983. Discovery of the clinical use of neuroleptics. In: Parnham, M.J., Bruinvels, J., (eds). Psycho- and Neuropharmacology, edn, vol. 1. Amsterdam, Elsevier. pp 163–180.

Desbonnet, L., O'Tuathaigh, C., Clarke, G., et al., 2012. Phenotypic effects of repeated psychosocial stress during adolescence in mice mutant for the schizophrenia risk gene neuregulin-1: a putative model of gene x environment interaction. Brain Behav. Immun. 26, 660–671.

Di Forti, M., Iyegbe, C., Sallis, H., et al., 2012. Confirmation that the AKT1 (rs2494732) genotype influences the risk of psychosis in cannabis users. Biol Psychiatr. 72, 811–816.

Dickerson, F., Schroeder, J., Stallings, C., et al., 2014. A longitudinal study of cognitive functioning in schizophrenia: clinical and biological predictors. Schizophrenia Res. 156, 248–253.

Drevets, W.C., Gautier, C., Price, J.C., et al., 2001. Amphetamine-induced dopamine release in human ventral striatum correlates with euphoria. Biol. Psychiatr. 49, 81–96.

Dudchenko, P.A., Talpos, J., Young, J., et al., 2013. Animal models of working memory: a review of tasks that might be used in screening drug treatments for the memory impairments found in schizophrenia. Neurosci. Biobehav. Rev. 37, 2111–2124.

Eells, J.B., Misler, J.A., Nikodem, V.M., 2006. Early postnatal isolation reduces dopamine levels, elevates dopamine turnover and specifically disrupts prepulse inhibition in Nurr1-null heterozygous mice. Neuroscience 140, 1117–1126.

Eells, J.B., Varela-Stokes, A., Guo-Ross, S.X., et al., 2015. Chronic *Toxoplasma gondii* in Nurr1-Null heterozygous mice exacerbates elevated open field activity. Plos One, 10.

Ellenbroek, B.A., 2012. Psychopharmacological treatment of schizophrenia: what do we have and what could we get? Neuropharmacology 62, 1371–1380.

Ellenbroek, B.A., Cools, A.R., 2000. The long-term effects of maternal deprivation depend on the genetic background. Neuropsychopharmacology 23, 99–106.

Ellenbroek, B.A., Cools, A.R., 2002. Apomorphine susceptibility and animal models for psychopathology: genes and environment. Behav. Genetic. 32, 349–361.

Ellenbroek, B.A., Lubbers, L.J., Cools, A.R., 1996. Activity of "seroquel" (ICI 204,636) in animal models for atypical properties of antipsychotics: a comparison with clozapine. Neuropsychopharmacology 15, 406–416.

Ellenbroek, B.A., Sluyter, F., Cools, A.R., 2000. The role of genetic and early environmental factors in determining apomorphine susceptibility. Psychopharmacology 148, 124–131.

Ellenbroek, B.A., van, L.G., Frenken, M., et al., 1999. Sensory gating in rats: lack of correlation between auditory evoked potential gating and prepulse inhibition. Schizophrenia Bull. 25, 777–788.

Farde, L., Wiesel, F.A., Halldin, C., et al., 1988. Central D2-dopamine receptor occupancy in schizophrenic patients treated with antipsychotic drugs. Arch. Gen. Psychiatr. 45, 71–76.

Farde, L., Wiesel, F.A., Nordstrom, A.L., et al., 1989. D1 and D2 dopamine receptor occupancy during treatment with conventional and atypical neuroleptics. Psychopharmacology 99, S28–S31.

Fatouros-Bergman, H., Cervenka, S., Flyckt, L., et al., 2014. Meta-analysis of cognitive performance in drug-naive patients with schizophrenia. Schizophrenia Res. 158, 156–162.

Feifel, D., Shilling, P.D., 2013. Modelling schizophrenia in animals. In: Conn, P.M., (ed.). Animal models for the study of human disease, edn. Amsterdam, Elsevier. pp 727–755.

Fischer, A.S., Whitfield-Gabrieli, S., Roth, R.M., et al., 2014. Impaired functional connectivity of brain reward circuitry in patients with schizophrenia and cannabis use disorder: effects of cannabis and THC. Schizophrenia Res. 158, 176–182.

Fosse, R., Joseph, J., Richardson, K., 2015. A critical assessment of the equal-environment assumption of the twin method for schizophrenia. Frontiers Psychiatr. 6, 62.

Friston, K.J., Frith, C.D., 1995. Schizophrenia: a disconnection syndrome? Clin. Neurosci. 3, 89–97.

Frith, C.D., Friston, K.J., Herold, S., et al., 1995. Regional brain activity in chronic schizophrenic patients during the performance of a verbal fluency task. Br.J. Psychiatr. 167, 343–349.

Gilmour, G., Arguello, A., Bari, A., et al., 2013. Measuring the construct of executive control in schizophrenia: defining and validating translational animal paradigms for discovery research. Neurosci. Biobehav. Rev. 37, 2125–2140.

Giusti-Rodriguez, P., Sullivan, P.F., 2013. The genomics of schizophrenia: update and implications. J. Clin. Investig. 123, 4557–4563.

Gottesman, I.I., Shields, J., 1967. A polygenic theory of schizophrenia. Proc. Natl. Acad. Sci. USA 58, 199–205.

Gottesman II, Shields, J., 1982. Schizophrenia: The epigenetic puzzle. edn. Cambridge University Press: Cambridge.

Gustavsson, A., Svensson, M., Jacobi, F., et al., 2011. Cost of disorders of the brain in Europe 2010. Eur. Neuropsychopharmacol. J. Eur.Coll. Neuropsychopharmacol. 21, 718–779.

Hafner, H., Maurer, K., Loffler, W., et al., 1993. The influence of age and sex on the onset and early course of schizophrenia. Br. J. Psychiatry 162, 80–86.

Haque, F.N., Lipina, T.V., Roder, J.C., et al., 2012. Social defeat interacts with Disc1 mutations in the mouse to affect behavior. Behav. Brain Res. 233, 337–344.

Harrison, P.J., Pritchett, D., Stumpenhorst, K., et al., 2012. Genetic mouse models relevant to schizophrenia: taking stock and looking forward. Neuropharmacology 62, 1164–1167.

Heinrichs, R.W., Zakzanis, K.K., 1998. Neurocognitive deficit in schizophrenia: a quantitative review of the evidence. Neuropsychology 12, 426–445.

Henquet, C., Rosa, A., Krabbendam, L., et al., 2006. An experimental study of catechol-O-methyltransferase Val(158) Met moderation of Delta-9-tetrahydrocannabinol-induced effects on psychosis and cognition. Neuropsychopharmacology 31, 2748–2757.

Hietala, J., Syvalahti, E., Vuorio, K., et al., 1995. Presynaptic dopamine function in striatum of neuroleptic-naive schizophrenic patients. Lancet 346, 1130–1131.

Hodgkinson, C.A., Goldman, D., Jaeger, J., et al., 2004. Disrupted in schizophrenia 1 (DISC1): association with schizophrenia, schizoaffective disorder, and bipolar disorder. Am. J. Human Genetic. 75, 862–872.

Howell, K.R., Pillai, A., 2014. Effects of prenatal hypoxia on schizophrenia-related phenotypes in heterozygous reeler mice: a gene x environment interaction study. Eur. Neuropsychopharmacol. 24, 1324–1336.

Howes, O., Bose, S., Turkheimer, F., et al., 2011. Progressive increase in striatal dopamine synthesis capacity as patients develop psychosis: a PET study. Mol. Psychiatr. 16, 885–886.

Howes, O.D., Kambeitz, J., Kim, E., et al., 2012. The nature of dopamine dysfunction in schizophrenia and what this means for treatment. Archiv. General Psychiatr. 69, 776–786.

Ibi, D., Nagai, T., Koike, H., et al., 2010. Combined effect of neonatal immune activation and mutant DISC1 on phenotypic changes in adulthood. Behav. Brain Res. 206, 32–37.

Jentsch, J.D., Roth, R.H., 1999. The neuropsychopharmacology of phencyclidine: from NMDA receptor hypofunction to the dopamine hypothesis of schizophrenia. Neuropsychopharmacology 20, 201–225.

Jiang, T.Z., Zhou, Y., Liu, B., et al., 2013. Brainnetome-wide association studies in schizophrenia: the advances and future. Neurosci. Biobehav. Rev. 37, 2818–2835.

Jones, C.A., Watson, D.J.G., Fone, K.C.F., 2011. Animal models of schizophrenia. Br. J. Pharmacol. 164, 1162–1194.

Kane, J.M., Honigfeld, G., Singer, J., et al., 1988. Clozapine for the treatment-resistant schizophrenic: a double-blind comparison with chlorpromazine. Arch. Gen. Psychiatry 45, 789–796.

Kantrowitz, J.T., Nolan, K.A., Sen, S., et al., 2009. Adolescent cannabis use, psychosis and catechol-o-methyltransferase genotype in african americans and caucasians. Psychiatric Quarterly 80, 213–218.

Karayiorgou, M., Simon, T.J., Gogos, J.A., 2010. 22q11.2 microdeletions: linking DNA structural variation to brain dysfunction and schizophrenia. Nature Rev. Neurosci. 11, 402–416.

Karl, T., 2013. Neuregulin 1: a prime candidate for research into gene-environment interactions in schizophrenia? Insights from genetic rodent models. Frontiers Behav. Neurosci. 7, 106.

Kay, S.R., Fiszbein, A., Opler, L.A., 1987. The positive and negative syndrome scale (PANSS) for schizophrenia. Schizophr. Bull. 13, 261–276.

Kegeles, L.S., Abi-Dargham, A., Frankle, W.G., et al., 2010. Increased synaptic dopamine function in associative regions of the striatum in schizophrenia. Arch. Gen. Psychiatr. 67, 231–239.

Kendler, K.S., 2015. A joint history of the nature of genetic variation and the nature of schizophrenia. Mol. Psychiatr. 20, 77–83.

Kety, S.S., 1968. The types and prevalence of mental illness in the biological and adopted families of adopted schizophrenics. J. Psychiatr. Res. 6, 345–362.

Kohl, S., Heekeren, K., Klosterkotter, J., et al., 2013. Prepulse inhibition in psychiatric disorders--apart from schizophrenia. J. Psychiatr. Res. 47, 445–452.

Kvajo, M., McKellar, H., Gogos, J.A., 2012. Avoiding mouse traps in schizophrenia genetics: lessons and promises from current and emerging mouse models. Neuroscience 211, 136–164.

Laborit, H., Huguenard, P., 1951. Artificial hibernation by pharmacodynamical and physical means. Presse. Med. 59, 1329.

Laborit, H., Huguenard, P., Alluaume R, 1952. A new vegetative stabilizer; 4560 R.P. Presse Med. 60, 206–208.

Laruelle, M., Abi, D.A., Gil, R., et al., 1999. Increased dopamine transmission in schizophrenia: relationship to illness phases. Biol. Psychiatr. 46, 56–72.

Laruelle, M., Abi-Dargham, A., van Dyck, C.H., et al., 1996. Single photon emission computerized tomography imaging of amphetamine-induced dopamine release in drug-free schizophrenic subjects. Proc. Natl. Acad. Sci. USA 93, 9235–9240.

Laskey, J., Klett, C., Caffey, E., et al., 1962. A comparison evaluation of chlorpromazine, chrloprothixene, fluphenazine, reserpine, thioridazine and ttriflupromazine. Dis. Nerv. Syst. 23, 8.

Leucht, S., Cipriani, A., Spineli, L., et al., 2013. Comparative efficacy and tolerability of 15 antipsychotic drugs in schizophrenia: a multiple-treatments meta-analysis. Lancet 382, 951–962.

Li, W., Yang, Y., Lin, J., et al., 2013. Association of serotonin transporter gene (SLC6A4) polymorphisms with schizophrenia susceptibility and symptoms in a Chinese-Han population. Prog. Neuropsychopharmacol. Biol. Psychiatr. 44, 290–295.

Lipska, B.K., Weinberger, D.R., 1995. Genetic variation in vulnerability to the behavioral effects of neonatal hippocampal damage in rats. Proc. Natl. Acad. Sci. USA 92, 8906–8910.

Long, L.E., Chesworth, R., Huang, X.F., et al., 2013. Transmembrane domain Nrg1 mutant mice show altered susceptibility to the neurobehavioural actions of repeated THC exposure in adolescence. Int. J. Neuropsychopharmacol. 16, 163–175.

Luby, E.D., Cohen, B.D., Rosenbaum, G., et al., 1959. Study of a new schizophrenomimetic drug - sernyl. Amer. Med. Assoc. Arch. Neurol. Psychiatry 81, 363–369.

Lustig, C., Kozak, R., Sarter, M., et al., 2013. CNTRICS final animal model task selection: control of attention. Neurosci. Biobehav. Rev. 37, 2099–2110.

Maibing, C.F., Pedersen, C.B., Benros, M.E., et al., 2015. Risk of schizophrenia increases after all child and adolescent psychiatric disorders: a nationwide study. Schizophr. Bull. 41, 963–970.

Malhotra, A.K., Pinals, D.A., Adler, C.M., et al., 1997. Ketamine-induced exacerbation of psychotic symptoms and cognitive impairment in neuroleptic-free schizophrenics. Neuropsychopharmacol. 17, 141–150.

Marder, S.R., Fenton, W., 2004. Measurement and treatment research to improve cognition in schizophrenia: NIMH MATRICS initiative to support the development of agents for improving cognition in schizophrenia. Schizophrenia Res. 72, 5–9.

Marder, S.R., Fenton, W., Youens, K., et al., 2004. Schizophrenia, IX - cognition in schizophrenia - the MATRICS initiative. Am. J. Psychiatr. 161, 25.

Markou, A., Salamone, J.D., Bussey, T.J., et al., 2013. Measuring reinforcement learning and motivation constructs in experimental animals: Relevance to the negative symptoms of schizophrenia. Neurosci. Biobehav. Rev. 37, 2149–2165.

Marsman, A., van den Heuvel, M.P., Klomp, D.W., et al., 2013. Glutamate in schizophrenia: a focused review and meta-analysis of (1)H-MRS studies. Schizophr. Bull. 39, 120–129.

McCarthy-Jones, S., Green, M.J., Scott, R.J., et al., 2014. Preliminary evidence of an interaction between the FOXP2 gene and childhood emotional abuse predicting likelihood of auditory verbal hallucinations in schizophrenia. J. Psychiatr. Res. 50, 66–72.

McEvoy, J.P., Lieberman, J.A., Stroup, T.S., et al., 2006. Effectiveness of clozapine versus olanzapine, quetiapine, and risperidone in patients with chronic schizophrenia who did not respond to prior atypical antipsychotic treatment. Am. J. Psychiatr. 163, 600–610.

Mesholam-Gately, R.I., Giuliano, A.J., Goff, K.P., et al., 2009. Neurocognition in first-episode schizophrenia: a meta-analytic review. Neuropsychology 23, 315–336.

Muntjewerff, J.W., Ophoff, R.A., Buizer-Voskamp, J.E., et al., 2011. Effects of season of birth and a common MTHFR gene variant on the risk of schizophrenia. Eur. Neuropsychopharmacol. 21, 300–305.

Murray, C.J.L., Vos, T., Lozano, R., et al., 2012. Disability-adjusted life years (DALYs) for 291 diseases and injuries in 21 regions, 1990-2010: a systematic analysis for the Global Burden of Disease Study 2010. Lancet 380, 2197–2223.

Narita, K., Sasaki, T., Akaho, R., et al., 2000. Human leukocyte antigen and season of birth in Japanese patients with schizophrenia. Am. J. Psychiatr. 157, 1173–1175.

Nicodemus, K.K., Marenco, S., Batten, A.J., et al., 2008. Serious obstetric complications interact with hypoxia-regulated/vascular-expression genes to influence schizophrenia risk. Mol. Psychiatr. 13, 873–877.

Nord, M., Farde, L., 2011. Antipsychotic occupancy of dopamine receptors in schizophrenia. CNS Neurosci. Ther. 17, 97–103.

Nuechterlein, K.H., Barch, D.M., Gold, J.M., et al., 2004. Identification of separable cognitive factors in schizophrenia. Schizophrenia Res. 72, 29–39.

O'Leary, C., Desbonnet, L., Clarke, N., et al., 2014. Phenotypic effects of maternal immune activation and early postnatal milieu in mice mutant for the schizophrenia risk gene neuregulin-1. Neuroscience 277, 294–305.

Oliver, P.L., Davies, K.E., 2009. Interaction between environmental and genetic factors modulates schizophrenic endophenotypes in the Snap-25 mouse mutant blind-drunk. Hum. Mol. Genet. 18, 4576–4589.

Paek, M.J., Kang, U.G., 2012. How many genes are involved in schizophrenia? A simple simulation. Prog. Neuropsychopharmacol. Biol. Psychiatr. 38, 302–309.

Palmer, A.A., Dulawa, S.C., Mottiwala, A.A., et al., 2000. Prepulse startle deficit in the brown Norway rat: a potential genetic model. Behav. Neurosci. 114, 374–388.

Porter, R.A., Dawson, L.A., 2015. GlyT-1 inhibitors: From hits to clinical candidates. In: Celanire, S., Poli, S., (eds.). Small molecule therapeutics for schizophrenia, edn., vol. 13. Heidelberg, Springer. pp 51–100.

Ripke, S., O'Dushlaine, C., Chambert, K., et al., 2013. Genome-wide association analysis identifies 13 new risk loci for schizophrenia. Nature Genetic. 45, 1150–U1282.

Ripke, S., Sanders, A.R., Kendler, K.S., et al., 2011. Genome-wide association study identifies five new schizophrenia loci. Nature Genetic. 43, 969–U977.

Ripke, S., Neale, B.M., Corvin, A., et al., 2014. Biological insights from 108 schizophrenia-associated genetic loci. Nature 511, 421-+.

Rojas, P., Joodmardi, E., Hong, Y., et al., 2007. Adult mice with reduced Nurr1 expression: an animal model for schizophrenia. Mol. Psychiatr. 12, 756–766.

Salamone, J.D., Cousins, M.S., Bucher, S., 1994. Anhedonia or anergia? Effects of haloperidol and nucleus accumbens dopamine depletion on instrumental response selection in a T-maze cost/benefit procedure. Behav. Brain Res. 65, 221–229.

Sams Dodd, F., 1996. Phencyclidine-induced stereotyped behaviour and social isolation in the rat: A possible animal model of schizophrenia. Behav. Pharmacol. 7, 3–23.

Schroeder, A., Bureta, L., Hill, R.A., et al., 2015. Gene-environment interaction of reelin and stress in cognitive behaviours in mice: Implications for schizophrenia. Behav. Brain Res. 287, 304–314.

Scully, P.J., Quinn, J.F., Morgan, M.G., et al., 2002. First-episode schizophrenia, bipolar disorder and other psychoses in a rural Irish catchment area: incidence and gender in the Cavan-Monaghan study at 5 years. Br. J. Psychiatr. 181 (Suppl. 43), S3–S9.

Seeman, M.V., 1986. Current outcome in schizophrenia: women vs men. Acta Psychiatr. Scand. 73, 609–617.

Seeman, P., Lee, T., Choa-Wong, M., et al., 1976. Antipsychotic drug doses and neuroleptic/dopamine receptors. Nature 261, 717–719.

Shifman, S., Johannesson, M., Bronstein, M., et al., 2008. Genome-wide association identifies a common variant in the reelin gene that increases the risk of schizophrenia only in women. PLoS Genet. 4, e28.

Siegel, S.J., Talpos, J.C., Geyer, M.A., 2013. Animal models and measures of perceptual processing in Schizophrenia. Neurosci. Biobehav. Rev. 37, 2092–2098.
Simons, C.J., Wichers, M., Derom, C., et al., 2009. Subtle gene-environment interactions driving paranoia in daily life. Genes Brain Behav. 8, 5–12.
Solis-Ortiz, S., Perez-Luque, E., Morado-Crespo, L., et al., 2010. Executive functions and selective attention are favored in middle-aged healthy women carriers of the Val/Val genotype of the catechol-o-methyltransferase gene: a behavioral genetic study. Behav. Brain Funct. 6.
Stefanis, N.C., Henquet, C., Avramopoulos, D., et al., 2007. COMT Val158Met moderation of stress-induced psychosis. Psychol. Med. 37, 1651–1656.
Stefanovics, E.A., Elkis, H., Liu, Z.N., et al., 2014. A cross-national factor analytic comparison of three models of PANSS symptoms in schizophrenia. Psychiatr. Res. 219, 283–289.
Stephan, K.E., Baldeweg, T., Friston, K.J., 2006. Synaptic plasticity and disconnection in schizophrenia. Biol. Psychiatr. 59, 929–939.
Sullivan, P.F., 2013. Questions about DISC1 as a genetic risk factor for schizophrenia. Mol. Psychiatr. 18, 1050–1052.
Sullivan, P.F., Kendler, K.S., Neale, M.C., 2003. Schizophrenia as a complex trait - evidence from a meta-analysis of twin studies. Arch. General Psychiatr. 60, 1187–1192.
Sun, J., Kuo, P.H., Riley, B.P., et al., 2008. Candidate genes for schizophrenia: a survey of association studies and gene ranking. Am. J. Med. Genet. B 147B, 1173–1181.
Szulwach, K.E., Li, X., Smrt, R.D., et al., 2010. Cross talk between microRNA and epigenetic regulation in adult neurogenesis. J. Cell Biol. 189, 127–141.
Tochigi, M., Ohashi, J., Umekage, T., et al., 2002. Human leukocyte antigen-A specificities and its relation with season of birth in Japanese patients with schizophrenia. Neurosci. Lett. 329, 201–204.
Tomimatsu, Y., Hibino, R., Ohta, H., 2015. Brown Norway rats, a putative schizophrenia model, show increased electroencephalographic activity at rest and decreased event-related potential amplitude, power, and coherence in the auditory sensory gating paradigm. Schizophr. Res.
van den Heuvel, M.P., Fornito, A., 2014. Brain networks in schizophrenia. Neuropsychol. Rev. 24, 32–48.
van Rossum, J., 1966. The significance of dopamine-receptor blockade for the mechanism of action of neuroleptic drugs. Arch. Int. Pharmacodyn. Ther. 160, 492–494.
van Winkel, R., van Beveren, N.J., Simons, C., 2011. AKT1 moderation of cannabis-induced cognitive alterations in psychotic disorder. Neuropsychopharmacology 36, 2529–2537.
van Winkel, R., Henquet, C., Rosa, A., et al., 2008. Evidence that the COMT(Val158Met) polymorphism moderates sensitivity to stress in psychosis: an experience-sampling study. Am. J. Med. Genet. B 147B, 10–17.
Varty, G.B., Geyer, M.A., 1998. Effects of isolation rearing on startle reactivity, habituation, and prepulse inhibition in male Lewis, Sprague-Dawley, and Fischer F344 rats. Behav. Neurosci. 112, 1450–1457.
Volkow, N.D., Wolf, A.P., Brodie, J.D., et al., 1988. Brain interactions in chronic schizophrenics under resting and activation conditions. Schizophr. Res. 1, 47–53.
Vuillermot, S., Joodmardi, E., Perlmann, T., et al., 2011. Schizophrenia-relevant behaviors in a genetic mouse model of constitutive Nurr1 deficiency. Genes Brain Behav. 10, 589–603.
Vuillermot, S., Joodmardi, E., Perlmann, T., et al., 2012. Prenatal immune activation interacts with genetic Nurr1 deficiency in the development of attentional impairments. J. Neurosci. 32, 436–451.
Wedenoja, J., Tuulio-Henriksson, A., Suvisaari, J., et al., 2010. Replication of association between working memory and Reelin, a potential modifier gene in schizophrenia. Biol. Psychiatr. 67, 983–991.
Yao, J., Pan, Y.Q., Ding, M., et al., 2015. Association between DRD2 (rs1799732 and rs1801028) and ANKK1 (rs1800497) polymorphisms and schizophrenia: a meta-analysis. Am. J. Med. Genetics Part B 168, 1–13.
Yilmaz, Z., Zai, C.C., Hwang, R., et al., 2012. Antipsychotics, dopamine D(2) receptor occupancy and clinical improvement in schizophrenia: a meta-analysis. Schizophr. Res. 140, 214–220.
Zalesky, A., Fornito, A., Seal, M.L., et al., 2011. Disrupted axonal fiber connectivity in schizophrenia. Biol. Psychiatr. 69, 80–89.
Zammit, S., Owen, M.J., Evans, J., et al., 2011. Cannabis, COMT and psychotic experiences. Br. J. Psychiatr. 199, 380–385.
Zammit, S., Spurlock, G., Williams, H., et al., 2007. Genotype effects of CHRNA7, CNRI and COMT in schizophrenia: interactions with tobacco and cannabis use. Br. J. Psychiatr. 191, 402–407.
Zhang, T.H., Koutsouleris, N., Meisenzahl, E., et al., 2015. Heterogeneity of structural brain changes in subtypes of schizophrenia revealed using magnetic resonance imaging pattern analysis. Schizophrenia Bull. 41, 74–84.

CHAPTER

10

Conclusions and the Road Ahead

INTRODUCTION

One in every three individuals in the developed world will be diagnosed with a brain disorder in their lifetime (Andlin-Sobocki et al., 2005), and as most of these develop relatively early in life and often lead to long-life disability, the impact is tremendous. According to the global burden of disease study as a group, brain disorders have by far the highest burden in terms of years lived with disability (Vos et al., 2012), reaching about 2×10^8 years in 2010 (Fig. 10.1). This high burden for the patients also translates into an enormous economic burden for society as a whole. In a study in the European Union, the overall costs for brain disorders was estimated at a staggering 800 billion Euro (Gustavsson et al., 2011), significantly more than the costs for lung disorders, diabetes, cardiovascular disorders, and cancer combined (Fig. 10.2). As we have seen in previous chapters on individual disorders, these costs are a combination of direct medical costs (296 billion, such as medication and hospitalization), direct nonmedical costs (186 billion, such as costs for adaptation in the house and social services) and indirect costs (315 billion, such as loss of productivity and early retirement). These figures emphasize the importance of a better understanding of the aetiology and pathophysiology of brain disorders in order to develop more effective treatment.

THE GENETIC REVOLUTION AND PSYCHIATRY

The elucidation of the structure of DNA sparked a revolution in biomedical research including psychiatry. Followed by multiple additional breakthroughs in both theoretical concepts [such as the universal code and the central dogma (see Chapter 2 for more details)] and in techniques [such as the polymerase chain reaction (PCR), high throughput sequencing, and the development of DNA and RNA microarrays], the analysis of the genetic basis of normal and abnormal aspects of life has accelerated to an enormous speed. However, at the same time we have seen throughout this book that the genetic basis of behavior and psychiatric disorders is still far from completely understood. Without exception, the disorders described are genetically complex and the results of many candidate gene and genome wide association

Gene-Environment Interactions in Psychiatry. http://dx.doi.org/10.1016/B978-0-12-801657-2.00010-0

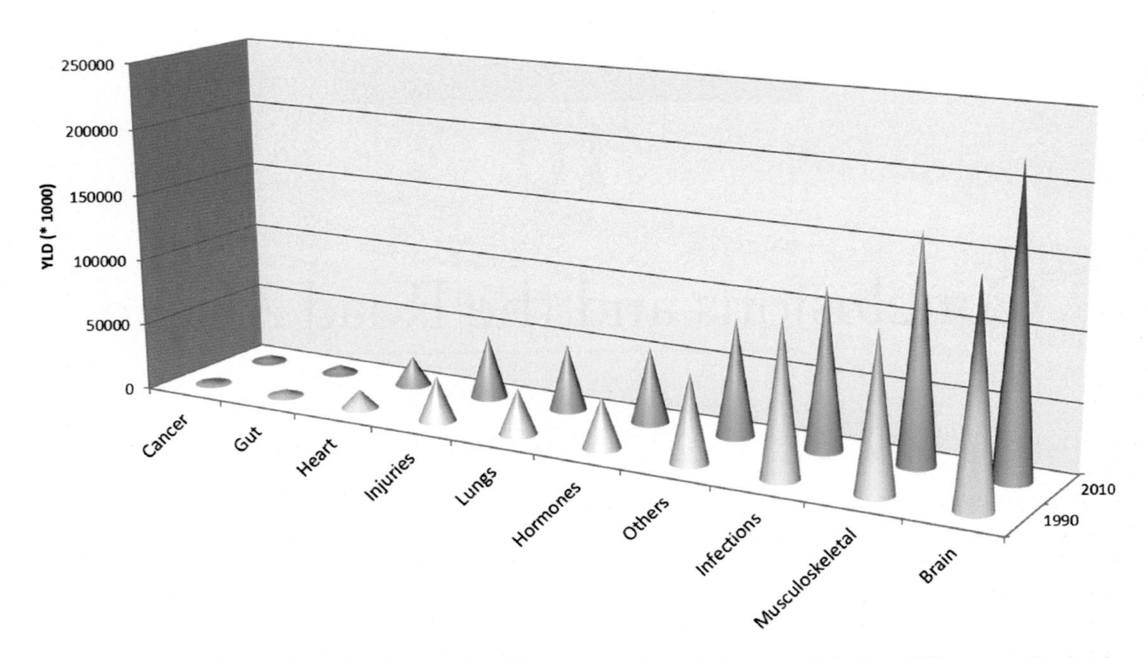

FIGURE 10.1 **Burden of disease in patients with brain disorders.** The Global Burden of Disease study clearly showed that brain disorder imposes the heaviest burden on patients, measured as years lived with disability (YLD). *Vos et al. (2012).*

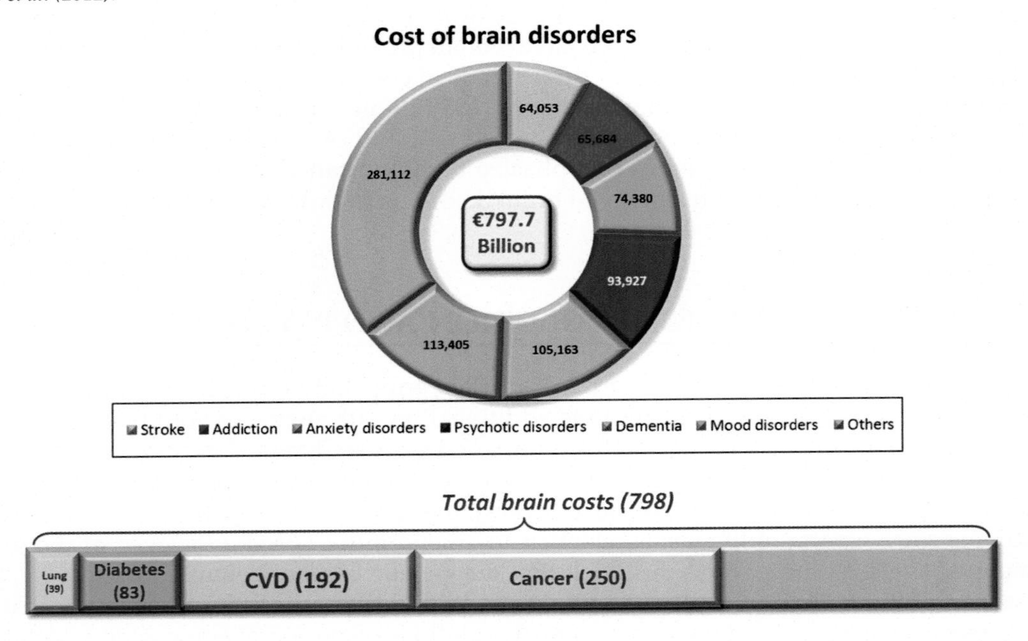

FIGURE 10.2 **The costs of brain disorders.** A recent study from Europe showed that the total costs for disorders of the brain was about 800 billion Euro, far exceeding the costs for lung diseases, diabetes, CVD, and cancer combined (Gustavsson et al., 2011). The numbers in brackets represent the estimated costs for these disorders.

studies (GWAS) have failed to find genes with a large impact. In fact, it seems safe to say that common genetic variants (ie, variants with a prevalence of more than 1% in the general population) only leads to odds ratios in the order of 1.00–2.00 (Fig. 10.3). Although some variants (especially copy number variants) can confer much larger risks, these are exceedingly rare and do not significantly contribute to the disease at a population level. Given that the heritability of most psychiatric disorders (Fig. 4.6) is substantial (between 45% and 70%), this begs the question why we have been unsuccessful in identifying genes with a major contribution at a population level so far.

In several neurological disorders we have been much more successful. Thus already in 1993 the gene responsible for Huntington's disease was identified (The Huntington's disease collaborative research, 1993), followed by several other disorders, also due to triplet repeat (Fondon et al., 2008). However, these diseases have a relatively simple, monogenetic origin. Other neurological disease, such as Parkinson's (PD) and Alzheimer's disease (AD) have a more complex origin with both 'familial' and 'sporadic' cases. With respect to PD, about 10% of all cases of PD have a positive family history and variations in several genes (such as *SNCA* and *LRRK2*) are responsible for autosomal-dominant (monogenetic) forms of PD (Klein and Westenberger, 2012). Similarly, familial AD is relatively rare (3%–5% of all cases) and is related

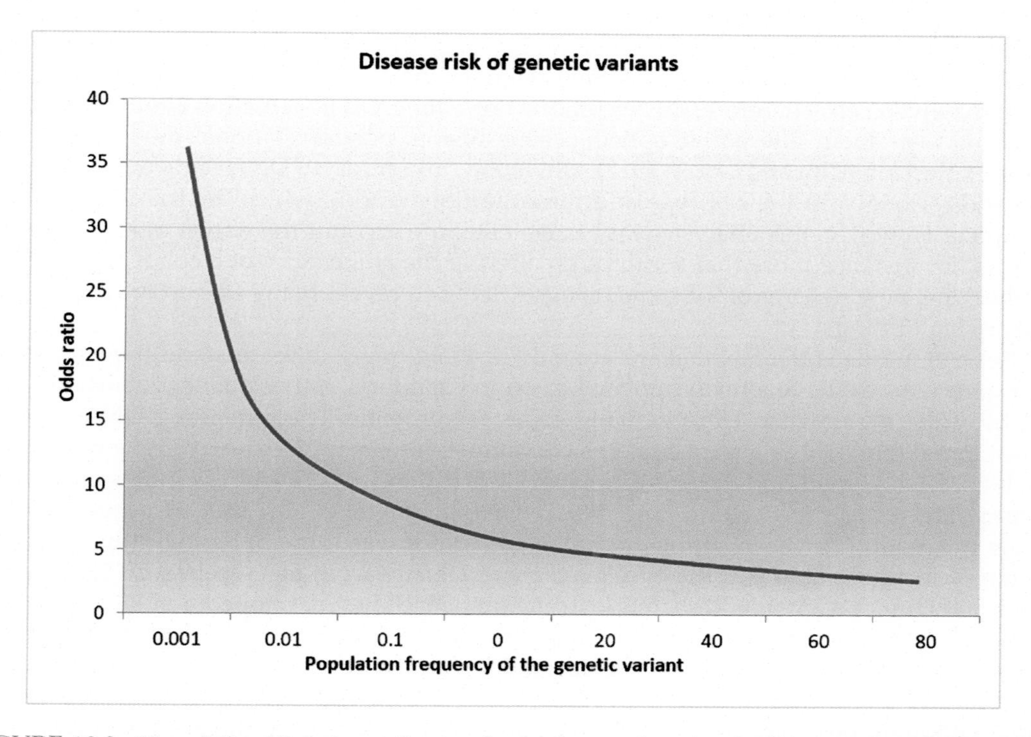

FIGURE 10.3 **The relationship between the prevalence of a genetic variant in the general population and its contribution to schizophrenia.** For schizophrenia, and indeed for most other psychiatric disorders, there appears to be a negative relationship between the prevalence of the gene and the risk it confers. Thus while some rare genetic variants have relatively high odds ratios, much more common variants only have a very small odds ratio (1.1–1.3).

to one of three genes (PS1, PS2 and APP) all related to the production of β-amyloid (Canevelli et al., 2014). Thus in contrast to psychiatric disorders, neurological diseases (when a familial basis is ascertained) have a relatively simple genetic origin. Several different theories have been proposed to explain the lack of "single genes with big impact". In the following sections we will touch upon these different theories, although the space and the aim of this chapter prohibit us from going into too much detail.

THE GENETIC ANALYSIS IS INCOMPLETE

The first, and perhaps most parsimonious explanation, is that such genes do actually exist, we just have not been able to find them yet. In this day and age where millions of genetic variants have been studied in tens of thousands of patients and controls, this becomes increasingly unlikely. Having said that, new techniques such as deep-sequencing and RNA sequencing have only recently been applied to psychiatry and their full potential has yet to be explored. Deep sequencing refers to the number of times a single nucleotide is read within the genome. With the latest generation of sequencing tools, it is now possible to read each nucleotide multiple times, thus significantly reducing sequencing errors (Goldman and Domschke, 2014), which is especially relevant for relatively rare variants. Recent so-called "ultra-deep sequencing" even allow coverage of more than 1000 times (Mirebrahim et al., 2015). RNA sequencing (RNA-Seq) refers to the analysis of the entire transcriptome within a given cell. Therefore, this technique not only gives information about genetic variants but also about alternative splicing, gene fusion, posttranslational modifications and gene expression (Chu and Corey, 2012). Obviously, in contrast to DNA sequencing, RNA-Seq depends strongly on the cell type and the conditions during which the transcriptome was sampled. However, this analysis would also take into account the effects of environmental effects, for instance those that act through altering the epigenetics of a cell (Chapter 5). Together, this new generation of sequencing is likely to reveal many important and perhaps surprising findings.

Related to this is the fact that we are only just beginning to understand the functions of our genes. As we have already touched upon in Chapter 2, only a small proportion (about 1.5%) of the mammalian DNA actually codes for proteins. The remaining DNA was long considered 'junk DNA', a term coined by Susumu Ohno in 1972. Even in a review in Nature in 1980, Leslie Orgel and Francis Crick concluded that junk DNA "*had little specificity and conveys little or no selective advantage to the organism*" (Orgel et al., 1980). However, over the years this notion has been seriously challenged and it was thought biologically and evolutionarily inconceivable that the vast majority of DNA was, in fact, useless. In 2003, the US National Human Genome Research Institute launched the Encyclopedia of DNA Elements (ENCODE) program aimed at elucidating all the functional elements of the human genome. Since approximately 90% of all single nucleotide variants related to various diseases identified in GWAS are found in the noncoding part of DNA, this exercise is extremely important. Although the project is still ongoing, some very important findings have already been published (Consortium, 2012). Most notably the authors concluded that, although indeed only about 1.5% of the DNA encodes for proteins, up to 80% is involved in gene transcription (among others through the production of long noncoding RNA and micro RNAs, see

Chapter 5). Although this finding has subsequently been criticized as being based on a very liberal definition of 'gene transcription' (Doolittle, 2013), it seems nonetheless safe to assume that the noncoding parts of DNA plays a much larger role in gene transcription than originally thought.

In addition to improving our understanding of the human genome, we also need to further our understanding of gene transcription. Many of the variants identified either in candidate studies or GWAS, are either in introns or represent synonymous mutations in exons. As discussed in Chapter 2, in contrast to missense and nonsense, neither intron variants nor synonymous mutations alter the amino acid sequence of the proteins coded by the gene. This leads to the question how (or if?) these changes can alter protein function. Evidence is accumulating that intron variants can alter gene splicing, affect the binding of enhancers or repressors and binding to the ribosomes. Likewise, synonymous mutations can alter gene transcription and translation as well as splicing and mRNA transport (Chamary et al., 2006), thus implying that synonymous mutations are not "silent" mutations. Finally it should be remembered that each triplet code is related to a single species of tRNA. By changing one nucleotide a different tRNA is required for RNA translation (even in the case of synonymous mutations). Given that not all tRNA levels are the same, this may affect protein formation (Dana and Tuller, 2014).

A final point, related to the idea that the analysis of DNA is incomplete is the existence of chimerism and mosaicism. As discussed in Chapter 3, both imply the existence of multiple forms of DNA within the same organism. While mosaicism involves different DNA from the same organism, chimerism refers to the presence of DNA from different organisms. Mosaicism can occur by a variety of mechanisms, for instance as a result of a mutation early in development which then can propagate to individual organs. Chimerism is usually due to the merging of multiple fertilized eggs. Originally thought to be a very rare condition in humans, it has recently been suggested to occur more frequently than originally thought (Chen et al., 2013b). The implication of chimerism and mosaicism is that analyzing the DNA from one cell or bodily fluid (such as plasma or saliva) may not fully capture an individual's genome, and it is, at least in theory, possible that specific genetic variants are present in the brain but not in plasma. However, in reality this seems more a theoretical possibility than a major concern, given the still relatively low incidence of chimerism and mosaicism.

GENETIC EFFECTS ARE MODERATED BY THE ENVIRONMENT

A second explanation why most genetic studies have only identified genes with small effects is related to the fact that virtually none of the GWAS or candidate gene studies take the role of environmental factors into account. As we have seen throughout the book none of the psychiatric disorders can be explained by genetic factors alone and many gene * environmental interactions have been found. Fig. 10.4 shows three of the most common pattern of interactions. In Fig. 10.4A, the traditional diathesis – stress interaction is displayed. In this model, the two genotypes by themselves do not confer a differential risk in relation to a particular symptoms of disease characteristics, but while one genotype (allele II) is relatively sensitive to a particular (usually negative) environmental factor, the other genotype (allele I) is resistant to

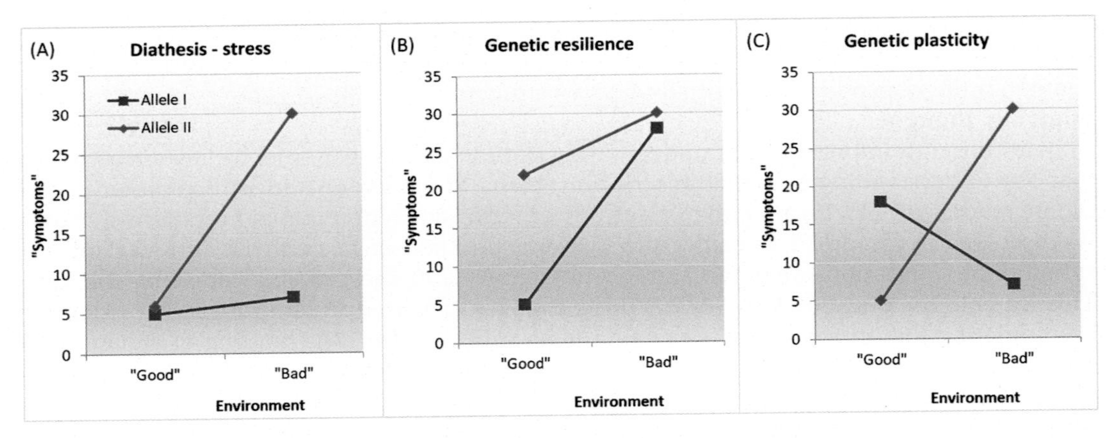

FIGURE 10.4 **Examples of gene – environment interactions.** Throughout the various chapters, several different types of gene – environment interactions have been described. The three main types of interactions are shown here. A: In the classical diathesis – stress model, both genetic variants have a comparable risk in a neutral (or good) environment. However, the risk is greatly increased in one genotype (II) when exposed to a bad environment. B: In the genetic resilience model, one genotype (II) confers a significant risk in a good environment, but, the other genotype (I) is much more sensitive to the negative influence of a bad environment. As a result, in a bad environment both genotypes have similarly (high) risks. C: In the genetic plasticity model, the genotypes differ in both the good and the bad condition. Whereas one (II) has a reduced risk in a good environment, the other genotype (I) has a reduced risk in a bad environment. See for more detail and concrete examples the main text.

its influence. This is probably the most common form of gene * environment interaction and a prime example is the classical study by Caspi and coworkers (Caspi et al., 2003) on depression (Chapter 7). In this study, the authors investigated the interaction between the variation in the promotor region of the serotonin transporter (*5-HTTLPR*) and stressful life events on the diagnosis of major depressive disorders (MDD). Their findings showed that in the absence of stressful life events the risk of developing MDD was similar between individuals with the short (s) and the long (l) allele. However, with increased exposure to stressful life events the risk increased for the s-allele carriers, but not for the l-carriers.

The second type of interaction (Fig. 10.4B) we refer to as the genetic resilience interaction. In this type of interaction, in a neutral (or good) environment one genotype (allele II) significantly enhances the risk compared to the second genotype (allele I). However, when exposed to environmental stressors, the risk selectively increases for the second genotype (I) while the original risk gene shows resilience. This kind of interaction can occur when the risk of a specific genotype is already quite high and additional exposure to environmental factors cannot increase the risk any further (due to a ceiling-type effect). An example of this type of interaction can be seen in a study investigating the interaction between peer smoking and genetic variations in the nicotinergic receptors (Johnson et al., 2010). In this study it was shown that the "AA" genotype of rs16969968 in the *CHRA5-A3-B4* gene cluster by itself significantly increased the risk of nicotine dependence compared to individuals with at least one "G" allele. However, with increased exposure to peer smoking this difference diminishes due to the fact that the risk of developing nicotine dependence significantly increased for the "G" allele carriers.

Finally the third form of gene environment interaction has been referred to as genetic plasticity (Belsky et al., 2009). This interaction is the strongest and basically leads to a reversal of risks. Thus whereas one genotype (allele I) has the highest risk in a good (or neutral) environment, the second genotype (allele II) confers the highest risk when exposed to environmental stressors ("bad" environment). An excellent example of this type of interaction is a study by Chen and coworkers who investigated the interaction between the well-known val/met variant in the *BDNF* gene (rs6265) and stressful life events (Chen et al., 2013a). In this study the authors found that in the absence of any stressful life events, individuals with the "AA" genotype (met/met) showed the greatest number of depressive symptoms in adolescence. However, in the presence of stressful life events this situation changed, and when exposed to 4 or more such events, carriers of the "G" allele showed more depressive symptoms.

These gene * environment interactions allow for a number of important conclusions. First of all, at least in the case of psychiatric disorders, the term 'genetic risk factor' should be used with great care, as often the risk only becomes apparent in the presence (Fig. 10.4A) or absence (Fig. 10.4B) of specific environmental factors. Moreover, in the case of plasticity genes neither (or both) of the genetic variants can be considered risk factors. Indeed in a meta-analysis of 77 studies investigating the *5-HTTLPR* it was shown that the detrimental effect of a negative environment and the beneficial effects of a positive environment on carriers of the 's' allele was equally strong (van Ijzendoorn et al., 2012).

Perhaps even more importantly, these gene * environment interactions indicate that studying genetic (or environmental) effects alone can lead to a substantial underestimation. For instance, a main effect of genotype was not found in either the study by Caspi et al. (2003) or the one by Chen et al. (2013a,b). In other words, if the environmental factors had not been included in the analysis, the conclusion would have been that neither the *5-HTTLPR* nor the *BDNF* genotype had an influence of the presence of depressive symptoms. Indeed, in many other studies that have identified a significant gene * environment interaction, no main effects of gene and/or environmental were found. This then implies that candidate gene and GWAS studies have the risk of systematically underestimating the significance of genetic factors and should be extended to include environmental factors as well.

An important additional factor that is of relevance to gene * environment interaction studies is the distinction between gene * environment interactions and gene * environmental correlations. So far we have primarily been referring to the traditional interactive effect of two independent factors. However, in certain circumstances the environment may be causally related to the genotype, a phenomenon referred to as gene * environmental correlation (usually abbreviated as rGE). Several different types of rGEs can be distinguished. Passive rGE refers to the association between the genotype of a child and the familial environment it is raised in. Reactive rGE refers to the genotype dependent reaction to specific environments. For instance certain genotypes (such as the s-allele of the 5-HTTLPR) enhances the stress response to a specific (negative) environmental factor. Thus, while objectively the environmental factor is identical for all genotypes, the perception (and thus reaction to it) may be different. Finally, active rGE refer to the active selection of specific environments by individuals with a specific genotype (Jaffee and Price, 2007). Although there are techniques available to distinguish between correlations and true interactions, animal studies can be very helpful in this respect as it is possible to much better restrict the exposure to the environment. Nonetheless, the reactive rGE is also in animal research an important factor to keep in mind.

PSYCHIATRIC DISORDERS ARE HETEROGENEOUS

A final important factor contributing to the lack of a single gene with big impact, is directly related to the diagnosis of psychiatric disorders. A diagnosis is generally based on the presence of a select number of symptoms from a larger checklist and as we have pointed out several times throughout the book, this leads to significant heterogeneity in patients. Indeed, inspection of Box 7.1 shows that patients with depression can have diametrically opposite symptoms (insomnia vs. hypersomnia, weight gain vs. weight loss). Considering this significant heterogeneity in the symptoms within a given patient population, it is more than likely that there is also significant heterogeneity in pathology and aetiology. Hence it is not surprising that so many genes with small effects have been linked to a specific psychiatric illness.

Related to this, is the fact that there is not only significant heterogeneity within a patient population with a specific diagnosis, but at the same time there is significant overlap in symptoms between patients with different diagnoses. Not only is comorbidity very common between different disorders (see for instance Figs. 7.10 and 8.2), but in addition comparable symptoms are part of different disorders, as is schematically illustrated in Fig. 10.5. One of the most obvious examples is of course MDD and bipolar disorder (BP), but many other similarities can be readily identified. For instance psychomotor agitation can occur in attention deficit hyperactivity disorders (ADHD), autism spectrum disorder (ASD), BP, MDD and schizophrenia. Likewise, alteration in reward anticipation and perception are an integral part of addictive disorders, MDD, BP and schizophrenia, and deficits in social cognition can occur in virtually all psychiatric disorders.

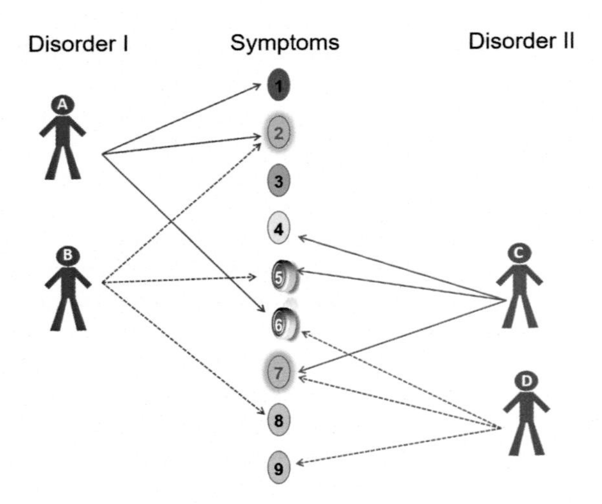

FIGURE 10.5 **The symptomatological relationship between different psychiatric disorders.** Since a psychiatric diagnosis is based on the occurrence of a set number of symptoms from a larger checklist, patients with the same diagnosis (such as the blue patients A and B, or the red patients C and D) can have both overlapping (illustrated by the ellipses with red numbers and a blue or red glow) and different symptoms. Moreover, certain symptoms can occur in multiple disorders and thus can be shared even by patients with different diagnoses (as illustrated by the 3D ellipses with white numbers).

Given the similarities in symptomatology between different psychiatric disorders, it is not surprising that similarities in underlying pathology are also apparent. This is already clear when inspecting the neurobiological substrates discussed in the previous chapters (see for instance Figs. 6.6, 7.3, 8.5, and 8.13). Structures such as the medial prefrontal and orbitofrontal cortex, and parts of the basal ganglia have been implicated in most psychiatric disorders. Likewise, neurotransmitters such as dopamine have been an integral part of the neurobiology of addiction, ADHD, MDD, BP and schizophrenia and serotonin has been implicated in ASD, MDD and BP. A recent study further investigated similarities in underlying pathology by performing a meta-analysis of all structural MRI scans published in patients with schizophrenia, BP, MDD, addiction, obsessive-compulsive disorder, or anxiety (Goodkind et al., 2015). Focusing on alterations in gray matter volume, the authors identified comparable changes in the anterior insular/dorsal anterior cingulate cortex network across diagnostic groups. Moreover they found that the combined gray matter volume was positively correlated with executive functioning, a prime example of a transdiagnostic cognitive deficit. Although these similarities are intriguing and clearly underscore transdiagnostic similarity, it would be even more interesting to investigate similarities at a functional level.

Finally, to further underscore the similarities between different psychiatric disorders, several recent papers have now started to use the large number of candidate gene and GWAS studies to identify similar genetic variants across diagnoses. One such initiative is the Cross Diagnostic Group of the Psychiatric Genomic Consortium and in a recent study they used the vast GWAS database to look for similarities between schizophrenia, MDD, BP, ASD and ADHD (Cross-Disorder Group of the Psychiatric Genomics et al., 2013). First, the authors calculated the heritability (h^2) based on single nucleotide variants, and these varied from 0.17 for ASD to 0.28 for ADHD. More importantly, they estimated the coheritabilities between disorders. Although for most pairs of disorders these were nonsignificant, five pairs had significant coheritability. As Fig. 10.6 clearly shows, there was substantial evidence that schizophrenia, BP and MDD have a (partly) shared genetic liability. This is in line with the familial studies, especially in relation to schizophrenia and BP (Figs. 7.9 and 7.10). In an effort to identify common genes, the same group analyzed GWAS data from 33,332 patients (suffering from one of the 5 disorders) and 27,888 controls (Cross-Disorder Group of the Psychiatric Genomics, 2013) and identified 4 variants (Table 10.1), three of which were found in all five disorders (variants in 3p21 (rs2535629), 10q24 (rs11191454) and in *CACNB2*) while the remaining variants (*CACNA1C*) was enriched in schizophrenia and BP. In addition, a number of other genes were highly associated with several disorders, but they did not reach the GWAS threshold. It should, again, be pointed out that all these GWAS identified variants only contributed a relatively small risk to all five disorders, with Odds Ratios varying from 1.07 to 1.13. Overlap in copy number variations (CNVs) between different disorders have also been reported. For instance, in one study, most of the CNVs associated with schizophrenia were also associated with ASD (Doherty and Owen, 2014), although some were shared with other psychiatric disorders as well. Among these latter were duplications of 7q11.2, associated with schizophrenia, ASD, ADHD and anxiety disorders, deletions and duplications of 16p11.2 and deletions of 22q11.2, both of which have been associated with schizophrenia, ASD, ADHD, mood and anxiety disorders.

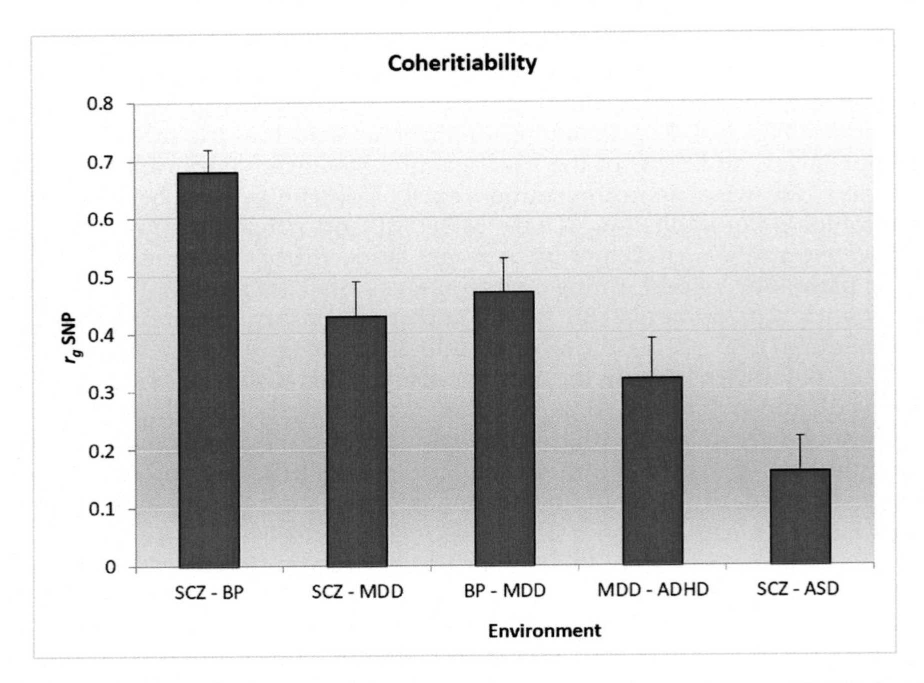

FIGURE 10.6 **Co-heritability between psychiatric disorders.** Based on a large database of GWAS data, the single nucleotide polymorphism heritability (r_gSNP) between different psychiatric disorders was calculated. The five pairs of disorders represented in this figure showed significant co-heritability. SCZ: schizophrenia, BP: bipolar disorder; MDD: major depressive disorder; ADHD: attention deficit hyperactivity disorder; ASD: autism spectrum disorder [data based on (Cross-Disorder Group of the Psychiatric Genomics et al., 2013)].

TABLE 10.1 Shared Loci Between Schizophrenia (SCZ), BP, MDD, ADHD, and ASD

Variant	Nearest gene	Variant	p value	Best fit model
rs2535629	*ITIH3* (+ many)	$G \rightarrow A$	2.5×10^{-12}	All five disorders
rs11191454	*AS3MT* (+many)	$A \rightarrow G$	1.4×10^{-8}	All five disorders
rs1024582	*CACNA1C*	$A \rightarrow G$	1.9×10^{-8}	BP, SCZ
rs2799573	*CACNB2*	$T \rightarrow C$	4.3×10^{-8}	All five disorders

An alternative approach to identify genetic similarities between different psychiatric disorders is to focus on candidate gene studies. One distinct disadvantage of using candidate gene studies, as we have remarked upon regularly in the previous chapters, is that many findings reported in one paper could not be replicated in subsequent papers. In an attempt to circumvent this problem, Gatt and coworkers focused only on genetic variants of candidate genes that were confirmed in meta-analyses (Gatt et al., 2015). Using this approach, the authors identified several genetic variants that were significantly associated with more than one disorder (Fig. 10.7).

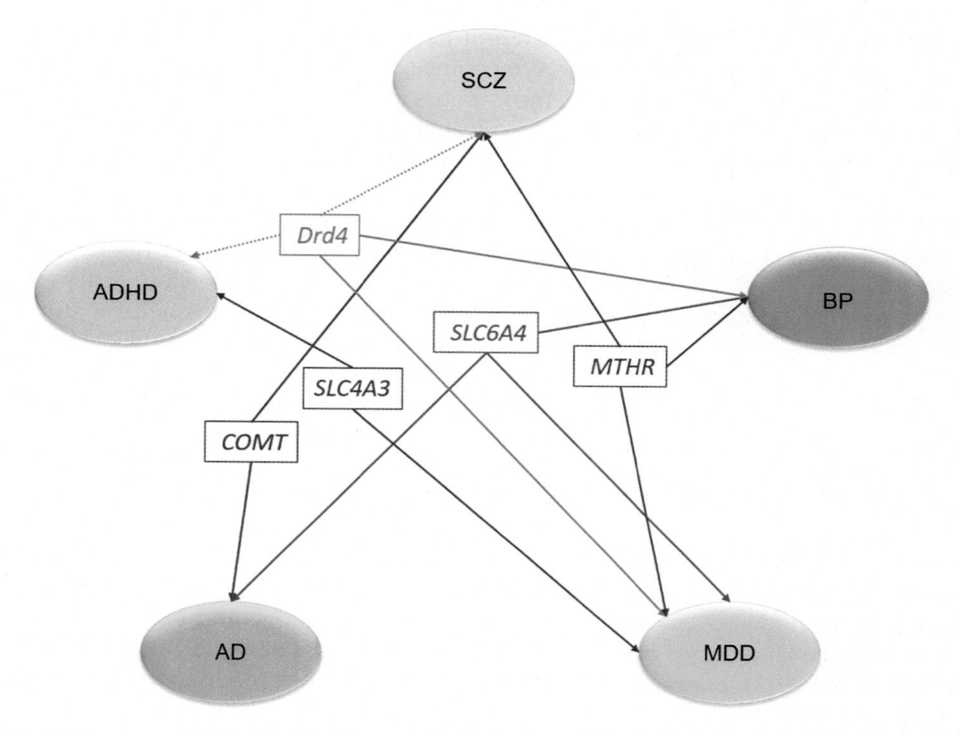

FIGURE 10.7 **The genetic relationship between different psychiatric disorders.** Comparing meta-analyses of candidate gene studies for psychiatric disorders, Gatt and coworkers identified several genetic variants that are shared by multiple disorders (Gatt et al., 2015). Solid lines represent similar genetic variants, while dashed lines represent different genetic variants. SCZ: schizophrenia, BP: bipolar disorder; MDD: major depressive disorder; ADHD: attention deficit hyperactivity disorder; AD: anxiety disorder.

LOOKING TOWARD THE FUTURE

The above discussion clearly shows that in recent years the boundaries between different psychiatric disorders are fading and although not incorporated in the DSM criteria, more and more research is focusing on symptoms rather than diagnostic classifications. One prime example, relevant for the present discussion is the study by Arnedo and coworkers (Arnedo et al., 2015) that focused on identifying the genetic architecture of schizophrenia. Although this paper was already discussed in Chapter 9, it is important to briefly recapitulate their approach. Rather than focusing on identifying single genetic variants related to schizophrenia, they first of all extensively characterized their patients in terms of symptoms and secondly investigated gene sets that could distinguish patients' symptoms (or symptoms clusters) from healthy volunteers. Their approach made it possible to identify more than 70% of all patients and the symptoms of certain patient clusters could be almost completely determined by only a handful of genetic variants.

In our opinion this study represents an important step forward in identifying on the one hand heterogeneity within a patient population with the same diagnosis, and on the other hand finding commonalities between different patient populations. Thus in the future, the emphasis should be on transdiagnostic symptoms (or symptoms clusters) and on endophenotypes or biomarkers. Indeed this was the impetus for the initiative by the National Institute for Mental Health in the US to develop the so-called Research Domain Criteria [RDoC, (Cuthbert, 2014)]. Endophenotypes were first introduced by Gottesman and Shield (Gottesman and Shields, 1973) and are considered quantifiable entities in the pathway between genes and behavior, or in the case of psychiatric disorders in the pathway between genes and symptoms. As discussed elsewhere, endophenotypes have to fulfill a number of criteria: [1] they are associated with the disorder; [2] they are heritable; [3] they are state-independent (ie, always present irrespective of whether the patient is ill or in remission); [3] within families they should co-segregate with the illness; [4] they should occur in family members of the proband to a higher degree than in the general population (Gould and Gottesman, 2006). Especially within the field of cognitive deficits in schizophrenia the concept of endophenotypes has received wide attention. For instance a recent study identified multiple cognitive and information-processing tasks that fulfil the criteria for endophenotypes (Braff, 2015), including deficits in prepulse inhibition and the California Verbal Learning Task. Interestingly, as we already mentioned, cognitive deficits are seen in many psychiatric disorders and thus these endophenotypes do not only offer the possibility of increasing the homogeneity within the schizophrenic patient population, but also could contribute to the transdiagnostic analysis of genetic factors. Indeed, deficits in working memory and executive functioning have also been considered as endophenotypes for BP (Raust et al., 2014), ASD, and ADHD (Rommelse et al., 2011). Likewise, multiple attempts have been made to identify neuroimaging endophenotypes for psychiatric disorders such as ASD (Mahajan and Mostofsky, 2015), drug addiction (Jupp and Dalley, 2014), schizophrenia (Ferrarelli, 2013), and MDD (Peterson and Weissman, 2011).

Much less success has been made regarding other (common) aspects of dysfunctioning in psychiatric patients. However, recent studies have suggested that heart rate variability (HRV) may be an interesting endophenotype for emotional dysregulation. HRV refers to the beat-to-beat variability in heart rate and can be easily assessed in humans (as well as in animals). Although, HRV is primarily determined by the balance between the sympathetic and parasympathetic nervous system, it is strongly regulated by several key component of the prefrontal cortex (Thayer and Lane, 2009). Given that many psychiatric disorders show altered activity (and/or connectivity) within the prefrontal cortical network, it is not surprising that many also show altered HRV (Moon et al., 2013). Again, this has been most extensively studied in schizophrenia and two recent meta-analyses found convincing evidence to classify a reduced HRV as an endophenotype (Clamor et al., 2014; Montaquila et al., 2015 Fig. 10.7).

Thus the concept of endophenotypes will be an important tool in the search for the aetiology of psychiatric symptoms (Fig. 10.8). RDoC, on the other hand, represent more 'theoretical' aspects of disease pathology, and include phenomena such as negative and positive valence and arousal /regulatory systems. Nonetheless, there is substantial overlap between the two concepts as well, and both are thought to represent intermediate aspects of disorders that are more closely related to specific genes and neurobiological abnormalities.

In addition, more effort should be put into simultaneously assessing genetic and environmental factors rather than focusing on either. Fortunately, more and more research initiatives are moving into this direction, but whereas GWAS studies have already formed multiple

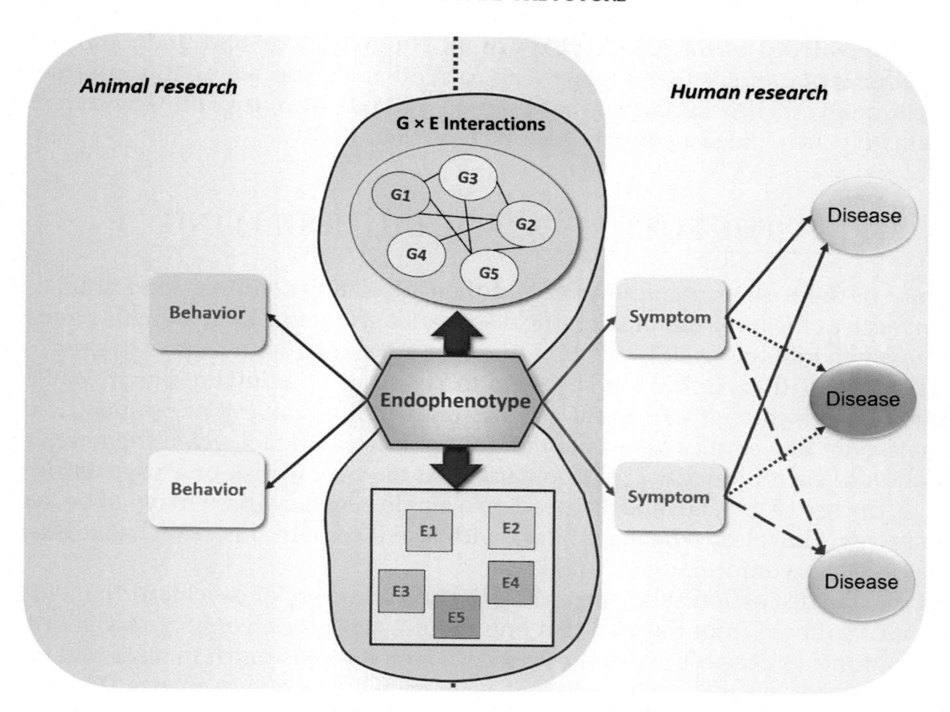

FIGURE 10.8 **Towards an improved approach to studying the aetiology of psychiatric disorders and symptoms.** Using endophenotypes (or Research Domain Criteria) as features on the pathway from genes to psychiatric symptoms, has the added benefit that they are more quantifiable and can often be assessed objectively in humans and animals with similar techniques. Thus animal and human research would intersect at the level of the endophenotypes. Genetic (G) and environmental (E) factors are thought to interact to shape the endophenotype, which would give rise to specific behavioral alterations in animals as well as clinical symptoms in humans that can occur in multiple disorders.

consortia in order to have enough statistical power to identify common genes with relatively small effects, most gene * environment interaction studies are done by individual research teams [see however (Arnedo et al., 2015)]. One distinct complication in studying environmental factors is the reliability of the information as we already discussed in Chapter 4. Thus most information is based on recall from either the patient or the family. Nonetheless, this information is important and often can be corroborated by other sources, such as hospital records, health-related databases or other independent sources. Moreover, the ever increasing popularity of social media on the internet, may represent a source of information hitherto unused. Although obviously information on social media is also biased and subjective, it is nonetheless preserved for a long time and can be assessed retrospectively with more confidence. In this respect it can serve similar functions as home videos that have already been used in the past, for instance to provide evidence for neurodevelopmental delays in children who later on develop schizophrenia (Walker et al., 1994).

Finally, an important development which has gained significant interest in recent years is the use of bio-informatics as a tool to identify commonalities between genes. Thus our ever increasing knowledge of the biochemistry of the (human) body and brain allows us to develop networks of proteins that interact with each other. By combining this information with the genetic analyses, functional networks can be build that can explain why different genes may ultimately have the same functional outcome (Fig. 10.7).

THE FUTURE OF ANIMAL MODELLING

Because of the ease of manipulation and their availability, animals and animal research has often been at the forefront of scientific discoveries and findings in animals often precedes similar research in humans. It is therefore perhaps surprising with respect to gene * environment interactions, that clinical studies seem to be ahead of animal research. While there is extensive research into either genetic or environmental modelling of psychiatric disorders in animals, only few studies have combined these factors. In fact, given the relatively small contribution of each individual genetic variant to the overall risk of a psychiatric disorder (Fig. 10.3), the use of genetic models based on a single genetic variant may not be very useful by themselves. However, when combined with specific environmental challenges they can nonetheless be very informative.

Similar to the discussion above regarding the heterogeneity of psychiatric patients, animal models have suffered from the idea that one should strive for an animal model in which 'as many symptoms as possible' are represented. In fact, it seems much more useful to develop animal models for specific symptoms that have transdiagnostic relevance (Fig. 10.8). Thus the endophenotype/RDoC concept is equally valid for animal modelling, especially when using technologies that are very similar in rodents and humans. This has been exemplified by techniques such as prepulse inhibition (Braff et al., 1978; Braff and Geyer, 1990) or P_{50} gating (Adler et al., 1982; Adler et al., 1986; Ellenbroek, 2004). Another relevant example may be HRV, which can also be assessed in humans and rats with identical techniques (Stiedl et al., 2009). Although not necessarily using identical procedures, several cognitive tasks have recently been identified that appear to map onto the specific (transdiagnostic and cross-species) domains such as attention, executive functioning, working memory etc. (see Neuroscience and Biobehavioural Reviews, 2013, 37B, pp. 2087–2103).

Thus, the same (or at least very similar) endophenotypes can be used in both humans and animals, offering the possibility of true translational validity, by assessing how genetic and environmental factors interact to shape the endophenotype and how this translates into behavioral changes in rats and clinical symptoms in humans.

CONCLUSION

Together, it seems that the technologies and theoretical concepts for better understanding the aetiology of psychiatric symptoms are now available. By further integrating clinical and preclinical research, especially through establishing cross-species designs we will be able to use the "best of both worlds". This will ultimately lead to important breakthroughs in the research into nature nurture neuroscience and will hopefully aid in the development of novel strategies for the treatment of patients suffering from psychiatric symptoms. Given the tremendous personal, social and economic burden brain disorders constitute, we owe it to the patients, their families and society as a whole to significantly increase our efforts into providing a better quality of life for all.

References

Adler, L.E., Pachtman, E., Franks, R.D., et al., 1982. Neurophysiological evidence for a defect in neuronal mechanisms involved in sensory gating in schizophrenia. Biol. Psychiatry 17, 649–654.

Adler, L.E., Rose, G.M., Freedman, R., 1986. Neurophysiological studies of sensory gating in rats: effects of amphetamine, phencyclidine and haloperidol. Biol. Psychiatry 21, 787–798.

Andlin-Sobocki, P., Jonsson, B., Wittchen, H.U., et al., 2005. Cost of disorders of the brain in Europe. Eur. J. Neurol. 12, 1–27.

Arnedo, J., Svrakic, D.M., del Val, C., et al., 2015. Uncovering the hidden risk architecture of the schizophrenias: confirmation in three independent genome-wide association studies. Am. J. Psychiatry 172, 139–153.

Belsky, J., Jonassaint, C., Pluess, M., et al., 2009. Vulnerability genes or plasticity genes? Mol. Psychiatry 14, 746–754.

Braff, D.L., 2015. The importance of endophenotypes in schizophrenia research. Schizophrenia Res. 163, 1–8.

Braff, D.L., Geyer, M.A., 1990. Sensorimotor gating and schizophrenia. Human and animal model studies. Arch. Gen. Psychiatry 47, 181–188.

Braff, D., Stone, C., Callaway, E., et al., 1978. Prestimulus effects of human startle reflex in normals and schizophrenics. Psychophysiology 15, 339–343.

Canevelli, M., Piscopo, P., Talarico, G., et al., 2014. Familial Alzheimer's disease sustained by presenilin 2 mutations: systematic review of literature and genotype-phenotype correlation. Neurosci. Biobehav. Rev. 42, 170–179.

Caspi, A., Sugden, K., Moffitt, T.E., et al., 2003. Influence of life stress on depression: moderation by a polymorphism in the 5-HTT gene. Science 301, 386–389.

Chamary, J.V., Parmley, J.L., Hurst, L.D., 2006. Hearing silence: nonneutral evolution at synonymous sites in mammals. Nature Rev. Genetics 7, 98–108.

Chen, J., Li, X.Y., McGue, M., 2013a. The interacting effect of the BDNF Val66Met polymorphism and stressful life events on adolescent depression is not an artifact of gene-environment correlation: evidence from a longitudinal twin study. J. Child Psychol. Psychiatry 54, 1066–1073.

Chen, K., Chmait, R.H., Vanderbilt, D., et al., 2013b. Chimerism in monochorionic dizygotic twins: case study and review. Am. J. Medical Genetics. Part A 161A, 1817–1824.

Chu, Y., Corey, D.R., 2012. RNA sequencing: platform selection, experimental design, and data interpretation. Nucleic Acid Ther. 22, 271–274.

Clamor, A., Hartmann, M.M., Kother, U., et al., 2014. Altered autonomic arousal in psychosis: an analysis of vulnerability and specificity. Schizophrenia Res. 154, 73–78.

Consortium, E.P., 2012. An integrated encyclopedia of DNA elements in the human genome. Nature 489, 57–74.

Cross-Disorder Group of the Psychiatric Genomics C, 2013. Identification of risk loci with shared effects on five major psychiatric disorders: a genome-wide analysis. Lancet 381, 1371–1379.

Cross-Disorder Group of the Psychiatric, Genomics, C., Lee, S.H., Ripke, S., et al., 2013. Genetic relationship between five psychiatric disorders estimated from genome-wide SNPs. Nat. Genet. 45, 984–994.

Cuthbert, B.N., 2014. The RDoC framework: facilitating transition from ICD/DSM to dimensional approaches that integrate neuroscience and psychopathology. World Psychiatry 13, 28–35.

Dana, A., Tuller, T., 2014. The effect of tRNA levels on decoding times of mRNA codons. Nucleic Acids Res. 42, 9171–9181.

Doherty, J.L., Owen, M.J., 2014. Genomic insights into the overlap between psychiatric disorders: implications for research and clinical practice. Genome Medicine 6.

Doolittle, W.F., 2013. Is junk DNA bunk? A critique of ENCODE. Proc. Natl. Acad. Sci. USA 110, 5294–5300.

Ellenbroek, B.A., 2004. Pre-attentive processing and schizophrenia: animal studies. Psychopharmacology 174, 65–74.

Ferrarelli, F., 2013. Endophenotypes and biological markers of schizophrenia: from biological signs of illness to novel treatment targets. Curr. Pharmaceutic. Des. 19, 6462–6479.

Fondon, 3rd, J.W., Hammock, E.A., Hannan, A.J., et al., 2008. Simple sequence repeats: genetic modulators of brain function and behavior. Trends Neurosci. 31, 328–334.

Gatt, J.M., Burton, K.L.O., Williams, L.M., et al., 2015. Specific and common genes implicated across major mental disorders: a review of meta-analysis studies. J. Psychiatric Res. 60, 1–13.

Goldman, D., Domschke, K., 2014. Making sense of deep sequencing. Int. J. Neuropsychopharmacol. 17, 1717–1725.
Goodkind, M., Eickhoff, S.B., Oathes, D.J., et al., 2015. Identification of a common neurobiological substrate for mental illness. JAMA Psychiatry 72, 305–315.
Gottesman, I.I., Shields, J., 1973. Genetic theorizing and schizophrenia. Br. J. Psychiatry 122, 15–30.
Gould, T.D., Gottesman, I.I., 2006. Psychiatric endophenotypes and the development of valid animal models. Genes Brain Behav. 5, 113–119.
Gustavsson, A., Svensson, M., Jacobi, F., et al., 2011. Cost of disorders of the brain in Europe 2010. Eur. Neuropsychopharmacol. J. Eur. Coll. Neuropsychopharmacol. 21, 718–779.
Jaffee, S.R., Price, T.S., 2007. Gene-environment correlations: a review of the evidence and implications for prevention of mental illness. Mol. Psychiatry 12, 432–442.
Johnson, E.O., Chen, L.S., Breslau, N., et al., 2010. Peer smoking and the nicotinic receptor genes: an examination of genetic and environmental risks for nicotine dependence. Addiction 105, 2014–2022.
Jupp, B., Dalley, J.W., 2014. Behavioral endophenotypes of drug addiction: etiological insights from neuroimaging studies. Neuropharmacology 76, 487–497.
Klein, C., Westenberger, A., 2012. Genetics of Parkinson's disease. Cold Spring Harb. Perspect. Med. 2, a008888.
Mahajan, R., Mostofsky, S.H., 2015. Neuroimaging endophenotypes in autism spectrum disorder. CNS Spectrum. 20, 412–426.
Mirebrahim, H., Close, T.J., Lonardi, S., 2015. De novo meta-assembly of ultra-deep sequencing data. Bioinformatics 31, i9–16.
Montaquila, J.M., Trachik, B.J., Bedwell, J.S., 2015. Heart rate variability and vagal tone in schizophrenia: a review. J. Psychiatr. Res. 69, 57–66.
Moon, E., Lee, S.H., Kim, D.H., et al., 2013. Comparative study of heart rate variability in patients with schizophrenia, bipolar disorder, post-traumatic stress disorder, or major depressive disorder. Clin. Psychopharmacol. Neurosci. 11, 137–143.
Orgel, L.E., Crick, F.H., Sapienza, C., 1980. Selfish DNA. Nature 288, 645–646.
Peterson, B.S., Weissman, M.M., 2011. A brain-based endophenotype for major depressive disorder. Ann. Rev. Med. 62, 461–474.
Raust, A., Daban, C., Cochet, B., et al., 2014. Neurocognitive performance as an endophenotype for bipolar disorder. Front Biosci. (Elite Ed) 6, 89–103.
Rommelse, N.N., Geurts, H.M., Franke, B., et al., 2011. A review on cognitive and brain endophenotypes that may be common in autism spectrum disorder and attention-deficit/hyperactivity disorder and facilitate the search for pleiotropic genes. Neurosci. Biobehav. Rev. 35, 1363–1396.
Stiedl, O., Jansen, R.F., Pieneman, A.W., et al., 2009. Assessing aversive emotional states through the heart in mice: implications for cardiovascular dysregulation in affective disorders. Neurosci. Biobehav. Rev. 33, 181–190.
Thayer, J.F., Lane, R.D., 2009. Claude Bernard and the heart-brain connection: further elaboration of a model of neurovisceral integration. Neurosci. Biobehav. Rev. 33, 81–88.
The Huntington's disease collaborative research, 1993. A novel gene containing a trinucleotide repeat that is expanded and unstable in Huntington's disease chromosome. Cell 72, 983.
van Ijzendoorn, M.H., Belsky, J., Bakermans-Kranenburg, M.J., 2012. Serotonin transporter genotype 5HTTLPR as a marker of differential susceptibility? A meta-analysis of child and adolescent gene-by-environment studies. Trans. Psychiatry 2, e147.
Vos, T., Flaxman, A.D., Naghavi, M., et al., 2012. Years lived with disability (YLDs) for 1160 sequelae of 289 diseases and injuries 1990-2010: a systematic analysis for the Global Burden of Disease Study 2010. Lancet 380, 2163–2196.
Walker, E.F., Savoie, T., Davis, D., 1994. Neuromotor precursors of schizophrenia. Schizophr. Bull. 20, 441–451.

Abbreviations

AC	Adenylate cyclase
ADHD	Attentional deficit hyperactivity disorder
ASD	Autism spectrum disorder
BDNF	Brain derived neurotrophic factor
BP	Bipolar disorder
BPM	Behavioral pattern monitor
BPRS	Brief psychiatric rating scale
CAR	Cortisol awakening response
CATIE	Clinical antipsychotics trial intervention effectiveness
CHO	Chinese hamster ovary
CLEQ	Childhood life events questionnaire
CNTRICS	Cognitive neuroscience treatment research to improve Cognition in schizophrenia
CNV	Copy number variations
COGA	Collaborative study on the genetics of alcoholism
CPP	Conditioned place preference
CREB	CAMP response element binding
CRH	Corticotrophin-releasing hormone
CSS	Chromosome substitution strains
CTQ	Childhood trauma questionnaire
CVD	Cardiovascular disorders
DNA	Deoxyribo nucleic acid
DALY	Disability adjusted life years
DBP	D-box binding protein
DDT	Delay discounting task
DFP	Diisopropyl fluorophosphate
DMI	Depressive-mania-interval
DMN	Default mode network
DR	Dorsal raphe
DSM	Diagnostic and statistical manual
DTD	Developmental trauma disorder
DTI	Diffusion tensor imaging
ENCODE	Encyclopedia of DNA elements
ERP	Event related potentials
ES	Embryonic stem
FAS	Fetal alcohol syndrome
FASD	Fetal alcohol spectrum disorder
FDA	Food and Drug Administration
FH	Fawn hooded
FST	Forced swim test
GAIN	Genetic association information network
GPC	Glial progenitor cells
GR	Glucocorticoid receptor
GWAS	Genome wide association studies
HAB	High anxiety behavior
HAT	Histone acetyl transferase
HD	Hyperkinetic conduct disorder
HDR	Homology dependent repair
HDRS	Hamilton depression rating scale
HEK	Human embryonic kidney

Gene-Environment Interactions in Psychiatry. http://dx.doi.org/10.1016/B978-0-12-801657-2.00016-1

HPA	Hypothalamic pituitary adrenal
HRV	Heart rate variability
ICD	International classification of disease
IGT	Iowa gambling task
ISBD	International society for bipolar disorders
KD	Knock-down
KO	Knock-out
LAB	Low anxiety behavior
LH	Lateral habenula
LIF	Leukemia inhibitory factor
MAO	Monoamine oxidase
MDD	Major depressive disorder
MDE	Major depressive episodes
MDI	Mania-depressive interval
MEG	Magnetic electroencephalography
NMDA	N-methyl-D-aspartate
NHEJ	Nonhomologous end joining
NPC	Neuronal progenitor cells
PANSS	Positive and negative syndrome scale
PCR	Polymerase chain reaction
PD	Personality disorder
PET	Positron emission tomography
PFC	Prefrontal cortex
PI	Polarity index
PTSD	Post-traumatic stress disorder
QTL	Quantitative trait loci
REM	Rapid eye movement
RISC	RNA induced silencing complex
RNA	Ribonucleic acid
SA	Splice-acceptor
SAM	S-adenosyl methionine
SD	Sprague dawley
SERT	Serotonin transporter
SFARI	Simons foundation autism research initiative
SHP	Stress hyposensitive period
SHR	Spontaneous hypertensive rat
SNI	Single nucleotide insertions
SNP	single nucleotide polymorphism
SNRI	Serotonin/noradrenaline reuptake inhibitors
SNV	Single nucleotide variant
SPECT	Single photon emission computer tomography
SSRI	Selective serotonin reuptake inhibitors
SST	Stop-signal task
SUD	Substance use disorder
TAF	Transcription associated factors
TAL	Transcription activator-like
TBP	TATA binding protein
TEPS	Temporal experience of pleasure scale
TF	Transcription factor
TFIIB	Transcription factor IIB
TFIIF	Transcription factor IIF
TNF	Tumor necrosis factor
TRAILS	Tracking adolescents individual lives survey
TRBP	HIV-1 TAR RNA binding protein

UNODC United Nations Office on Drugs and Crime
VNTR Variable number tandem repeat
VP Ventral pallidum
VTA Ventral tegmental area
WES Whole exome sequencing
WMH World mental health
YLD Year lived with disability
YMRS Young mania rating scale
ZF Zinc finger
ZFN Zinc-finger-nuclease

Index

E

F

G

X

Y

Z

Printed in the United States
By Bookmasters